T0344995

Complexity Science

Ecosystems, the human brain, ant colonies and economic networks are all complex systems displaying collective behaviour, or emergence, beyond the sum of their parts. Complexity science is the systematic investigation of these emergent phenomena, and stretches across disciplines, from physics and mathematics, to biological and social sciences. This introductory textbook provides detailed coverage of this rapidly growing field, accommodating readers from a variety of backgrounds and with varying levels of mathematical skill. Part I presents the underlying principles of complexity science, to ensure students have a solid understanding of the conceptual framework. Part II introduces the key mathematical tools central to complexity science, gradually developing the mathematical formalism, with more advanced material provided in boxes. A broad range of end-of-chapter problems and extended projects offer opportunities for homework assignments and student research projects, with solutions available to instructors online. Key terms are highlighted in bold and listed in a glossary for easy reference, while annotated reading lists offer the option for extended reading and research.

Henrik Jeldtoft Jensen is Professor of Mathematical Physics at Imperial College London, and leads the Centre for Complexity Science. He is a prominent expert in complexity science and is involved in a variety of high-profile research projects including the application of co-evolutionary dynamics to the modelling of socio-economical sustainability, finance, cultural evolution, innovation and cell diversity in cancer tumour growth, and has also worked with the Guildhall School of Music and Drama to identify differences in the neuronal response of audience and performers depending on the mode of performance. He has published two books on self-organised criticality and complex systems.

Complexity Science
The Study of Emergence

Henrik Jeldtoft Jensen
Imperial College London

CAMBRIDGE
UNIVERSITY PRESS

Shaftesbury Road, Cambridge CB2 8EA, United Kingdom

One Liberty Plaza, 20th Floor, New York, NY 10006, USA

477 Williamstown Road, Port Melbourne, VIC 3207, Australia

314–321, 3rd Floor, Plot 3, Splendor Forum, Jasola District Centre,
New Delhi – 110025, India

103 Penang Road, #05–06/07, Visioncrest Commercial, Singapore 238467

Cambridge University Press is part of Cambridge University Press & Assessment,
a department of the University of Cambridge.

We share the University's mission to contribute to society through the pursuit of
education, learning and research at the highest international levels of excellence.

www.cambridge.org
Information on this title: www.cambridge.org/highereducation/isbn/9781108834766

DOI: 10.1017/9781108873710

First published 2023

A catalogue record for this publication is available from the British Library.

ISBN 978-1-108-83476-6 Hardback

Additional resources for this publication at www.cambridge.org/jensen

To Vibeke, Barbara and Rebecca

Contents

Acknowledgements   *page* xi
Nomenclature   xiii
Preface   xvii

I Conceptual Foundation of Complexity Science   1

Introduction to Part I   3

1 The Science of Emergence   5
   1.1 The Importance of Interaction   9
   1.2 Past Views on Emergence   15
   1.3 Further Reading   18
   1.4 Exercises and Projects   19

2 Conceptual Framework of Emergence   21
   2.1 Emergence of a Characteristic Scale or Lack of Scale   23
   2.2 Emergence of Collective Robust Degrees of Freedom   26
   2.3 Structural Coherence   28
   2.4 Evolutionary Diffusion   31
   2.5 Breaking of Symmetry   33
   2.6 Emergence of Networks   35
   2.7 Temporal Mode   37
   2.8 Adaptive and Evolutionary Dynamics   39
   2.9 Further Reading   40
   2.10 Exercises and Projects   41

3 Specific Types of Emergent Behaviour   46
   3.1 Ising-Type Models: Transitions and Criticality   48
   3.2 Network Models and Scale vs. No Scale   52
   3.3 Emergence of Coherence in Time: Synchronisation   57
   3.4 Evolutionary Dynamics: Adaptation   60
   3.5 Mean-Field Modelling: Dimensionality and Forecasting   64
   3.6 Further Reading   69
   3.7 Exercises and Projects   70

4   The Value of Prototypical Models of Emergence   75
     4.1   The Need for Simplification of Models   76
     4.2   O'Keeffe–Einstein Propositions at Work   78
     4.3   Further Reading   82
     4.4   Exercises and Projects   83

II   Mathematical Tools of Complexity Science   87

Introduction to Part II   89

5   Branching Processes   93
     5.1   Generator Functions: Sizes and Lifetimes   97
            5.1.1   Size of the Progeny   99
            5.1.2   Time to Extinction   102
     5.2   Branching Trees and Random Walks   103
     5.3   Further Reading   106
     5.4   Exercises and Projects   107

6   Statistical Mechanics   110
     6.1   Probabilities and Ensembles   110
     6.2   The Ising Model   119
     6.3   The Peculiar Nature of the Critical Point   125
     6.4   Fluctuations, Response and Correlations   127
     6.5   Examples of Correlation Functions: Brain, Flocks of Birds, Finance   132
     6.6   Diverging Range of Correlations   133
            6.6.1   Correlation Function – Exact Approach   134
            6.6.2   Correlation Function – Intuitive Discussion   139
     6.7   The Two-Dimensional XY Model   143
            6.7.1   2d XY: Some Mathematical Details   148
            6.7.2   Vortex Unbinding   153
            6.7.3   The Vortex Unbinding Transition in Other Systems 154
     6.8   Further Reading   156
     6.9   Exercises and Projects   156

7   Synchronisation   163
     7.1   The Kuramoto Model: The Onset of Synchronisation   164
     7.2   Chimera States   170
     7.3   Further Reading   174
     7.4   Exercises and Projects   175

8   Network Theory   177
     8.1   Basic Concepts   178
     8.2   Measures of the Importance of Nodes   179

	8.2.1	Degree Centrality	179
	8.2.2	Eigenvector Centrality	184
	8.2.3	Closeness Centrality	187
	8.2.4	Betweenness Centrality	187
	8.2.5	How Well Does it Work?	188
8.3	Community Detection		188
8.4	Spreading on Networks – Giant Cluster		196
8.5	Analysis of Dynamics of and on Networks		203
	8.5.1	Generating Networks	204
	8.5.2	Random Walk on Networks	212
	8.5.3	Synchronisation on Networks	216
8.6	Further Reading		224
8.7	Exercises and Projects		225

9　Information Theory and Entropy　　230
9.1	Information Theory and Interdependence		232
9.2	Entropy and Estimates of Causal Relations		237
9.3	From Time Series to Networks		241
9.4	From Entropy to Probability Distribution		245
9.5	Measures of Degrees of Complexity		256
	9.5.1	Lempel–Ziv Complexity Measure	256
	9.5.2	Information-Theoretic Approach to Emergence	259
	9.5.3	Group Entropy Measure of Complexity	272
9.6	Further Reading		274
9.7	Exercises and Projects		275

10　Stochastic Dynamics and Equations for the Probabilities　　279
10.1	Random Walk and Diffusion	280
10.2	First Passage and First Return Times	293
10.3	Correlations in Time	297
10.4	Random Walk with Persistence or Anti-persistence: Hurst Exponent	302
10.5	Stationary Diffusion: Ornstein–Uhlenbeck Process	307
10.6	Evolutionary Dynamics and Clustering	309
10.7	Master Equation, Coarse Graining and Free Energy	313
10.8	Further Reading	318
10.9	Exercises and Projects	319

11　Agent-Based Modelling　　324
11.1	Flocks of Birds or Schools of Fish	325
11.2	Models of Segregation	328
11.3	The Tangled Nature Model	337
11.4	Further Reading	349
11.5	Exercises and Projects	350

12 Intermittency   356
    12.1 Self-Organised Criticality   357
         12.1.1 Sandpile Models   358
         12.1.2 Mean-Field Analysis   361
         12.1.3 Lessons from Sandpile Models   364
         12.1.4 Forest Fire Model   367
    12.2 Record Dynamics   370
         12.2.1 Statistics of Records   371
         12.2.2 Spin Glasses, Superconductors, Ants and Evolution   375
    12.3 Tangent Map Intermittency   379
    12.4 Further Reading   382
    12.5 Exercises and Projects   383

13 Tipping Points, Transitions and Forecasting   387
    13.1 Externally Induced Transitions   387
    13.2 Intrinsic Instability   389
    13.3 Further Reading   395
    13.4 Exercises and Projects   395

14 Concluding Comments and a Look to the Future   397
    14.1 Further Reading   399

Glossary   401
References   411
Index   436

Acknowledgements

My perception of what complexity science is, and my understanding of how it is best pursued, have been developed over many years of fortunate exchanges of views, ideas and expertise through collaboration and dialogue with a wealth of people. Some scientists, some artists and musicians, and some with other backgrounds. It is not possible for me to pinpoint the exact effect of each conversation and each person, but I know that my thinking and knowledge have grown from this lush environment of inquisitive curiosity. Around the world I have been lucky to be with people sharing the curiosity to understand, for the sake of contributing to our comprehension of our surroundings. The kind of people convinced that investigating basic *why and how* questions can lead to multitudes of valuable applications of a nature that we were not able to imagine at the outset.

It is probably unrealistic to mention everyone who has been formative in my scientific outlook, but I will try – despite the embarrassment if I leave someone out.

I am consciously aware of specific valuable incidences of influence from the following during their affiliation with Imperial College London: Tomaso Alarcon, Saoirse Amarteifio, Jørgen Vitting Anderson, Elsa Arcaute, Li Bai, Gil Benkoe, Thibault Bertrand, Lydéric Bocquet, Luís Borda-de Água, Katharina Brinck, Kris Broga, Andrea Calandruccio, A. David Caplin, Andrea Caroli, Seng Cheang, Yang Chen, Kim Christensen, Lesley Cohen, Simone Avogadro di Collobiano, Geraldine Cox, Bjön Crütz, Kajsa Dahlstedt, Ari Datta, David Edwards, Daniel Erikson, Murat Erkurt, Tim Evans, Paul Expert, Pavel Vazquez Faci, Max Falkenberg, Ignazio Gallo, Michael Gastner, Cedric Gaucherel, Peyman Ghaffari, Andrea Giomette, Hayato Goto, Jelena Grujic, Matt Hall, Dominic Hamon, Vasilis Hatzopoulos, Kerstin Holmström, David Jackson, Dominic Jones, Rishi Kumar, Simon Laird, Renaud Lambiotte, Daniel Lawson, Nathan Lindop, Stefano Lise, Chuan Wen Loe, Juan M. Lopez, Pedro Mediano, Miguel Molina-Solana, Nicholas Moloney, Mario Nicodemi, Dominic O'Kane, L. F. Pereira de Oliveira, Lorenzo Palmieri, Andrew Parry, Roozbeh Pazuki, Adele Peel, Garry Perkins, Ole Peters, Giovanni Petri, Duccio Piovani, John W. Polak, Gunnar Prussner, Hardik Rajpal, Shama Raman, Fatimah Adul Razak, Chris J. Rhodes, R. M. del Rio-Chanona, Juan David Robalino, Fernando Rosas, Jacob Runge, Anand Sahasranaman, Giovanni Sena, Maheen Siddiqui, Proshun Sinha-Ray, Steven Spencer, Marianne Storey, Clemens von Strengel, Denise Thiel, Jimmy Totty, Eduardo Viegas, Xiaogeng Wan, Nanxin Wei, Geoffrey West, Galen Wilkerson, Nicola Wilkin, Alastair Windus, Nicky Zachariou, Jaleh Zand.

From beyond Imperial College London I am very much aware that I have benefitted from knowing: Lucilla de Arcangelis, Baris Bagci, Per Bak, John Berlinsky, Thomas

Bohr, Steven T. Bramwell, Andrew Brass, Yves Brechet, Sandra C. Chapman, Dante Chialvo, Norma B. Crosby, Francisco de la Cruz, Marina Diakonova, Alvaro Diaz-Ruelas, David Dolan, Benoit Doucot, Deniz Eroglu, Thomas Fiig, Hans Fogedby, Nigel R Franks, Torsten Freltoft, Geoffrey Grinstein, Rudolf Hanel, John Hertz, Peter C. W. Holdsworth, Vincent A. A. Jansen, Mogens Høgh Jensen, Catherine Kallin, Kunihiko Kaneko, Dimitris Kugiumtzis, Alan Luther, Hildegaard Meyer-Ortmanns, Ole G. Mouritsen, Pietro Panzarasa, Tiago Pereira, Christopher Pethick, Oscar Pla, Tom Richardson, Alberto Robledo, Ana B. Sendova-Franks, Paolo Sibani, John Anthony Sloboda, Edouard Sonin, Didier Sornette, Nico Stollenwerk, Hideki Takayasu, Misako Takayasu, Piergiulio Tempesta, Stefan Thurner, Ugur Tirnakli, Constantino Tsallis, Mats Wallin, Nicholas W. Watkins, Hans Weber, Roseli Wedemann.

I have received valuable comments on the manuscript from Vibeke N. Hansen, Barbara N. J. Jensen, Nishanth Kumar, Santiago Musalem Pinto, Fernando Rosas, Madalina Sas, Eduardo Viegas and Gezhi Xiu, and from a number of anonymous reviewers. Thank you very much for your suggestions, comments, corrections and your time.

I am grateful to Simon Capelin for suggesting, as commissioning editor at Cambridge University Press, that I should write this book and to my two CUP editors Nicholas Gibbons and Melissa Shivers for their assistance.

My worldview in the broadest sense, and the scope and aim of the book, have been strongly influenced by my daily interactions with my inner circle: Rebecca N. J. Jensen, the teacher, Barbara N. J. Jensen, the theologian, and Vibeke N. Hansen, the medical physicist.

Nomenclature

The following table contains, in alphabetical order, the Greek alphabet followed by the Roman, mathematical symbols, a brief description and the page of their first use.

Symbol	Description	Page
α	Phase lag	171
β	Inverse of the temperature times Boltzmann's constant	112
β	Parameter in the transition probability	314
β	Critical exponent for order parameter	126
γ	Drift velocity	285
γ_{Tot}	Total restriction ratio	261
γ_{mar}	Marginal restriction factor	267
γ_T	Product of reduction factors	268
γ_R	Relative reduction factor	269
$\gamma_{\neg}$	Complement restriction factor	267
δ	Size of change caused by mutation	32
δ	Asymmetry of random walker	284
$\delta_{i,j}$	Kronecker delta function	96
$\delta(x - x_0)$	Dirac delta function	287
$\Delta h(x, y)$	Change in function h	115
$\Delta(AB)$	Group entropic complexity measure	273
Δu_{ag}	Change in agent utility	330
ΔU_{BA}	Change in potential function	314
ϵ	Coupling strength between Kuramoto rotors	164
η	Coefficient of restoring force	307
θ_i	Angle of rotor or arrow	143
λ_q	Eigenvalue number q	185
μ	Carrying capacity-like coefficient	338
μ	Average number of branches	93
ν	Lack of conservation parameter	361
ξ	Correlation length	130
π_Q	Cluster size probability	198
π_{ij}	Probability walker moves from node i to node j	213
Π	Transition matrix	213
ρ	Vector of local densities	316
ρ_i	Component of vector of local densities	316
ρ_Q	Probability of size of cluster at the end of link	199
σ	Random velocity coefficient	307
σ_i	Component of occupancy vector	315
σ	Occupancy vector	315
σ_X	Standard deviation	232
Σ	Structural entropy	270
τ	Time to extinction	102
τ	Correlation time	130
χ	Magnetic susceptibility	127
Υ	Helicity modulus	147

Symbol	Description	Page
$\Phi(x, y)$	Composition function	251
$\psi_k(t)$	Deviation from the average of rotor k	165
$\chi(t)$	Stochastic force	297
$\bar{\omega}$	Average rotor velocity	165
ω_k	Speed of rotor number k	164
Ω	Number of possible events or states	113
Ω	Organisation entropy	270
$\Omega(E, V)$	Total number of microstates possible under these constraints	111
$\vec{\nabla} f$	Gradient of function f	114
a	Random walker step size	280
a_{ij}	Element of adjacency matrix	179
A	Adjacency matrix	179
B	Applied magnetic field	120
$\tilde{B}_i$	Effective field	122
C	Heat capacity	127
C	Connectance	179
$C(r, t)$	Correlation function	129
C_q	Network cluster number q	190
C_{XY}	Correlation coefficient	232
$C_{LZ}(s)$	Lempel–Ziv measure	258
D	Diffusion constant	285
$DTE_{X_i \to X_j}$	Direct transfer entropy	243
E	Energy	111
$E_{\mathbf{B}}$	Energy of bath	111
E_{Tot}	Energy of bath and system	111
E_{s}	Energy of system	111
$E(\mathbf{r})$	Dynamical variable of the Zhang model	359
$\langle E \rangle$	Average energy	117
$F(\boldsymbol{\rho})$	Stochastic process free energy	317
$F(x)$	Accumulated probability density	371
$\hat{f}(k)$	Fourier transform of function $f(x)$	286
$g(x, t)$	Source term	297
$g_0(s)$	Generator function for the degree distribution	197
$g_1(s)$	Generator for the excess degree distribution	197
$g_X(s)$	Generator function for the stochastic variable X	94
$g(\omega)$	Density of Kuramoto rotors	165
$G(t)$	Group generator function	252
$G(\boldsymbol{\rho})$	Stochastic process potential	331
H	Hurst exponent	303
$H(X)$	Shannon entropy	237
$H(X|Y)$	Conditioned entropy	239
$H[S_1, \ldots, S_N]$	Hamiltonian of the Ising spins S_i	120
$H(\mathbf{S}, t)$	Offspring probability weight function	338
$H_{\neg 1}$	Entropy conditioned on complement to $\neg 1$	267
$I(X; Y)$	Mutual information	233
J	Strength of interaction in the Ising model	120
J_{ij}	Strength of interaction in the Ising spin glass	375
$J(x, t)$	Current	293
$J(\mathbf{S}, \mathbf{S}')$	Interaction strength between type $\mathbf{S}$ and type $\mathbf{S}'$	338
k	Degree of node	179
k_i	Degree of node i	180
k_B	Boltzmann's constant	9
K	Kuramoto order parameter	165
L_{C_q}	Number of links in cluster C_q	190

Symbol	Description	Page
L	Number of links in network	179
$L(x)$	Lambert function	254
$n(\psi)$	Rotor density	167
$n_{as}(\psi)$	Density of non-synchronised rotors	167
$n_s(\psi)$	Density of synchronised rotors	167
$n(x,t)$	Number of agents of type x at time t	309
N	Number of nodes in network	179
N	Number of particles	9
m	Magnetisation per spin	120
$\langle m \rangle$	Thermal average of magnetisation per spin	120
$\frac{\partial f}{\partial x}$	Partial derivative of function f with respect to x	114
p	Pressure	9
p_+	Right step probability	280
p_-	Left step probability	280
p_i	Probability of event i	113
$p(s)$	Probability that the system is in a particular state s	111
p_k	Branching probability	93
p_{mut}	Mutation probability	309
$P_{\text{deg}}(k)$	Degree distribution	179
$p_{\text{off}}(\mathbf{S},t)$	Offspring probability	338
p_{kill}	Killing probability	340
$P_{\text{mut}}^{(0)}$	Probability that no 'genes' mutate	342
$P_{1\text{pass}}(t)$	First passage time probability	293
$P_{1\text{ret}}(t)$	First return time probability	296
$P_i(t)$	Probability system is in state i at time t	282
$P(x,t)$	Probability system is at x at time t	281
$P(k)$	Probability of event k	53
$P_X(n)$	Probability that $X = n$	94
$P_{X_i, X_j}(x_i, x_j)$	Joint probability	231
$P_{\mathbf{X}}(x_1, x_2, \ldots, x_N)$	Simultaneous probability	231
$P_{X\mid Y}(x, y)$	Conditioned probability	238
$P(Q\mid k)$	Probability of size of cluster of k neighbours	199
q	Number of neighbours	122
q_i	Number of agents on node i	213
$\boldsymbol{Q}$	Vector of occupancies	213
Q	Number of configurations each component can occupy	261
$Q[C]$	Modularity of partitioning C	191
s_k	Step size	302
$s(\rho_q)$	Entropy of state ρ_q	316
$s(l_1, l_2)$	Sequence factor	257
S	Thermodynamic entropy	112
S	Strength of giant cluster	201
S	Avalanche size	361
$\mathbf{S}_\alpha$	Agent label	338
S_α	Rényi entropy	252
$S(N)$	Sequence	257
$S[p]$	Group entropy	252
$S[\mathbf{p}]$	Shannon entropy	113
S_q	Tsallis's q entropy	250
$\langle S \rangle$	Thermodynamic average of an arbitrary spin S	121
$\langle S_i \rangle$	Thermodynamic average of a specific spin S_i	121
S_i	Ising variable at position i equal to $+1$ or -1	48
t_i	Record time	372
T	Temperature	9

Symbol	Description	Page
T	Avalanche duration	361
T_{BA}	Transition probabilities	314
T_{ij}	Transition probabilities	282
$TE_{Y \to X}$	Transfer entropy	236
$\text{Tr } A$	Trace of matrix A	135
$u(\rho_q)$	Coarse-grained potential function	316
u_{ag}	Agent utility	330
U_{gl}	Global utility	329
$U(A)$	Potential function	314
v_q	Eigenvector number q	185
V	Volume	9
W_{Tot}	Effective total number of allowed configurations	261
W_i	Effective number of allowed configurations of i	267
$W_{\neg i}$	Effective number configurations in the complement of i	267
x	Agent position along label axis	32
x_d	Position at which agent dies	32
x_0	Initial position of population of agents	32
x_{off}	Position of offspring	31
x_n	Probability that branching process stops at generation n	98
X_i	The ith component of a time series vector	230
$\mathbf{X}(t)$	Time series vector	230
Y_n	Total number of decedents up to generation n	97
Y_∞	Total size of the progeny	97
$z(\mathbf{r})$	Sandpile dynamical variable	358
Z	Partition function	112
Z_n	Size of generation n	96
$Z_n^{(k)}$	Number of nodes in generation n originating from node k in the first generation	96

Preface

Many good books on complexity science currently exist. Some discuss complexity science from the perspective of a specific methodology such as network theory, analysis of power laws or use of agent-based simulations, while others discuss real systems which are considered to be complex, such as finance, sociology or ecology. References will be given where relevant to the context throughout the following pages.

This textbook is different in its aim and format from existing books. The book will present complexity science as a science in its own right, which focuses on the systematic study of emergent phenomena. Figure 1 is included here to indicate from the onset what our focus will be. The figure is best read with a concrete example of the components,

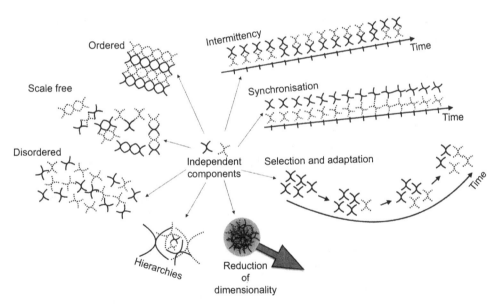

Figure 1 Emerging structure. Two components, agents, are shown in the centre. We now contemplate what will happen when collections of such components are made to interact and form collective structures. Surrounding the agents are sketches of structures and patterns that may emerge in time and space at the aggregate level. Types of static structures are to the left. Order and disordering tendencies may compete and produce a richness of properties with no equivalence amongst the individual agents. The dynamics of the interacting components can produce a wealth of different collective dynamical modes. For example, intermittent rearrangement, synchronisation and selection leading to adaptation. Or the interaction between components may produce a rigidity that allows the description of the time dependence in terms of just a few parameters, allowing us not to specify the dynamics of the individual components. This corresponds to a reduction of dimensionality.

or agents, in mind. They might, for example, be molecules forming rigid materials, neurones exhibiting synchronous firing patterns, biological cells building up hierarchies of structure, social agents segregating, etc. We will return to this figure in much more detail in Chap. 3; for now we simply suggest that the reader takes a careful look while thinking of possible manifestations of the examples sketched in this figure to obtain a feeling for what we will mean by the term 'emergence'.

We will not try to define a complex system, because systems in general cannot be divided into two exclusive classes: the complex and the not complex. The situation is the same as if we try to define a physical system or a biological system. Most systems will have both physical and biological aspects. Think of a bird flying through the air. Its motion is subject to the physical laws of aerodynamics but its metabolism and physiology are part of biology. This is the reason why we focus on the concepts and methodology of complexity science and take the viewpoint that its subject matter is emergence rather than specific types of systems. We will, however, as we go along mention many examples of applications of complexity science to real systems.

This textbook is intended for students, researchers and others with a wide variety of backgrounds. The book is separated into two parts. The first part is dedicated to a non-mathematical exposition of basic ideas and concepts used in complexity science. The second part develops a broad range of mathematical tools often used in complexity science. The hope is that the mathematics is introduced at a level that will be manageable for readers with just a basic high school background in calculus, vectors and matrices and probability theory. The presentation tries in a gentle way to gradually introduce the mathematical formalism and strives towards being self-contained, with some elaboration included as separate 'technical boxes'. Concepts are boldfaced the first time they are introduced and a glossary presents brief definitions for rapid consultation.

The book's structure is intended to make it useful for self-study as well as an accompanying text for a taught course. Each chapter is framed by a brief synopsis and summary. References to other textbooks and particularly relevant scientific papers are included throughout the text, and an annotated list of recommended Further Reading can be found at the end of each chapter. An online updated copy of the Further Readings can be found on the book's webpage at Cambridge University Press. Exercises and projects are included at the end of each chapter. Some of these exercises are intended as brief discussions, perhaps with fellow students, to help digest the material. Others are more comprehensive and can, for example, be used as take-home assignments or even be developed into small research projects. In both cases, thinking through this material is expected to greatly help the reader to obtain a working knowledge of the field of complexity science.

The hope is that the reader will develop a clear understanding of the subject matter and the methodology of complexity science. The book will explain the focus of complexity science, its aim and how understanding complexity science can be helpful to people from various backgrounds and with different objectives. Think of the environment, our society, the economy, the mind or advanced IT systems. A biologist, a sociologist, a psychologist, an economist, a neuroscientist, a mathematician, a physicist and so forth may want to know about complexity science because of certain emergent aspects encountered while analysing a particular phenomenon, such as species diversity, social segregation, mental health, financial crashes, mental wellbeing or non-supervised machine learning.

Faced with the same kind of phenomena, the engineer, the decision maker, the politician and, for example, the journalist may want some familiarity with complexity science, its purpose and the kinds of analysis it can offer, for the simple reason that although complexity science cannot predict accurately the behaviour of large complex systems, it is able to help sort out what kind of behaviour can be expected.

The aspiration is that the book may serve as a guide to at least one possible path through the enormous and ever-growing terrain of concepts, methods, applications and literature of complexity science. Of course, the best way to read the book is to start at the beginning and work through to the end. But since for many it may be difficult to allocate the time needed for this, the book is written to make it possible to dive in and out of chapters and sections. This means that concepts will be considered multiple times from different perspectives throughout the book. Cross references should help to connect the discussion, but to make a non-sequential reading easier a certain amount of repetition occurs. The sequential reader can just make a nod of recognition and move on.

For easy reference, a list of mathematical notation is included at the front of the book and a glossary at the end of the book.

I

Conceptual Foundation of Complexity Science

Introduction to Part I

The chapters of Part I will discuss why **complexity science** is important, how this science relates to other sciences and also a little bit about its philosophical status. The aim is to make clear what makes complexity science special and in which way it contributes to our understanding of the surrounding world. Part I concentrates on the conceptual level with two intentions. Firstly, the part will contribute to the discussion of the precise demarcation of complexity science. Secondly, a description of concepts, results and perspectives of this science is one way to illustrate its important relevance as a meta-science to subject fields which at present span from linguistics and economics to biology and physics. It will become clear that complexity science in this sense is comparable to mathematics, but although complexity science when fully at work may need to make use of mathematics, its conceptual basis can to a large extent be presented without mathematical formalism. We can, for example, discuss collections of agents and the kind of collective behaviour to expect at the aggregated systemic level without specifying the specific identity of the individual agents. They may represent certain aspects of people settling in a city or they may represent molecules moving on a surface. Both situations share aggregated behaviours, which can be captured by general concepts such as segregation, ordering or mixing.

Any science will use words from daily life and through refinement try to focus and sharpen the meaning in order to develop specific concepts that form the subject matter of the particular scientific activity. In Part I we discuss the way complexity science uses words such as **complex, complexity** and **emergence** to build up our understanding of the behaviour of systems consisting of many interacting components. It is important to be aware of the terminology and its distinct meaning, which sometimes can be different from the use encountered in other situations. For example, we may intuitively think of 'complex' and 'complicated' as being synonymous. Complexity science makes the distinction that a complicated phenomenon is quantitatively difficult to keep track of. It might be that we try to compute the properties of many different independent components, such as the particles in a gas. This will be computationally demanding but conceptually easy. Complexity arises when the collection of components interact and new collective phenomena emerge possessing properties entirely different from those of the individual components.

Part I will also discuss the modelling approach of complexity science. Contrary to what we may at first expect, the use of well-chosen simple models is needed to improve our understanding. The simplicity and transparency of the models are particularly important because we are trying to capture the behaviour of phenomena that are both complicated and complex. We will discuss how conceptually simple models allow us to identify behaviour shared across very different systems from sociology, physics and economics, for example.

We consider complexity science to be the investigation of emergent phenomena. This focus is behind what is included in Part I. We will explain the necessity of *interactions* between the constituent parts and try to classify a number of different types of emergent behaviour encountered in very different systems. We will discuss general aspects of modelling complexity and what features can make a model particularly useful. We will try to make our presentation concrete by relating ideas and concepts to applications and include references to further discussions.

1   The Science of Emergence

> **Synopsis:** The subject matter of complexity science will be identified and placed in a historical perspective.

What is *complexity science*? There exists no universally agreed definition. Is complexity science a well-defined discipline with its own subject matter, or is it essentially just another term for science?

Let us consider what might be a useful and constructive working definition of the term 'complexity science'. Consider traditional disciplines such as mathematics, biology or physics. What such disciplines encompass appears more or less uncontroversial. Although with time the focus and methodology have changed, we have some fairly clear idea of which kind of problems biologists and physicists study. We are also broadly familiar with the methods employed by biologists or by physicists. Nor are we in doubt that mathematics, physics and biology are existing disciplines, each with a specific focus and subject matter and well-established institutions and educational traditions.

We can meaningfully talk about a specific scientific field without a very precise definition. In fact, it is important to realise that the scope and focus of a science will change as it develops and so will our understanding of the part of the world explored by the scientific activity. Accordingly, classifications of scientific fields should be flexible and accommodating and not restrictive. New methods will be launched and new phenomena included. Obviously, before the realisation that DNA carries the genetic code, genetics was very different from what it is now. Or before physics discovered quantum mechanics, it was an entirely different discipline with a completely different view of what constitutes matter and of the applicability of deterministic predictions. So we are looking for a flexible and informative description to define complexity science. One commonly used description of biology is that it studies animate matter in contrast to physics, which then is seen as the study of inanimate matter. This definition asks us to define animate and inanimate. Along the same lines, mathematics is the discipline concerned with a systematic and abstract study of patterns. So, we need to try to define what we mean by patterns. Of course these definitions can be deliberated endlessly, which we will not do; our purpose is simply to point out that it is possible in straightforward terms to make a reasonably useful demarcation of such scientific fields. It is helpful to think of biology as concerned with animate matter and physics as dealing with inanimate matter and mathematics as the study of patterns, even if we are unable to define rigorously the terms animate, inanimate and patterns.

When we turn to complexity science, a similar consensus for a manageable brief description in terms of some salient aspects of the activity does not exist. Often complexity science is considered as synonymous with cross-disciplinary activities such as studies of brain dynamics involving, say, neuroscientists, biologists, mathematicians and physicists. Or complexity science is seen as identical to one of the many methodologies it makes use of. Some might see complexity science as essentially identical to data science; others might think of **game theory** as the essence of complexity science. Indeed when complexity scientists study a complicated system composed of very many individual components, they do make use of methodology from data science, and various versions of game theory are used to analyse complex systems ranging from internet dynamics to sociology.

Sometimes complexity scientists may even contribute to the development of strategies and approaches in data science, or refine the methods and approaches of game theory, but the complexity scientist's main aim is not to find efficient ways to analyse large data sets or to refine the theory of games. The complexity scientist focuses on extracting general patterns and essential behaviours from the data sets or understanding common consequences of the rules of games. This is done with the aim to improve our general understanding of how systemic structure emerges from the interaction and dynamics of the constituent parts.

Other times complexity science may be identified with **network** science. Again, to analyse many component systems, complexity science does indeed make use of, and often contributes strongly to, the development of the science of networks and mathematical graph theory. This is absolutely natural, since commonly various features of a complex system can be represented as sets of nodes connected by various types of links. The natural relationship between network science and complexity science is even more clear when one thinks of the view of mathematician and philosopher Alfred North Whitehead (1861–1947) 'that scientists should concentrate on multi-perspective networks of relationships, rather than on the behaviour of the aggregated atomic unit' [343, 493]. More about Whitehead and complexity science in a moment.

So, clearly complexity science participates very often in cross-disciplinary research and it makes use of, and contributes to, the theory of networks, data science and other disciplines. But its raison d'être goes beyond these activities. The viewpoint of the present book is that the **systematic investigation of the general patterns and structures of emergent phenomena** is what makes complexity science a distinct scientific activity. One may ask if this statement makes it clear what complexity science is about, since we will have to agree on what emergent phenomena are. And true enough, care is needed. We may use terms like 'animate' and 'inanimate' matter to pinpoint biology and physics, but obviously that does not answer all philosophical questions concerning the nature of biology and physics. Likewise when we start to think carefully about 'emergent phenomena' a wealth of philosophical questions suddenly present themselves [155].

While being aware of existing important philosophical concerns, we will for the moment make do with a pragmatic description of an emergent phenomenon. We will stress the importance of interactions between the components that generate properties of the system as a whole which are not found among the properties of the individual building blocks. Our viewpoint is that complexity science sees the dynamics of interrelated processes as its subject matter, so it looks for the shared common behaviours between totally different parts of the world. This is why complexity science is eager to enter into collaborations with subfields such as finance, economics, neuroscience, ecology, etc.

The sketch in Fig. 1.1 tries to indicate the difference between the focus of complexity science and that of the specific subject fields. To the left is a stereotypical representation of the structure as seen from the various subject fields. Each subject field is interested in a specific component at a certain level of the structure. For example, the main focus of cell biology consists in the internal workings of the cell. Similarly, psychology's subject matter is the dynamics of the human mind and typically the two fields are studied more or less independently of each other. The two columns to the right in Fig. 1.1

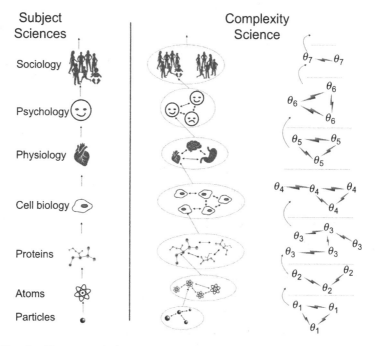

Figure 1.1 Sketch of how complexity science focuses on general features of the interaction between components and tries from these interactions to identify generalities shared across different systems and across the different levels of organisation. For example, psychology focuses on the individual. The interactions between individuals lead to sociology. Complexity science will investigate if similar processes and types of emergence can be identified at different levels. We can, for example, look for similarities at the level of sociology and the level of molecules. This may happen despite humans being very different from molecules since, viewed schematically, the interaction between humans, and between molecules, can share properties such as being attractive or repulsive, and this can in both cases, at the aggregate level, lead to phenomena such as segregation or mixing.

indicate the perspective of complexity science. The middle column emphasises that complexity science is concerned with the behaviour of aggregates, and the rightmost column is meant to show that complexity science commonly extracts at a given level a few salient features of the interactions between the building blocks and represents these in a schematic way by a few parameters depicted as the θs.

The rationale behind the approach of complexity science is to be able to shift the focus from the processes inside the components to the processes between the components. The inside at one level is 'the between' of the level below. For example, the inside of cell biology is the between of proteins. This shift in which one puts the focus on the interaction between components may allow us to identify general aggregate-level aspects of emergent dynamics. And what complexity science has found is that such generalities may be shared between different levels of the hierarchy in Fig. 1.1, For example, segregation or mixing can occur in a population of people and this behaviour can exhibit similarities with segregation or mixing of molecules. Sometimes the focus on interaction also allows us to understand, at least to some extent, the level above as emerging from the level below. We will discuss a mathematical example of this in Sec. 6.7, where we discuss how vortices that appear as structures at the aggregate level can arise from the spatial arrangement of the components.

Here follow a few examples which without mathematics illustrate how processes at the aggregate level, also often referred to as the systemic or collective level, may arise.

Let us first think about 'thoughts'. Ponder the processes occurring in our brain while we think, for example when thinking about emergence. We do not know exactly what the processes we call thoughts are, but we do know that they involve different brain regions and zillions of neurones firing in some sort of coordinated manner. The individual participating neurone does not 'think', it simply undergoes a process of loading, firing and reloading. The thought process is a property of the collective interacting dynamics of all the participating neurones.

As our next example we will look at the phenomenon of colour. Perhaps at first we consider colours to exist in some objective sense. We may correctly define the colour red in physical terms as electromagnetic waves of frequency around 430 THz. This is the great physical insight of Isaac Newton (1642–1726), inspired by his observation that white light from the sun can split into different colours when passed through a prism. But this definition does not properly grasp the multitude of attributes the colour red possesses when we think of the colour experience. Wolfgang Goethe (1749–1832) developed a theory of colour, published in 1810, which studied colour as a combined physical, physiological and psychological phenomenon [476].

The mental experience of red is formed in our mind when light of the appropriate frequency passes through our eyes to generate a signal in the visual cortex that further propagates up through the hierarchies of cognitive processing. The qualities of the colour red as warm, or as the contrast colour to green, cannot be deduced from the value of its frequency as an electromagnetic wave. Even less can the emotional character of red, such as to do with romance or warmth, be reduced to a property of the wave. So where reside the properties of colours which stir us? Not in a single physical property of the electromagnetic field, but rather in the collective effect of processes generated by the light absorbed through our eyes generating a hierarchy of other processes in our brain's neural system. All this leads in some not very well understood way to processes which we sense as thoughts and emotions in our mind. This example is simply meant to illustrate that some of the most familiar phenomena surrounding us are very much emergent in their nature. They exist in the form of some kind of amalgamated collaborative state across many participating components and not at all as some tangible *thing* that can be isolated and understood as an independent component of reality.

To highlight the aspects of emergence that are most important for complexity science, in the next section we consider two simple examples taken from physics and sociology, respectively, namely the **ideal gas** and **social segregation**.

1.1   The Importance of Interaction

Physics has a tradition for developing very schematic and simplified representations of our surrounding world. Physics describes the matter we encounter in our daily life as composed of molecules which interact more or less strongly. If interactions between molecules are relatively weak compared to the available thermal energy, which is the case at high temperatures, one may ignore that the molecules interact and can then describe the matter as an ideal gas consisting of independent non-interacting particles (see Fig. 1.2). In this situation the product of the pressure p and the volume V is proportional to the temperature T. This relation was established during the seventeenth and eighteenth century first as an empirical fact and then later understood in terms of the statistical behaviour of the molecules. At fixed temperature the product pV is proportional to the number of molecules N within the volume:

$$pV = Nk_BT. \tag{1.1}$$

The factor k_B denotes Boltzmann's constant, which we do not need to worry about right now. The mathematical description of the ideal gas makes it clear that each individual

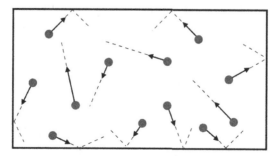

Figure 1.2 Sketch of an ideal gas. Each molecule is considered to move about independently of the others. When a molecule hits a wall, momentum is transferred between the molecule and the wall, leading to a force on the wall. The pressure on the wall is given by the total amount of force transferred from all the molecules hitting a unit area during a unit time interval.

particle contributes the same amount to the pressure against the walls of the container. A single particle in the container and N particles behave identically. The only difference between having one particle in the container and N particles is that the walls are hit more frequently by the molecules, leading to the pressure growing proportionally with the number of molecules. In this sense the entire gas has the same properties as do the individual molecules. The pressure originates in the force transfer between the molecules and the walls when the molecules collide with the walls. The pressure is the sum of the force from the particles exerted on a unit area per unit time and therefore the measured pressure will fluctuate more if only a few molecules are inside the container. For a macroscopic number on the order of Avogadro's number 6×10^{23}, the relative fluctuations are negligible and therefore the measured pressure appears as independent of time.

In this sense one might be tempted to consider the pressure as an emergent property of the gas. But this is misleading. The pressure is simply the direct sum of properties of the individual parts and nothing more. The reason the pressure becomes a negligibly fluctuating and therefore well-defined quantity for a gas containing Avogadro's number of molecules is not that some new state foreign to the individual particles has formed, but because of a simple and universal mathematical fact following from the **Central Limit Theorem** [139]. Namely that the fluctuations in a sum of *independent* terms vanish compared with its average as the number of terms increases. This ensures that the pressure, being the net effect of the bombardment of the walls by the molecules in the container, is very stable for a volume containing a macroscopic amount of gas.

The lack of interaction between the components of the ideal gas prohibits emergent systemic behaviour qualitatively different from the behaviour of the individual particles. Perhaps this is a little bit surprising for anyone who has used a bicycle pump. The piston does feel like it is compressing some sort of elastic medium and one might imagine that this has to do with squeezing the air molecules together. But the effect would also be there if the pump only contained a single molecule – though the pressure exerted by a single molecule would be minute and strongly fluctuating. The pressure increases because the molecules fly around inside the container and will hit the walls more frequently when the distance to the walls decreases as the volume becomes smaller. So even with one molecule inside the container, the pressure will increase when we decrease the available volume because the molecule will have a shorter distance to travel between collisions with the wall and hence hit the wall more often, so the time-averaged force exerted on the wall increases.

The situation becomes very different if we take into account that real atoms, or molecules, do interact. The description in terms of an ideal gas, i.e. in terms of non-interacting particles, eventually breaks down when the density in the gas becomes sufficiently high. This density depends on the specific material. Let us think of water. Not too dense water vapour behaves like an ideal gas. But water in liquid or ice form obviously behaves in entirely different ways. Liquid water or solid ice have properties that are completely different from the properties possessed by individual water molecules. The wetness of water or the hardness of ice are the result of the interactions between

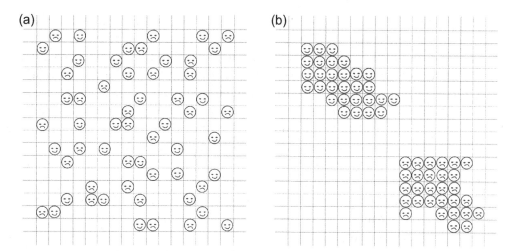

Figure 1.3 Lattice population of two types of agents: smilies and grumpies. In panel (a) the agents do not interact, i.e. they are indifferent to how their neighbour sites are occupied and as a result we see a homogeneous mix of types across the lattice. In panel (b) the agents are conscious about their neighbours. They prefer to be surrounded by agents of the same type as themselves, i.e. an attraction between like types is present. This interaction leads to a structured arrangement, namely a segregated distribution of the two types on the lattice.

the water molecules and are examples of emergent phenomena with no equivalence amongst the properties of the individual components. It is the forces acting between the water molecules that allow water to feel wet, which for example makes water able to climb up the wall of the container, and similarly at lower temperature the solid and elastic behaviour of ice is a result of the ability of the water molecules to interact with each other.

Next we consider emergent, and perhaps surprising, behaviour in models of sociology. Game theory is concerned with the resulting behaviour when **agents** interact according to a given set of rules. In it simplest form, two participants follow a table of pairs of actions in the form of player one does that and player two does that. A table allocates a payoff for each possible pair of actions and the theory then studies the accumulated payoff for a player choosing a certain string of actions. The rigorous mathematical analysis of such games was pioneered by von Neumann [478]. The approach has been generalised in various ways to include multiple players and the evolutionary aspects included by Maynard Smith [194].

The approach was generalised by combining ideas from game theory with dynamic action (see Fig. 1.3) in the 1960s by the American sociologist and game theoretician Schelling[1] [386, 387]. Schelling wondered why American cities were often found to segregate entirely into White and Black regions. Schelling had noticed that when asked,

---

[1] The model Schelling made use of in his study is now known as the Schelling model, although it is a specific instance of a more general model introduced earlier by James M. Sakoda [376]. The story of why the segregation model became known solely as the Schelling model is presented in detail in [188].

people were not very concerned about the exact composition of the colour of people in their neighbourhood. The percentage of co-inhabitants of the opposite colour an individual feels comfortable with, Schelling calls the individual's tolerance level. If the ethnic composition of the population in a neighbourhood is directly determined by the tolerance levels of the individual inhabitants, one would expect cities to consist of a mix of White and Black with a specific ratio that might fluctuate somewhat from place to place, but as long as the individual tolerance levels are larger than zero there is no reason to expect total segregation into White and Black neighbourhoods.

However, nearly complete segregation was very often observed in American cities. The discrepancy between the expectation based on the preferences of the individual and the actual behaviour at the systemic population level led Schelling to consider collective dynamical effects induced by the fact that each individual contributes to the environment of the others. We will return to Schelling's model in detail later in Sec. 11.2, all we need to be aware of here is that Schelling's simple mathematical model demonstrated that the dramatic difference between behaviour expected from an extrapolation of the inclinations of individuals and the actual aggregate behaviour may be related to what one might call the synchronised collective dynamics of the individuals. This tendency towards herd behaviour can be seen as caused by a kind of trend setting. When an individual, say White, moves out from a region, the density of White goes down and hence it becomes more likely that the tolerance level of the remaining Whites is breached, motivating more Whites to leave. The systemic dynamics is a result of a collective magnification of the effect of the individual's limited tolerance.

That the collective of many interacting components can possess properties that are different and richer than those of the individual components is surely not a big surprise. But how different can the systemic level be? And will the properties of the aggregate be a specific and unique reflection of the components of this particular system? If this is the case, we will not be able to determine general principles and identify classes of behaviour. Each case would be totally unique and a science of emergence would not be possible. But this is not how aggregates behave.

We considered the examples of how the wetness of water and the hardness of ice are emergent features with no equivalence at the level of the individual water molecules. And we discussed how Schelling's model of social segregation demonstrates that segregation may appear at the collective level although the inclinations of the individuals do not suggest a preference for separation.

Moreover, fortunately it turns out that very different systems can share the same kind of emergent properties that are understood as manifestations of general principles or 'laws'. The observation that all matter is normally found to exist in three forms – gas, liquid and solid – is just one such example of regularities of very general scope. Segregation is another example of similar collective systemic behaviour in very different situations, such as amongst human populations and molecules in materials. Complexity science studies these regularities and how they are related to the different classes of interactions amongst components. In later chapters we will elaborate on conceptual and mathematical descriptions.

Right now it is useful to consider how regularities, and even something like laws, can exist at different levels of organisation amongst entirely different types of components. Think again of matter. Water molecules, H_2O, are certainly very different building blocks compared with, say, a piece of metal made up of copper atoms. Nevertheless, in both cases the three phases – gas, liquid and solid – can be observed, though the specific properties of the solid phase are very different for a metal compared to ice. Or think of macro-evolution as recorded in the fossil record, which is found to exhibit long periods of little activity separated by relatively brief eruptions of hectic extinction and creation. How can this dynamics possess similarities with the dynamics of e.g. financial systems? These questions relate to the huge discussion concerning the ontological structure of the world and how ontology and epistemology are related [155], and can easily be overwhelming. For our purposes it suffices to remind ourselves that the world does not really consist of 'things'.

Already Aristotle (385–323 BC) in his *Metaphysics* speculated about the relationship between the collection of components and the totality they form. This is often summed up as, for a complex system: 'The whole is greater than the sum of its parts'. As a contrasting example, we can again think of the ideal gas where the whole, i.e. the pressure, is simply the sum of the contributions of each molecule.

Perhaps 'the invisible hand' of Adam Smith (1723–1790) can be seen as a mechanism generating a greater systemic level. It is the invisible hand that in Smith's view turns the multitude of self-centred activities of the individual participants in the economy (i.e., the merchant, the farmer, the craftsman, etc.) into a directed force for the common good.

The aim of complexity science is to investigate and develop an understanding of such 'invisible hands' and how they produce the regularities, commonalities and general properties of the greater 'whole'. But how can the systemic-level 'whole' of two very different systems have anything in common, say the precipitation of rain in the atmosphere and neuronal activity in the brain [325]? How likely one finds it that general emergent behaviour may exist in very different systems depends very much on how one envisages the world at its fundamental ontological level.

Assume the world consists of 'things' stacked on top of each other in some mechanical manner, resembling how we construct using Lego bricks. The choice of colour and shape of the little plastic pieces will very much determine what we can then assemble. Similarly, if the world fundamentally is made of 'things' it seems unlikely that the brain, consisting of biological cells, can share much with a system like the atmosphere, whose building blocks are molecules: oxygen, nitrogen, water, etc.

But the most fundamental building blocks of our world are not really hard material 'things', but rather processes nested within processes and similar processes may be at play in different situations. The world perceived as a nested hierarchy of processes is an ontological viewpoint resonating with Alfred North Whitehead's process philosophy [271, 401, 494]. Biological cells are made up of an entanglement of molecular processes. Molecules are made up of quantum processes. The brain is made up of neuronal processes and the atmosphere is made up of electrochemical and thermodynamical processes.

Take as an example an ant colony. At one level the colony possesses integrity and may be viewed as an 'organism'. At another level the ants are of course themselves complicated beings with an internal set of organs, and the organs are composed of interacting cells, etc. The ant colony can be viewed as an intricate composite of processes, e.g. involving foraging for food by ants laying out pheromone trails [90]. A very accurate description of the detailed physiology of the individual ant is not needed in order to realise that the foraging is a collaborative activity that can be investigated in its own right and lessons can be learned about task organisation without central control [336]. On the other hand, when we focus on the individual ant, we will have to describe it in terms of other processes at a different hierarchical level, including e.g. sensing, metabolism, etc.

When we think of the world as consisting of nested processes, it becomes understandable that different complex systems may, at given levels of their hierarchy, be controlled by processes also found in otherwise different complex systems. As an example, consider **evolutionary adaptive dynamics** in which the agents, or components, modify their strategies while subject to some selective pressure, e.g. competition for customers or resources. This kind of dynamical process may be relevant to a collection of micro-organisms competing for the same nutrition in a small pond, as well as to companies competing for customers.

Complexity science will help to figure out which features of a phenomenon emerge as general consequences of many components evolving together and which are specific and unique to a certain system. As an illustration think of the **normal distribution**, also frequently referred to as the bell curve; for an example see panel (a$_2$) in Fig. 2.5 later. This distribution is encountered when looking at the height of people, scores in exams, blood pressure, average number of goals scored by football players, sum of eyes when rolling dice, etc. This does not establish that the same mechanisms are responsible for blood pressure and dice rolling, but it is an example of how totally different phenomena can share the same features because they do share some similarity at a more abstract general level, often what one might call mathematical processes. In the case of the normal distribution, this is to do with the mathematics of adding independent random numbers and the so-called Central Limit Theorem [139] mentioned above.

This example elucidates how general behaviour may be produced simply by multitude, but since the normal distribution is found even when the components do not interact, it is not an example of what complexity science considers to be new emergent aggregate behaviour. In many different types of systems where interactions are shaping the collective behaviour, we will see in the chapters below that the normal bell-shaped probability distribution may be replaced by very broad distributions which correspond to situations where no *typical* event characterises what to expect. We are very familiar with such highly varied behaviour, for example from earthquakes where we know all too well that it is impossible to predict the size and time of the next earthquake.

We can turn these consideration around and use the observation of probability distributions that differ from the normal distribution as an indication that interactions are at play and have generated new emergent behaviour. The observation of probability distributions that follow broad power law-like behaviour, rather than the bell-shaped normal distribution, is often taken as a first indication of new collective behaviour at the aggregated systemic level. We discuss this at great length in the following chapters.

1.2   Past Views on Emergence

Let us close this chapter with a few examples of what influential scientists in the past have said about emergence. This is mainly thought of as indicative of how emergence is an integrated part of science and has been considered long before the invention of the term 'complexity science'.

Probably the earliest recorded thoughts about emergence are those mentioned above by Aristotle from about 350 BC, paraphrased as 'The whole is greater than the sum of its parts'. This articulates that the whole, i.e. the systemic level, can exhibit behaviour which is not in any way a property of the parts when these are considered in isolation.

From Ancient Greece we jump to one of the fathers of the scientific revolution of early 1900, namely the atomic physicist and key person in the development of quantum mechanics: Niels Bohr (1885–1962). Bohr is supposed to have pointed out that the wetness of water emerges out of the proper combination of hydrogen and oxygen and cannot be found or predicted by analysing those chemicals individually [225].

The American scientist and mathematician Warren Weaver (1894–1978) was a very successful science administrator and among many achievements wrote the book *The Mathematical Theory of Communication* together with Claude Shannon (1916–2001), the father of **information theory**. Already in 1948 Weaver discussed two types of complexity [487]. The first he called *disorganised* complexity, which he ascribes to collections of large numbers of independent particles and mentions that this is where the uncorrelated statistics is very powerful. The ideal gas case mentioned above is a prototype example, essentially no new macroscopic behaviour can emerge since the components do not interact. Mathematically, statistics like the normal distribution will be relevant since the collection of particles will be described by sums of uncorrelated random processes.

Weaver calls the second type *organised* complexity and points out that this can relate to particles undergoing highly correlated dynamics. This is the situation we are concerned with in this book. Weaver's organised complexity entails emergent collective behaviour and consists of any kind of phenomenon that involves interacting components which create complicated behaviour: cancer, exchange rates, etc. Weaver makes a point of emphasising that these classes of problems are of extreme importance. As mentioned above, in this book we suggest that it is useful to use the term *complexity science* to describe the systematic study of the commonalities shared across many different instances of organised complexity. This is in agreement with Weaver when he stresses the coherent dynamics amongst interdependent components. Weaver did not focus particularly on large collections of components. Later on in the book we will discuss how certain types of emergent behaviour, such as phase transitions, only occur in the limit of very large numbers of components.

We now move on to the physics Nobel laureate Philip W. Anderson (1923–2020). Anderson was hugely influential in the field of physics, using quantum mechanics to understand properties of materials. The field makes use of **quantum many-body theory** to understand theoretically how the interactions between the atoms and electrons in a material produce the observed macroscopic properties. Anderson's interest in the emergence of physical properties led him in 1972 to write an article in the journal *Science* with the title 'More is different' [18]. The article clearly articulates his view on complexity, which Anderson continued to elaborate extensively during the following years; see the collection of essays *More and Different: Notes from a Thoughtful Curmudgeon* [19]. The phrase 'more is different' is, as Anderson is very well aware of, of course an elegantly terse condensation of the second law of materialist dialectics as formulated by Engels [129]: 'The transformation of quantity into quality' or in other words 'the conversion of quantitative changes to into qualitative changes'.

Being a theoretical physicist, Anderson is obviously inspired by how the statistical descriptions of matter have been developed mathematically to capture emergent phenomena such as **phase transitions** that have absolutely no equivalent at the level of the individual components and strictly speaking can only take place when infinitely many components are present. The mathematical process, namely the so-called **breaking of symmetry**, involved in the transition, say, from ice to water (Fig. 1.4) is similar to those thought to be involved in the early stages of the formation of the universe, and possibly of relevance to sociological processes such as consensus formation.

Anderson's main aim is to demonstrate that fundamental and entirely new concepts are needed to understand the world of many interacting components. Anderson points out that after reducing a system or phenomenon to its components, one needs to understand how the components influence each other in order to understand the

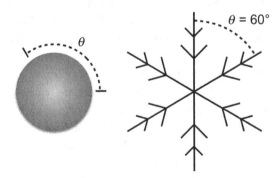

Figure 1.4 Water drop and snowflake. We can turn the water drop through any angle θ and it will look the same. The snowflake has lower symmetry, since we need to turn it through an angle of $60°$, $120°$, $180°$, etc. to ensure it lands on itself. We say that the high symmetry of the water drop is broken into a lower symmetry in the snowflake.

systemic level. As one seeks to reconstruct and understand the systemic level, new scientific principles and laws are needed at each level. This might be less of a surprise to chemists, biologists, sociologists, economists, etc. than apparently it is to some physicists.

But besides arguing against naive reductionism, Anderson also cautiously suggests an optimistic position that sees complex phenomena as accessible to systematic modelling, including mathematical analysis. Anderson rightly emphasises that methods and concepts from physics will have to be developed and replaced as one moves into biology, sociology, etc. Nevertheless, he does speculate that useful conceptual similarities may exist across very different fields of science, and highlights the mechanism of symmetry breaking as an example.

The economist, political scientist and cognitive psychologist Herbert A. Simon (1915–2001) published in 1962 a very comprehensive view entitled *The Architecture of Complexity* [412]. He discussed how his experience in behavioural sciences had led him to develop his view that hierarchical organisation is the essential feature shared amongst very different types of complex systems. He discusses processes found in physics, biology and social systems that are able to generate hierarchical structures and argues that it is possible to identify general mechanisms of relevance across entirely different types of complex systems.

Emergence and complexity are also of interest to philosophers. The philosopher and sociologist Edgar Morin (1921–), in his book *On Complexity* [302], puts emphasis on self-organisation and stresses the importance of von Neumann's introduction of cellular automata. Von Neumann (1903–1957) was a mathematician who made influential contributions across many fields of natural science and invented the notion of **cellular automata** as a mathematical device to study self-replicating entities. Morin, like Anderson, is also particularly interested in collections of huge numbers of interacting components. He does argue that a science of complexity is possible and that eventually a paradigm for this science will be established and possess principles of distinction, conjunction and implication.

Let us briefly mention the discussion, especially amongst philosophers [155], concerning degrees of emergence. The idea is that certain emergent phenomena are only weakly emergent because they somehow in terms of causality can be reduced to their components. In contrast, other phenomena are strongly emergent and possess a different type of irreducible relation to their components, which seems to imply that they will never be explained or understood in terms of their constituent parts. Consciousness is, according to Chalmers [81], the prime example of strong emergence. Fortunately we do not need to resolve these difficulties. All we have to bear in mind is that any emergent phenomenon in our sense possesses properties that are not possessed by the individual components. The wetness of water may to some extent be understood in terms of surface tension that can be analysed from the physics of the H_2O molecules, but this does not imply that a single water molecule in any reasonable sense can be said to possess wetness. Similarly the Schelling model discussed above shows that the dynamics at population level may entail responses beyond the behaviour ascribable to the individual.

Summary: We have noted that complexity, complex systems and complexity science and emergence can mean different things to different people. We have also indicated that conceptually people have investigated and theorised about complexity and emergence for a very long time. As a demarcation of complexity science to help focus and clarify our discussion, we use the following two working definitions:

Emergence denotes the occurrence of properties or phenomena at the aggregate level, which the individual parts do not possess.

Complexity science involves the systematic investigation of emergent phenomena. The aim is to characterise and understand commonalities shared by emergent phenomena in different systems and fields of research.

1.3   Further Reading

General audience:

The establishment of complexity science as a field of modern enquiry is very much related to the Santa Fe Institute. The book *Complexity: The Emerging Science at the Edge of Order and Chaos* [479] by M. Mitchell Waldorp offers an engaging overview of this story.

Melanie Michell's book *Complexity: A Guided Tour* [298] presents a non-mathematical grand scientific overview by one of the members of the Santa Fe Institute.

A succinct explanation of basic ideas, with an emphasis on adaptive systems and agents, is given in *Complexity: A Very Short Introduction* [196] by computer scientist and complexity science pioneer John Holland.

Intermediate level:

The paper 'Science and Complexity' [487] by Warren Weaver is both of great historical interest and greatly inspiring.

Philip Anderson's paper entitled 'More Is Different' [18] introduced this catch phrase and discusses the role of symmetry breaking and how science is a hierarchical enterprise.

Complexity and emergence are discussed from the viewpoint of a sociologist and philosopher in Edgar Morin's elegant book *On Complexity* [302].

Emergence and complexity are discussed with a focus on games by the computer scientist John Holland in his very clearly written book *Emergence From Chaos to Order* [195]. With respect to conceptual issues, the last chapter 'Closing' is of special interest.

The physicist Murray Gell-Mann's 1994 book *The Quark and the Jaguar: Adventures in the Simple and the Complex* [153] is intended for a general audience but offers nevertheless many important and scientifically thought-provoking discussions.

Advanced level:

Several chapters in Part I of *The Routledge Handbook of Emergence* [155] are of great interest to anyone speculating about basic philosophical aspects of emergence. In particular, Chapter 3 by Kerry McKenzie is helpful when thinking about what we mean by the term 'fundamental'.

1.4   Exercises and Projects

Exercise 1
Make a summary of the hierarchies of processes, nested within each other, you can identify when you think of a human being from the level of atoms, through cells to organs, to the level of brain and mind.

Exercise 2
Discuss similarities and differences between the views concerning complexity and emergence expressed by Weaver [487] and Anderson [18].

Exercise 3
Discuss emergence from the viewpoint that reality acually consists of processes and not of 'things' [271, 401, 494]. Then, from this position, consider the notion of strong and weak emergence [81, 155].

Project 1 – Different degrees of emergence
See the Introduction chapter of the *Routledge Handbook of Emergence* [155].

(a) Define the two concepts: epistemic emergence and ontological emergence.
(b) Can you think of methodologies, experiments and data sampling that will allow one to decide whether an observed emergent phenomenon is epistemic or ontological?
(c) Even if emergence is 'only' epistemic, why is it then still of importance to characterise its nature and structure?
(d) How do the emergence discussed by Herbert Simon [412] and by Phil Anderson [18] fit into the classification in terms of epistemic and ontological emergence.

(e) Consider how a viewpoint along the lines of Whitehead's emphasis on processes at the basic ontological level impacts on the classification in terms of epistemic and ontological emergence.

Project 2 – Evolutionary ecology and sociology

Consider in general terms similarities and differences between the evolution and emergence of structure in (1) biological ecosystems and (2) sociological structures and building blocks of society.

(a) Which emergent structures can we identify with confidence?
(b) To what extent can the evolutionary dynamics be reduced and explained in terms of the dynamics of these emergent building blocks?
(c) Is it possible to describe the functioning and development of biological and social systems without the use of concepts such as biological species or social classes?
(d) Are biological species or social classes examples of epistemic emergence or ontological emergence, see Project 1 above.

Project 3 – Emergence and interactions

We mentioned around Eq. (1.1) and in Sec. 1.1 that without interaction between component, new emergent properties cannot be generated.

Nevertheless, sometimes people do mention pressure as an emergent property, like in this quote by J. A. Wheeler (1911–2008) from p. 341 in [143]:

In thinking about the world in the large, I have another phrase that I like, borrowed from my Princeton colleague Philip Anderson: "More is different". When you put enough elementary units together, you get something that is more than the sum of these units. A substance made of a great number of molecules, for instance, has properties such as pressure and temperature that no one molecule possesses. It may be a solid or a liquid or a gas, although no single molecule is solid or liquid or gas.

(a) Reproduce the argument given in this chapter for why the pressure of a gas is *not* an example of an emergent property.
(b) Can you think of a sense in which, as Wheeler proposes, pressure can be seen as an emergent property?
(c) According to the viewpoint of Anderson mentioned in Sec. 1.2, would Anderson agree with Wheeler and classify pressure as an emergent phenomenon?
(d) Assume with Wheeler that pressure is an example of emergence and make a list of similar 'emergent' properties, such as mass, height, population of a country, etc.
(e) Let us assume we go along with Wheeler and accept the pressure of a non-interacting gas as an emergent phenomenon. From this perspective, can you think of any property of a system containing multiple components that are not emergent?

2 Conceptual Framework of Emergence

> **Synopsis:** Many types of emergence exist. This chapter will discuss some of the most prominent, and broadly occurring, examples of emergent structure in space and time.

In this chapter we will present some of the concepts most commonly used to classify and analyse the collective aggregated behaviour of many interdependent components.

We will first discuss in general terms some of the ways to identify, describe and classify emergent structures and behaviours. We will consider a number of particularly important 'pillars' of emergent behaviour. The following list consists of phenomena ubiquitously occurring across a range of different types of many-component systems and their description is widely discussed in the literature, both conceptually and mathematically. These examples of emergent behaviour will also help us to develop our intuition of the difference between what happens at the level of components in contrast to the collective systemic level.

Some Prototypical Types of Emergence

- Temporal mode: **intermittency**, non-**equilibrium** or lack of stationary state.
- **Characteristic scale** or no particular scale.
- Robust **collective degrees of freedom**.
- Structural coherence in space and time.
- **Symmetry breaking** and transitions.

From the outset we should be aware that the identification of components and their interaction, which will produce the emergent phenomena of the aggregate, is often far from self-evident but has to be considered as the first step towards understanding the phenomena, or research problem, at hand. We will find it helpful to distinguish emergence in time from emergence in space, although mixtures and overlaps between the two will often be found. **Synchronisation** between fireflies is an example. Fireflies are small insects that rhythmically emit brief flashes of visible light. When fireflies are exposed to each other's blinking, they are able to gradually adjust and produce an emergent synchronised state in which the entire river bank periodically lights up and blacks out as a whole. How the synchronised state is reached can depend on the spatial arrangement of the fireflies [358].

The time variation observed in the fossil record is another example of temporal emergence. The observed intermittency has been termed **punctuated equilibrium** by

Eldredge and Gould, see Chap. 9 in [169]. Observations show that extinction and appearance events documented in the fossil record exhibit long stretches of quiescence where relatively few extinctions of existing **species** or appearance of new species are recorded. These long quiet intervals are abruptly interrupted by bursts of extinction-appearance episodes where during relatively short time spans the content of the fossil record changes markedly [142]. One may ask if that is in contradiction to Darwin's 'gradual process of improvement' [94] or if punctuated equilibrium is related to a general intermittent mode of emergent dynamics relevant to the level of species and higher taxonomic structures. We will discuss modelling in Sec. 11.3 that suggests the latter.

For emergent structures in space one may, for example, think of patterns such as the stripes of the zebra or, to mention an all important and far from understood example, the shaping of an embryo from a single cell to a fully formed organism consisting of zillions of cells. Or, an example already touched upon in Chap. 1, segregation in cities and the model of Schelling.

Usually there are more than just the two levels consisting of individual components and the collective systemic level. For example, consider a biological organism in which several layers of aggregation can easily be identified. We can change our viewpoint from the scale of macromolecules to the level of cells and consider the cells as components that, through their interaction, allow organs to appear. At the next level we can consider the organs as components forming the entire organism. One may rightly feel overwhelmed and perhaps doubt that it is possible to identify simple recurring themes and typical behaviours among such intricate and convoluted arrangements of components within components.

However, in the following chapters we will see many examples that demonstrate that it is possible to focus on certain schematic aspects of the interactions, ignoring large amounts of detail, and thereby be able to identify some typical outcomes produced by the most important features of the *interactions* between components. Right now we just mention one concrete example. Namely, we will see that intermittent behaviour similar to that depicted in Fig. 2.1 is a general consequence in many ways of competing tendencies, or frustration, between the components. Moreover, we will see that the intermittency, which is a consequence of interaction, can be captured by a simplistic representation of interactions. For example, to obtain model similarity with real earthquakes one does not need a full description of the fault dynamics.

Often a deliberate simplified representation of the interactions allows an identification of what is of greatest importance. How can this be the case? Because certain aspects of the emergent collective behaviour depend on so-called many-body effects, namely the cooperative collaborative effects generated by the interactions *between* the components and to a much lesser degree on the internal properties of the components. The balance between how much a certain phenomenon is controlled by interactions between components and their collective behaviour, in contrast to the intrinsic specific properties of the individual components, will determine if one can develop a description

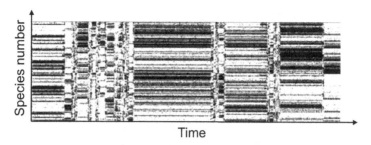

Figure 2.1 Sketch of the time dependence of the composition of species in the Tangled Nature model. Time is along the *x*-axis and at each time instance the label for the species present is marked by a black dot. Continuous occupation corresponds to the horizontal lines. Whenever a species goes extinct the line stops and when a new species comes into being a line starts. The dynamic is intermittent since we see extended periods where the pattern of lines changes little, interrupted by sudden transitions to a new pattern of lines. Based on simulation by Dr Matt Hall.

Figure 2.2 Example of a collection with a characteristic scale and one without a scale. In panel (a) all ducks are more or less of the same size and we can collectively describe them by the average size of the 13 ducks, represented by the black duck. In panel (b) no single scale captures the variety of different sizes of the 12 ducks.

with some degree of general applicability, or universality, by use of generic and stylistic schematic models.

In the following sections we will expand on some of the details of a few very general types of emergence.

2.1 Emergence of a Characteristic Scale or Lack of Scale

To try to illustrate the difference between a situation that can be captured by one characteristic scale and one without a scale, Fig. 2.2 shows two collections of ducks. In panel (a) all ducks vary little in size, some are smaller and some are larger than the average size, indicated by the black duck, but the average is typical of the sizes of the ducks in this collection. Whereas in panel (b) we see ducks spanning a broad range of sizes and the black duck of average size is not representative of the sizes.

We will now consider how either one characteristic scale or a broad range of scales can be generated when components influence each other. The scale of a human biological cell is 20 μm, in comparison to the scale of 1–2 m of the human body. Understanding the scales of a problem is an important step towards an overall understanding of how a system functions, and which properties it may have. It is the characteristic time scale of the human visual processing system that makes us see frames projected at a rate less than about 10 frames per section as individual static pictures, whereas when the rate goes above 10 frames per second we experience the frames as motion on the screen.

We will look at spreading from one component to another as an example of how the interaction between components can produce a specific scale for the average dynamics of the components. To be concrete, we will think of an infection spreading in a population.[1] Imagine that a person becomes infected and during a certain time span is able to infect other members of the population. We denote by μ the average number of people infected. Each of the infected will also cause μ new infections on average. So we expect that the first person infected generates μ new infected, and that these in turn will generate a total μ^2 and so forth. In round n of infections, called generation n, we expect the average number of infected individuals $N(n)$ to be $N(n) = \mu^n$. The mathematical details are discussed in Chap. 5, here we just point out that the average number of infected μ sets a scale for how the subpopulation of infected individuals grows in time. This is seen in the following way. We note that $\mu = 1$ is a very special value for the problem. If $\mu < 1$, then μ^n rapidly becomes smaller and smaller as n is increased and the number of individuals infected during the n generations of infections becomes rapidly small. The exact opposite happens if $\mu > 1$. In this case the number of people infected in round n explodes rapidly. We can ask how many rounds of infections will double the number infected in a given round. We focus on a round n_0 in which $N(n_0) = \mu^{n_0}$ are infected and ask how many additional rounds of infection Δn it takes before $N(n_0 + \Delta n) = 2N(n_0)$. We substitute the expression $N(n) = \mu^n$ and get an equation for Δn:

$$\mu^{n_0 + \Delta n} = 2\mu^{n_0}.$$

The factor μ^{n_0} can be divided out and we obtain

$$\mu^{\Delta n} = 2 \Rightarrow \Delta n = \frac{\ln 2}{\ln \mu}.$$

This tells us that the number of new infections $N(n)$ during round n doubles after every $\Delta n = \ln 2 / \ln \mu$ rounds of infections.

If we are able to determine the value of μ we have an average estimate of the scale of the speed with which the infection spreads through the population. As $\mu > 1$ is decreased

---

[1] We could at this level of schematic description equally well have thought of a rumour spreading from person to person, or a forest fire spreading from tree to tree.

towards 1 the number of infection rounds required for a doubling of $N(n)$ goes to infinity, because $\ln 1 = 0$.

We can make a similar calculation for $\mu < 1$ by asking how many infection rounds are needed to half $N(n)$. So we will have to solve

$$\mu^{n_0 + \Delta n} = \frac{1}{2} \mu^{n_0},$$

and now get $\Delta n = \ln \frac{1}{2} / \ln \mu = -\ln 2 / \ln \mu$. We recall that $\ln \mu < 0$ when $\mu < 1$.

The conclusion is that the scale of the infection dynamics, described by how fast a doubling or a reduction to the half occurs – given by the average number of new infections caused by one infected individual – is given by the numerical value of $\ln 2 / \ln \mu$ for all values of μ different from one.[2] The value $\mu = 1$ is special and needs more careful mathematical analysis, but we can already see that $\mu = 1$ is a very important value because the characteristic scale of the considered problem diverges when the value of μ approaches one and because $\mu = 1$ separates opposite types of behaviour, namely fast **exponential growth** for $\mu > 1$ and fast reduction for $\mu < 1$. As μ approaches one, a change occurs in the behaviour of distributions describing the size of the number of infected individuals at round n or the total number, S, of infections induced by a single infected person. When $\mu < 1$ the probability distribution describing the size of the total number of people infected by a single person follows essentially an exponential function. By which we mean that the probability of S assuming a specific value, Prob$\{S = s\}$, depends on s like Prob$\{S = s\} \propto \alpha^{-s}$, for some α. We just discussed that an exponential function has a characteristic scale and therefore for $\mu < 1$ the burst of infections originating from one person will also have a characteristic scale.

A particular important and simple spreading process consists of the **branching process**, or chain reaction, or avalanche process, where any infected agent infects $k = 0, 1, 2, \ldots$ new agents with the same set of probability p_k, see Chap. 5. In this idealised version one can show that as μ approaches one, the total size, S, of the chain of infected agents starting from one single infected agent becomes power law distributed,[3] which means that Prob$\{S = s\} \propto s^{-a}$ with some exponent a. In the case of the branching processes $a = 3/2$. This value of the exponent is, at least approximatively, observed in nature for very different spreading phenomena: for example the spreading of brain activity both at the level of individual neurones [56] and at the macroscopic level between brain regions measured by an **fMRI scanner** [433]. Another example is the area burned down by forest fires [420].

We will later encounter many cases where a characteristic scale of a system diverges for a specific value of some parameter and where this value separates different regimes of behaviour, see in particular Sec. 6.6. Obviously more than one scale may be relevant, such as when we consider the dynamics of an ecosystem and we have to take into account variation on a daily, seasonal, yearly and even much longer time scales, like the

---

[2] In epidemiology, μ is called the basic reproductive number and denoted by R_0.

[3] A function $f(x)$ is said to exhibit power law-like behaviour if for a range of x the function can be approximated as $F(x) \approx x^a$ for some exponent a.

Milankovitch cycles of order 100,000 years, see e.g. [230]. Frequently, complex systems are found in states where essentially all scales are relevant, in which case we talk about a **critical state**. A concrete example consists in the time variation monitored by EEG measurements on the brain. Although neuroscientists identify particular frequency bands,[4] or time periods, to be of particular interest in specific situations, it has been found that the brain exhibits time variation spanning periods from milliseconds up to many seconds [281].

When all scales are of importance, time or spatial patterns will be repeated such that if we magnify a small part it will resemble the whole. The broccoli plant can be used as an illustration, see Fig. 3.4. These structures are called **fractals**. The importance of all scales is manifested in the occurrence of fractal structures in space and time and relates to correlations emerging between all parts of the system. We will see in Sec. 6.6 how such a state can be reached by the divergence of length scale over which correlations vanish. For an engaging and wide-ranging discussion of the role of scale in biology, sociology and economics, see the excellent book by Geoffrey West [491].

2.2   Emergence of Collective Robust Degrees of Freedom

We will first look at how it is sometimes possible to describe the behaviour of huge numbers of components by just a few parameters. Even without being explicitly conscious about it, we often describe huge collections of components by referring to their overall behaviour, captured by a few parameters. We can describe a floating iceberg by following its centre of mass, see Fig. 2.3. Instead of keeping track of the trajectories of all the water molecules in the iceberg, we just need to know the three coordinates of the position and the three coordinates of the velocity of the centre of mass. But if the ice melts, the iceberg will lose its rigidity and the trajectory of each water molecule

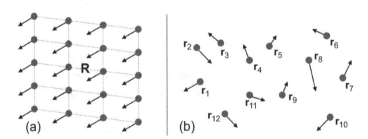

Figure 2.3 Independent vs. coordinate collective motion. In panel (a) particles move as a solid body and their motion can be described simply by following their centre of mass **R**. In panel (b) the particles move around independently and to specify their motion we will need to keep track of all the positions $r_1, r_2, \ldots, r_{12}$.

---

[4] The so-called delta, theta, alpha, etc. bands.

will not be determined by the centre of mass motion. The freezing of water responsible for the solid integrity of the iceberg is an example of the emergence of rigidity at the systemic level. The change from liquid to solid is a change in the collective behaviour of the components from a regime, the liquid, of fairly independent components to a regime, the solid, of strong coherence between the components. The components are even more independent in the gas form and the three phases of solid, liquid and gas are separated by phase transitions. Right now we are not focusing on the transitions, but want to emphasise the fact that under certain conditions, huge numbers of components may interact so strongly that they act as one coherent object as macroscopic rigidity emerges [21].

We are interested in the conceptual framework of the emergence of collective rigidity. This is of great relevance to complexity science since we often tacitly assume that a collection of components exhibit sufficient collective rigidity to allow a description in terms of broad sweep generalisations. At the informal level this is the sort of thing we do whenever we make statements like 'Norwegians are good skiers' or 'Danes like beer and akvavit'. The analogy with the iceberg, water and centre of mass consists in that, if the collection of people with Norwegian passports were in the frozen rigid coherent state – like the iceberg – then it makes sense to describe the ensemble of all Norwegians in terms of a few characteristics they all share, but if the term 'Norwegian' is more similar to the term 'liquid water', then a description will need to account for individual heterogeneity. It is certainly true that many Norwegians are wizards with skis and that beer and akvavit can often be found on lunch tables in Denmark. But there are exceptions. It is probably possible to find a Norwegian unable to ski and certainly not all Danes are fond of akvavit.

When dealing with the scientific analysis of complex systems it is important to keep in mind that the robustness and accuracy of terms and definitions describing entities at the macroscopic emergent level may depend on circumstances. The centre of mass of a set of molecules may capture the motion of a solid[5] but is far from enough to tell us how a liquid is moving about. The solid is an example of an emergent coherent structure encompassing essentially all the components allowing one to ignore differences in the behaviour of the individual components. This is often referred to as a situation where fluctuations can be ignored and the behaviour understood in terms of the on-average or majority behaviour. When this is the case even systems consisting of large numbers of individual components can be described in terms of a few parameters. The accuracy of such a description may vary depending on the controls imposed on a system from its surroundings; like in the example of a solid that becomes liquid as the temperature is increased. As the melting temperature is approached, the description in terms of the centre of mass behaviour will not continue to capture the dynamics.

The important point is to be aware that a given description can cease to be adequate and that a given conceptual framework may not be able to foresee its own limitations.

---

[5] To entirely specify the motion of a solid body, we of course also need to specify the angles that determine the orientation of the body as it moves along the trajectory of the centre of mass.

If, for example, one decides to describe the financial system in terms of smooth, slowly varying statistics, as traditionally has been common practise, then it is no big surprise that the theory fails to forecast sharp sudden crashes, though such abrupt events do occur in reality.

Biological evolution and ecology are prominent and important scientific examples where the dynamics is understood not by tediously following each individual. Instead, one tries to identify groups of components that are assumed to possess sufficient rigidity to be the building blocks of the systemic dynamics. To do this, evolutionary theory and ecosystems theory rely on the concept of species. Despite the crucial importance of the species concept in biology, the debate continues concerning how one should define and identify species [11, 275]. Is it best to define a species by listing a set of specific characteristic properties, say foraging preferences, social patterns, anatomic features, etc. or can species be effectively defined by looking at the genomes [11, 275]?

Sociology is in a similar situation. We cannot understand the forces responsible for the dynamics of history and society just by adding together the world views, preferences and actions of the individual citizens. Sociology needs to identify how the interaction between the individuals leads to the formation of groups with shared interests or shared outlooks, or whatever may generate the formation of rigid groups which can be treated as building blocks. The Marxist identification of classes according to how a person is related to the means of production was one such attempt. According to the classification, two main groups exist. One group consists of workers that do not own any machinery or other means of production. The other group, denoted the capitalists, contains the owners of the means of production. Despite its limitations, this classification was an attempt to identify the emergent sociological building blocks.

The formation of macroscopic rigid structures which exhibit some degree of stable robustness and which can possess some level of integrity is an open and active research field in many areas of complexity science. There exist a number of possible mechanisms which can be formulated and analysed mathematically and we will return to these in Part II. Some principles of general relevance can be identified and here we will outline the principles involved in a couple of scenarios.

First we look at the formation of coherent structures controlled by some effective energetic relation between the components. Next we consider synchronisation and finally we turn to clustering induced between agents that undergo evolutionary dynamics, which can be described as a kind of **diffusion process**.

2.3   Structural Coherence

The rigidity which is brought about when a liquid changes into a solid is an example of how coherence can emerge to span across huge numbers of components, even when direct interaction only involves nearby neighbours. The liquid can become solid when the erratic motion of the individual molecules at low temperature becomes so weak that the interaction between neighbouring molecules is able to restrict the motion. The change from relatively unhindered molecular motion in the liquid, to the constricted dynamics in the solid, happens without any change to the properties of the molecules, meaning no chemical reactions take place when water becomes ice. The dramatic change happens because at the lower temperature the displacement of one molecule in between its neighbours will necessitate an essential displacement of the neighbour molecules. In the liquid this is also the case but the disturbance, and the corresponding interaction energy, is negligible compared to the **kinetic energy** of the motion of the molecules at the higher temperature. In this sense the formation of the solid is a collective effect in which the molecules mutually restrict the motion of each other. Somewhat like a traffic jam. A truly fascinating aspect is that despite the molecules only interacting by direct forces with those nearby, the collective effect impedes the motion of molecules separated by zillions of molecules. This is what makes the solid rigid.

The competition between diversifying single-component behaviour and collective unifying behaviour is in one form or another responsible for many types of coherence, which may comprise a significant fraction of all components. For example, the formation of a consensus amongst people of different opinion. The diversifying effect would be the tendency to develop your own unique worldview with a distinctive opinion about whichever topic. The restrictive mechanism could be peer pressure in the sense that you feel an urge to align your opinion with your friends and maybe even more broadly your acquaintances. But even if peer pressure only works 'locally' amongst immediate acquaintances, consensus can be induced across the entire population.

Structural coherence can also take the form of synchronisation when temporal organisation is essential. Different types of synchronisation exist. It may involve many components executing the exact same dynamics, like sets of metronomes going tick-tock beat by beat, or synchronisation can involve a more elaborate but strictly coherent time evolution. For example, a symphony orchestra where the individual instruments of course do not play the exact same notes in unison, but still the different notes played by each musician are synchronised carefully with all the other members of the orchestra, see Fig. 2.4.

Collections of very different types of components are frequently found to be able to synchronise. The functioning of our brains is considered to depend strongly on patterns of synchronisation amongst the neurones [466]. Certain types of financial crashes involving herding, where large collections of traders suddenly behave uniformly, is also a form of synchronisation. A spectacular example of synchronisation from the world of insects consists of the gradual onset of fireflies emitting pulses of light in synchrony.

There are many different kinds of mechanisms responsible for coordinated synchronous dynamics. Of course a central controller can synchronise the dynamics, think of the conductor of a symphony orchestra or the air traffic controllers at airports. But fireflies are not centrally conducted and, as far as we know, nor is there a central controller directing the neurones of our brains. It is, however, possible for synchronisation to spread by travelling from one neighbour to the next in ways similar to the generation of spatially extended structure we mentioned above. In this way the entire set

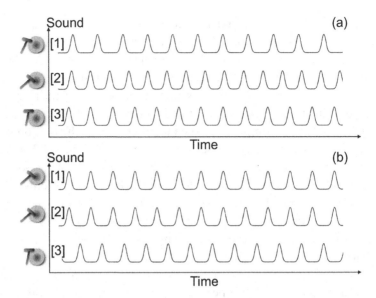

Figure 2.4 Synchronisation of beating gongs. In panel (a) the three gong players pay no attention to each other and their beats accordingly occur in a disorganised manner. In panel (b) gong players [1] and [2] are fully synchronised and their beats occur at exactly the same time. Player [3] likes syncopation and carefully arranged beats to appear exactly between the beats of [1] and [2]. We still consider [3] to be synchronised with [1] and [2], though with a constant lag.

of components can end up in a globally synchronised state even when interaction only exists amongst neighbours. And as for spatial structure, the existence or non-existence of synchronisation can depend on a competition between the internal dynamics of the individual component, say the firing rate of a firefly, and the strength of the influence of one component on the other. It is not fully understood how the blinking of one firefly influences the rate of firing of another firefly, but the fundamental importance of the interaction between the dynamical components has been known for a very long time.

The first scientific analysis of how synchronisation can be induced by interaction was given by Huygens in 1673, see e.g. [345]. Huygens was able to establish that his two pendulum wall clocks always ended up moving their pendulums synchronously whenever they were mounted in a way that made it possible for the vibrations produced by one clock to reach the other clock. If the two clocks are unable to 'communicate' via the elastic vibrations produced by their clockwork, no synchronisation is possible. In large ensembles of oscillators, it is found mathematically that the onset of synchronisation between the components happens at a sharp value of the strength of interaction and that the number of components which move synchronously then grows rapidly as the strength is further increased. The way the components synchronise can exhibit very intricate structures, especially through the formation of so-called chimera states, where the components arrange themselves in varying subgroups which internally synchronise while no synchronisation takes place between different groups.

The organisation of the groups may change with time. This kind of synchronisation has been suggested to be relevant to the dynamics of the brain [43]. It has long been known that epileptic seizure is related to an abnormal strong synchronisation that can involve large parts of the brain [415].

2.4   Evolutionary Diffusion

When we think of diffusion we tend to have in mind a process like the spreading of a droplet of cream in a cup of coffee. This kind of ordinary diffusion is similar to a **random walk** in which agents walk around at random in space, choosing the direction of the next step completely unrelated to the previous step. In the top panel of Fig. 2.5 an agent makes a move to the left. As time passes agents will make approximately the same number of steps to the left as to the right, so their trajectory will remain centred about their starting position. When we look at a collection of random walkers all initially starting from the same position, some of the walkers will make a much larger number of steps to the right, say, than to the left and will therefore end up far from their starting

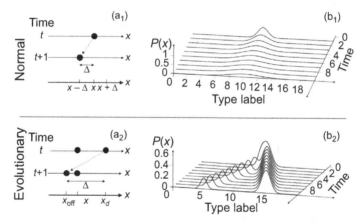

Figure 2.5 A sketch of the difference between ordinary diffusion and diffusion generated by evolutionary dynamics. Top row refers to ordinary diffusion. The agent has made a step from position x at time t to position $x - \Delta$ at time $t + 1$, depicted in (a_1). Bottom row refers to evolutionary birth–death dynamics. An agent at x duplicated at time t, a mutation occurs and at time $t + 1$ the offspring ends up at position $x_{\text{off}} \neq x$. The agent at position x_d dies at time t. The net effect is that an agent moves from position x_d to position x_{off} from time t to $t + 1$, as depicted in (a_2). The histograms of a population of agents are shown in (b_1) and (b_2). In both cases the agents are initially distributed about a narrow range of types, the central peak at type label 10 at time zero. As time passes the histogram corresponding to normal diffusion stays centred about the initial position, type level 10. The width of the peak increases and its height decreases. The effect is that a wider range of types becomes occupied with time. The peak describing evolutionary diffusion may, as depicted in (b_2), break up into more peaks which move around while they maintain their height and narrow width. This leads to separate clusters of types becoming occupied. See Fig. 10.10 for a simulation realisation of the process.

point while other walkers will do the opposite. Most of the walkers will make more or less the same number of steps to the left and to the right. For mathematical details see Chap. 10.

Assume that initially a collection of such agents are placed in a narrow region about position x_0, this could be the initial location of the droplet of cream and is illustrated by the peak centred about the position $x_0 = 10$ in panel (b$_1$) of Fig. 2.5. The peak indicates the probability that an agent picked at random initially has position x, so the area of the peak is equal to one. As time passes the agents will disperse and spread out because they each follow their random trajectory, but if steps to the left and to the right occur with equal probability, the peak will remain symmetric while it becomes broader. As the peak broadens it also becomes lower since the area of the peak remains equal to one as it gives us the total probability of finding an agent somewhere along the x-axis at a given time. The important conclusion is that ordinary diffusion will smear out the structure rather than maintaining or creating the structure. This is indicated in panel (b$_1$) of Fig. 2.5.

Now let us consider diffusion combined with evolutionary reproductive dynamics [256, 292], a phenomenon of broad relevance for instance to ecology [91] and evolution (see [169], p. 901) as well as nuclear reactors [199]. Consider agents of different types and assume for simplicity that they are characterised by one single parameter which we denote by the variable x, see panel (a$_2$) in Fig. 2.5; it could, for example, be the body weight of organisms of agents. We now assume that the agents at random instances duplicate asexually (somewhat like bacteria) in a process that consists of an agent producing an offspring with the same value of the property x except when a 'mutation' occurs, which then leads to a slight change in the x value of the offspring. This means that if the parent agent is characterised by the value x, the offspring will assume the value $x_{\text{off}} = x$ if no mutation occurs and $x_{\text{off}} = x - \delta$ or $x_{\text{off}} = x + \delta$ if a mutation of size δ takes place. The agents are also assumed to die randomly. Now look at the combination of the death of one agent with parameter x_d and the duplication of another agent at x producing an offspring agent with x-value x_{off}; we notice that effectively an agent has made a random walk step Δ from x_d to x_{off}, see Fig. 2.5.

Assume that initially all the agents are characterised by x-values in a narrow region about $x_0 = 10$, which as in the case of ordinary diffusion is described by a probability peak around x_0, see panel (b$_2$) in Fig. 2.5. As time passes, one finds that the evolutionary peak does not remain centred about x_0, as it does in the case of ordinary diffusion. The initial collection of agents will tend to break up into separate peaks that move about along the x-axis. In Fig. 2.5, panel (b$_2$) this is indicated by the two smaller solid-line peaks.

In this sense, evolutionary diffusion can maintain some degree of structure during one realisation of the dynamics of a collection of agents, which is of course what is typically done in observations. The degree of robustness of the peaks, or clusters of agents, is caused by the effective attraction towards regions with an increased population density. The move of the evolutionary diffusion process from one time step to the next consists effectively, as described in Fig. 2.5, in an agent being moved to the position of the offspring, which of course will have an x-value close to the parent agent and therefore increase the probability that agents effectively are moved towards the vicinity of a region with high population density. This behaviour corresponds to the dynamics of one specific realisation of the evolution of a population initially placed at x_0.

When we, say in computer simulations [256], average over many realisations of the stochastic evolution starting from the dashed peak about x_0, we will average over the many small peaks observed in the individual realisations. This averaging will eventually produce one broad peak about x_0 like the one observed at the top in Fig. 2.5. This does not describe the time evolution of the distribution of *one* population starting initially from the dashed peak. The averaging over the small peaks describes the distribution of many repeated realisations of the dynamics. The difference between the statistics of *one* realisation of the evolutionary dynamics – i.e. the moving about of sub-peaks in panel (b_2) of Fig. 2.5 – and the collapsing central peak describing the statistics of many realisations is an example of what is called lack of **ergodicity** or self-averaging, and has attracted much interest in economics [340] and in physics, describing an entirely new type of thermodynamic behaviour denoted as spin glass physics [112], which is also of importance to neural networks (see e.g. Kappen in [304]).

2.5 Breaking of Symmetry

In Sec. 1.2 we mentioned that symmetry can be broken (see Fig. 1.4 which illustrates how the formation of ice when water is cooled below the freezing temperature breaks the isotropic structure of the liquid form and replaces it by the intricate crystal structure of ice). Let us now elaborate on this concept and use evolutionary diffusion as an example.

The difference between the ordinary diffusion and the evolutionary diffusion, see Fig. 2.5, can be viewed as a difference in symmetry. The solid peak about x_0 of the ordinary diffusion preserves the left–right symmetry of the individual random walkers. The two solid-line peaks of the evolutionary diffusion break this symmetry.

The ordinary diffusion is described by symmetric peaks at all times. The symmetry originates in the symmetry between moving to the left and moving to the right whenever an agent chooses a new step. The symmetry at the microscopic level, the level of the dynamics of the individual agent, or component, is also present at the level of the collection of agents described by the probability distributions. The evolutionary diffusion is also symmetric at the level of the individuals. The mutant offspring is equally likely to end a little distance to the left as it is to end the same distance to the right of the parent. Nevertheless, a given realisation of the dynamics of a collection of agents initially distributed according to the central peak in panel (b_2) of Fig. 2.5 will at later times be described by distributions – the two solid peaks – which are not left–right symmetric about x_0. This is called symmetry breaking, because the macroscopic dynamics *breaks* the symmetry present at the microscopic level of the individual agents. As mentioned above, averaging the evolutionary diffusion process over many realisations of the stochastic reproductive and death processes will restore the symmetry by producing a peak of the same form as the solid line in the top panel. But this is not what one will observe by following a single realisation of the dynamics of a collection of reproductive agents initially distributed according to the narrow peak about x_0. We say that a single realisation of the dynamics breaks the symmetry and that averaging over many realisations restores the symmetry.

Emergence of structure associated with the breaking of symmetry is a fundamental phenomenon which is seen as underlying the emergence of structure in hugely different situations. The Big Bang theory of how our universe acquired its present structure is explained as consecutive events at which some symmetry was lost or, in other words, broken [489]. The transitions to a less symmetric situation made the four different fundamental forces (strong and weak nuclear forces, electromagnetism and gravity) separated out as distinct forces. Before these transitions the various forces did not have their own individual identity, symmetry made it impossible to distinguish one type of force from another.

Symmetry and breaking of symmetry is also deeply involved in our understanding of the properties of materials surrounding us [21]. Magnetic materials, for example, become magnetic because the interaction between the microscopic **magnetic moments** inside the material is able to establish a direction for the majority of the microscopic moments, i.e. a preferred direction which will break the directional symmetry of the individual magnetic moments. See Fig. 2.6. Individually the microscopic magnetic moments are as likely to point in one direction as in another. The total additive effect of moments pointing in random directions is zero since, say, one moment pointing north will be cancelled by another pointing south. This is a symmetric situation since no particular direction is distinguishable. The interactions between the microscopic moments can change this. The interaction favours alignment of pairs of moments and this can at low temperature lead to a majority of the moments pointing in a common direction. When this happens the effect of all the microscopic moments adds up and the material becomes magnetic.

Precisely at the temperature where a preferred common direction of the individual moments first appears, the system exhibits an exceptional sensitivity. This kind of special critical state is also found at the aggregate level for interacting components of many different types and in many different situations. Self-organised criticality, to be discussed in Sec. 12.1, studies such states as they occur broadly in complex systems. A local

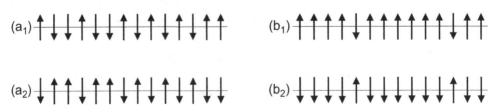

Figure 2.6 An example of up–down symmetry breaking. The configurations (a$_1$) and (a$_2$) possess the symmetry between up and down in the sense that there is no preference for up or down and when each arrow is flipped the overall structure of the configurations remains the same. This is clearly not the case for configurations (b$_1$) and (b$_2$) and we say they break the symmetry between up and down.

perturbation at a given position will have effects which can be sensed at a distance very far away. The spatial arrangement of the components will have fractal properties.

Symmetry breaking and new emergent behaviour of the electrons flowing in wires is behind the phenomena of **superconductivity**. When a material becomes superconducting the electrons can flow without any electrical resistance. This is, for example, used in the electric coils needed to generate the strong magnetic fields of MRI scanners used for medical diagnostics. The superconducting state becomes possible because a quantum-mechanical so-called gauge symmetry is broken [349] and allows in this case the electrons to move collectively as a coordinated body that is not slowed down by collisions with obstacles in the wire.

The different phase transitions observed in materials when they become magnetic, freeze or become superconducting are all examples of symmetry breaking which does not relate to any changes to the individual properties of the components. The chemistry of H_2O molecules is the same whether the molecule is found in a bucket of water or in an ice cube. The dramatic changes to the system properties caused by symmetry breaking are solely due to the mutual influence amongst the components, which allow the motion of the individual components to become restricted by collective effects. This kind of collective symmetry breaking is strictly speaking only possible mathematically if infinitely many components are involved. Which of course is never the case for real systems, but we expect only to observe the abrupt sharp change from one type of macroscopic property to another when the number of components is large. Systems with few components may still show smoothed-out versions of such behaviour.

Occurrences in our daily lives can also be considered related to the breaking of symmetry, regardless of whether the populations are infinite or not. Take the example of being for or against an issue. The population is symmetric in this respect if there are equal numbers of opposite opinion and this symmetry is broken when a majority of one viewpoint develops. Looked at in this perspective the formation of a preferred opinion can then be related to the methods used to analyse symmetry breaking [251], though care must be taken to account for details of social dynamics and social networks [149, 357].

Perhaps the most spectacular, and for us most important, symmetry breaking is the one that occurs after the first few cell divisions of the zygote on its morphogenic journey towards the embryonic state. This process is not fully understood, but we do of course know that chemical changes are involved, as suggested by Turing in his 1952 paper [458]. Turing used diffusion combined with chemical reactions to development a mathematical description. Very recent simulations which further develop Turing's approach and make use of dynamically induced symmetry breaking [131] are able to produce remarkable structures through morphogenesis.

2.6   Emergence of Networks

Like any other science, complexity science also progresses by cataloguing typical behaviours and developing methodology for analysis and modelling. We have already established that the interaction between components of a complex system can lead to

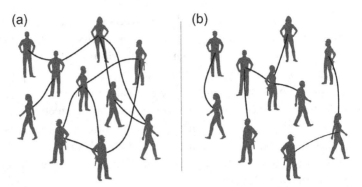

Figure 2.7 A population can be represented by different networks depending on which aspect is investigated. In panel (a) the links indicate shared work place; in panel (b) the links indicate friendship relations.

properties at the collective systemic level that have no parallel amongst the properties of the components. Our analysis will need to identify the essential aspects of the interactions and at the same time will need to identify which are the relevant building blocks or components in a given situation. When we think of the components as nodes and the interactions between the components as links, we can use network theory to describe our complex system. What we identify as nodes and connecting links will depend on which aspects we are interested in and therefore we can expect that a given complex system may be described by many different networks relevant to various emergent structures in the system, see Fig. 2.7. The structure and temporal dependence of the network may then help our classification, analysis and modelling. As an example, we will briefly look at neuroscience.

The scientific study of the brain is a very successful example of using network theory as an analysis tool. We know that the brain consists of a network of about 10^{11} neurones linked together through about 10^{14} connections [10]. The size of this network is, at least for the moment, too big to analyse directly. And even if we obtain computational power sufficient to directly make a model including all neurones and connecting axions, we still need to find a simplifying representation to facilitate our understanding of the functionality of this vast structure.

To reduce the numbers to a more manageable level, one can focus on networks of brain regions and their collaboration during specific tasks, i.e. identifying the effective components for that task. The identification of the components and their interaction are often equally subtle. For the brain the attempts to identify how the functioning of the entire brain is related to certain regions responsible for specific functionalities has a long history. In 1909 Brodmann [71] published maps of the brain identifying a number of areas classified according to differences in cellular structure. The relation of the Brodmann areas to function has been discussed ever since.

In recent years researchers have made use of brain scanners such as functional magnetic resonance imaging (fMRI), electroencephalograms (EEGs) and magnetoencephalography (MEG) in combination with a range of mathematical, statistical and probabilistic, or information-theoretic, techniques. These are used to identify the relationship between the time series from the scanners measuring activity at different positions in the brain. This has, for example, led to the identification of the resting state network [398], consisting of a network of brain regions which are found to be interdependent in a statistical manner even if an actual hard wiring between the nodes of this network in terms of axion connections has not been established and may not exist. As the name suggests, this brain network is identified as active even if the brain is not carrying out any particular explicit task. The dynamics of the networks activated during activity has been related to consciousness and creativity [356].

The description of the emergent structure of a complex system in terms of a network helps to identify commonalities with other systems and to formulate a mathematical analysis of functionality and stability. Networks emerging in different complex systems frequently exhibits similar topology. For example the degree distribution, which describes the number of nodes with a specific number of links, or the way different parts of the network are connected can be classified and related to how easy it is to pass information or energy across the network. It is often found that networks emerging in complex systems such as the brain have very broad degree distributions [197] and a well-balanced connectivity; this can be a way to obtain a well-connected structure without spending too many resources on unnecessarily many links [72]. Changes to the network structure of the brain are found to influence mental health. For example, the topological structure of the brain has been suggested to link to the onset of pathological behaviour such as psychosis [269] and excess connectivity of the frontal cortex may be related to obsessive–compulsive disorder [24].

Networks are now used across many different disciplines as a means of analysis. This opens up the possibility of using the concepts and language of network science to look for the general mechanism behind emergent behaviour [417]. In Chap. 8 we will discuss such details as we develop some of the mathematics of network science. The connection between the stability of processes unfolding on networks and the topology of the networks is obviously of great importance in many cases; besides the brain one may think of communication networks or ecology (e.g. the relationships between population stability and food-web topology [300]).

2.7 Temporal Mode

The difference in character between dynamics of individual components and the changes with time at the systemic level can be one of the most spectacular manifestations of emergence. The relevant time scales of the two levels can often differ by a very big factor. Just think of the brain. The firing dynamics of individual neurones happens at the scale of milliseconds. Reaction times of the brain are around hundreds of milliseconds and the contemplative thought processes can of course span anything from seconds to years. Similar separation is seen between the generation times of organisms

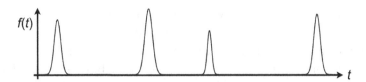

Figure 2.8 A sketch of intermittent activity. Time is along the x-axis and the y-axis is meant to indicate the level of activity at a given moment. One might think of earthquake activity or neuronal activity in the brain.

and the evolutionary time scales of species. In addition to the difference in the rate between the individual and the systemic level, the mode of the dynamics is also often very different. Even when the individual components tick along at a more or less even average pace, patterns of abrupt collective activity can suddenly happen. At the systemic level this can lead to long stretches of quiescence separated by sudden bursts of hectic activity. An example of this kind of intermittent dynamics is observed in the brain [433] and has been suggested to be associated with consciousness [331]. Other examples of intermittent systemic dynamics are, for example, the abrupt disruptive events in finance with ever-returning crashes. We can also think of how the slow motion of the earth's tectonic plates gives rise to intermittent earthquakes, or mention how mass extinctions have happened intermittently during the history of life on earth. The intermittent macro evolution in the form of extinction and creation events, as documented by the fossil record, has been denoted by the punctuated equilibrium of Eldredge and Gould [128, 169] (Fig. 2.8).

Similarities in the intermittent activity of very different systems have been found, for example, in terms of the distribution of sizes of the abrupt events and the time interval between them. The size of events, for example in terms of the energy released during an earthquake or the number of species disappearing during a mass extinction, is observed to follow very broad power law-like distributions in large classes of systems and this observation has been suggested to be a consequence of a general type of dynamical behaviour called self-organised criticality [34, 214, 354].

The distribution of the time intervals between the abrupt events is found to exhibit similar behaviour in very different systems such as earthquakes [462], granular piles [148] and ants leaving their nest [364]. When the interactions between the components can trap and store some kind of stress, the abrupt events will release the energy stored and decrease the stress in the system. As a result, the frequency of the abrupt events will decrease with time as the systemic stress level diminishes [406]. This mechanism will make the dynamics non-stationary and be accompanied by a change in the properties of the system. The details of the change will depend on the specific system but will in all cases correspond to the gradual development of more stable configurations; on average, the upheavals caused by the abrupt intermittency will, with time, happen less often. Examples of such slowing down of the dynamics might be seen e.g. in the Omori law [462] for the frequency of earthquake aftershocks or the decrease of rate of mass extinctions [310]. The precise mathematical description will be discussed in Chap. 12.

2.8   Adaptive and Evolutionary Dynamics

Adaptive evolutionary dynamics can be viewed as both an example of emergence and a very effective motor for generating emergent collective structure. The most famous and important example must be evolution by natural selection, as first advocated by Darwin and Wallace [93]. The entire fantastic taxonomic structure around us we now presume is a consequence of co-evolution and adaptation, and this demonstrates that evolution and adaptation is a dynamical mechanism able to generate all the emergent features mentioned in this chapter:

(1) Emergence of a characteristic scale, like the size of a certain type of animal.
(2) Emergence of robust macroscopic structures describable by a few collective parameters, e.g. a bird flying through the air.
(3) Symmetry breaking, e.g. the formation of distinct types of organism possessing nearly identical properties, like a species (cf. Sec. 2.4).
(4) Emergence of networks, e.g. ecosystems.
(5) Emergence of intermittent dynamics, e.g. punctuated equilibrium.

To further illuminate how evolutionary dynamics encompasses all these forms of emergence, let us for argument's sake think of the evolution of life on earth in the following schematic way. Assume life started out from the proverbial primordial soup, see e.g. Chap. 3 in [284]. Initially we have some collection of self-reproducing molecules with little taxonomic structure in comparison with later organismic life. Somehow the reproduction, mutation and adaptation of the molecules leads to the formation of cells from which multicellular organisms develop. These organisms will, as soon as they appear on the evolutionary stage, influence each other and thereby contribute to the co-evolutionary selective pressure which will contribute to the adaptation resulting from selection upon random mutations. Groups of similar types appear and we can think of these groups as the formation of species. The entire taxonomic structure is emergent. We classify the different members of the taxonomy by a few parameters (number of legs, spine or no spine, etc.). The lowest level in this hierarchy, just above the level of the individuals, we denote species and the species do possess a degree of robustness (although they eventually may go extinct).

The dramatic mass extinctions motivating the concept of punctuated equilibrium [169] are examples of emergent intermittent transitions and the web of interactions in the ecosystems, e.g. the food webs, are emergent networks. The very fact that evolution creates structure in the form of multicellular organisms and ecological networks is an example of symmetry breaking in a number of ways. The primordial soup was presumably spatially symmetric in the sense that the reproducing macromolecules floated around in some kind of broth without the formation of spatial structures. Already at this stage the symmetry of time was broken in the sense that evolution self-organised a direction towards a higher degree of organisation, represented by higher forms of life. At the level of random mutations there is, as far as we know, no particular preferred direction to the arrow of time. Good and bad mutations are equally likely to occur in this sense, as the genome is equally likely to evolve in a detrimental as in a beneficial direction. But the collective adaptation of the co-evolutionary process breaks this symmetry and higher-order structure evolves.

Of course we do not know in exact detail how the evolution of life on earth proceeded, but it is certainly possible to make abstract models that support the scenario just sketched, see e.g. [216] and Sec. 11.3.

Summary: We have discussed a variety of systemic properties which have no counterpart amongst the properties of the individual components. These properties are generated by the interactions between the components and can lead to new structures in space and time. Because emergent properties are an accumulated effect of many interacting components rather than determined by the inherent properties of the individual components, aggregates of entirely different components can possess the same type of emergent properties.

We looked at the emergence of:

- Characteristic scales.
- A critical state where all scales are important.
- Macroscopic robustness.
- Structural coherence in space and time.
- Symmetry and its breaking.
- Network structure.
- Intermittency.

2.9   Further Reading

General audience:

In the popular science book *Critical Mass: How One Thing Leads to Another* [40], Philip Ball argues that the analysis of emergent phenomena in political science, sociology, economics, biology, etc. can benefit from ideas and mechanisms, appropriately modified, for collective behaviour first developed in physics.

A non-technical explanation of various conceptual aspects of emergence can be found in the high-quality popular science book *Emergence: The Connected Lives of Ants, Brains, Cities and Software* [224] by Steven Johnson.

Intermediate level:

The introductory chapter by Robin Findlay Hendry, Sophie Gibb and Tom Lancaster in *The Routledge Handbook of Emergence* [155] is an up-to-date discussion of the philosophical aspects of emergence with a historical dimension.

The concept of symmetry breaking is lucidly explained in the 1972 paper 'More is different' [18] by Philip Anderson.

The emergence of scale-free phenomena is discussed enthusiastically in the book *Scale: The Universal Laws of Life and Death in Organisms, Cities and Companies* [491] by Geoffrey West.

A quantitative discussion, requiring only little mathematics background, of complexity and lack of scale is given in the book *Complexity and Criticality* [84] by Kim Christensen and Nicholas R. Moloney.

Advanced level:

The discussion by R. Noble and D. Noble in Chap. 31 of [155] is a wonderfully readable explanation of how emergence is fundamental and tangible in biology, with examples of circular, upwards and downwards causation.

It is worthwhile to be aware of David Chalmers's suggested classification in terms of strong and weak emergence [81]. It is an open question how to do this in a precise and quantitative way.

2.10 Exercises and Projects

Exercise 1

List examples from biology, sociology and neuroscience/psychology which involve the emergence of some kind of:

(1) Clustering or grouping amongst components.
(2) Intermittency in time.

Exercise 2

Make a list of examples of synchronisation across biology, physics, economics, neuroscience, etc.

Exercise 3

Consider the emergence of social class and the emergence of the way we perceive ourself. Discuss if it is an example of two interacting emergent phenomena, as seems to be implied in [277].

Exercise 4

Discuss and compare the views of complexity as expressed in Herbert Simon's 'The architecture of complexity' [412] and Phil Anderson's 'More is different' [18].

Exercise 5

Discuss the different scales relevant to the organisation of:

(a) Human society.
(b) A forest ecosystem.
(c) The brain.
(d) The financial system of a country like Italy and like China.

Exercise 6

Discuss what might be considered the robust emerging degrees of freedom of:

(a) Human society.
(b) A forest ecosystem.
(c) The brain.
(d) The financial system of a country like Italy and like China.

Exercise 7

Consider whether the concept of symmetry breaking can be defined in a way to make it of value to our understanding of:

(a) Social transitions such as industrialisation.
(b) Political revolutions.
(c) Shifts in moral code.
(d) Changes in accepted fashion.

Exercise 8

The zoology of emergence. In this chapter we listed the following examples of well-established classes of emergence:

(1) Characteristic scale.
(2) Collective degrees of freedom.
(3) Transitions.
(4) Networks.
(5) Temporal mode.
(6) Adaptation.

Choose two very different types of complex systems, e.g. the human brain and a biological ecosystem. Now think through the following:

(a) Identify for each of the two complex systems specific instances of the above six examples of emergence.
(b) How can each of the six classes of emergence be interrelated? For example, how may adaptation shape networks within each of the considered complex systems and further, how may the network structure link to transitions, etc.?
(c) List examples of emergence not included amongst the six types listed above.

Keeping in mind the two complex systems you chose above, now think of the following very general questions:

(d) Try to list, from as many different disciplines as possible (biology, economics, neuroscience, etc.), examples of emergence of emergence, for instance something like the emergence of interaction (competition and cooperation) between companies that themselves are emergent economical structures.

(e) For each of the examples of emergent structures you listed under (d), make lists of intrinsic processes (mainly) operating within the structure and extrinsic processes (mainly) operating between the structures. For example, think of a biological organism. Intrinsic processes within the organism can include transport of nutrition and cell division. Extrinsic processes connecting different emergent structures can be the trophic interaction defining a food web.

Project 1 – A taxonomy of emergence
Discussion of the paper 'How emergence arises' by the ecologist de Haan [102]. De Haan makes a number of suggestions for how we understand what emergence is and how we can distinguish between different types of emergence.

(a) What does de Haan mean by the 'conjugate'? Do you see the concept as useful. Can you think of examples of conjugates in some different complex systems, such as for example the brain, the economy, ecology?

(b) Do you agree with de Haan that it is beneficial to involve the observer in the analysis of emergence?

(c) Describe in as few words as possible de Haan's three types of emergence: discovery, mechanistic emergence and reflective emergence.

(d) Do you consider the three types of emergence in (c) as a useful classification and do you find them exhaustive or would you distinguish other types?

(e) Choose a complex system you feel somewhat familiar with. Now use the framework developed by de Haan to write a brief plan for a research project aimed at identifying and analysing the system's emergent properties.

Project 2 – Adaptation and intermittency
Eldredge and Gould have argued that the fossil record exhibits intermittent dynamics, which they named punctuated equilibrium [169]. That adaptive evolutionary dynamics is able to generate intermittent abrupt upheavals was argued in [473] to be of relevance also to financial crashes.

(a) Both organisms and companies viewed individually are born and die at a more or less constant rate. How does intermittent macrodynamics co-exist with smooth gradual dynamics at the level of the individuals?

(b) What might be the mechanism in adaptive dynamics that generates intermittency at the macro level?

(c) If adaptive dynamics inevitably leads to intermittency, will it then be impossible to avoid or mitigate financial crashes?

(d) List examples of intermittent dynamics and discuss to what extent they may contain elements of adaptive dynamics. For example, the development of human societies through revolutions and wars.

Project 3 – Scale and scale invariance
This project is somewhat open ended. It studies several papers to develop some insights into the importance of the co-existence of brain structures possessing characteristic scales with scale-invariant (power law dependence on separation) correlation function and power-distributed bursts of brain activity.

Recall the discussion in Sec. 2.1. In [136], analysis of fMRI scans of the brain found that the BOLD activity in the brain lacks a characteristic scale. Scale invariance and criticality is described in [281] as crucial for a healthy brain.

From the same type of scans it is routinely concluded that the brain possesses well-defined structures, see Fig. 2 in [463] consisting of brain regions particularly involved in certain tasks such as vision and motion. These regions appear to form collaborative networks, see Fig. 3 in [463] and [72].

(a) Try to think about how the BOLD signal of fMRI, when averaged over time and space, can lead to a scale-invariant correlation function while when analysed as in [463] it allows the identification of characteristic spatial regions.

In agreement with the observed behaviour of the BOLD correlation function [136], indications of scale-invariant brain dynamics were also seen at the microscale of bursts of neuronal firing [56] and at the macroscale of BOLD signal activity [433].

(b) Try to think about how the burst activity needs to be organised in space and time in order to be able to be described by a scale-free correlation function and also at the same time to exhibit spatial structures and networks when analysed as in [72, 463].

Project 4 – Synchronisation
Here are a number of questions which are meant to stimulate thinking and should be illuminated by web search and open access journals.

(a) Why is synchronisation a problem in finance (herding), the brain (epileptic seizure) and pedestrians passing bridges?
(b) Why and how do fireflies synchronise?

Let us now focus in more detail on synchronisation amongst humans. The following is concerned with a critical reading of two important synchronisation papers that discuss human behaviour and on the basis of these, looking closer at how the state of mind during improvisation can be further investigated.

(c) Make a short summary of the review article on synchronisation-coordinated human activity [316] and include a brief discussion of how useful you find the concept of synchronisation as a tool to understand collective human activity.

(d) Now do the same for the more specialised study of collaborative learning [317].

(e) On the basis of your considerations in (c) and (d), consider the suggestion that a special improvisational state of mind exists and is utilised by classical musicians when interpreting a composed peace in the more creative let-go performance style [109].

(f) Imagine you had access to brain scanners such as EEG and MEG equipment or other devices you would find useful and a set of top trained musicians, how would you design an experiment that would be able to check if human synchronisation of some form takes place during the improvisational state of mind?

Project 5 – System spanning structures and connectivity

This project is concerned with the review article 'Connectivity and complex systems: Learning from a multi-disciplinary perspective' [459]. The article is rather long, but a very detailed reading is not needed. It suffices to skim read to become familiar with the authors' use of the terms

(1) Emergence.
(2) Structural connectivity vs. fundamental connectivity.
(3) Fundamental unit.

Next we want to think about the way these three concepts are developed and used in the review in relation to the discussion in Chaps. 1 and 2.

(a) Consider the definition of emergence and complexity science given in the summary to Chap. 1 and discuss a comparison with the review article's notion of item (1) above.

(b) Discuss item (2) above in light of the discussion of processes in Sec. 1.1.

(c) Relate item (3) to the notion of robust degrees of freedom discussed in Sec. 2.2, structural coherence discussed in Sec. 2.3 and breaking of symmetry discussed in Sec. 2.5.

The authors say on page 6: 'However, deriving quantitative descriptions of emergent behaviour is challenging. Theoretical tools to relate emergent behaviour to structure and function are lacking, and by separating SC from FC we make it even more difficult to analyse emergent behaviour.'

We can use this somewhat pessimistic view as a motivation to take a look at what the present book presents as available tools for the analysis of emergence. Go through the Synopsis paragraph at the beginning of each of the chapters of Part II.

(d) Now think for a moment about how the view and conclusions of this multi-disciplinary review might have changed had it included mathematics, statistical mechanics and information among the disciplines.

3   Specific Types of Emergent Behaviour

> **Synopsis:** We present a few paradigmatic modelling strategies selected due to their generality in addressing the emergence of spatio-temporal structure, including criticality, synchronisation, intermittency, adaptation and forecasting.

We continue the discussion of concepts and processes of importance to emergence. We will not attempt to make a comprehensive list of examples of complex systems, but will mention a few central specific models and some of their applications as typical examples of systems-level structure relevant to emergence in space and time. The sketch in Fig. 3.1 presents a brief summary.

Spatial organisation may possess overall features which do not change with time. Such stationary behaviour is characteristic for systems in thermal equilibrium. In the following sections we will first consider the very minimalistic, but exceptionally efficient, modelling approach captured by the **Ising model**,[1] which originates in the analysis of equilibrium systems. Starting from binary components, the model studies how competition between opposite tendencies can generate either ordered or disorder configurations. When the two tendencies are in balance, particularly fascinating configurations without any specific scale or pattern are possible. These configurations are related to fractals and correlations that can reach across the entire system.

This kind of modelling has been very successfully applied to the physics of phase transitions, such as when a magnetic material loses its magnetic moment above a certain temperature. Although the careful scientific study of phase transitions originates in physics, similar competition between opposing tendencies goes beyond physics and analogous changes between different types of systemic behaviours are frequently found in other fields of science.

From sociology we can mention the segregation transition of the Schelling model [386, 423], concerned with the distribution in space of a population consisting of two types of agents. The disordering tendency is caused by agents searching their location of residence at random. In addition, there is a tendency towards the formation of structure because agents have a tendency to reject a location in a neighbourhood which is sparsely inhabited by agents of their own kind.

The **error threshold** of Eigen and co-corkers [105, 122] is an example of an order–disorder transition from evolutionary biology. Order and coherence are imposed by the

---

[1] The simple stylistic description was introduced by Wilhelm Lenz in 1920 and solved in the simplest one-dimensional case by Ernst Ising in 1924. See e.g. [392] for a discussion of magnetism and the Ising model.

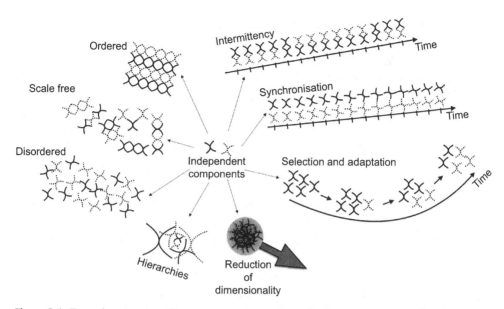

Figure 3.1 Emerging structure. Two components are shown in the centre, surrounding them are depicted typical types of structures which may be generated when components interact. The static structures to the left, i.e. ordered, scale-free and disordered, can arise as a result of competition between ordering and disordering processes. When the ordering is strongest, patterns reaching across the system can arise. When disorder wins, randomness may appear. When there is a balance between ordering and disordering processes, many different kinds of patterns of many different sizes may appear, leading to so-called scale-free configurations. Structures in time generated by the dynamics of the interacting components include examples such as intermittent rearrangement, where periods of little change are interrupted by sudden dramatic reorganisation. To the right, various temporal behaviours are depicted. Different forms of spontaneous, or driven, synchronisation occur commonly. When components are able to multiply and are subject to selection and adaptation, the composition of the population will change with time. The interaction between the components may produce a rigidity that allows the description of the time dependence in terms of just a few parameters, allowing us not to specify the dynamics of the individual components. This corresponds to a reduction of dimensionality. Although we list the various mechanisms for emergent structure separately they may act simultaneously, for example as is often the case in biology, and the result can be the emergence of hierarchies.

inherence of adaptively selected traits. Random mutations tend to work against this ordering mechanism. When the error in the inherited traits becomes too frequent due to a high mutation rate, beneficial traits disappear from the population as fast as they appeared and no adaptation can occur.

Dynamics and evolution are integral to the understanding of most complex systems, assuming equilibrium and stationarity is too restrictive when dealing with, for example, evolutionary ecology, economics, sociology or the dynamics of the brain. Equilibrium and stationarity is commonly replaced by intermittent dynamics consisting of long periods of little activity separated by abrupt, relatively short, hectic activity. This type of scenario was denoted as punctuated equilibrium in palaeontology [169]; it is a general feature of complex systems dynamics, which makes it important to study models of transitions between different structures and how to forecast approaching sudden changes. We will therefore also discuss model strategies that allow analysis of how intermittency may emerge as a consequence of interaction amongst components and how the associated abrupt changes may be forecasted.

Different types of synchronisation is another possible consequence of interaction between dynamical components. For example, fireflies may synchronise completely, in which case all the fireflies emit a flash of light periodically at the same time. Or the synchronisation may only be partial. This will involve some degree of synchrony amongst groups of fireflies and can also involve synchronised groups flashing regularly but with a time lag between different groups [358]. This is similar to when musicians make use of syncopated rhythms and comparable to a time lag between a periodically occurring cause and its effect, such as the relationship between the number of hours of daylight and the seasons.

As the strength of interaction between dynamical components changes, transitions between synchronised and non-synchronised dynamics can occur. Such transitions can be observed amongst collections of fireflies and appears also to be related to transitions in information processing in the brain [262]. The Kuramoto model offers a simple and very useful description of such transitions [345].

Finally we look at how the behaviour of collections of many components can sometimes be captured by a few parameters. This is similar to when we use the centre of mass to describe the motion of a body consisting of many parts.

3.1   Ising-Type Models: Transitions and Criticality

A collection of many components can be structured in multiple ways. If the components, whether molecules or agents, move around totally oblivious to each other, no coordinated patterns or structure will be produced. Although atoms at high temperature in the gas phase may bump into each other, and thereby interact, the collisions are too random and dominated by the kinetic energy of the individual particles to cause any formation of spatial structure. The interaction between the atoms has to be strong enough to curb the kinetic energy. Similarly for people settling in an urban area; if they choose their residence without relating to their fellow inhabitants, no structure in composition of the population will appear. Structure and interdependence, or interaction, are intimately related.

The Ising model [392] is probably the simplest possible description of how interaction between components can lead to structure. The model is schematic and consists of components placed on a lattice, or if one prefers on a network, as indicated in Fig. 3.2. The components are numbered $i = 1, 2, \ldots, N$ and can be in two states denoted by a label S_i, perhaps a natural starting point in the age of binary information processing, though originally the choice of two states $+1$ and -1 was motivated by the quantum mechanics of magnetic materials. We can, for example, imagine the two choices $S_i = 1$

Figure 3.2 A section of a lattice of binary components. The plus or minus signs indicate that the variable at the site assumes the value $+1$ or -1, respectively. The interaction energy between S_i and S_j is at a minimum when $S_i = S_j$.

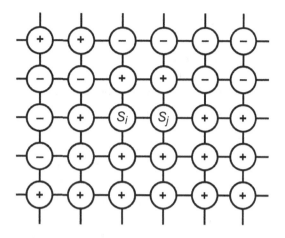

or $S_i = -1$ as denoting the microscopic magnetic moment of an atom. Or it could denote the decision of being for or against a certain issue, like staying or leaving a certain neighbourhood as exemplified in Schelling's model, or it could be describing whether a location labelled i is occupied by a molecule or not, etc.

In the physics interpretation of the Ising model, the choice $S_i = 1$ or $S_i = -1$ is determined by energy. Neglecting interactions $S_i = 1$ or $S_i = -1$ will have the same energy, when we assume no external field is present, but including interactions with a given set of neighbours will change this. We need to consider the energy stored in the interaction between two neighbours S_i and S_j, see Fig. 3.2. This energy represents how one of the components S_i influences another component S_j. Let us assume the lowest energy is obtained when $S_i = S_j$. That is, the interaction energy is lower when the two members of the pair assume the same value in comparison to assuming opposite values. When the lowest energy is favoured, the preference would be to let all S_i assume the same value. Either all $S_i = 1$ or all $S_i = -1$. In which case we are left with only two configurations for the entire set of N components. If we think of the model as describing opinions, this uniform configuration might be a consequence of the desire for agreement being stronger than the importance paid to a specific choice of opinion. We can think of being for or against smoking or alcohol consumption as examples where being aligned with the opinion of peers can be a crucial factor and as a result everyone might end up either being for or against, leading to a total consensus.

Evidently, the world is in general more complicated. When dealing with the physics of magnets, thermal energy can supply the energy needed to make a pair of S_i and S_j assume opposite values.

When using the binary variable S_i to describe the opinion of voters, opposite signs of the S_i and S_j may perhaps occur with a certain probability because two neighbours are incapable of agreeing on a specific issue, even if etiquette would encourage aligned opinions.

If thermal energy for the magnet, or various individual preferences in the case of the opinion formation, dominates, the configuration of the S_i variables will of course not consist of uniform allocations of the same value to all the variables S_i. The 'individuality' of the components will dominate and each S_i will essentially assume the value $+1$ or -1 at random, independently of its neighbours. There are as many as 2^N of these configurations.

When looking at the magnetic interpretation of the S_i, or the opinion representation, a competition between different preferences is relevant. When the magnet is heated to a certain temperature, the interaction between the components S_i will have to find a balance between minimising the interaction energy between neighbour pairs and storing the thermal energy by allowing certain neighbours to have opposite signs for their variable S. Equivalently, when thinking in terms of opinions, a competition between the wish to agree and some other concerns may be relevant. This kind of competition between opposing tendencies is of broad relevance.

Having realised that the state of our system is influenced by a competition between two opposing tendencies, we notice that the ordering tendency points the system towards one of only two configurations: all S_i equal to $+1$ or all equal to -1. In contrast, the disordering tendency pushes our system towards one of the large number of configurations in which values of S_i are assigned at random. From this consideration we see that it is to be expected that a change between order and disorder may happen if we are able to tune the balance between the two opposing tendencies. What we cannot know from this conceptual discussion is whether the change happens abruptly or not. Consider the magnetic interpretation: as we make more thermal energy available by raising the temperature, will the change from a completely ordered configuration of the S_i at low temperature to a totally disordered one at high temperature happen at one particular value of the temperature or will it happen gradually as the temperature increases? Experience and mathematical modelling, discussed later in Secs. 6.2 and 6.3, show that often the disorder gradually becomes more important up to a specific value of the temperature, or some other relevant control parameter, after which the disorder dominates. The poised competition, precisely at the transition where the ordering loses to the disordering, brings about properties of surprising relevance to many complex systems in many different situations. Perhaps because there is some kind of balance between opposing tendencies at the transition, one talks about a *critical point* and a *critical phenomenon*.

In later chapters we will use mathematical formalism to discuss all these aspects in more depth, right now we simply emphasise that competing interactions between components can lead to emergent macroscopic configurations with totally different regimes separated by a critical transition. The structure at the critical point is special by not possessing a particular characteristic scale. We say that we deal with a **scale-invariant** phenomenon. We are familiar with this kind of situation when we look at a photo and cannot decided whether we are looking at pieces of rock or entire mountains, see e.g. Fig. 3.3. The absence of a particular characteristic scale is related to a repetition of the same structures at different scales. A small part of a mountain, when magnified, looks like the entire mountain. And a little piece of broccoli, see Fig. 3.4, looks like the entire broccoli. Since each part looks similar to the whole structure, we say the structure is **self-similar**. This corresponds to the emergent configurations having fractal properties.

Figure 3.3 What is the scale? Are we viewing the coast of mountainous Greenland and the waters of the North Atlantic Ocean? Or is this a photograph of a few metres of exposed bedrock adjacent to a small lake or stream? Answer.[2]

Figure 3.4 Fractal and self-similar structures. Photographs of (a) broccoli and (b) a magnified part of the same floret. The magnified part appears essentially identical to the whole section it was cut from.

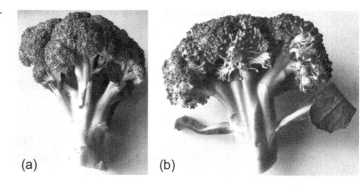

(a)   (b)

We will make use of self-similarity later in our mathematical analysis, see for example Chap. 5.

Mathematical phenomena without a characteristic scale have statistical properties that are not captured by the familiar normal, or Gaussian, distribution. In contrast, away from the critical point one will find that the system is described by a certain length and time scale and the normal distribution will commonly describe the statistical properties.

We will elaborate on the difference between a scale or no scale being present in the next section, Sec. 3.2, using a specific example from networks and return to the discussion with more precision in Part II, but here we can try to make intuitively clear the significant difference between the existence of one characteristic scale and the lack of a particular scale.

For this we will use the production of shoes as an example. Although different people have different size feet, the variation is not very big and shoe factories can cater for most people by producing a relatively small range of shoe sizes. Had the size of the human foot not followed a fairly narrow and peaked distribution, it would have been very difficult to provide for everyone through industrially produced shoes. If there were a significant number of people with feet of length anywhere between, say, 10 cm and

[2] Photo from Kinugawa River near Kawji-Onsen Station taken by the author on 14.4.2018.

100 cm, or even longer, factories and shops would find it difficult to make the right shoe size available for everyone.

Despite the fact that we may tend to think in terms of narrow distributions and hence typical sizes, many important exceptions exist. Most dramatic is probably the size distribution for earthquakes given by the Gutenberg–Richter law. Precisely because earthquakes of extremely different magnitudes do occur every so often, it is difficult to estimate how strong a building needs to be to withstand the shock of all the earthquakes it may be exposed to.

Across fields like biology, economics, finance, geophysics, astrophysics, neuroscience, etc. phenomena with essentially no typical scale emerge from the processes between the interacting components [34, 88, 435, 491]. This has made critical systems relevant far beyond the field of physics where they were initially studied. In the next section, as an example we will consider models of how networks with or without a characteristic scale can be generated.

3.2   Network Models and Scale vs. No Scale

In the preceding section we imagined components to be placed on a given existing structure like the one indicated in Fig. 3.2, and that the value of the variable on a particular site was determined either by thermodynamics, for the magnet, or by some social consideration in the case of opinion or choice dynamics. The focus was on the order vs. disorder of the configurations emerging on the lattice or network hosting the components. The lattice was supposed to represent either the crystal structure of the magnetic material or the social connections between the agents forming opinions. We will now change our focus and discuss how different structures of networks representing the connections between components may emerge.

First we notice that the same set of components may very well be connected in many different ways. Think of a group of people. A person may be related to another person because they share the same interest in, for example, music. Or the person could be related to other people because they share the same employer, etc. To describe such relationships, networks are very useful. Here we would think of people as the *nodes* of the network, representing existing specific relationships, shared interest or shared employer as a *link* between two nodes symbolising two persons.[3] In some cases it is straightforward to decide how to identify links between nodes. If we think of a train network, the stations are natural to identify as the nodes and we could decide to connect two stations by a link if there is a direct rail track between them. In other cases the identification of nodes and the links between them may be less straightforward. Consider the brain as an example. At a very microscopic level we could decide to consider the cell bodies of the neurones as the nodes and the axons connecting neurones as the links. Or we might consider local brain regions, say the volume of a

---

[3] There is a huge literature on networks, one very good place to start is [312].

voxel of an fMRI scanner or an anatomically identified region, as the nodes and link these nodes together if there are measurable correlations or some identifiable causal interdependence between two nodes. This is just to point out that the identification of nodes and relationships between them can in itself be a challenge. In Chap. 9 we will discuss how, for example, the use of information theory can be useful in this respect.

Let us proceed here in the same abstract and simplified way as in the previous section, where we tried to focus on the effect of interactions by representing the properties of a component, the microscopic magnetic moment of an atom or the opinion of an agent, by a binary variable. Similarly we may start out with very simple models in order to develop some understanding of the possible network structures that may emerge when different nodes and links are put together according to given sets of rules. As always when analysing complex systems, we are interested in understanding how the macroscopic properties are related to the interactions and interdependencies between components. For networks we will consider how different procedures for connecting nodes relate to networks with different properties.

It is useful to start by considering network structures that may arise in cases where components are paired at random. We can imagine people bumping into each other as they try to navigate a busy shopping street or a train station. We consider each person to be a node and a link is established between two persons if they collide (one or more times). In abstract form we can represent this as N nodes, the number of people considered, and place a link between any two persons with probability p representing the chance that collision occurs between the two.

In the literature the word 'graph' and 'network' are often used synonymously, and the random procedure we have just introduced is the Gilbert model of a random graph, see [157]. It is very much related to the Erdös–Rényi random graph model, see Sec. 8.5.1 below and e.g. [312], Sec. 11.1.

Deferring the mathematical details to Chap. 8, we can nevertheless notice here that the resulting network will be characterised by a specific 'scale' in the sense that there is a most likely number of links connected to a node. We see this in the following way. A given node can be linked to any of the other $N - 1$ nodes. Since a link is placed between each of these pairs with probability p, the average number of links starting from the node will be $(N - 1)p$. This average degree can be seen as the typical number of links one will find connected to a node chosen at random, and in this sense it is a characteristic scale of the network. Our randomly produced network is an example of a system with a well-defined average and typical scale, and therefore it does not possess scale-free or critical properties in the sense discussed above in Sec. 3.1.

Let us next describe a procedure for constructing a network which generates a scale-free structure in the sense that the nodes do not have a typical degree. The degree of a node is the number of links attached to the node. The probability, $P(k)$, that a node selected at random has degree equal to a certain number k is called the degree distribution. A network is said to be scale-free if the function $P(k)$ does not exhibit a specific scale. Figure 3.5 depicts the difference between a distribution with a scale and one without. The peaked dashed curve possesses a scale, namely the most likely degree

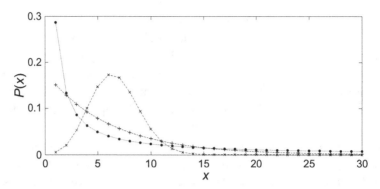

Figure 3.5 A binomial (crosses), an exponential (plus signs) and a power-law (dots) distribution on the integers $1, 2, \ldots, 30$. The three distributions have the same average of about 6.6. The most striking difference is the presence of the peak in the **binomial distribution**. Essentially all probability is concentrated in this peak. Both the exponential and the power law have their maximum at one and are monotonously decaying. The essential difference between the exponential and the power law is that the probability for events that are exponentially distributed rapidly becomes vanishingly small above the average value, whereas the power law allows a significant probability even for very large event sizes.

given by the location of the maximum of the peak. The smooth decreasing dotted curve does not make one specific degree stand out, the functional form is a power law, i.e. of the form $P(k) \propto 1/k^\tau$, and power laws do not exhibit any scale in the following sense. We look at relative change when we compare the probability of finding a degree-k node with the probability of finding a node with a degree which is x times bigger, i.e. equal to xk. The ratio between these two probabilities is given by $P(xk)/P(k) = 1/x^\tau$. The relative change in probability only depends on x. This means that the relative change in degree probability between degree $k = 10$ and $k = 100$ is the same as the change between 100 and 1000. So independent of where we are on the degree scale. The power-law distribution also has a most likely size, and it can also have an average, but neither of these will characterise the typical size drawn at random from the power-law distribution. This was indicated in Fig. 2.2.

Networks without a scale appear, at least in approximate form, in many different contexts in the world around us [393]. There are several different algorithms able to generate scale-free degree distributions, the one we describe here is simple and much celebrated and can be view as an implementation of the so-called Matthew or 'more begets more' principle (see the *New Testament*, Matthew 25:29 or Luke 19:26). As an example, think of the number of acquaintances of a person. One may expect that the more a person is already known, the more likely it is that even more people will get to know about that person. The following procedure for growing a network follows this philosophy. At present this is in parts of the literature related to preferential attachment. The mechanism has a long history and is sometimes denoted the Yule–Simon mechanism [411, 499], the Simon mechanism [411] or cumulative advantage processes [352]. The term 'preferential attachment' was used by Barabási and Albert [45].

The procedure is as follows. We start to build a network from M initial nodes, e.g. $M = 5$. Next, new nodes are brought along one by one. Each node is connected to Q, e.g. $Q = 3$ of the existing nodes, but the link between a newcomer and an existing node is placed with a probability that is larger the larger, the degree of the existing node. So if an existing node already has accumulated e.g. a degree of 20, newcomers will attach a link to this node with a higher probability than to another existing node that only has e.g. a degree of 10.

We notice two differences compared to the totally random allocation of links between nodes assumed by the Gilbert model described above. Firstly, links and nodes will be correlated since nodes receive links depending on their degree at a certain stage in the growth process. This may obviously make the generated network structure more relevant to the analysis of real networks, such as friendship or online social networks, where links are not just allocated completely randomly. It will also make the statistical analysis somewhat more difficult since one will not be dealing with independent probabilities.

The second observation we can make without introducing detailed mathematical formalism is that the distribution of the degrees of the nodes is expected to be very different from the Gilbert case. In the random Gilbert network a link is attached to a node with probability p.

Conceptually we have that a peak is produced in the degree distribution of the Gilbert network because of competition between combinatorics and probability. This is similar to when we throw two dice and it is more likely that the sum of the faces equals 7 than 2. The probability of each throw of the two dice has the same probability, namely 1 out of 36. The reason the sum being equal to 7 is more likely is that we obtain 7 from all the following configurations of faces of first and second dice: 1+6, 2+5, 3+4, 4+3, 5+2 and 6+1. Whereas to get a sum of 2, only the configuration 1+1 is relevant.

This paragraph sketches the combinatorial and probability arguments behind the degree distribution of the Gilbert network. To determine the probability that k links are attached to a node, we imagine picking a node and then going through the $N - 1$ other nodes and with probability p adding, and with probability $1 - p$ leaving out, a link. The total number of ways of choosing the k nodes to be linked out of all the $N - 1$ nodes is given by the binomial coefficient $\binom{N-1}{k} = \frac{(N-1)!}{k!(N-1-k)!}$. So the probability that a node is connected to k other nodes is the product of the probability factor $p^k(1-p)^{N-1-k}$ and the combinatorial fact $\binom{N-1}{k}$. This gives us the expression $P(k) = \binom{N-1}{k}p^k(1-p)^{N-1-k}$, which is simply the binomial distribution, see Fig. 3.5 and for more mathematical detail, Secs. 8.2.1 and 8.5.1. The probability factor $p^k(1-p)^{N-1-k}$ will, for $p < 1/2$, be largest for $k = 0$ and decrease as k is increased from 0 to $N - 1$; for $p > 1/2$ this factor will be smallest for $k = 0$ and then grow as k is increased from 0 to $N - 1$. The peak in the degree distribution originates in the factor $\binom{N-1}{k}$ describing the number of ways we can allocate the k links. This number is strongly peaked about k equal to half of all the possible links, i.e. $k = (N - 1)/2$. The result of multiplying the probability factor and the combinatorial factor is to produce a degree distribution with a peak at $p(N - 1)$, which is the average degree. Let us recall that the binomial distribution approaches the ubiquitously encountered normal distribution for large N. Already for N around 10 the binomial distribution is well approximated by a normal distribution with average $\mu = pN$ and variance $\sigma^2 = pN(1 - p)$. So we can safely think of the Gilbert network as being described by a normal degree distribution.

In the case of preferential attachment, the attraction of the nodes with high degree overcomes the rapid decrease in the number of ways the k links can be allocated for k larger than $(N - 1)/2$. The result turns out to be a degree distribution that essentially follows a decreasing power law in the degree $P(k) \propto k^{-a}$, see Sec. 8.5.1. This means the degree distribution for preferential attachment is ever decreasing from its value at $k = 1$, as sketched in Fig. 3.5.

In this sense we have a situation similar to the one we encountered at the critical point described in Sec. 3.1, but in the present case the lack of a characteristic scale does not relate to space or time, but concerns the topological nature of the network and can be seen as a property of the probability distribution describing the number of links attached to the nodes.

To get a feeling for the consequences of the different types of degree distributions, we plot in Fig. 3.6 three networks, each with 100 nodes and 200 links. We have only allowed degrees $k = 1, 2, \ldots$, i.e. disconnected degree-zero nodes have been excluded. The network to the left in the figure has a binomial degree distribution, the one in the centre has an exponential distribution and the network to the right has a power-law degree distribution. We have just discussed that the binomial distribution has a characteristic scale and so does, as mentioned in Sec. 2.1, the exponential function. When looking at Fig. 3.6 we notice that the nodes in the network to the left and the one in the centre have a preferred typical degree and that no high-degree nodes are present in these networks. This is different for the network to the right. We see plenty of degree-one nodes, because the power law has a maximum at degree $k = 1$. This is the same

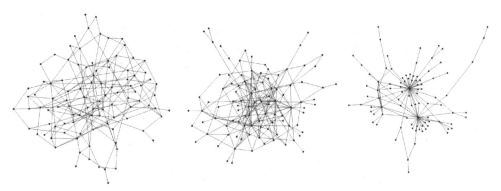

Figure 3.6 A plot of three networks. The degree distribution of each network is, from left to right, given by a binomial, an exponential and a power law. Each network has $N = 100$, $L = 200$, and $<k> = 4$. Figure courtesy of Madalina Sas.

for the exponential distribution, but where the exponential decays so fast that no large-degree nodes are seen, the network with the power-law degree distribution contains both degree-one nodes and several nodes of high degree.

3.3   Emergence of Coherence in Time: Synchronisation

So far we have considered some examples of emergence of structures either in ordinary physical space as in the case of the magnetic system in Sec. 3.1, or in some more abstract sense as when we looked at structures of networks. Now we turn to structure in time. There are many different ways the temporal behaviour at the systemic level may differ dramatically from the dynamics of the components. Let us first mention that a range of different types of synchronisation can occur. A famous example of such non-intended synchronisation happened when the new pedestrian Millennium Bridge across the Thames opened in London in June 2000. Any mechanical structure is bound to move when subject to forces, this is a consequence of Newton's laws and designers and engineers cannot avoid buildings or bridges bending when hit by winds or people moving about. So the aim is to make designs which render the motion of a bridge so insignificant that no one notices. The engineers responsible for the structural calculations for the Millennium Bridge had to estimate how much one person walking would make the bridge move and, in order to adjust for many people moving at the same time, they added the effect of the maximum number of people that can possibly be squeezed together on the bridge. They then designed the supporting beams strong enough that even if the bridge was crowded with pedestrians, the motion of the bridge resulting from all the forces from all the busy feet would not be noticeable.

This was the intention. But as soon as the bridge opened, it was closed again. In contrast to the engineers' expectations, the bridge swayed considerably, greatly scaring the people navigating their way across. The bridge remained closed for about one and a half years while the engineers tried to figure out why it did not behave as their careful calculations told them it should. The answer turned out to be that when accounting for the combined effect of pedestrians simultaneously walking on the bridge, the engineers had assumed the motion of the bridge induced by one pedestrian should be added at random to the effect of another pedestrian. Therefore the expectation was that when the walking of one pedestrian made the bridge sway slightly in one direction, the effect of another pedestrian would typically make the bridge sway in the opposite direction, so the net effects of the two pedestrians would be expected to cancel each other out, by and large.

But adding the effects at random ignores the way the sway of the bridge caused by one pedestrian influenced the gait of another pedestrian, and this turned out to be wrong. In reality, as soon as a pedestrian feels the slightest sway of the bridge, they will adjust their walking in a way that leads to synchronisation between different pedestrians and suddenly everyone is walking in step and the forces from all their feet, rather than nearly cancelling out, add up to an appreciable strength that makes the bridge move in a very noticeable way. This is a spectacular example of getting the systemic-level dynamical behaviour wrong by neglecting the effect of one component on the others. As soon as the engineers realised the importance of the feedback amongst the pedestrians, the assumption of independent random effects was discarded and they redesigned the dampers on the bridge, causing the swaying to go away. The bridge reopened in February 2002 after an example of practical complex systems emergence engineering was completed.

We notice that the first and simplest assumption made by the engineers, namely just adding the forces from each pedestrian as independent random contributions, would have been correct if the motion of one pedestrian did not influence the motion of the others. The tiny swaying of the bridge generated by the forces from the feet of one pedestrian leads to an effective interaction between the pedestrians and they end up walking in step, i.e. in a coordinated manner as if some central control conducted them, like when an officer counts to keep parading soldiers walking in step.[4]

Let us mention in passing that the emergence of unexpected effective interactions is not only of relevance to the design of pedestrian bridges. It is, for example, also at the heart of the mechanism that allows electrical resistance to disappear at low temperatures where the state of superconductivity may occur. In this case it is the electrons tugging at the ions of the material the electrons flow through that is able to change entirely the forces between the electrons. When one electron moves through the electrical wire, it is able to pull in the positive charged ion of the crystal structure composing the wire. This produces locally an excess of positive charge, which will act as a region of attraction on other electrons. In this way the electrons are able to attract each other by means of the motion of the crystal structure they move through and this turns the repulsive forces, which normally act between charges of the same sign, into an effective attractive force. The effective attraction mediated by the crystal structure allows quantum-mechanical mechanisms to make the electrons flow through the wire without suffering the usual resistance due to collisions with obstacles, such as impurities, in the conducting wire, and the net effect is that electrical resistance vanishes [444].

The importance of structure in time can be found in many very different settings. We will just mention two from biology. Probably everyone has either observed or seen movies of the remarkable dynamical structures formed by flocks of birds or schools of fish. We do not know all the details of how thousands of birds or fish manage to move around as one big swarm, resembling a single organism. But several mechanisms are being investigated [189, 428] and all, of course, consider the mutual influence between the animals and try to deduce in what way one animal responds to and influences the animals surrounding it. Mathematical details will be discussed in Sec. 11.1.

At a very different time scale, a subtle relation among the mutual influence between species, their abundance dynamics and the stability of ecosystems is found to be of importance. It has been suggested that higher than otherwise expected diversities can

---

[4] The curious fact is that when a marching regiment of soldiers is to cross a bridge they are commanded to fall out of step, precisely to avoid swaying the bridge more than necessary.

be sustained by mechanisms such as **out-of-phase** oscillations in the population size of different species [322], or the effect of multiple species interactions in contrast to just pairwise influences [260].

Synchronisation of the walking pedestrians on the bridge is an example of synchronisation consisting of individual oscillatory motion becoming coordinated across many components. Similarly, when the electrons moving through a wire enter the superconducting state they undergo a transition to a globally coherent state. The term 'synchronisation' is often used quite broadly to cover the emergence of temporal coherence at the systemic level. We may in this sense think of all the musicians of a symphony orchestra being synchronised. Although the different instruments play different notes, the musicians do very much coordinate their playing with everyone else and we can think of the changes in the music as a change in the structure of how the musicians synchronise with each other. This is at least qualitatively reminiscent of the flock of birds or the temporal variation in species abundance.

We looked at the Ising model and saw how, despite it being very abstract and simple, it offers a remarkably fruitful framework to study the formation of and transition between spatial structures. The Kuramoto model [345] is a similar modelling approach for studying the onset of synchronisation. A detailed mathematical discussion will be given in Chap. 7. Here we summarise the conceptual content of the model and its predications, see Fig. 3.7.

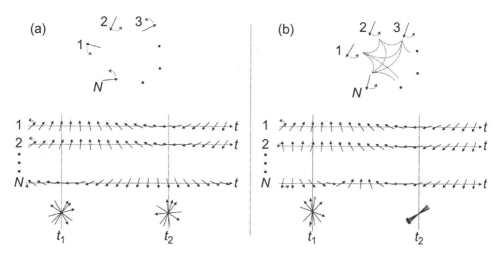

Figure 3.7 Synchronisation of Kuramoto rotators. Rotors labelled $1, 2, \ldots, N$ are depicted at the top. In panel (a) they rotate independently of each other and accordingly they point in many different directions. In panel (b) the dotted lines connecting the rotors indicate a mutual influence amongst the rotors. The rotation of the rotors is indicated by the three time axes in the middle of each panel. In panel (a) the rotors keep pointing in a broad range of directions as time passes, as seen clearly when all the rotors are placed on top of each other (see the bottom figures). The interaction between the rotors, assumed in panel (b), establishes synchronised rotation as time passes. This narrows the range of directions covered by the superimposed rotors at the bottom of panel (b).

The Kuramoto model considers a number of rotors, as depicted at the top of Fig. 3.7. Each rotor is assumed, when left alone, to rotate at a distinct velocity. In panel (a) each rotor turns around at its specific velocity, independently of the other rotors. One could think of this as members of a very enthusiastic audience desperately applauding with no mind for the clapping of their fellow audience members. In this case the set of rotors will point in all sorts of directions and the overlay of all the rotors at the bottom of the panel will resemble a two-dimensional curled up hedgehog. Panel (b) depicts the same rotors, still each possessing a preferred velocity of rotation, but now they are paying attention to each other and will, as time passes, attempt to adjust their velocities to try to achieve a degree of collective synchronised behaviour. The interaction between the rotors encourages them to speed up if they are behind the rotors they are coupled to, or slow down if they are ahead. This would correspond to the audience members realising that their effect may become more convincing if they coordinate their clapping. In this case the rotors will tend to point in the same direction while they rotate, and the rotors placed on top of each other at the bottom of the panel now look more like a bunch of flowers than a hedgehog.

The Kuramoto model analyses the competition between the disordering effect of a set of different individual rotor velocities and the ordering effect of interaction between the rotors. The interaction is implemented pairwise and different from zero if the two rotors are not aligned. In this case, an amount will be added to the speed of the rotor which is behind and the rotor which is ahead will be slowed down a bit. If the spread of the individual rotor speeds is too large, the pairwise interaction will not be able to synchronise them. The analysis of the Kuramoto model derives the value of the interaction strength at which onset of synchronisation happens for a given spread in the velocities.

Although individual components of course may follow dynamics very different from simple two-dimensional rotors, the collective effects which control the onset of synchronisation give a nice intuitive insight of broad general interest. A good understanding of how and when changes in the arrangement of coordinated synchronised behaviour occur is often very important. It has, for example, been conjectured that the state of our mind is strictly related to how the different brain regions synchronise [466], and it is a fact that the onset of epileptic seizure is accompanied by a very high degree of neuronal synchronisation.

Sudden changes in systemic behaviour are of very general importance and we will discuss attempts to forecast such abrupt changes in the next section, after we have looked at evolutionary dynamics.

3.4   Evolutionary Dynamics: Adaptation

Development driven by a combination of random changes and adaptation comes in many formats. We will consider how simple versions of random mutations amongst a set of reproducing agents can lead to observed network structures encountered in different types of complex systems; e.g. biological or economical ecosystems. As we have stressed, interaction amongst the agents is a prerequisite for the emergence of systems level collective structures.

Since Darwin and Wallace, biological evolution has of course been considered to be the prime example of evolutionary dynamics. However, this kind of dynamics is relevant to a large number of situations, e.g. the dynamics of competing companies, traders or even the dynamics of opinions in a population share some similar elements. Of course, companies, traders and opinions do not reproduce nor have a genome that undergoes mutations in the same way as biological organisms do. But the assets of a company or the number of traders following a certain strategy or the popularity of an opinion can increase and decrease and may be compared to the number of organisms of a specific biological type. And companies may start up new companies and in this sense produce 'offspring' with properties different from the mother company, which will correspond to mutations. A trader may decide to change strategy, which will correspond to a mutation from one type of trader (species) to another. And in the case of opinion dynamics, a person will move from one opinion group to another when deciding to change opinion.

When considering company dynamics, it is important to include merger between companies as a process creating new entities [168, 472]. A process which has been argued also to be crucial for the dynamics of the evolution of biological life. The types of biological cells which constitute our bodies are well known to have a cell nucleolus supposed to originate in a merger event, see e.g. Chap. 5 in [279].

In all these cases the viability, whether of a company, a trading strategy or an opinion, will depend on the properties of the surrounding companies, strategies or opinions. A company gains or loses income depending on the commercial landscape of other companies it is involved with, either through the production chains or by competing for customers. The success of a strategy depends on the strategies followed by other traders and the attractiveness of an opinion is vey much measured against the current trends and view points. And when we consider biological evolution, we realise that the so-called fitness of a type of organism is not an intrinsic property like weight or height, but is highly dependent on the surrounding ecosystem. A koala bear may be fit and thriving in a forest of eucalyptus trees in Australia but would not fit in well if transported to the Artic.

We will now look at the network structures that can be generated as a result of evolutionary dynamics amongst interdependent agents. We include the most basic evolutionary processes which are, in the case of biological evolution, reproduction, mutation and death. If we consider for example cultural evolution, processes will be growth or decline and innovation. This can be the growth or decline of the popularity of an idea and the invention of new concepts and ideas. In the case of economy we might for instance look at the growth of the size of companies, merger, bankruptcy and new startups.

To be specific, we will here consider the processes of most relevance to biological evolution. The reproduction rate will be made to depend on the interactions with other types of agents present. Mutations will make the occupancy on types move about, since when a mutation occurs during a reproduction event, an agent will be added to the occupation on a node different from the one which the reproducing parent occupies. Ultimately, death will reduce the population of the node containing the deceased agent.

To fix our language, we will use the term 'reproducing agents', which may represent biological organisms, companies, traders, etc. The crucial aspect we want to include in our modelling is, as usual, that one type of agent can influence another type of agent. It is likely that such influences can have many different manifestations. Think of a tree, insects and birds. The bark, or the leaves of the tree, may create an attractive habitat for certain insects which, on the other hand, may serve as a food source for some insects. The birds eating the insects may help check the number of insects so the tree does not succumb to the burden parasites can put on its metabolism. For any collection of agents there will be zillions of possible influences between them, which we want to simplify by assuming sets of random effects of one type of agent on another and sometimes no effect at all.

We consider a large collection of potential agent types, indicated by the black dots, the nodes of the network in Fig. 3.8, and we assume that a mutation may put an agent on neighbour nodes near the parent node. An example of such a process is indicated by the curved arrow in Fig. 3.8. The links of the network represent the effect one type of agent can have on another type and are indicated by the straight lines in Fig. 3.8. Of course, only nodes occupied by agents can influence agents on connected nodes since, say, although bananas are always good food for monkeys, if the banana plants have gone extinct, the monkeys cannot benefit from the bananas' possible healthy effect. To

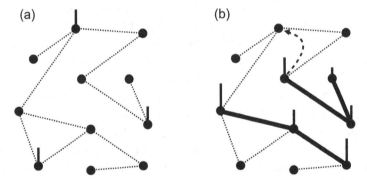

Figure 3.8 Newtworks of co-evolving agents. The network in panel (a) is the hardwired structure on which the agents evolve: a black dot, a node, represents one type of agent and the links between the nodes represent that some form of interaction exists between a pair of types. The interaction is not symmetric. A flu virus has a very different effect on humans than humans have on the virus. We imagine that initially a random set of agent types are occupied. The vertical columns represent the size of the population of agents on a given node. Initially, only three types are extant. Reproduction and mutation will move the population around on the network. The dashed arrow indicates how a mutant may arrive on a nearby node due to a reproduction event. The network in panel (b) indicates the occupancy on the hardwired network after the evolutionary dynamics has run for some time. Another subset of nodes are now occupied. We are interested in the properties of the subnetwork of occupied types. This network is indicated by the thick lines.

incorporate this in the model, the dotted lines depicted in the network to the left indicate potential effects between types which only become effective (solid lines) when agents of these types are actually present. For example, there would clearly be a very strong effect on humans if dinosaurs were still roaming the earth, but since they went extinct long before we arrived, we do not need to worry about this potential interaction. On the other hand, some existing strain of bacteria may in their present form be harmless, but that could change if a mutation occurred, which in Fig. 3.8 would correspond to a nearby node with a different set of interactions becoming occupied, as indicated by the curved dashed arrow.

Representing the world of evolutionary possibilities by the 'hardwired network' as sketched to the left in Fig. 3.8, the nodes in the network represent, in an admittedly naive and simplistic way, all the possible genomes, or blueprints, for the collection of all possible agents. The dotted lines connecting the nodes indicate that if agents of those two types are present simultaneously they will affect each other with a certain strength. In the figure we have just drawn single lines, in reality the links are not symmetric since often the relationship between two types of agents has very different consequences for the two types, just think of predators and prey.

Not all possible ways of populating the nodes will be equally sustainable. The number of agents on the various nodes will have to be balanced according to the influences between the types. The evolutionary dynamics of reproduction, mutation and death will tune the occupancies in order to establish this balance. Accordingly, only some configurations of occupation will be viable, as indicated by the network to the right in Fig. 3.8.

As the evolutionary process runs its course, the population of agents of the various types will grow or decline; unoccupied nodes may become occupied as a result of mutations placing an offspring from a reproduction event on a nearby node. Other nodes may become unoccupied, i.e. that type goes extinct, if their reproduction rate is unable to balance the decline due to the ever-acting death. In terms of networks, this means that the subset of occupied nodes will change over time scales corresponding to species creation and extinction. This corresponds to the evolutionary time spans such as studied by palaeontology in the fossil record. One can think of the occupied set of nodes and the links between them as forming an ecosystem of interdependent species. As time passes this network will evolve and we can use this modelling framework to address questions concerning how the topological structure of the occupied network, the ecosystem, is related to the evolutionary process.

We can compare how the hardwired network and the occupied network differ, for example in terms of the degree distributions, the density of links and the formation of communities. And we can investigate how these network properties depend on time and on parameters such as the mutation rate, the level of resources and the topology of the hardwired network. We may also look at how fast the occupancy spreads across the hardwired network and from this get some idea about the speed with which the evolutionary dynamics is able to produce diversity. All these questions are of relevance wherever we encounter evolutionary dynamics in the real world [216], and we will return to a more detailed comparison when we include mathematics in our discussion (see Sec. 11.3). Here we just mention that the occupied evolved network produced by this simplistic version of evolutionary dynamics can generate occupied networks with degree distributions that follow the exponential function and the distribution changes to a power law when the rate of mutation goes towards zero [250]. This behaviour is comparable say to ecological networks [116]. The diversity, i.e. the number of different types generated due to the mutations, can be studied as a function of time. Such studies may be of value when analysing the growth of cancer tumours [307].

The message at this stage is that a range of observed features can be captured by a very simple model that focuses on the very basic ingredients of multiplication, mutation and death together with the fact that co-evolving agents create an environment of interdependencies for each other. This is yet another example of how minimalistic modelling may help us to identify some basic processes that can generate network structures of relevance across e.g. biology, sociology and economics. It is the simplicity and focus on just a few essential mechanisms that helps us to understand the systemic effects of basic evolutionary processes.

3.5   Mean-Field Modelling: Dimensionality and Forecasting

Besides understanding how systemic coherent dynamics may emerge, we also want to understand typical mechanisms responsible for switching between one structure and another and, in particular, it can be crucial to know how to forecast if or when such changes may happen. We would like to develop techniques that can enable us to handle cases like ecosystem collapses, financial crashes, the onset of epileptic seizure, earthquakes and similar abrupt events. Different types of dynamics will be relevant at different time scales and the importance of stochastic elements in the dynamics may also be important. To illustrate this, for concreteness we will think of ecosystems.

When we look at time scales over which the composition in terms of species of an ecosystem remains unchanged, we are dealing with population dynamics and will have to keep track of the variation in the number of individuals belonging to each of a fixed number of species. Mathematically speaking, this means that we have a fixed set of variables to care about. For example, in the standard **Lotka–Volterra model** of predator–prey dynamics one looks at e.g. foxes and rabbits and assumes that the abilities of these two species remain unchanged in time.

Such models of population dynamics may be studied using smooth and deterministic differential equations when one is dealing with large population sizes. In reality, the offspring production of foxes or rabbits at the level of individuals changes fairly stochastically, but these fluctuations become relatively less pronounced when there are large numbers of individuals present. However, when a population is near extinction the stochastic aspects of the dynamics of the individual reproduction events are important.

Additionally, at time scales spanning many generations the effect of the evolutionary processes (mutation, selection, adaptation) will have to be considered. For example, the selection of rabbits that are better adapted to avoid the foxes perhaps by developing faster speed or a better camouflage.

Moreover, it may not be sufficient just to look at rabbits and foxes in isolation. Co-evolving species may of course influence the ability to sustain the populations. For example, if grass disappears, rabbits will have to move elsewhere. Including too few components may be a severe limitation in many other situations in ecology, or economics, or even when looking at earthquakes.

Earthquake dynamics involve many blocks of elastically coupled sections of the earth's crust and in such cases the breakdown of stability may *not* be captured appropriately by just following the average behaviour. We expect that earthquake prediction is difficult precisely because a minor local slip between two interlocked parts of the fault can have crucial consequences and will not be captured by an on-average description. It is likely that sometimes such a local breakdown rapidly generates a catastrophic slip involving large portions of the fault. A similar kind of chain reaction, or snowballing, is also at play in mountain avalanches and can also maybe play a role in ecological mass extinctions when a keystone species goes extinct and pulls other dependent species into extinction. Equivalent scenarios may be important in finance and perhaps also for the mechanism behind the onset of epileptic seizures.

Now we are well aware of the possible limitations of an on-average description, let us neglect these difficulties for a moment and assume that we are dealing with dynamics of many interdependent components which can be captured by the on-average time dependence; one talks about a **mean-field** description. The name indicates that instead of following each and every detail of all the components, one develops a mathematical description of the average influences, such as e.g. forces acting between components. This type of description has been used systematically for a very long time by physicists and engineers. For example, when designing an aeroplane one does not keep track of the collisions of the plane with each individual air molecule. Instead, a hydrodynamic description by use of the Naiver–Stokes equations, which are based on Newton's mechanics, is used. The motion of the individual molecules is replaced by equations for the average velocity and density of molecules and the pressure.

As we mentioned above, comparable approaches are also used in ecology. Since in ecology we do not have basic equations to start from, such as Newtonian mechanics, the applicability or limitations of the mean-field approach can be difficult to judge in any other way than by trial and error. But if it turns out that the on-average description suffices, then a large array of mathematical techniques are ready to be applied. An application of ideas from the branch of mathematics called **dynamical systems theory** [383] has attracted much attention. We will present a mathematical discussion in Sec. 13.1 and try here to explain the essence of the approach. One focuses on a variable that is expected to capture the overall stability of the entire system, this could be the size of the total population. It is assumed that, left to itself, the dynamics will be stable and if for some reason the configuration is perturbed by some external force, the variable, such as the total population, will return to the equilibrium values. One can visualise the postulated dynamics by thinking of a ball in a two-dimensional bowl, see Fig. 3.9.

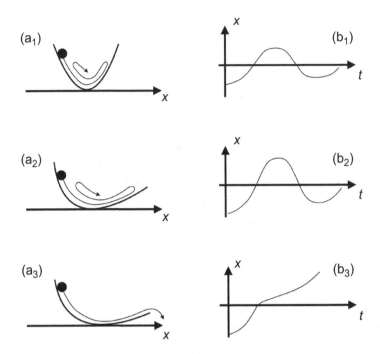

Figure 3.9 Ball rolling in a bowl. The panels (a_1), (a_2) and (a_3) depict the motion of the ball starting from the same position with three different heights of the right rim. The panels (b_1), (b_2) and (b_3) indicate the corresponding behaviour of the position $x(t)$ of the ball as a function of time. As the bowl is made more shallow, the ball is able to make larger excursions and eventually to escape over the rim.

If one kicks the ball in order to give it a certain amount of kinetic energy, friction will after a while return the particle to the bottom of the bowl. Since the ball climbs the wall to a height determined by its energy, steeper walls will confine the particle in a smaller interval than a more shallow bowl. Now assume some external influence is able to change the shape of the bowl to make it gradually more shallow on one side. We will observe that as the bowl becomes more shallow, the size of the ball's excursions away from the position at the bottom will increase and it will take a longer time to return to the bottom. Eventually the ball may be able to reach the rim of the bowl, which will allow the particle to leave the bowl and never return. So the system has become unstable and one says that a tipping point has been reached. If this description in terms of a single, or a few, system parameters is relevant, one should be able to forecast the approach of a tipping point by monitoring the fluctuations of the system. The scenario has been suggested to be relevant to ecology, finance and other multi-component systems [383], though as we will discuss further below the applicability of the approach is far from obvious in all cases. Some crucial abrupt transitions are not preceded by an increase of fluctuations in any obvious parameter. Earthquakes are one example, and other example is the sudden disruption due to a new successful invasive mutant.

For the Millennium Bridge, this would correspond to the oscillations induced by the pedestrians walking in synch becoming so large that the beams of the bridge break. This was never a possibility, as the engineers had made the bridge strong enough for this to be impossible. Phrased in the language of Fig. 3.9, the engineers had made sure that the rim of the bowl was so high that there would be no way to escape over it, but they had not made the bowl sufficiently steep and narrow to reduce the oscillations below a comfortable level.

One example where a sudden change might be preceded by an increase in the fluctuations of a single parameter could be the contamination of Lake Erie by fertilisers from the surrounding farmland [293]. The total algae population was monitored. Without external perturbation of the lake's environment, the algae population oscillates but remains checked, so the size of the algae population behaves like the ball in the steep bowl. The effect of the increasing level of nitrogen from fertilisers corresponds to making the bowl more shallow and eventually the algae population will not return from its excursions but explode to such high levels that the oxygen consumed by the algae becomes detrimental for other life forms in the lake. Hence, it should be possible from a careful monitoring of the oscillations in algae level to discover an approaching breakdown of the stability of the lake's ecosystem.

If all approaching abrupt changes in complex systems were foreshadowed by an increase in fluctuations, we would not repeatedly be taken by surprise when we are hit by financial crashes, severe floods or devastating earthquakes. In the remainder of this section we will think about the shortcomings of the mean-field description and also consider possible ways to forecast when the mean field is insufficient.

We will expect the mean-field approach to be applicable *if* the system is well described in terms of what we may call robust macroscopic variables. That will, for instance, be the case if we consider an aeroplane. The centre of mass and two angles is enough to determine the motion of the entire plane. It may also be sufficient to monitor the total population of algae in Lake Erie to predict future tipping points, but we could imagine other dependencies to develop that make the total algae population insufficient as predictor. Just for argument's sake, we can imagine that the larger algae population encouraged other species to thrive, which might lead to different changes in the population dynamics of the lake. For example, an increase in the shellfish population could perhaps co-exist with a larger algae population and a new kind of stable, but very different, environment could be established in the lake.

One example of the importance of detailed tracking of multiple species is seen in the synchronisation effects found in a study of 17 years of data from Little Rock Lake in Wisconsin [232]. We are faced with a similar situation when we consider a flock of birds or a school of fish. Their centre of mass might not help us much in predicting how the crowd of individuals stay together or break up in their motion around obstacles or as a response to predator attacks. The same holds for groups of pedestrians moving through a crowded city centre, or voters participating in an election. In such cases one will need a mathematical description that provides more detail. For example, a description of how the motion of an individual bird depends on the interaction with the surrounding birds.

A further complication is encountered when the individual components are able to change and adapt. Such changes to the composition may trigger sudden effects. The extinction of the dinosaurs and the ensuing establishment of large numbers of new species of mammals, including humans [179, 278], may be an example and a result of combinations of internal (co-evolutionary pressure) and external factors (perhaps a meteorite collision).

Disruptions to our societies' economic or financial life may be triggered in a similar way. We can think of the collapse of Lehman Brothers in 2008 and the following financial crash, or we can think of the arrival of a completely new type of business and the consequential restructuring. Think for example of the arrival of Google or Amazon, which generated a tipping point of seemingly no return. The world of advertisement and retail shopping changed dramatically. Before 1997, the year Google and Amazon went public, one would not have been able to forecast their arrival by looking at the oscillations in some market variable. To capture such 'mutation'-induced abrupt changes where in a short time span a hitherto non-existing agent suddenly influences the dynamics significantly, one needs to go beyond monitoring the slow time variation of the large aggregated variables and include variations and fluctuations in less conspicuous agent types. The Tangled Nature framework, mentioned in Secs. 4.2 and 11.3, which studies different types of interacting co-evolving agents, is a prototypical model in which dramatic transitions keep occurring intermittently.

The transitions in the Tangled Nature model consist of one configuration of co-existing type becoming unstable due to the effect of new mutants. At the level of the individuals, the dynamics always runs at a steady pace. Agents reproduce and die. The reproduction produces an unceasing production of new types and once in a while new mutants may be created that interact with the existing set of types in a way that destabilises the existing balance. Some types will go extinct and some mutants will succeed and become part of the new ecosystem. At the level of the individuals, the dynamics is more or less steady as birth and death continues ceaselessly. But when looking at the level of the entire ecosystem we observe intermittent dynamics with abrupt rearrangement of the species composition; this is similar to punctuated equilibrium [169].

These transitions may be thought of as tipping points in the sense that there is no going back when a transition first starts to unfold. We cannot bring back the time before the dinosaurs disappeared, or the time before Google and Amazon arrived. Nor is there any warning of the approaching transition in terms of an increase in oscillations of e.g. the total population. Nevertheless, it turns out that if one keeps making instantaneous stability analysis of the population of the different types of agents, the approaching transitions can typically be forecasted a couple of generations before they happen. The methodology is similar to the type engineers routinely use to ensure stability of mechanical structures, the important difference being that since we are dealing with evolutionary dynamics, the stability keeps changing as the features of the agents change due to mutations and therefore the analysis of the stability has to be updated as the evolution unfolds. The method consists in looking for perturbations which go from being harmless to detrimental as the composition of the population keeps changing. In terms of Fig. 3.9, the analysis corresponds to investigating if the internal dynamics causes the bowl to become shallow. The arrival of Amazon and Google was not a factor external to the market economy, they were simply new types with strong influences on the other co-evolving companies.

Luckily it is possible to make the above discussion much more detailed and specific when it is turned into a more precise mathematical formulation; we will return to this in Chap. 13.

Summary: We have looked at the Ising model of interacting binary components, models of network formation, the Kuramoto model of onset of synchronisation and the Tangled Nature model of co-evolutionary adaptive dynamics. We have emphasised a few very general mechanisms generating structure at the aggregate level:

- Competition between ordering and disordering mechanisms, which when at balance, can bring about critical states without a particular dominating scale. In such cases fractal structures and power-law distributions are relevant.
- Random allocation of links will produce networks with degree distributions with a scale. When a more-begets-more mechanism is used, the degree distribution may follow a power law, i.e. be without a scale.
- Forecasting disruptive macroscopic events is more likely to be successful when the dynamics involves robust emergent on-average variables.
- In contrast, evolutionary dynamics involves the appearance of new entities, such as mutants, which make forecasting challenging.

3.6 Further Reading

General audience:

In his popular science book *Critical Mass* [40], Philip Ball gives many specific examples of analysis and modelling of concrete emergent phenomena across engineering, biology, economics, sociology and network design.

Intermediate level:

The book *Think Complexity* [113] by Allen Downey is a nice hands-on Python-based exploration of models with emergent complex behaviour. *Note*: Concerning Chap. 9 and 1/f it should be kept in mind that the claim that 1/f exists in the sand pile is incorrect, see Sec. 12.1.

Advanced level:

In the book *Basic Notions of Condensed Matter Physics* [21], P. W. Anderson explains carefully symmetry breaking and generalised rigidity to an audience interested in physics.

Manfred Eigen's monumental opus *From Strange Simplicity to Complex Familiarity* [121] is scientifically high level but accessible to anybody with a high school science background. It addresses some of the most profound questions concerning how matter, life and thought emerges from energy and information.

3.7   Exercises and Projects

Exercise 1
Find an online movie on fireflies such as David Attenborough's 'Talking to Strangers' in the BBC series *Trials of Life*. Notice how Malaysian fireflies are able to gradually synchronise along large portions of the river bank. Try to think of why the synchronisation sets in gradually. Is it likely that the more fireflies are pulsating in synch, the easier it is for the others to notice and respond to the synchronisation?

Exercise 2
In Sec. 3.5 we discussed various aspects of relations between dynamics at different levels of organisation in a complex system.

(a) Think of the human body and try to list the different time scales relevant to the organisation at levels from the subcellular molecular dynamics, through the level of organs up to the dynamics, including motion, of the entire body.
(b) Try the same exercise, this time concentrating on the brain from neuronal firing to the dynamics of thought processes.

Exercise 3
Make a list of as many transitions as possible in as many different complex systems as you can think of. For each transition try to suggest ways to check if the transition involves a diverging length scale (Sec. 2.1) and some type of scale invariance (Sec. 3.1).

Exercise 4
Recall the discussion around Fig. 3.9.

(a) List examples of abrupt changes, or tipping points, which may be preceded by an increase in the fluctuations of some observable parameter.
(b) Now list examples of abrupt changes, or tipping points, which are not preceded by an increase in the fluctuations of some (known) observable parameter.

Exercise 5
Try to think of networks in different systems such as food webs, interconnected airports, social networks, etc. Use a web search to investigate which type of distribution the number of links attached to a node follows for each of the networks. This is the so-called degree distribution, see Sec. 8.2.1.

Project 1 – Characteristic scale
We discussed in Sec. 2.1 and the end of Sec. 3.1 how dynamics can establish a characteristic scale and we also briefly mentioned that complex systems may have many scales, even infinitely many, equally relevant scales.

(a) Try in general terms to list the characteristic spatial and temporal scales relevant to humans. There will be different scales for the intrinsic workings of a human body (e.g. cell sizes and division times) and other scales for the body as a whole (e.g. body weight and height, lifetime, response times).
(b) Take a look at some photos of clouds and forest canopy and try to identify characteristic scales, recall the discussion around Fig. 3.3 and Fig. 3.4.
(c) Now take a look at some Jackson Pollock drip paintings and look for characteristic scales or the absence of these.
(d) In the light of your analysis in (b) and (c), would you agree that Taylor, Micolich and Jones's findings [437] make Pollock a naturalistic painter?

Project 2 – The Ising model
Recall the discussion around Fig. 3.2. Though the model was introduced to describe magnetic phase transitions, it has been used much more widely. Applications include, for example, gas theory, neuroscience, spin glasses and sea ice.

(a) Do a literature search and identify applications of the Ising model to as many different systems as possible.
(b) Compare the similarities and differences in the emergent features of these different versions of the Ising model.
(c) Discuss strengths and limitations of the Ising model approach.

Project 3 – Intermittency and forecasting
We discussed in Sec. 3.5 how the dynamics of complex systems commonly involve intermittent bursts of activity and the importance of developing methods enabling forecasting of the events. Without going into the details of the mathematics, which we will return to in Chap. 13, we will in this project consider attempts at forecasting earthquakes and financial crashes and compare in broad terms to the approach suggested in [383].

(a) Briefly review the approach to forecasting earthquakes based on machine learning [223].
(b) Briefly summarise the review 'Forecasting financial crises in emerging market economies' [360].

(c) Consider whether the machine learning approach of (a) can be generalised to apply to (b).

(d) Think of how the approaches in (a) and (b) compare with the forecasting in terms of growing fluctuations suggested in [383] and [385].

Note that the big financial crash relating to the collapse of Lehman Brothers occurred in 2008 seven years after Scheffer and collaborators first published the ideas that abrupt transitions can be forecasted by the increase in fluctuations preceding the transition.

(e) What does the fact that the 2008 financial crash took the world by surprise imply concerning the relevance of the forecasting approach suggested by Scheffer and collaborators?

(f) By tracing the citations of the papers [383, 385] discuss if cases can be found where transitions have been predicted identifying an increase in fluctuations.

Project 4 – Networks and their structure: Ecology as an example
This project is intended as an example of how the structure of a network is related to its stability and at the same time as an illustration of how difficult it can be to settle the actual structure of a real network.

(a) Think of an ecological food web. Try to think about how the number of links attached to a node, i.e. a species, will influence its viability.

A keystone species is one that if removed will trigger a major upheaval in the ecosystem. Bees are examples of keystone species.

(b) Try to consider how you will imagine the relation between stability of a species and the number of links between this species and other species in the food web.

(c) How do you expect a keystone species to be placed in terms of network structure in a food web?

Now we turn to two ecological papers [226, 416], which attempted to address the the link between structure and stability of food webs.

(d) Make a list of the most important points in [416] and summarise in particular how the authors describe the relation between the structure, or topology, of the food webs and a food web's stability.

(e) Summarise briefly the focus of reference [226].

(f) What is the main difficulty according to [226] in identifying keystone species from food web data.

(g) Now think of a network describing social interactions. Consider how we can reliably identify key persons in such a network. You can e.g. focus on decision making in a parliamentary democracy.

Project 5 – Models of brain dynamics
As summarised in the *New Scientist* article [49] there is substantial evidence that neuronal dynamics exhibits emergent scale invariance, see Sec. 2.1 and Project 3 in

Chap. 2. Some of the evidence comes from the shape of the probability distribution describing bursts, avalanches, of neuronal activity being approximately power law, recall Fig. 3.5. Observations in [56] and [433] find such evidence. The present project studies how far simple models are able to capture these emergent aspects of the immensely complex web of interacting neurones. We will look at a number of papers and use skim reading to get an overview

(a) Describe in bullet points the spatial structure and dynamics of the model used in [178].
(b) Summarise your evaluation of how well the model captures the experimental observations.

Although the model used in [178] captures some aspects of neuronal dynamics, it does neglect the fact that the brain contains both excitatory and inhibitory neurones. This challenge is taken up in a series of papers by de Arcangelis and collaborators.

(c) Describe in bullet points the spatial structure and dynamics of the model used in [101].
(d) What are the most important differences between the model studied in (a) and the one studied in (c)?

In Chap. 4 we introduce the third O'Keeffe–Einstein proposition:

Einstein: **P3** *A theory is the more impressive the greater simplicity of its premises, the more different kinds of things it relates, and the more extended its area of applicability.*

With this statement in mind, it is important to study how wide a range of phenomenology a given model approach is able to capture. After all, the brain does more than just support intermittent bursts of activity. Read the abstract, summary and skim read rapidly through the following papers: [267, 268, 294, 295, 373].

(e) Make a list of the emergent features of brain dynamics recorded in the above list of papers.
(f) Would you consider the model introduced in (b) as able to capture aspects of consciousness?

Project 6 – The Gaia hypothesis and Daisyworld
The earth's ability to maintain a stable environment far out of equilibrium is a spectacular example of large-scale homeostasis and requires an accurate balance between the influx of energy received from the sun and the loss of energy by radiation leaving the earth. The Gaia hypothesis, first suggested by James Lovelock in around 1970, considers the balance to be a result of feedback loops between the biotic and non-biotic environment. The idea is illustrated by the simple 'Daisyworld' model.

(a) Summarise the effect of the equations for Daisyworld as e.g. presented in [485] and Box 1 in [259].

(b) Check how the body maintains a constant temperature. Do you see any similarity between the body's temperature regulation and the Daisyworld mechanism?

(c) Can you think of similar stabilising relationships in other complex systems?

One may ask how the regulatory mechanisms suggested by the Gaia hypothesis have come about. Hence, relating Gaia to evolution is of great importance.

(d) Skim read Lenton's review in [259] and make a brief summary of attempts to generalise the Daisyworld model to include evolutionary aspects.

Lenton mentions 'Daisyworld ignores important levels of the environment between the individual and the global levels' but concludes that 'Gaian models suggest that we must consider the totality of organisms and their material environment to fully understand which traits come to persist and dominate'. With these comments in mind, we now turn to recent attempts at modelling evolution and Gaia within one framework. Familiarise yourself with [27, 28].

(e) Explain what is meant by an *entropic hierarchy*.

Clearly the step from the original Daisyworld in [485] to the Tangled Nature framework used in [28] involves more complicated modelling.

(f) In the light of the O'Keeffe–Einstein proposition (see introduction to Chap. 4), discuss if moving to the model in [28] is worthwhile.

4　The Value of Prototypical Models of Emergence

> **Synopsis:** Conceptual and mathematical models can serve many purposes. We will discuss why simple stylistic models are particularly useful in complexity science since they can help to identify the most essential mechanisms amongst the profusion of interdependencies at play.

For some readers, a chapter on the value of models may seem unnecessary and, of course, they are perfectly welcome to skip ahead. Although readers may decide not to dive into this chapter, they are encouraged to periodically think about why we build models and how models may relate to reality and the phenomena we like to describe and understand.

Those familiar with research approaches in physics will most likely recognise the term 'toy model'. It is not a very appropriate name, but was presumably coined as a way to acknowledge that a certain model deliberately only includes a few of the aspects of a specific phenomenon. The term is often applied to the Ising model of magnetism mentioned in Chap. 3 and to be considered in more detail in Sec. 6.2. In reality, quantum-mechanical phenomena are responsible for the magnetic moment of a material, e.g. of a compass needle. However, the famous Ising model replaced the complicated effects of quantum mechanics with a model consisting of variables that can only assume the values plus or minus one. In sociology, Schelling's model of segregation [386] employs a similar strategy. Schelling's model considers two types of people settling in a city and represents each type in terms of their tolerance level only.

It is unfortunate to call this type of model a toy model, since it can be misunderstood to imply that the models are not really meant to be taken seriously. The situation is in fact exactly the opposite. When we develop models of the world around us, we always need to go stepwise. It is instructive here to recall one quote by the American painter Georgia O'Keeffe (1887–1986) and two quotes attributed to Albert Einstein (1875–1955). We will, for ease of reference, denote them as the three O'Keeffe–Einstein propositions, which are as follows:[1]

O'Keeffe: **P1** *Nothing is less real than realism. Details are confusing. It is only by selection, by elimination, by emphasis, that we get at the real meaning of things.*

---

[1] For **P1** see p. 22 in [430]. **P2** Roger Sessions, *New York Times*, 8 January 1950, attributes the quote to Albert Einstein. **P3** is from the third paragraph on p. 31 of Einstein's *Autobiographical Notes* [125].

Einstein: **P2** *Make everything as simple as possible, but not simpler.*

Einstein: **P3** *A theory is the more impressive the greater simplicity of its premises, the more different kinds of things it relates, and the more extended its area of applicability.*

We are reminded by O'Keeffe and Einstein about the very nature of how we comprehend our overwhelmingly confusing and complicated surroundings. It is perhaps of particular importance that people dealing with the science of complexity are conscious about the need for a deliberate abstract and simplifying approach. Obviously the degree of detail that needs to be included will depend on the context. But it is important to bear in mind that if one from the onset tries to include all possible aspects of a phenomenon, one rapidly formulates a description that is beyond comprehension. Starting out with an oversimplification will most likely help in one way or another. This strategy can help to identify the essentials and to distinguish between the specific and the general.

If the abstract simplified model one starts out from does not manage to reproduce the wanted behaviour, the lesson is that the initial understanding is flawed. Perhaps the right ingredients were not included or perhaps one got the interactions between the included components wrong. This is also a lesson of sorts. If, on the other hand, one is successful and able to explain some of the observed behaviours, it may signify that one is on track to sort out which aspects are the most important.

We will in later chapters return to examples of how this kind of modelling works and how the aspiration to do simple modelling is of the greatest importance, not least when analysing convoluted complex systems. Right now we emphasise that by starting out simple one may also be able to understand that an observed phenomenon is not specific to the particular situation studied, but rather a general phenomenon shared across a variety of systems that, despite their differences, do share some underlying similarities.

Since we are dealing with systems of intricate hierarchies of nested components and their interrelationships, it may be assumed that starting with a richly detailed description is essential. However, the reason complexity science works is precisely because systemic complex behaviour at the aggregate level often can originate from simple mechanisms. One might say that this is the very reason science at all works, namely that simple regularities underlie the wealth of bewildering and convoluted behaviour we are surrounded by. Of course this has been the lesson of science through the centuries, but it is worth stressing that complexity science boldly studies how far this route of investigation applies to complex phenomena such as the brain, the economy, the climate, etc.

In the following two sections, we will elaborate on the need for simplicity and in the light of the O'Keeffe–Einstein propositions look at two specific models of biological evolution.

4.1   The Need for Simplification of Models

When investigating complex systems one is faced with large numbers of often heterogeneous components. Even when modern computer capacity is able to construct a near realistic replica to simulate, we still need simplification to help us identify possible emergent structures that make manageable description possible. Here are a couple of well-known concrete examples from different sciences of attempts to identify the effective structure responsible for behaviour at systems level.

Consider sociology and economics. We might try to understand the history of human society solely as an effect of personal actions. History was once taught mainly as a record of kings and their personalities. Though the precise course of events clearly will involve individuals and be influenced by the idiosyncrasies of the most powerful protagonists, it does seem superficial to reduce world history to a battle between kings, emperors and generals. A very eloquent and critical discussion of the shortcomings of reducing history to a battle between a few influential personalities is given by Leo Tolstoy in his second epilogue to *War and Peace*. Tolstoy argues that one needs to go beyond the few famous names and see history as a result of the contributions of huge numbers of small contributions from ordinary people adding together with the actions of leaders. In Chap. 7 of the second epilogue, Tolstoy attempts to identify the forces that produce the movement of the nations. Namely, he says:

The movement of the nations is caused not by power, nor by intellectual activity, nor even by a combination of the two as historians have supposed, but by the activity of all the people who participate in the events, and who always combine in such a way that those taking the largest direct share in the event take on themselves the least responsibility and vice versa. [447]

Tolstoy does also mention that the total effect of all the individual contributions is like the addition of infinitesimals in the mathematics of calculus. Each infinitesimal contribution is in itself vanishingly small but nevertheless integrating all the contributions leads to a non-zero value for the integral. But Tolstoy does not identify any emergent structure at the level of society. In contrast, Karl Marx did just that. As is probably very well known, Marx suggested that the dynamics of society is the result of the struggle between the classes and he tried to define and identify the classes present in a society. More recently Tony Blair, while Prime Minister of the UK, argued that social classes are not the important emergent components, but rather we have to understand society as the dynamics of *stake-holders*. The discussion continues. For our purpose we just want to note that these are examples of attempts to make a complex phenomenon, the history of society, comprehensible by identifying the essential emergent structures.

We see this strategy whenever dealing with complex many-component systems, although often no explicit mention is made of emergence and complexity science. Just two other examples will suffice. Freud's theory of neurosis is one example of trying to identify structure, or in this case processes, governing the dynamics of the mind, which presumably amounts to trying to understand aspects of the high-level dynamics of our brains. See e.g. [39] for a recent discussion of the history and the scientific controversies surrounding the term 'neurosis'. A similarly much used, but difficult to define, systems-level concept is sitting at the centre of ecology. At first one might think that the notion of a species cannot be too contentious. We suppose we all know a dog when we see one. But perhaps not. Are wolves dogs? Is a coyote a dog? They can interbreed but are considered to be separate species, despite the so-called biological species concept defining species as consisting of individuals that can produce fertile offspring. It turns out that it is difficult to make a robust, sharp definition of biological species, see e.g. Chap. 24 in [77]. This would seem to be a significant short-coming, since evolutionary theory, or ecosystems theory, heavily makes use of species as the building blocks, see e.g. Gould's great opus *The Structure of Evolutionary Theory* [169].

The important point in the present context is that when considering complex systems, the building blocks are not necessarily self-evident but the identification of the hierarchy of structures is part of the development of an understanding. Some will emphasise (e.g. Dawkins [99]) that selection foremost acts at the level of the reproducing entity, i.e. the individual organism, and that genes are the essential entities determining biological evolution. Others (e.g. Gould [169], Jablonka and Lamb [252] and more recently Goodnight [166] and Walsh [480]) will point out that the mechanisms behind selection in Darwinian evolution are much more involved than Dawkins's catchy selfish gene picture suggests, and may often involve selection pressures acting at several levels. We will return later in Sec. 11.3 to a discussion of how simple models can be helpful in this long-standing discussion.

Let us keep in mind Whitehead's viewpoint that processes rather than things are what really exist while we consider some concrete models in the light of the three O'Keeffe–Einstein propositions listed above.

4.2   O'Keeffe–Einstein Propositions at Work

It is natural to start with what has become the prototypical model of simple effective modelling, namely the Ising model mentioned above. Ising and Lenz were very well aware that representing a magnet in terms of a set of variables that can only assume two values, plus and minus one, is a tremendous simplification. Real magnetic material contains atoms and electrons moving about according to the laws of quantum mechanics. The only aspect of quantum dynamics included in the Ising model is the assumption that working with a binary variable will suffice, corresponding to the relevant quantum variables only assuming two values. The phenomenon that the Ising model can address is the formation in ferromagnetic materials of a magnetic state at low temperature. At high temperature no overall magnetic field is produced by the material, but as temperature is lowered below a certain level, the material undergoes a transition and becomes magnetic. It was this specific phenomenon that the Ising model was designed to address.

The explanation and modelling of phase transitions has become one of the pivotal achievements in modern science, as the underlying concepts apply to a huge range of phenomena including the emergence of the universe as we see it from the Big Bang to the understanding of the formation of opinions. We will return to the Ising model in mathematical detail in Sec. 6.2, right now we just want to look at the model in the light of our three propositions. Though the Ising model leaves out most of the complications of real material, it includes two crucial aspects. The first is that the individual components on their own assume the value -1 and $+1$ with equal likelihood and therefore their average value will be zero. The second crucial assumption is that the individual components interact and therefore the actual average value may differ from 0. The components interact with their neighbours and the energy of each of these interactions is lowest if a component and its neighbour have the same value, either both equal to -1 or both equal to $+1$.

It is this dependence, or interaction, which allows the success of the Ising model. And it is a spectacular success. The fact that components of the Ising model can have average values different from zero, which happens at low temperature, is an example of symmetry breaking, see Sec. 2.5. It is this symmetry breaking that turned out to be of immense importance across science. So although the Ising model is unable to explain many materials science problems related to magnetic materials (such as self-induction domain formation), the model is successful in explaining why very different materials and different types of phase transitions may behave in the same way for temperatures near the phase transition. The simple structure of the Ising model is of much broader relevance than magnetic materials. It also captures aspects of melting transitions, opinion formation and neuronal networks models of brain dynamics. This kind of common, or universal, behaviour is not at all obvious when one looks at the microscopic properties of materials and is an example of how the collective effect of interactions may make the macroscopic behaviour of large numbers of components behave in a simple structured way that falls into a few distinct classes.

So how does the Ising model fare in terms of the three propositions? It clearly does very well. **P1** is very much satisfied since the model is formulated making use of the most extreme abstraction, focusing on just the absolutely most minimal characterisation of the mechanisms involved, while neglecting many complications of real quantum-mechanical atoms and electrons. **P2** is also beautifully satisfied since the model is simple but not so simple that it leaves out the essential phenomenology. And finally **P3** is satisfied in a very impressive way. The Ising model explains how the interaction between components can lead to the observed change in the macroscopic magnetic properties, but it also turns out that the model is able to describe, with numerical accuracy, mathematical exponents describing a large varied family of phase transitions that falls into the so-called Ising class. The fact that the behaviour of complex systems can be grouped into classes of emergent behaviour is a recurrent theme throughout this book. That such classification is possible is an essential reason why complexity science can be considered a well-defined scientific endeavour.

We want to discuss three more models from the viewpoint of our three O'Keeffe–Einstein propositions. First the Schelling model of racial segregation, then two models of punctuated equilibrium.

The Schelling model of residential racial segregation is as bold as the Ising model in its neglect of detail in order to identify what may be the most important mechanism behind the process of segregation. The inspiration for the model is based on the observation that people express fairly tolerant views when asked how much they are concerned about residing in a neighbourhood where their colour, or race as it used to be misleadingly called, only constituted a minority. If the individual person felt content as long as their ethnic group constituted more than some $X\%$ of the local population, one might expect that an urban area would show a heterogeneous mix with a fractional representation of the ethnic group in all neighbourhoods of at least $X\%$.

But Schelling – and probably very many others – noticed that this was not what one observed. Now there can be many reasons for this discrepancy between what the inclination of the individual person permits in terms of ethnic mix and the actual composition of urban areas. One can think of laws, traditions, economics, cultural and many other effects. But like in the case of the Ising model, the Schelling model demonstrated that the observed phenomenon can be reproduced by a model that neglects many plausible effects, some of which presumably may be relevant to some degree. And similar to the Ising model, the crucial mechanism in Schelling's model is the interaction, or influence, between the components, i.e. the 'agents' of the model representing the inhabitants of the city. In Schelling's model the interaction occurs through the relative density of each ethnic group, imbalance can breach an agent's tolerance level and drive the agent to change neighbourhood.

As we have mentioned before, it is a recurrent theme in complexity science that models with simple components with some minimal interaction can capture emergent processes at the systemic level and in doing so help to elucidate the generality of the underlying processes. Schelling's model, like the Ising model, also does well in terms of our three propositions. Among the zillions of possible influences on a person's choice of residence, the model focuses on one single mechanism: whether the tolerance level $X\%$ is breached or not. So **P1** is satisfied. The model is remarkably simple, yet it establishes a possible explanation for the observed behaviour, hence **P2** is satisfied. Finally it turns out, as we shall return to in Sec. 11.2, that the framework of the model is of relevance to a range of segregation phenomena, which make **P3** satisfied. It is also interesting that there is even a link between the Schelling model and the Ising model, as will be explained in Sec. 11.2.

Finally, let us turn to two models of punctuated equilibrium. We will argue that the first, the **Bak–Sneppen model**, is an example of a model that breaks the requirement of **P2**. In contrast, the **Tangled Nature model** manages to satisfy all three O'Keeffe–Einstein propositions though at the cost of a less abstract and simplified modelling framework compared to the Bak–Sneppen model.

Both models are meant to be mathematical models of intermittency in the form of Eldredge's and Gould's punctuated equilibrium of macroevolution [169]. People familiar with mathematical modelling in statistical mechanics became aware of punctuated equilibrium to a great extent through the elegant paper published by Bak and Sneppen in 1993 [36]. The model takes a dramatic step and represents the collection of evolving species by a set of numbers, $B_1, B_2, \ldots, B_N$ between zero and one placed on a circle, see Fig. 4.1. The idea is that these numbers correspond to the '**fitness**'. It is important to keep in mind that fitness is a difficult concept to define precisely. In biology, fitness is related to the reproductive success but that is not really an intrinsic feature, such as the

Figure 4.1 A sketch of the Bak–Sneppen model. The height of the vertical lines indicates the different values of the random numbers B_i allocated to the N sites $i = 1, 2, \ldots, N$ arranged on the ring.

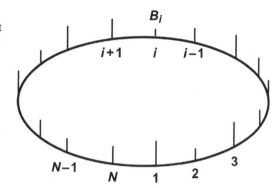

number of legs of an organism. Reproductive ability will always be intimately related to the environment. Penguins may be very fit in Antarctica but very unfit in the Sahara.

Evolutionary dynamics is represented by choosing the species with the smallest fitness value and removing it. The assumption is that low fitness somehow will correspond to being doomed to extinction. The disappearance of the least-fit species may drive the species it interacts with to extinction. This is represented by removing the two species neighbouring the one with the lowest fitness value. The model represents the creation of new species by simply replacing three new random numbers on the sites where the extinction took place.

This is obviously a very simple model and easily fulfils proposition **P1** above. The question is whether it also satisfies **P2** and **P3**. One can notice that the rate of activity is always the same, namely in each update three species go extinct and three are created. This means that the processes defined by the model are smooth and not at all intermittent. So since we wanted to explain intermittency, one might feel that proposition **P2** is not satisfied and that we are dealing with a model which is too simple to capture the intermittency of punctuated equilibrium. This conclusion may to some extent be a matter of taste. The authors note that the interaction between the species, represented by the removal of the two neighbours, leads to a structure in the distribution of fitness numbers present in the system. The authors argued that this structure would relate to a real time, not the updated time, in such a way that the smoothness of the updates will become intermittent in real time because, it is argued, real time is exponentially related to the updated time. However, even if one accepts the argument for this exponential connection to real time, it seems that the model's simplicity makes it unable to satisfy **P3**.

So the model seems to break proposition **P2**, since the phenomenon of punctuation does not readily emerge from the defined process, but has to be added through a not altogether transparent interpretation of time. Moreover, the model's elegant simplicity also prevents it from addressing the observed gradual decay since the Cambrian explosion in the number of mass extinctions per time [310] and the variation in the number of species and other taxonomic structures with time. The drastic simplicity makes the model unable to relate to the effect of evolution on ecological measures such as species abundance distributions (SAD), species area relations (SAR), ecosystem network structures, etc. For this reason the model does not fulfil **P3** very well. In Sec. 11.3 we will look at the Tangled Nature model. This model is simple compared to any real evolving biological system, though significantly more complex than the Bak–Sneppen model. The Tangled Nature model is more complicated because it considers reproducing individuals. It is simple because it only assumes that individuals are able to reproduce and die, an abstraction that makes the model qualify for **P1**. The reproduction probability of an individual depends on how the species it belongs to is related to other existing species. This set of processes turns out to capture punctuated equilibrium as the emergent mode of dynamics at the macro level. Furthermore, the simplistic individual-based evolutionary dynamics generates the decay in the overall extinction rate and a range of ecosystem properties including SAD and SAR, together with aspects of ecosystem network structures. Hence the Tangled Nature model can be seen as an example of a simplified model of a complex system undergoing evolution and adaptation, attaining **P1** without sacrificing **P2** and **P3**.

Now that we are aware of the strengths and limitations of approaching complex phenomena by use of simple models, in the next chapters we will elaborate on the mathematical modelling of emergent behaviour.

Summary: We have discussed how the level of detail impacts on the usefulness of a model. Adding or subtracting features in a model can help identify the most relevant mechanism. We notice that:

- Including every little detail may hide the generality of underlying mechanisms.
- Too little detail can make a model irrelevant.
- The ideal model will explain the most and be based on the fewest assumptions.

4.3   Further Reading

Introductory level:

The article 'The truth about scientific models' [198] by Hossenfelder contains many important observations concerning the nature of models.

Intermediate level:

The very enjoyable book *Evolution in Four Dimensons* [252] by Jablonka and Lamb contains many instances of discussions of what makes models useful and which level of description is needed. See e.g. the section Selfish Genes and Selfish Replicators in Chap. 1 and Chap. 9 Lamarckism Evolution: The Evolution of the Educated Guess.

The book *Introduction to the Modeling and Analysis of Complex Systems* [379] by Hiroki Sayama gives in Chap. 2 a hands-on discussion of what models are and how to construct them.

Advanced level:

The short book *Scientific Models: Red Atoms, White Lies and Black Boxes in a Yellow Book* [154] by Philip Gerlee and Torbjörn Lundh offers a thorough discussion of conceptual and practical aspects of what scientific models are about and how to construct them.

The paper 'Values and evidence: How models make a difference' [333] by Wendy Parker and Eric Winsberg presents a critical discussion of how models can influence our assessment of evidence.

4.4   Exercises and Projects

Exercise 1
Consider the financial system of a country like Japan for example and discuss which ingredients we may have to include in a model that can address the reason for the observed recurrent financial crashes.

Exercise 2
Discuss how one decides whether a model is too simple or unnecessarily complicated.

Exercise 3
Imagine that the EU flagship *Human Brain Project* has succeeded in producing a high-performance computer that can simulate the entire network of 8.6×10^{10} neurones and their 10^{15} connections.

(a) Will it be easier to understand this artificial brain function than to understand the human brain?
(b) Can such an artificial brain help to identify the relevant emergent mental structures? Or put differently, can we *understand* how the brain gives rise to mental activity without identifying the relevant macroscopic collective building blocks?
(c) Will an artificial one-to-one representation of the brain in itself explain cognition and mental processes?
(d) Could we 'educate' this machine brain in order to develop a human mind? Think in terms of system and its environment.

Exercise 4

The term 'fitness' is frequently used without any attempt to define it carefully. Discuss how to define fitness and how to measure it. Is it reasonable to assume fitness to be an inherent property of an agent or does fitness always depend on context and environment?

Project 1 – The robust components

The stakeholder concept has become popular in management theory. The publisher's description of the book *Stakeholder Theory* [146] says:

The stakeholder perspective is an alternative way of understanding how companies and people create value and trade with each other. Freeman, Harrison and Zyglidopoulos discuss the foundation concepts and implementation of stakeholder management as well as the advantages this approach provides to firms and their managers. They present a number of tools that managers can use to implement stakeholder thinking, better understand stakeholders and create value with and for them. The Element concludes by discussing how managers can create stakeholder oriented control systems and by examining some of the important stakeholder-related issues that are worthy of future scholarly and managerial attention.

This quote makes use of the concept of 'stakeholder' in relation to management theory of firms. The concept has also been used in the form 'stakeholder society'. The online dictionary Wiktionary explains:

A society whose members have both rights from it, and duties or responsibilities to it; a concept within the New Labour movement.

Consider the concept 'stakeholder' from the perspective of establishing a prototypical model.

(a) What do 'stakeholders' represent: individuals or groups?
(b) How can the concept help to establish a description, i.e. model, of the dynamics of the management of either a firm or society?
(c) At the level of firms, discuss if the stakeholder concept represents robust structures possessing independent integrity.
(d) At the level of society, discuss if the stakeholder concept represents robust structures possessing independent integrity.
(e) Can the stakeholder concept help us to establish robust prototypical models?

Project 2 – The brain

The discipline of neuroscience makes use of network theory to identify the structures relating to functionality of the brain. Consider the paper by Vértes et al. [469]. You do not need to study all the details of the paper.

Note that the authors write: 'We therefore anticipate that many aspects of brain network organization, in health and disease, may be parsimoniously explained by an economical clustering rule for the probability of functional connectivity between different brain areas'.

(a) From the perspective of prototypical models, how will you characterise the modelling described in the paper?

(b) Check which later papers have cited Vértes et al. [469]. Does this indicate that the modelling presented in [469] was robust and captured the most important aspects of the brain functionality?

(c) With an eye on the paper 'Brain networks and cognitive architectures' [342] by Petersen and Sporns, consider what a prototypical model of human cognition could look like.

(d) Imagine we have established a simple model of the brain network. Could such a model possess emergent structures resembling aspects of the human mind?

Project 3 – The O'Keeffe–Einstein propositions

The project is concerned with a model of ant foraging. The observational background for the model consists of ants leaving their nets to go and collect food and is described in [365]. The theoretical model is presented in [405].

(a) Make a brief summary of the salient features of the observations described in [365].

(b) Without going into the mathematical details, make a brief summary of the structure and focus of the model [405]. Try to sketch the flow of the mechanisms included in the modelling.

Now think of the model in terms of the three O'Keeffe–Einstein propositions.

(c) Is the description used by the model viable? Does it include too much or too little detail to be considered 'realistic' in the sense of O'Keeffe **P1**?

(d) Can you think of a way to make the model simpler, as asked by Einstein **P2**, or would any simplification make the model 'too simple'?

(e) Reading the Summary, how would you evaluate the model in terms of the third proposition **P3**. Can you think of potential applications not mentioned by the author?

II

Mathematical Tools of Complexity Science

Introduction to Part II

The next chapters are dedicated to mathematical approaches of central relevance to the analysis and modelling of emergent behaviour amongst many interacting components.

As with any other language, it is not a matter of being either, on the one hand, an expert or, on the other hand, completely unwilling to learn even the basics. That is, although I am not a master of English like Shakespeare, I am still able to use the language as an aide to my own thinking and as a way to benefit from presenting my ideas in discussions amongst peers. And the more we use a language, the greater our ability to think in a structured and reasonably consistent manner. Mathematics is not just formalism [191, 492]. Mathematics consists of *ideas* that can be captured by symbols and communicated by formalisms. This view of mathematics is not very different from observing that music consists of emotional landscapes captured by sound that can be communicated by the written score. Music is much more than the written score and mathematics is much more than the symbols in equations.

Everyone knows that science and engineering make use of mathematics to compute and predict. But it is still worth emphasising that formulating scientific models in mathematical terms is essential in order to be able to identify accurately the consequences of ideas and assumptions. A verbal model, a notion sometimes used e.g. about Darwin's theory of evolution, is a first step to identify the ingredients needed for an explanation. But to identify the relative relevance of various mechanisms, mathematics is needed to handle the details. As models are developed mathematically, the astonishing observation is that the study of the mathematical structure can itself lead to discoveries. A spectacular example is the investigation in the late 1930s by the theoretical physicist Wolfgang Pauli (1900–1958) of the connection between statistics and the **spin quantum number** of particles [335]. Pauli was able to predict very profound and completely new types of behaviour of collections of particles from the details of the mathematical structure of the model describing them. This is an example of how we are able to use models that are formulated in sufficient mathematical detail to uncover mechanisms and relations that we otherwise would not have noticed. Mathematical formulations not only make it possible to draw the precise consequences of the assumptions we make, the mathematical formulation of a model is also able to point backwards and help us to discover new phenomena and behaviours, which were not explicitly included when first defining the model. The mathematician Eugene Wigner has written a very readable and stimulating essay 'The unreasonable effectiveness of mathematics in the natural sciences' [495]. The developments since 1960, when Wigner wrote his article, have seen mathematics become an essential tool in fields such as linguistics, psychology and musicology, outside the traditional natural sciences. Had Wigner written his article today, he would probably have dropped the last four words of the title.

This part of the book is written with the intention of striking a balance between verbal explanation of ideas and presentation of the mathematical formalism. The hope is that working through the chapters will help readers who are not already fluent in mathematics to develop sufficient mathematical literacy to be able to appreciate and use mathematics as a language to formulate and express ideas.

With this in mind, we will try to present the conceptual basis, including their scope and purpose, behind the formalism as we go along. This means that readers already acquainted with a topic covered in one of the following chapters can skim through the text to become familiar with the notation and terminology, while readers less trained in mathematics may use the exposition as a springboard to develop their mathematical skills.

The topics included are, of course, not entirely exhaustive of all the kinds of mathematics used in complexity science. They have been selected because they are all widely used in the literature and because they are crucial to the characterisation, analysis and classification of a broad range of emergent phenomena arising as components are aggregated. The various topics are interrelated and it is not crucial to read the chapters in the order they appear. It may be beneficial first to skim through to obtain a rough overview and then to return for a more in-depth study.

Since complex dynamics commonly consists of bursts of activity spreading across the components, such as an epidemic, a forest fire or a burst of neuronal activity, we start out with the simplest mathematical version of this kind of dynamics, namely the branching process in Chap. 5.

After this look at spreading processes of crucial importance to complex systems, we turn in Chap. 6 to the very general formalism of statistical mechanics. A formalism developed to analyse macroscopic phenomena such as phase transitions in physical systems in terms of the properties of the components. We will be particularly interested in ideas and concepts which are helpful when analysing emergence in general, i.e. beyond the traditional situations considered by physics. The formalism offers examples where one can relate the emergent structures and their dynamics in a very precise way to the collective behaviour of the underlying microscopic components. The formalism of statistical mechanics is particular well suited when dealing with equilibrium or stationary phenomena. The analysis of phenomena with an explicit time dependence needs other methods.

In Chap. 7 we look at synchronisation as an example of deterministic dynamics that involves some degree of coherence amongst the motion of the individual components.

To be able to analyse and classify structures, static as well as dynamic, across components embedded in a web of interdependencies, we turn to network theory in Chap. 8. Because complexity science is concerned with the phenomena rooted in the interaction between components, network theory is a natural mathematical tool that enables the identification of which kind of systemic behaviour to expect when components are connected in specific ways.

The mathematical analysis of the interdependence between observables based on joint probability functions is called information theory. This is the focus of Chap. 9, where we present methodologies of particular relevance to data analysis. Rather than starting from a mechanistic model, we may be presented with a number of **observables** in the form of data series, such as time series from a brain scanner, ecological records, financial records or climatic recordings. Information theory can then help us to understand the nature of relationships between the observables, for example whether they might be causally related or just correlated.

Where Chap. 9 starts from joint probability distributions, Chap. 10 looks at how, from a knowledge about the stochastic processes, we can derive equations from which we determine the probability distributions.

Chapter 10 addresses agent-based modelling. In this approach we want to know what the consequences are when a specific set of actions are made available to the individual components, now called agents.

The last two chapters, Chap. 12 and 13, focus on the emergence of systemic-level jerky dynamics, i.e. intermittency. Chapter 12 presents some of the many frameworks focusing on how intermittency arises and what kind of statistics to expect for the intermittent events. Chapter 13 focuses on methods to forecast disruptive sporadically occurring events.

5 Branching Processes

> **Synopsis:** The process of entities successively splitting into two or more is of great relevance: biological reproduction, infection spreading, rumour spreading, nuclear reactions and much more.

A branching process consists of replicating entities called agents, or parents and offspring, see Fig. 5.1. The parent may, with certain probabilities, produce a number of offspring or none. This stochastic mathematical description summarises the dynamics of many different real processes. In Sec. 2.1 we discussed the connection to epidemiology. Here are a few other relevant examples. The population dynamics of asexual organisms, where a parent produces a number of offspring and each of these may in turn produce new offspring, etc. Nuclear reactors make use of branching processes. Here the disintegration of one nuclei may induce disintegration of other nuclei and so on. Or we can think of a disease spreading through a population. In this case an infected person may cause infection in a number of susceptible people. Likewise we can think of a piece of information, or a rumour, that spreads through a population by one person telling a number of other persons.

We will call the reproducing agents 'nodes' and can then describe the dynamics in terms of an initial root node, the very first parent, which may give rise to a number $k = 0, 1, 2, \ldots$ of new nodes. In the simplest situation we will assume that the probability p_k that a node produces k new nodes is the same for all nodes at all times, i.e. that the root node at generation 0 in Fig. 5.1 gives rise to k offspring with the same probability as e.g. each of the nodes at generation 3 does.

Since p_k are probabilities, normalisation demands

$$\sum_{k=0}^{\infty} p_k = 1 \tag{5.1}$$

and the average number of branches, or offspring, is given by

$$\mu = \sum_{k=0}^{\infty} k p_k. \tag{5.2}$$

The average branching ratio μ is well known from e.g. the Covid-19 pandemic. In epidemiology, μ is called the basic reproductive number and denoted R_0. In this chapter we will see why it is significant whether this number is smaller or greater than 1. We will analyse how μ determines the number of descendants of the root node after n generations, the total number of descendants before the lineage goes extinct, and the

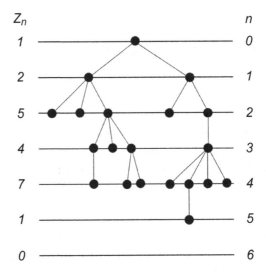

Figure 5.1 Branching process. A parent is represented by a node (black dot) placed on the horizontal lines indicating generations labelled by n. The offspring of a parent, if any, are placed as nodes on the line below, connected to the parent by thin solid lines called branches. The node in generation 0 has two branches, representing two offspring. In generation 2 we see a node with three and one with two branches.

time to extinction. To discuss the statistical properties of the branching process, we will make use of the formalism of generator functions. This formalism is introduced in Box 5.1.

Box 5.1 Generator Functions for Probability Distributions

Generator functions are convenient mathematical tools that simplify the computation of the statistical properties, such as the average and higher moments of a probability distribution. We will consider a discrete **stochastic variable** X which can assume the values $X = 0, 1, 2, \ldots$. The probability that the X assumes a specific value is denoted by

$$P_X(n) := \text{Prob}\{X = n\}. \tag{5.3}$$

The **generator function** for X is defined as

$$g_X(s) = \sum_{n=0}^{\infty} P_X(n)s^n. \tag{5.4}$$

We notice that the generator function is a function of a variable s. This variable does not have any particular significance besides allowing a way to keep track of the properties of X in the form of some simple algebraic manipulations. For example, we have

$$g_X(1) = \sum_{n=0}^{\infty} P_X(n) = 1,$$

$$g_X'(1) = \sum_{n=0}^{\infty} n P_X(n) = \langle X \rangle, \tag{5.5}$$

$$g_X''(1) = \langle X^2 \rangle - \langle X \rangle.$$

Furthermore, the generator function helps to analyse sums of stochastic variables of the following type:

$$S_N = \sum_{k=1}^{N} X_k, \tag{5.6}$$

where X_k are independent and identical, also denoted by i.i.d., meaning that all the X_k are described by the same generator function given by Eq. (5.4). The number of terms in the sum, N, is another stochastic variable that assumes values in $\{0, 1, 2, \ldots\}$ and is independent of the X_k. The generator for N is

$$g_N(s) = \sum_{n=0}^{\infty} P_N(n) s^n. \tag{5.7}$$

In this case the following holds for the generator function for the sum S_N (see e.g. [140]):

$$g_{S_N}(s) = g_N(g_X(s)). \tag{5.8}$$

The proof is as follows.

Proof
We have $g_{S_N}(s) = \sum_{q=0}^{\infty} P_{S_N}(q) s^q$ and

$$P_{S_N}(q) = \text{Prob} \left\{ S_N = \sum_{k=1}^{N} X_k = q \right\} \tag{5.9}$$

$$= \sum_{n=0}^{\infty} \text{Prob}\{N = n\} \text{Prob} \left\{ \sum_{k=1}^{n} X_k = q \right\} \tag{5.10}$$

$$= \sum_{n=0}^{\infty} P_N(n) \sum_{q_1} \cdots \sum_{q_n} \delta_{q_1 + \cdots + q_n, q} P_X(q_1) \cdots P_X(q_n). \tag{5.11}$$

In the last equation we made use of the probability that $X_1 = q_1, \ldots, X_n = q_n$ is the product of the individual probabilities and to be able to rewrite the sum we introduced the Kronecker delta function

$$\delta_{i,j} \begin{cases} 1 & \text{if } i = j \\ 0 & \text{otherwise} \end{cases}. \tag{5.12}$$

We can then proceed as follows:

$$g_{S_N}(s) = \sum_{q=0}^{\infty} \sum_{n=0}^{\infty} P_N(n) \sum_{q_1} \cdots \sum_{q_n} \delta_{q_1 + \cdots + q_n, q} P_X(q_1) \cdots P_X(q_n) s^q. \tag{5.13}$$

$$= \sum_{n=0}^{\infty} P_N(n) \sum_{q=0}^{\infty} \left(\sum_{q_1} P_X(q_1) s^{q_1} \right) \cdots \left(\sum_{q_n} P_X(q_n) s^{q_n} \right) \delta_{q_1 + \cdots + q_n, q} \tag{5.14}$$

$$= \sum_{n=0}^{\infty} P_N(n) \left(\sum_{q=0}^{\infty} P_X(q) s^q \right)^n \tag{5.15}$$

$$= \sum_{n=0}^{\infty} P_N(n) (g_X(s))^n \tag{5.16}$$

$$= g_N(g_X(s)), \tag{5.17}$$

which is the expression in Eq. (5.8).

To analyse the statistics of the dynamics of the progeny as the reproductive process proceeds through consecutive generations, we look at the size of the population Z_n at generation n. We want to discuss the following properties:

(1) Extinction corresponding to $Z_n = 0$ for some n and onwards. We will look at Prob$\{Z_n = 0\}$ for some $n < \infty$.
(2) How the size of the average of the nth generation, i.e. $\langle Z_n \rangle$, depends on time.
(3) The probability distribution of the total progeny Y_∞, i.e. the total number of descendants of the root node.
(4) The probability distribution of the time to extinction, τ, defined by $Z_n > 0$ for $n < \tau$ and $Z_N = 0$ for all $n \geq \tau$. We will look at properties of the distribution $P(\tau)$.

We have looked at scale invariance and self-similarity before, see e.g. Sec. 3.1. Since each node produces offspring according to the same probabilities p_k, the branching process can be considered to repeatedly start afresh from every node. This makes the tree structure generated in Fig. 5.1 self-similar. The self-similarity of the tree structures is key to our analysis. At generation n we separate the nodes into groups according to their parent at generation 1, as sketched in Fig. 5.2. This allows us to write

$$Z_n = \sum_{k=1}^{Z_1} Z_n^{(k)}. \tag{5.18}$$

Here $Z_n^{(k)}$ denotes the number of nodes in generation n that originate from node k in the first generation. There are Z_1 nodes in generation 1 and one node, per definition, in generation 0.

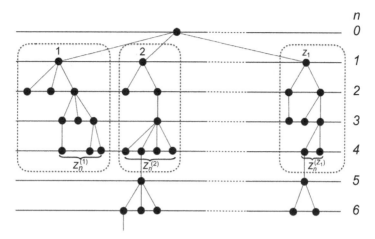

Figure 5.2 Branching process. The tree is broken up into Z_1 trees according to the progeny of the first generation, $n = 1$. Each of these subtress follows the same statistics as the entire tree.

In the next section we will analyse the statistics of the branching process by use of the self-similarity of branching trees to find equations for the generator functions for the following stochastic variables: the size Z_n of generation n, the total number Y_n of descendants up to generation n and the total size $Y_\infty = \lim_{n \to \infty} Y_n$. In case the process started at generation 0 continues forever, the total size Y_∞ will be infinite.

5.1 Generator Functions: Sizes and Lifetimes

The size and duration statistics of the process initiated by the seed node at generation 0 are contained in the probability distributions $P(Z_n)$ and $P(Y_n)$ of Z_n and Y_n, so we should compute these distributions. This can be done but the mathematics is somewhat advanced, see [183, 324]. A good understanding of the branching process is fortunately provided by a study of the moments of Z_n and Y_n, and these can be analysed in relatively simple ways by use of the generator formalism.

The self-similarity of the branching process ensures that it can be separated into subtrees, as done in Fig. 5.2, each subtree having the same statistical properties as the entire process, except that the subtree runs for one generation less. It follows that the probability that $Z_n^{(k)} = q$ is equal to the probability that $Z_{n-1} = q$. We use this observation to analyse the size of generation n. As in Eq. (5.18), we have $Z_n = \sum_{k=1}^{Z_1} Z_n^{(k)}$ and the self-similarity tells us that the generator for each of the terms $Z_n^{(q)}$ is identical to the generator for Z_{n-1}:

$$g_{Z_n^{(k)}}(s) = g_{Z_{n-1}}(s). \tag{5.19}$$

Next we notice that the number of terms Z_1 in the sum for Z_n is equal to the number of offspring of the root node in generation 0. This means that $\text{Prob}\{Z_1 = k\} = p_k$, where p_k are the branching probabilities defining the process and therefore the generator function $g_{Z_1}(s)$ for Z_1 is equal to the generator function $g(s)$ for the offspring probabilities:

$$g_{Z_1}(s) = \sum_{k=0}^{\infty} p_k s^k = g(s). \qquad (5.20)$$

We can use the expression in Eq. (5.8) in Box 5.1 to analyse Z_n in Eq. (5.18). We conclude that

$$g_{Z_n}(s) = g(g_{Z_{n-1}}(s)). \qquad (5.21)$$

Here we made use of the fact that the number of terms N in the sum in Eq. (5.8) is $N = Z_1$ and its generator is $g(s)$.

This equation makes it possible to determine how the probability, x_n, that the branching process stops at generation n depends on the branching ratio μ. We have that

$$x_n := \text{Prob}\{Z_n = 0\} = g_{Z_n}(0), \qquad (5.22)$$

and we notice

$$x_n = g_{Z_n}(0) = g(g_{Z_{n-1}}(0)) = g(x_{n-1}). \qquad (5.23)$$

We also know how the function $g(t) = \sum_{k=0}^{\infty} p_k t^k$ behaves. Namely, since $g'(t) > 0$ for $t \in [0, 1]$, the function $g(t)$ is an increasing function on $[0, 1]$ and has the graph depicted in Fig. 5.3. The intercept with the y-axis is determined by $g(0) = p_0$ and the slope at $t = 1$ is $g'(1) = \mu$. The probability that for some n the process stops is given by $x := \lim_{n \to \infty} x_n$. According to Eq. (5.23), this limit will satisfy $x = g(x)$.

So to determine whether there exists an n at which the process stops, we need to look for the roots of the equation $t = g(t)$ which is where $g(t)$ crosses the identity line $y = t$ in Fig. 5.3. For $\mu \leq 1$ this only happens at $x = 1$. Since x is the stopping probability, this means that when the branching ratio is smaller than or equal to one, the process will stop at some point. In contrast, when $\mu > 1$ the generator $g(t)$ will cross the identity line at a value of $x < 1$. For $\mu < 1$ the probability that the process stops is less than

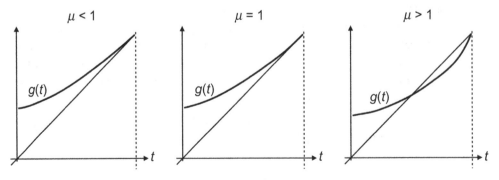

Figure 5.3 Roots of the equation $s = g(s)$ for different values of the average branching ratio μ.

one, which means that there is a probability larger than zero that the process continues forever.

We summarise:

- For $\mu \leq 1$ the branching process will stop.
- For $\mu > 1$ the process may continue forever with a probability larger than zero.

The generator function for Z_n contains the information about the average size of generation n, namely $\langle Z_n \rangle = g'_{Z_n}(1)$. By use of Eq. (5.21) we can write

$$
\begin{aligned}
\langle Z_n \rangle = g'_{Z_n}(1) &= \frac{d}{ds} g(g_{Z_{n-1}}(s))|_{s=1} \\
&= \left[g'(g_{Z_{n-1}}(s)) \frac{d}{ds} g_{Z_{n-1}}(s) \right]_{|s=1} \\
&= \left[g'(g_{Z_{n-1}}(s)) g'(g_{Z_{n-2}}(s)) \frac{d}{ds} g_{Z_{n-2}}(s) \right]_{|s=1} \\
&= \left[g'(g_{Z_{n-1}}(s)) g'(g_{Z_{n-2}}(s)) \cdots g'(g_{Z_1}(s)) g'(s) \right]_{|s=1},
\end{aligned}
\tag{5.24}
$$

which together with the observation that

$$
g_{Z_n}(1) = g(g_{Z_{n-1}}(1)) = \cdots = g(g(\cdots (g(s))))|_{s=1}
\tag{5.25}
$$

and that $g(1) = 1$ and $g'(1) = \mu$ leads to the conclusion that

$$
\langle Z_n \rangle = \left[g'(1) \right]^n = \mu^n.
\tag{5.26}
$$

That is, the average size of the nth generation will decay exponentially when $\mu < 1$ and explode exponentially when $\mu > 1$. This, of course, explains why during an epidemic outbreak it is important to restrict the contact between people sufficiently to bring the average number of new infected $R_0 = \mu$ below 1.

We know from the analysis of Eq. (5.22) that for $\mu > 1$ there is a larger than zero probability that the process never stops, leading to the total progeny Y_n becoming infinite, and therefore the average $\langle Y_n \rangle$ will also diverge for $\mu > 1$. In the next section we will look at how $\langle Y_n \rangle$ behaves for $\mu < 1$.

5.1.1 Size of the Progeny

How many descendants will eventually be produced by the replication process starting from an agent at generation 0? This will depend on the specific realisation of the stochastic process described by the branching probabilities p_k. The total number of progeny produced during the n consecutive generations is given by the sum

$$
Y_n = 1 + Z_1 + Z_2 + \cdots + Z_n,
\tag{5.27}
$$

where each of the terms Z_k counts the size of generation k and is itself a stochastic variable, meaning that Z_k is a random number. Very often when we sum many random numbers, the sum, when adjusted appropriately (subtracting the average and dividing by the square root of the number of terms) turns out to be Gaussian distributed. But in the case of a branching process the terms in the sum Z_k are not at all independent, since Z_{k+1} is very much related to the size of Z_k. If the generation number k is large, it is more likely that generation $k + 1$ is also large.

The distribution of $Y_\infty = \lim_{n \to \infty} Y_n$ can be determined for large values of Y_∞. Namely, for $\mu < 1$ it can be shown that for the leading k and μ, the dependence of the probability that $Y_\infty = k$ has the form

$$\text{Prob}\{Y_\infty = k\} = Ak^{-3/2}e^{-k/k_0(\mu)} + \epsilon, \quad \text{for } k \gg 1. \tag{5.28}$$

Here A is a factor independent of k and the term denoted by ϵ is smaller than the first term in the equation by a factor k. The scale of the exponential is given by $k_0(\mu) \sim (1 - \mu)^{-2}$. The detailed computation by use of the generator formalism is given in [183, 324]. We want to stress that when the average branching ratio μ is approaching 1, the characteristic scale of the exponential factor in Eq. (5.28) diverges and the distribution for Y_∞ will follow a power law with the exponent $-3/2$. Although we will not derive the expression in Eq. (5.28), it is instructive to look at how the generator formalism in a simple way lets us analyse the dependence on μ of the average $\langle Y_\infty \rangle$.

As we did above, see Fig. 5.2, we divide the process up according to the ancestor at generation 1 and make use of the self-similarity. Let $Y_n^{(k)}$ denote the total number of nodes up to generation n, which originates from node k in generation 1:

$$Y_n = 1 + \sum_{k=1}^{Z_1} Y_n^{(k)}. \tag{5.29}$$

And therefore, according to Eq. (5.8), we know that the generator function for $\sum_{k=1}^{Z_1} Y_n^{(k)}$ is given by

$$g_{Y_n}(s) = g(g_{Y_{n-1}}(s)). \tag{5.30}$$

Note that if $A = 1 + B$, where A and B are stochastic variables with $A, B \in \mathbb{N}$, then $\text{Prob}\{A = k\} = \text{Prob}\{B = k - 1\}$ and we have, since $\text{Prob}\{A = 0\} = 0$, that

$$\begin{aligned}
g_A(s) &= \sum_{k=0}^{\infty} \text{Prob}\{A = k\}s^k \\
&= \sum_{k=1}^{\infty} \text{Prob}\{A = k\}s^k \\
&= \sum_{k=1}^{\infty} \text{Prob}\{B = k - 1\}s^k \\
&= s \sum_{q=0}^{\infty} \text{Prob}\{B = q\}s^q \\
&= s g_B(s).
\end{aligned} \tag{5.31}$$

So we can conclude that

$$g_{Y_n}(s) = s g(g_{Y_{n-1}}(s)). \tag{5.32}$$

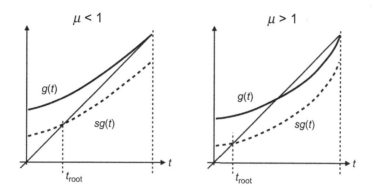

Figure 5.4 Roots of Eq. (5.33) for different values of the average branching ratio μ.

The limit $g_{Y_\infty}(s) := \lim_{n \to \infty} g_{Y_n}(s)$ is given as a root in the equation

$$g_{Y_\infty}(s) = sg(g_{Y_\infty}(s)). \tag{5.33}$$

We can again graphically see, cf. Fig. 5.4, that there are two regimes depending on the value of μ. Recall that the root $t_{root} = g_{Y_\infty}(s)$ and that

$$g_{Y_\infty}(1) = \sum_{k=0}^{\infty} \mathrm{Prob}\{Y_\infty = k\} \tag{5.34}$$

is the normalisation of the variable Y_∞. That is, when we find the t_{root} that satisfies $t = sg(t)$ and let $s \to 1$, the solution t_{root} will give us the left-hand side of Eq. (5.34) and thereby the value of the sum on the right-hand side. Recall that the slope of $g(t)$ at $t = 1$ is $g'(1) = \mu$, as indicated in Fig. 5.4. For $\mu > 1$ we have $t_{root} < 1$ as $s \to 1$. This implies that for $\mu > 1$ the sum $\sum_{k=0}^{\infty} \mathrm{Prob}\{Y_\infty = k\} < 1$ and therefore the stochastic variable Y_∞ is not normalised for $\mu > 1$, corresponding to what we found above, that for $\mu > 1$ the branching process can continue forever with a probability larger than zero, so a subset of the generated trees will have infinite size. This is in contrast to the situation for $\mu < 1$ where t_{root} approaches 1 as $s \to 1$, corresponding to the fact that all trees stop for some value of $n < \infty$ when $\mu < 1$. The two cases are separated by the critical value $\mu_c = 1$.

We saw above that for $\mu < \mu_c$, i.e. the subcritical case, the average generation size decays exponentially and for the supercritical case, $\mu > \mu_c$, we have exponential growth of $\langle Z_n \rangle$ with n. This behaviour is also reflected in the average size of the progeny $\langle Y_\infty \rangle$, which is given by $g'_{Y_\infty}(1)$. We have

$$
\begin{aligned}
g_{Y_\infty}(s) &= sg(g_{Y_\infty}(s)) \Rightarrow \\
g'_{Y_\infty}(s) &= g(g_{Y_\infty}(s)) + sg'(g_{Y_\infty}(s))g'_{Y_\infty}(s) \Rightarrow \\
g'_{Y_\infty}(s) &= \frac{g(g_{Y_\infty}(s))}{1 - sg'(g_{Y_\infty}(s))}.
\end{aligned}
\tag{5.35}
$$

Let $s = 1$, and use the fact that for $\mu < 1$, Y_∞ is normalised, meaning that $g_{Y_\infty}(1) = 1$ and therefore

$$g'(g_{Y_\infty}(1)) = g'(1) = \mu.$$

We conclude that

$$\langle Y_\infty \rangle = g'_{Y_\infty}(1) = \frac{g(1)}{1 - g'(1)} = \frac{1}{1 - \mu}. \tag{5.36}$$

As expected, $\langle Y_\infty \rangle \to \infty$ as $\mu \to 1$ and Eq. (5.36) describes in detail how the size of the progeny diverges.

Up to this point in our discussion we have concentrated on the size of the individual generations Z_n, the accumulated number of individuals Y_n during the first n generations and the total number of descendants Y_∞ of the parent in generation 0. In the next section we will look at temporal aspects by analysing the probability that the process stops, i.e. goes extinct, after a certain number of generations.

5.1.2   Time to Extinction

Whenever the probability that a node does not produce any offspring, p_0, is larger than zero, the branching process may stop because none of the Z_n nodes present in generation n manage to produce any offspring. This will happen with probability $p_0^{Z_n}$. However, this observation does not allow us to compute the probability that the process for n generations has $Z_k > 0$ for $k = 0, 1, 2, \ldots, n - 1$ and then $Z_n = 0$. The value of n is called the **time to extinction**, denoted by τ, and we write

$$\tau := \min\{n \geq 1 \mid Z_n = 0\}. \tag{5.37}$$

The time to extinction, τ, is the generation number at which, for the first time, no offspring are produced; of course $Z_k = 0$ for all $k > \tau$. In Fig. 5.1, $\tau = 5$.

We will use the generator formalism to sketch an analysis[1] of the asymptotic behaviour for the accumulated distribution $P(\tau > n) := \text{Prob}\{\tau > n\}$ for the branching processes with $\mu \leq 1$ and finite standard deviation[2] $\sigma < \infty$. We find that

$$\begin{aligned} \mu < 1 &: P(\tau > n) \sim \mu^n \text{ for large } n \\ \mu = 1 &: P(\tau > n) \sim \frac{2}{\sigma^2 n} \text{ for large } n. \end{aligned} \tag{5.38}$$

Recall that the generator for the size of the nth generation is $g_{Z_n}(s)$. And that according to Eq. (5.22) the probability of stopping is

$$x_n = \text{Prob}\{Z_n = 0\} = g_{Z_n}(0) = 1 - \text{Prob}\{Z_n > 0\}$$

---

[1] The discussion is inspired by lecture notes on stochastic processes made available online by Professor Steven Lalley, Department of Statistics and the College, University of Chicago. https://galton.uchicago.edu/~lalley/.

[2] Here σ refers to the standard deviation of the offspring process described by the branching probabilities p_k.

and therefore

$$P(\tau > n) = \text{Prob}\{Z_n > 0\} = 1 - g_{Z_n}(0).$$

For $\mu < 1$ we found that $x = \lim_{n\to\infty} x_n = 1$. So for large n we have $g_{Z_n}(0) \simeq 1$. We want to make use of $g_{Z_n}(s) = g(g_{Z_{n-1}}(s))$ from Eq. (5.21) and Taylor expand the outer function $g(s)$ on the right-hand side about $s = 1$. Since $g(1) = 1$ and $g'(1) = \mu$, we have

$$P(\tau > n + 1) = 1 - g_{Z_{n+1}}(0) = 1 - g(g_{Z_n}(0)) = 1 - g(1 - (1 - g_{Z_n}(0))$$

$$= 1 - [g(1) - (1 - g_{Z_n}(0))g'(1)] = 1 - [1 - (1 - g_{Z_n}(0))\mu] \qquad (5.39)$$

$$\mu(1 - g_{Z_n}(0)) = \cdots = \mu^n.$$

Hence, for $\mu < 1$ the probability that the branching process is active for more than n generations decreases exponentially as n increases.

For $\mu = 1$ we need to Taylor expand to second order, while we make use of $g(1) = 1$ and $g'(1) = \mu = 1$. Hence

$$1 - g_{Z_{n+1}}(0) = 1 - g(g_{Z_n}(0)) = 1 - g(1 - (1 - g_{Z_n}(0))$$

$$= 1 - [g(1) - g'(1)(1 - g_{Z_n}(0)) + \frac{1}{2}g''(1)(1 - g_{Z_n}(0))^2]$$

$$\Downarrow \qquad\qquad (5.40)$$

$$1 - g_{Z_n}(0) = 1 - g_{Z_n}(0) - \frac{1}{2}g''(1)(1 - g_{Z_n}(0))^2.$$

We conclude that $x_n = 1 - g_{Z_n}(0)$ satisfies the iterative equation

$$x_{n+1} = x_n - bx_n^2, \qquad (5.41)$$

with $b = g''(1)/2$. We can check by substitution into this equation that asymptotically (i.e. for $n \gg 1$), $x_n = \alpha/n$ is a leading-order solution to the equation. Namely, substitute $x_n = \alpha/n$ into Eq. (5.41) and notice that when $\alpha = 1/b$, the equation reduces to $n^2/b = n^2/b - 1/b$ so right and left hand is asymptotically equal. We conclude that $P(\tau > n) = 1 - g_{Z_n}(0) \sim 1/n$.

In summary, we have for the probabilities

$$\text{Prob}\{\tau = n\} = \frac{d}{dn}(1 - P(\tau > n)) \sim \begin{cases} n\mu^{n-1} & \text{for } \mu \leq 1 \\ 1/n^2 & \text{for } \mu = 1 \end{cases}. \qquad (5.42)$$

We will end this chapter by making in the next section a link between branching processes and random walks.

5.2 Branching Trees and Random Walks

The branching process produces branching trees and by designing a random walk moving from node to node on the branching tree we can make a link between the two stochastic processes.

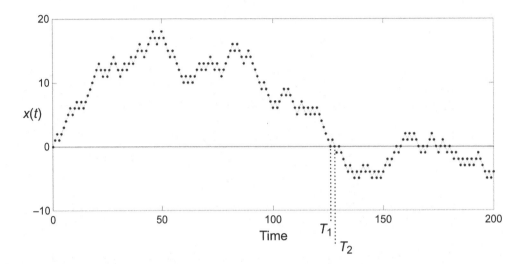

Figure 5.5 Trajectory of a random walker. At time $t = T_1$ the walker returns to the position $x(0) = 0$ where it started at time zero. The walker returns again at time T_2 and continues to do so at later times.

We saw in Eq. (5.28) that the tree size distribution Prob$\{Y_\infty = k\}$ behaves like a power law $k^{-3/2}$ multiplied by an exponential factor and that the latter approaches 1 when the average branching ratio approaches the critical value, i.e. when $\mu \to 1$. The distribution for the first return time for a random walker is also described by the exponent $-3/2$, see Sec. 10.2 and e.g. Sec. III.7 in [139]. Is this a coincidence or is it an example of similar behaviour in different systems? The following construction relates the branching tree to a random walk and thereby shows that the branching process and the random walk are mathematically similar processes.

Let us first recall what exactly is meant by the return time of a random walker, for details see Sec. 10.2. Think of a random walker with position $x(0)$ at time zero. See Fig. 5.5. Assume that $x(0) = 0$. In each time step, the position of the walker changes randomly according to

$$x(t + 1) = x(t) + \Delta.$$

Let the walker increments Δ be distributed according to some distribution for which the first two moments exist and let $\langle \Delta \rangle = 0$, so the walk is unbiased. We can consider the times $0 < T_1 < T_2 < \ldots$ when the walker has returned to the value $x(0)$ for the first time T_1, for the second time T_2, etc., i.e. $x(T_1) = x(T_2) = \cdots = 0$. The probability that $T_1 = T$ is (asymptotically) given by $T^{-3/2}$.

We can make a map between the branching process and the random walker in the following way. We think of walking on the tree generated by a branching process, for example the tree in Fig. 5.1. We enumerate the nodes $i = 1, 2, 3, \ldots$, see Fig. 5.6. Next we need to relate the increments of the walk to the offspring production. We let the increment Δ_i be given by the number of offspring emerging from a node, k_i minus 1,

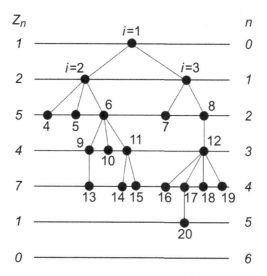

Figure 5.6 A branching tree seen as a random walk. The random walker starts from the root node at time $i = 1$. At each time step it moves to a new node in the tree, sweeping through the nodes of consecutive generations. Every time the walker moves to a new node, the time i of the walker is incremented by 1. The position of the walker $x(i)$ is not indicated on the figure. See the text for how $x(i)$ is related to the number of offspring produced by each node.

i.e. we define $\Delta_i = k_i - 1$. We now ascribe a 'position' to the walker which is updated according to $x_{i+1} = x_i + \Delta_i$, so we have

$$x_{i+1} = x_i + k_i - 1. \tag{5.43}$$

We can think of the walk in the following way. Whenever we arrive at a node, the offspring shooting off from that node is uncovered to us. If we per definition assume $x_1 = 1$, after i steps, x_i is equal to 1 plus the accumulated number of nodes uncovered but not yet visited. We of course increase the number of visited sites by 1 whenever we step onto the new node. And if no offspring sprouts from that node we lower x_i by 1. In Fig. 5.6 we have per definition $x_1 = 1$ and $k_1 = 2$ therefore

$$\begin{aligned}
x_1 &= 1 \\
x_2 &= x_1 + k_1 - 1 = 1 + 2 - 1 = 2 \\
x_3 &= x_2 + k_2 - 1 = 2 + 3 - 1 = 4 \\
&\vdots \\
x_{19} &= x_{18} + k_{18} - 1 = 3 + 0 - 1 = 2 \\
x_{20} &= x_{19} + k_{19} - 1 = 2 + 0 - 1 = 1 = x_1 \\
x_{21} &= x_{20} + k_{20} - 1 = 1 + 0 - 1 = 0.
\end{aligned} \tag{5.44}$$

We see that the random walk variable x_i returns to its start value after 20 steps, equal to the number of nodes in the branching tree.

If $k_i > 0$, offspring sprout from the node, the number of non-visited sites increases by k_i and the net change to x_i is $k_i - 1$. The process continues until for some i the random walk variable has returned to 1 (i.e. $x_i = 1$) and in the next step we have $x_{i+1} = 0$. The value of i will then equal the total number of nodes, Y_∞, in the tree generated by the branching process.

Perhaps it is illuminating to think of the process in terms of a literature search. We want to scan the literature on a topic of interest starting from one particular publication. Say, in a paper on sociology, we see the *Thomas theorem* mentioned and that the paper gives k_1 references of relevance to this theorem. We decide to check all the references to the Thomas theorem we can reach by going through these references and the references that these papers mention and so forth. The variable x_i is equal to the number of papers we still have to check after having visited the first i papers. Every time we check a paper we can tick one paper off as inspected, hence we can lower x_i by 1, so $x_i \rightarrow x_i - 1$. On the other hand, if we find k_i new references, we will need to add these to the list of 'still-to-be-checked' papers; to account for this we have to make the change $x_i \rightarrow x_i + k_i$. The process stops when the length of the 'still-to-be-checked' list is equal to 1 and we do not find new references in the last paper on the list, meaning x_i will become 0 in the next time step.

The increments of the random walk $\Delta_i = k_i - 1$ are given in terms of the number of offspring, which are independent and identical stochastic variables. This means that the random walk performed by x_i is of the type mentioned above and the return time distribution for x_i, and therefore for Y_∞, will be characterised by the exponent $-3/2$.

Summary: Offspring production, fire spreading from one tree to others, infections spreading from one person to others, rumours passed from one person to others, etc. are all examples of branching processes. We analysed the simplest version where each node, with the same set of probabilities p_k, produces k offspring and saw that the average number of offspring, μ, determines the nature of the spreading with

- $\mu < 1$: exponential decay of activity.
- $\mu = 1$: critical power law behaviour.
- $\mu > 1$: exponential increase in activity.

5.3   Further Reading

Introductory level:

The paper 'Criticality and self-organization in branching processes: Application to natural hazards' [86] by Corral and Font-Clos gives a clear explanation of branching processes and generator formalism.

Intermediate level:

Chapter 12 in Feller's phenomenally useful book *An Introduction to Probability Theory and Its Applications* [139] contains a very readable introduction to the mathematics of branching processes.

A long list of applications of branching processes to biological population dynamics is given in the very readable collection *Branching Processes: Variation, Growth, and Extinction of Populations* [176], edited and authored by Haccou, Jagers and Vatutin.

Advanced level:

Theodore E. Harris's book *The Theory of Branching Processes* [183] is a standard mathematical introduction to the field.

The book by Marek Kimmel and David E. Axelrod [235] carefully presents the mathematics behind branching processes applied to molecular and cellular biology, as well as to human evolution and medicine.

The specific application of branching processes to the modelling of cancer can be found in Richard Durrett's book [117].

5.4 Exercises and Projects

Exercise 1

Consider a probability distribution for event sizes s given by

$$P(s) = \begin{cases} As^{-\tau} & \text{if } s \in \mathbb{N} \\ 0 & \text{otherwise} \end{cases}$$

where A is a normalisation factor.

(a) Explain in what sense power laws are scale invariant.
(b) Derive the normalisation constant.
(c) Compute the moments of the distribution and derive criteria on τ for the existence of the ath moment.

Exercise 2

Let $S_N = \sum_{k=1}^{N} X_k$, where the stochastic variables $X_k \in \mathbb{N} \cup \{0\}$ are identical and independent, each described by the generator function $g_X(s)$. Furthermore, the stochastic variable $N \in \mathbb{N}$ is described by the generator function $g_N(s)$ and $g_{S_N}(s)$ denotes the generator function for S_n.

(a) Show that

$$g_{S_N}(s) = g_N(g_X(s)).$$

Consider a branching process. Let $g(s)$ denote the generator function for the branching probabilities and $g_{Z_n}(s)$ denote the generator function for the size of the generation number n.

(b) Show that

$$g_{Z_n}(s) = g(g_{Z_{n-1}}(s)).$$

(c) Let the generator for the branching probabilities be given by

$$g(s) = \alpha + \beta s.$$

Use the expression in (b) to determine the generator function for the size distribution of generation number n.

(d) From the result in (c) determine $\mathrm{Prob}\{Z_n = k\}$ for $k = 0, 1, 2, \ldots$.

Exercise 3

Consider a branching process for which the generator function for the branching probabilities is given by

$$g(s) = \frac{1-p}{1-ps}.$$

(a) What is the probability that an individual produces k offspring?
(b) What is the average branching ratio?
(c) For which value of p is the process a critical branching process?
(d) Show that the generator function for the size of the progeny Y_n after n generations, $g_{Y_n}(s)$, is given by

$$g_{Y_n}(s) = \cfrac{1-p}{1 - \cfrac{\alpha}{1 - \cfrac{\alpha}{1 - \cfrac{\alpha}{\ddots \cfrac{\alpha}{s}}}}}$$

where $\alpha = p(1-p)$ and there are a total of $n-1$ factors α in the nested fraction n on the right-hand side.

Project 1 – From generator function to statistics

Consider a general branching process. Let $g(s)$ denote the generator function for the branching probabilities and let $g_{Z_n}(s)$ denote the generator function for the size Z_n of the nth generation. Furthermore, let

$$x_n = \mathrm{Prob}\{\text{extinction by generation number } n\}.$$

(a) Show the relation

$$g_{Z_n}(s) = g(g_{Z_{n-1}}(s)).$$

(b) Show that

$$x_n = g(x_{n-1}).$$

(c) Show that the probability $x = \lim_{n \to \infty} x_n$ that the process eventually goes extinct is therefore given by the smallest root in $s = g(s)$.

Consider a branching process with the following branching probabilities:

$$\begin{aligned} p_0 &= q \\ p_1 &= 0 \\ p_2 &= p \\ p_k &= 0 \ \forall k \geq 2. \end{aligned} \tag{5.45}$$

(d) For which value of p is this process critical?
(e) Determine how the average size of the nth generation depends on p.
(f) For arbitrary $p \in (0, 1)$ determine the probability that the process eventually goes extinct.

Now consider a branching process with branching probabilities given by

$$p_k = qp^k \ \forall \ k \in \{0\} \cup \mathbb{N}. \tag{5.46}$$

(g) Let μ denote the average branching ratio, consider $1 - g(s)$ and show that for the process in Eq. (5.46) the generator function satisfies the following relation:

$$\frac{1}{1 - g(s)} - \frac{1}{\mu(1 - s)} = 1.$$

(h) For the process in Eq. (5.46) determine x_n as a function of the average branching ratio.
(i) Use the result in (h) to show that

$$\lim_{n \to \infty} x_n = \begin{cases} 1 & \text{if } \mu < 1 \\ 1/\mu & \text{if } \mu > 1 \end{cases}.$$

(j) Consider the process introduced in Eq. (5.45). Let $P(T)$ denote the probability that extinction occurs at generation T. By direct simulation, consider the limit $\mu \to 1^-$ and estimate the value of the exponent of the power law $P(T)$ is approaching. Explain how you perform your simulation and show the plots you used to reach your conclusion. How does your numerical estimate of the power-law exponent compare with what you expect theoretically?

6 Statistical Mechanics

> **Synopsis:** Properties and behaviours at the systemic aggregate level are derived as statistical averages from probability distributions describing the likelihoods of the various states available to the components.

The overall behaviour of systems with many components can be estimated if we have access to the probabilities for each of the possible configurations of the constituents. Of course it is very difficult in general to obtain these probabilities. However, for more than a century we have had a formalism which works very successfully in the special case of thermodynamic equilibrium. The formalism is known as statistical mechanics and ascribes a probability to each microscopic configuration, even when components are interacting and the joint probability is not just given as a product of the single component probabilities.

The joint probability is determined by the energy of the configuration and the temperature. A very large number of good books have been written during the last 100 years, but we will just refer to two books which are both very readable and contain comprehensive presentations of formalisms, see [362] and [392]. Since the aim of complexity science is to find ways to compute the systemic-level behaviour for any type of many-component system, and since this agenda is still work in progress, it is worth being aware of how statistical mechanics approaches the challenge.

In the next section we discuss how statistical mechanics ascribes probabilities to each of the microscopic configurations of a system. A micro configuration consists of specifying the state of each component, such as the position and the velocity, when we deal with particles. The following sections in this chapter will describe how it is possible, from the probabilities of the micro configurations, to derive the behaviour at the systemic level. We will in particular look at how being able to compute the systemic properties as averages over the micro configurations makes it possible to analyse transitions between regimes of entirely different systemic structures. In Sec. 6.7 we will discuss how new emergent structures can be identified.

6.1 Probabilities and Ensembles

To describe how statistical mechanics is able to identify structures emerging at the macroscopic level, we look at how macroscopic, or systems-level, quantities are obtained through averaging procedures. The reason that equilibrium systems can be analysed in particular detail is because the equilibrium, where the systems of interest can be considered as in thermal equilibrium with a heat bath, makes it possible to determine the probability weights of the individual microstates. One starts out with the following fundamental assumption concerning isolated, i.e. closed, systems:

- **Microcanonical ensemble**
  For a closed system it is assumed that *all* microstates, consistent with the macroscopic constraints, occur with *equal* probability.

The macroscopic constraints can, for example, be the total energy E (which is constant for a closed system) and the volume V. Denote by $\Omega(E, V)$ the total number of microstates possible under these constraints, meaning that the components or particles of the system have to be located within the given volume V and that the energy of all the particles taken together must equal E. The probability $p(s)$ that the system is in a particular state s is then, according to the assumption of the microcanonical ensemble, given by

$$p(s) = \frac{1}{\Omega(E, V)}. \tag{6.1}$$

Closed systems are not very interesting in the sense that one is unable to interact with them. It is much more interesting to consider a system $\mathbf{S}$ in contact with its surroundings. The generic example analysed by statistical mechanics is a system brought in contact with a heat bath $\mathbf{B}$, also called a heat reservoir. The heat bath is a system sufficiently big that even when it exchanges energy with the small system of experimental interest, the heat bath remains unchanged. Pictorially we can think of a cup of tea in contact with the Pacific Ocean. In reality, cryostats are used to maintain the temperature of the experimental sample constant with high precision. The heat bath is characterised by its temperature T. We can now use the fundamental hypothesis above to determine the probabilistic weights for the states of $\mathbf{S}$. We assume the combined system $\mathbf{B} + \mathbf{S}$ to be closed, so the combined system is described by the microcanonical ensemble, i.e. all microstates of the combined system are equally likely. To proceed we note that if the system is in a state of energy E_s, the bath will be in one of its states of energy $E_{\mathbf{B}} = E_{\text{Tot}} - E_s$. The number of those states is denoted by $\Omega_{\mathbf{B}}(E_{\text{Tot}} - E_s)$. Here we make use of the assumption that interactions between the heat bath and the system are unable to alter the microstates available to either system.

We conclude that the system will be found in a specific microstate s_0 with probability

$$p(s_0) = \frac{\Omega_{\mathbf{B}}(E_{\text{Tot}} - E_{s_0})}{\sum_{s \text{ of } \mathbf{S}} \Omega_{\mathbf{B}}(E_{\text{Tot}} - E_s)}, \tag{6.2}$$

where the denominator ensures normalisation. In order to introduce the temperature into the mathematical formalism, it turns out that we should consider the logarithm of $p(s)$. We have

$$\ln[p(s_0)] = \text{constant} + \ln[\Omega_{\mathbf{B}}(E_{\text{Tot}} - E_{s_0})] \tag{6.3}$$

$$= \text{constant} + \ln[\Omega_{\mathbf{B}}(E_{\text{Tot}})] - \frac{\partial \ln[\Omega_{\mathbf{B}}(E_{\text{Tot}})]}{\partial E_{\text{Tot}}} E_{s_0} \tag{6.4}$$

$$= \text{constant} - \frac{1}{k_B T} E_{s_0}. \tag{6.5}$$

Here we Taylor expanded to linear order to obtain the second equality. The third equality follows, because it can be shown by use of the first and second laws of thermodynamics that the temperature is given by

$$\frac{1}{k_B T} = \frac{\partial \ln[\Omega_{\mathbf{B}}(E_{\text{Tot}})]}{\partial E_{\text{Tot}}}. \tag{6.6}$$

The constant k_B on the left-hand side is called Boltzmann's constant. The identity in Eq. (6.6) is derived in the following way. The first and second laws of thermodynamics lead to the thermodynamic identity $dE = TdS - pdV$, where the thermodynamic entropy $S = k_B \ln[\Omega(E)]$. Since the thermodynamic identity takes the form of an **exact differential**, we conclude that $\partial E/\partial S = T$ or $1/T = \partial S/\partial E$, from which Eq. (6.6) follows and we can conclude from Eq. (6.5) that

$$p(s_0) = \exp\left(\text{constant} - \frac{1}{k_B T} E_{s_0}\right) = \exp(\text{constant}) \exp\left(-\frac{1}{k_B T} E_{s_0}\right).$$

We rename the factor $\exp(\text{constant})$ as $1/Z$ and write

$$p(s_0) = \frac{e^{-\frac{E_{s_0}}{k_B T}}}{Z}. \tag{6.7}$$

The constant Z, called the **partition function**, is obtained from the normalisation condition

$$\sum_{s \text{ of } S} p(s) = 1, \tag{6.8}$$

to be given by

$$Z = \sum_{s \text{ of } S} e^{-\frac{E_s}{k_b T}} = \sum_{s \text{ of } S} e^{-\beta E_s}, \tag{6.9}$$

where we introduced the standard notation $\beta = 1/k_B T$. The partition function plays a central role in statistical mechanics because statistical average quantities can be obtained from it. For example, the average energy $\langle E \rangle$ is given by

$$\langle E \rangle = \sum_{s \text{ of } S} E_s p(s)$$

$$= \sum_{s \text{ of } S} E_s \frac{e^{-\frac{E_s}{k_b T}}}{Z} \tag{6.10}$$

$$= -\frac{\partial \ln(Z)}{\partial \beta}.$$

The probability in Eq. (6.7) is called the Boltzmann weight of the specific microsate s_0 and represents the **canonical ensemble** in statistical mechanics. The derivation presented here is based on the fundamental assumption of the microcanonical ensemble, namely that for a closed system at a specific energy E all possible micro configurations of energy E are equally likely to occur. It is not possible to derive this assumption rigorously from the basic equations of physics. But if we step outside physics and take the broader view of probability and information theory, the assumption appears reasonable as a least-biased assumption. Moreover, since the derivation starting from a closed system and the assumed equilibrium established by contact with a heat bath is difficult to generalise beyond physical systems, it is of interest to look at how the same probabilistic weights can be obtained from the principle of **maximum entropy** [211].

The information-theoretic version of entropy is very much used to extract information about system behaviour that is concealed in time series or similar data sets. It is worth keeping in mind that the term *information* as used in information theory does not have the same meaning as when in daily life we talk about the information contained in, say, a manual for how to operate a certain device or a cooking recipe. We will return to information theory in more depth in Chap. 9, here we concentrate on the alternative derivation of the statistical mechanics probabilities.

Information theory is concerned with the analysis of probabilities, and the information one has in mind is not the meaningful, so-called semantic information contained in a verbal explanation. In information theory one considers a narrow probability distribution to be more informative than a broad distribution. The idea is that if we know a certain quantity, say the weight of an animal, is distributed in a narrow range, we know more about what to expect if we meet a member of that species than if the weight is known to be distributed across a very broad range. Shannon introduced his entropy [394, 427] to capture the fact that probabilities carry information about what to expect. If a certain event i occurs with probability p_i, Shannon considers $-\ln p_i$ to be a measure of how surprised we become when we actually observe i. Think of throwing two dice. The probability that the sum of the eyes equals 2 is 1/36. So we get surprised when we see the combination one–one. Whereas the probability for the sum of the eyes to add up to 7 is 1/6. So we are less surprised when we find that the sum of our throw equals 7. In Shannon's formalism, our surprise at seeing the sum being equal to 2 is $-\ln(1/36) \approx 3.6$ and our surprise at an outcome summing to 7 is only $-\ln(1/6) \approx 1.8$. So what is the overall surprise we will experience when dealing with a certain probability distribution? As a measure of this, Shannon took the average surprise carried by each of the realisations of the variable described by the distribution $\boldsymbol{p} = (p_1, \ldots, p_i, \ldots, p_\Omega)$, namely[1]

$$S[\boldsymbol{p}] = -\sum_{i=1}^{\Omega} p_i \ln p_i. \tag{6.11}$$

---

[1] To use notation as much in agreement as possible with the usual tradition in the literature, we will sometimes denote the entropy by S and sometimes by H. This is certainly not ideal, but makes it easier to connect to other books and papers.

This expression turned out to be equivalent to the expression physicists, especially Boltzmann and Gibbs, had considered around 1900 when developing a statistical description of thermodynamics. For this reason the expression in Eq. (6.11) is also called entropy, though of course it is not immediately clear how exactly it relates to the entropy introduced in terms of heat exchange by Clausius in 1855.

Box 6.1 Lagrange Multipliers

To illustrate the ideas behind the method of Lagrange multipliers, we look at a function of two variables $f(x, y)$. We want to locate the values (x^*, y^*) of x and y that maximise the value of $f(x, y)$. When x, y are allowed to assume values everywhere, i.e. we are looking for the point in the entire plane that maximises the value of $f(x, y)$, we simply need to find the x, y that makes the gradient of the function vanish:

$$\vec{\nabla} f = \left(\frac{\partial f}{\partial x}, \frac{\partial f}{\partial y} \right) = (0, 0). \tag{6.12}$$

Perhaps we are looking for the highest point in some landscape. If there are no restrictions on where we are allowed to ramble about, the solution of Eq. (6.12) will give us the location of the highest point we are looking for. See Fig. 6.1. The contours given by $f(x, y)$ = constant will correspond to positions of equal height above sea level.

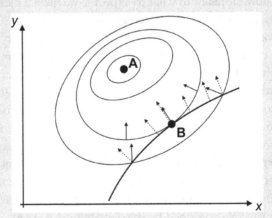

Figure 6.1 Sketch of the ideas behind Lagrange multipliers. The curves circulating the point **A**-represent contours of the function $f(x, y)$. The point **A** is the global maximum. This point cannot be reached when we are confined to the thick curve. On this path, the highest point we can reach is **B**. The arrows indicate the gradients of the function $f(x, y)$ (solid arrows) and of the function $g(x, y)$ (dashed arrows). These are always perpendicular to the contours and for this reason **B** is characterised by the t for which they are parallel.

However, it might be that we are searching the highest point in a park where the delicate alpine vegetation is protected by restricting people to stick to certain paths. This could be the thick curve in Fig. 6.1 specified by the mathematical expression $g(x,y) = c$, where $g(x,y)$ is a function of two variables and the relation between x and y along the curve is given by $g(x,y)$ being equal to the constant c. When we have to stay on the path, we will not be able to reach the overall highest point **A** but will need to be satisfied with the highest point **B** along the path specified, $g(x,y) = c$.

To pinpoint **B**, we make use of the gradient ∇g. We know from the theory of Taylor expansions that the change, $\Delta h(x,y)$, in a function $h(x,y)$ when we go from one location (x,y) to another nearby $(x + \delta_x, y + \delta_y)$ is given by

$$\Delta h(x,y) = \vec{\delta} \cdot \vec{\nabla} h. \tag{6.13}$$

We make use of this for the two functions $f(x,y)$ and $g(x,y)$. When we stay on the curve $g(x,y) = c$, the function value of g remains equal to c, so if we let $\vec{\delta}$ point along this curve, i.e. along the tangent of the curve, we will have $\Delta g(x,y) = 0$. Equation (6.13) tells us that for $\vec{\delta}$ parallel to the tangent, $\vec{\delta} \cdot \vec{\nabla} g = 0$ and therefore the gradient $\vec{\nabla} g$ is perpendicular to the direction of the tangent $\vec{\delta}$ of the curve $g(x,y) = c$. The same is, of course, also true for the gradients of $f(x,y)$, they will also be perpendicular to the curves $f(x,y)$ along which f is constant, which will be the height contours of the map if $f(x,y)$ were to indicate the height of the landscape. This is indicated in Fig. 6.1. To locate the point **B**, we simply need to find the point along the curve $g(x,y) = c$ where the gradients of f and g are parallel:

$$\vec{\nabla} f \parallel \vec{\nabla} g,$$

which means that a number λ exists such that $\vec{\nabla} f = \lambda \vec{\nabla} g$, or expressed differently:

$$\vec{\nabla}(f - \lambda g) = 0. \tag{6.14}$$

Here the number λ is called a Lagrange multiplier and by its introduction we have managed to handle the constraint in a simple way. We just need to find the unconstrained maximum of the function $J(x,y) = f(x,y) - \lambda g(x,y)$ by solving $\nabla J = 0$. While we solve Eq. (6.14) we treat λ as a free unknown parameter. The solution of Eq. (6.14) will depend on λ, so we denote this solution by $(x^*(\lambda), y^*(\lambda))$. To complete the computation we need to determine λ by finding the value of λ for which the point $(x^*(\lambda), y^*(\lambda))$ is located on the curve $g(x,y) = c$. So the final step is to solve the equation

$$g(x^*(\lambda), y^*(\lambda)) = c. \tag{6.15}$$

The strength of the Lagrange method to find maxima under constraints is that what we have just described can be generalised to functions of any number of variables, i.e. $f(x_1, x_2, \ldots, x_N)$ and constraints of the form $g(x_1, x_2, \ldots, x_N) = c$. The condition in Eq. (6.14) is derived from the geometrical condition that the gradients of the two functions f and g are parallel: $\vec{\nabla} f \parallel \vec{\nabla} g$. And this argument is also valid in higher dimensions, i.e. when we deal with many variables.

To apply the method to find the maximum entropy for the probability distribution $P(x)$ of an observable X which can assume the values $x_1, x_2, \ldots, x_\Omega$, we look at the problem in the following way. We are looking at the set of probabilities $p_1 = P(x_1), p_2 = P(x_2), \ldots, p_\Omega = P(x_\Omega)$ which corresponds to the largest value of the entropy in Eq (6.11). While we are looking for the maximum of the entropy, we will consider the probabilities $p_1, p_2, \ldots, p_\Omega$ to be our variables and we are looking for the set of values of these variables $p_1^*, p_2^*, \ldots, p_\Omega^*$ that makes the function

$$S(p_1, p_2, \ldots, p_\Omega) = -\sum_{i=1}^{\Omega} P_X(x_i) \log P_X(x_i) \tag{6.16}$$

maximal under the certain constraints of normalisation:

$$g_1(p_1, p_2, \ldots, p_\Omega) = \sum_{i=1}^{\Omega} p_i = 1 \tag{6.17}$$

and perhaps the constraint given by a specified average:

$$g_2(p_1, p_2, \ldots, p_\Omega) = \sum_{i=1}^{\Omega} x_i p_i = \langle X \rangle. \tag{6.18}$$

The Lagrange method tells us that we simply need to determine the maximum of the function

$$J(p_1, p_2, \ldots, p_\Omega) = S(p_1, p_2, \ldots, p_\Omega) - \lambda_1 g_1(p_1, p_2, \ldots, p_\Omega)$$
$$- \lambda_2 g_2(p_1, p_2, \ldots, p_\Omega). \tag{6.19}$$

The Lagrange multipliers are to be found by making sure the solution $p_1^*, p_2^*, \ldots, p_N^*$ of $\vec{\nabla} J = 0$ satisfies the constraints given by g_1 and g_2 in Eqs. (6.17) and (6.18).

Let us now see how Shannon's entropy can lead to the Boltzmann weights [211]. One makes the assumption that the probability, p_s, with which the system is in a microstate s will be given by the probability distribution which makes the entropy in Eq. (6.11) largest. To do this we consider the Shannon entropy as a function of many variables $p_1, p_2, \ldots, p_\Omega$. One variable p_i for each of the Ω microstates. The set of values $\boldsymbol{p}^* = (p_1^*, p_2^*, \ldots, p_\Omega^*)$ of the probabilities that maximise $S[\boldsymbol{p}]$ is determined by differentiation. Namely, each of the partial derivatives of $S[p]$ must vanish when evaluated on p^*. However, while we maximise $S[p]$ we cannot neglect that the sum of all the p_i must equal 1 since we are dealing with probabilities. To ensure we respect this constraint, we make use of the standard technique of Lagrange multipliers, see Box 6.1, and incorporate the constraint $\sum_i p_i = 1$ by maximising the quantity

$$J_1[p] = -\sum_s p_s \ln p_s + \lambda \sum_s p_s, \tag{6.20}$$

where $\lambda \in \mathbb{R}$ is a real constant to be determined later from the constraint $\sum_s p_s^* = 1$. We need to solve the equations

$$\frac{\partial J_1}{\partial p_s} = -\ln p_s - 1 + \lambda = 0, \tag{6.21}$$

for all $s = 1, 2, \ldots, \Omega$. This is straight forward. All the p_s are equal and given by

$$p_s = \exp(\lambda - 1) = \frac{1}{\Omega}. \tag{6.22}$$

In the last step we determined λ so that the sum of all the probabilities equals 1. We notice that, when we substitute this expression for the p_s into the expression for the entropy Eq. (6.11) we obtain

$$S[\mathbf{p}] = -\sum_{s=1}^{\Omega} p_s \ln p_s = -\sum_{s=1}^{\Omega} \frac{1}{\Omega} \ln \frac{1}{\Omega} = \ln \Omega. \tag{6.23}$$

This expression is very famous in statistical mechanics and information theory and is engraved on Boltzmann's gravestone.

We notice that when we maximise the Shanon entropy under the single constraint that the probabilities must add up to 1, we get that all microsates are equally likely, i.e. the result is the same as the fundamental assumption made for the closed system above and denoted as the microcanonical ensemble.

Let us now go one step further and assume that we are able to control the average energy of the system, $\langle E \rangle$, and keep it at a certain fixed value. We then need to ensure that when we find the probabilities p_s^* for the microstates, besides adding up to 1, they also correspond to the correct value of the average energy. That is, we need to respect the constraint $\sum_s E_s p_s = \langle E \rangle$ for the given value of $\langle E \rangle$. We do this by adding one more Lagrange multipliers:

$$J_2[p] = -\sum_s p_s \ln p_s + \lambda_1 \sum_s p_s + \lambda_2 \sum_s E_s p_s. \tag{6.24}$$

In this case we have

$$\frac{\partial J_2}{\partial p_s} = -\ln p_s - 1 + \lambda_1 + \lambda_2 E_s = 0, \tag{6.25}$$

from which we find

$$p_s = \exp(-1 + \lambda_1) \exp(\lambda_2 E_s). \tag{6.26}$$

The factor $\exp(-1 + \lambda_1)$ can be eliminated by the normalisation condition, so we can write

$$p_s = \frac{\exp(\lambda_2 E_s)}{Z} \text{ and } Z = \sum_s e^{\lambda_2 E_s} \tag{6.27}$$

and find p_s to have the same exponential form as the canonical ensemble in Eq. (6.7). In principle we would now have to tune the Lagrange multiplier λ_2 such that the average energy computed using the derived expression for p_s matches the imposed value $\langle E \rangle$ of the average energy:

$$\sum_s \frac{\exp(\lambda_2 E_s)}{Z} E_s = \langle E \rangle, \tag{6.28}$$

but in practice this equation is not possible to solve for λ_2, so instead one makes contact with thermodynamics and concludes that $\lambda = -1/k_B T$, see e.g. [213].

The important point we want to make is that the derivation of the probability weights from considerations in terms of the microcanonical and canonical ensemble is equivalent to an information-theoretic derivation in terms of maximising the Shannon entropy under constraints. If all we know about the system is that the p_s are probabilities and have to add up to 1, we get all p_s equal, which is equivalent to the closed system microcanonical ensemble. If we also impose a value for the average energy, we get the Boltzmann weights (when making appropriate contact with thermodynamics). The equivalence between the information-theoretic approach of maximum entropy (MaxEnt) and the ensemble approach suggests it is worthwhile trying to generalise the MaxEnt approach to complex systems not usually considered in statical physics. We will return to this in Chap. 9.

Let us conclude and summarise. The probabilistic weights, needed to calculate the average macroscopic behaviour of a system in contact with a heat bath at temperature T, is given by the Boltzmann weights in Eq. (6.7). A large number of average quantities can be calculated from the sum in Eq. (6.9). This important sum is called the *partition function* or partition sum. The average and higher moment of variables such as the energy can be obtained by differentiating the partition function.

Some states, or configurations of the microscopic degrees of freedom will contribute more to the partition sum than others, because the exponential factor $\exp(-E_s/K_B T)$ depends on the ratio between the energy of the state and the temperature. Such important configurations can sometimes be identified as macroscopic collective excitations, see Sec. 6.7. These may possess a degree of robustness and stability and can then be identified as macroscopic emergent objects with specific properties that can be considered essential building blocks.

Perhaps it is instructive to have the following picture in mind. Think of a pool table. To describe the motion of the balls, we can either follow the trajectories of all the individual molecules making up 15 coloured balls or we can notice that some of the molecules move together in a coordinated way and thereby form each of the 15 balls. We can therefore instead simply follow the trajectories of the centre of mass (COM) of each of the balls. Obviously we lose a lot of information since we can't go from the COM of the balls to the motion of all the molecules; whereas we can derive the COM motion if we know the motion of all the molecules. Hence we note that emergence involves a loss of some information; in this specific case we lose track of the motion of the individual particles. On the other hand, realising that we only need to follow the COM of the 15 balls, rather than following each of the 15×10^{26} molecules forming the 15 balls, represents an enormous jump in the level of our understanding. In this case we of course identify the 15 balls right away, because that is what we can see. But in other cases the identification of how the many microscopic components form some few important macroscopic structures may not be so easy. The vortices considered in Sec. 6.7 are one well-analysed example. Understanding how the dynamics of the neurones combine to form the components of our mind is another important example which is still very much work in progress.

Before we return to the emergence of collective excitations with some individual integrity in Sec. 6.7, we will in the next section study a very simple model of interacting components and use the statistical mechanics formalism to discuss transitions between regimes with different properties in which the correlations between different parts of the system differ dramatically.

6.2   The Ising Model

We already introduced the motivation and idea behind the Ising model in Secs. 3.1 and 4.1 and saw that the model is designed to describe the effect of interactions amongst collections of components, each of which can be in one of two states.

The Ising model originates in physics, where it is a model of magnetism. The magnetism of a material is produced by the alignment of atomic magnetic moments and is fundamentally a quantum phenomenon. Nevertheless, it has turned out that many macroscopic properties of magnetic materials can be understood by use of very simple statistical mechanics models that do not explicitly refer to quantum mechanics. We will introduce the Ising model and describe how it is easily handled in the mean-field approximation.[2] The model considers N variables, or degrees of freedom, S_i with $i = 1, \ldots, N$. These are called spins and we assume they are placed on a d-dimensional hypercube. See Fig. 6.2 for a two-dimensional sketch (we did already see this figure in Sec. 3.1).

Each spin can assume two values, e.g. up $S_i = 1$ and down $S_i = -1$. The model is defined in terms of the energy assigned to each configuration of the spin variables, which is given by the Ising Hamiltonian[3]

---

[2] We recall that 'mean field' is a general term for an approximation where, instead of following the exact evolution of each component, one develops a description of the 'on-average' behaviour; see also the comments in Secs. 3.5 and 12.1.2.

[3] It is unfortunate that the entropy in Eq. (9.16) and the Hamiltonian in Eq. (6.29) are denoted by the same letter H, but this is the most common notation used in the literature. The context will make it clear which quantity we are dealing with.

Figure 6.2 Sketch of two-dimensional Ising model. The plus and minus sign indicate the value, $S_i, S_j = \pm 1$, of the variable on that site.

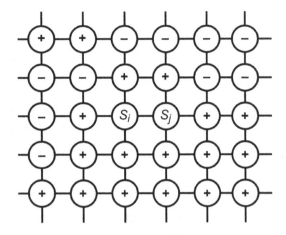

$$H[S_1, \ldots, S_N] = -\frac{1}{2}J \sum_{i=1}^{N} \sum_{j \in \text{nb}(i)} S_i S_j - B \sum_{i=1}^{N} S_i. \tag{6.29}$$

Here, J describes the strength of the influence, i.e. the coupling, between the spins. The set of neighbours of spin i is denoted by $\text{nb}(i)$ and B represents the effect of an externally applied magnetic field on each of the individual spins.

Since we are interested in emergent behaviours, we want to calculate the mean of the total collection of the spin variables:

$$m = \frac{1}{N} \sum_{i=k}^{N} S_k. \tag{6.30}$$

This quantity is called the total magnetisation per spin. It is somewhat like a centre of mass of a collection of particles, or if we think of the variables S_i as representing opinions for or against some option, it would give us a measure of the population's opinion with $m = -1$ corresponding to a consensus with everyone against, $m = 1$ everyone for and $M = 0$ would represent total disagreement or polarisation. We want to know the behaviour of m when averaged over all the possible configurations. In our present statistical-mechanics analysis we assume that the configurations occur with the probabilities given by the Boltzmann weights of the energies in Eq. (6.29). We denote averaging over these probabilities by angular brackets and write

$$\langle m \rangle = \frac{1}{N} \left\langle \sum_{k=1}^{N} S_k \right\rangle = \frac{1}{N} \sum_{k=1}^{N} \langle S_k \rangle = \langle S_i \rangle, \tag{6.31}$$

for all $i = 1, 2, \ldots, N$. The last equality is correct since we are dealing with a homogeneous system where all the sites statistically are subject to the same environment. For an inhomogeneous system J might depend on the site label i and j or B might depend on i, in which case we would have $\langle S_i \rangle \neq \langle S_j \rangle$ for $i \neq j$. Inhomogeneous systems are obviously of great relevance to real systems: all atoms in a material many not experience the same interaction strength to their neighbours or be influenced by the same strength of the magnetic field. Or if we think of the Ising model as relating to opinions and i being the label of the agents, clearly all agents may not influence each other in the same way or respond to external information (represented by B) in the same manner. The inhomogeneous Ising model is called a **spin glass** because its dynamical properties are similar to those of ordinary glass. The spin glass is also closely related to optimisation problems such as encountered when studying protein folding [426] and used in artificial intelligence and machine learning, see e.g. [184, 185].

Here we will ignore the inhomogeneity, but in Sec. 12.2 we will return and discuss ways of dealing with the dynamics of systems which are similar to the inhomogeneous Ising model. Now we want to make use of the fact that for the homogeneous Ising model, statistically all the sites $i = 1, 2, \ldots, N$, are equivalent, so to compute $\langle m \rangle$ we just need to consider the average of an arbitrary spin S_i and we can suppress the index i and write $\langle S \rangle = \langle S_i \rangle$, i.e.

$$\langle S \rangle = \langle S_i \rangle = \sum_{s \text{ of } \mathbf{S}} S_i p_{\text{state}} = \frac{1}{Z} \sum_{S_1 \in \{-1, 1\}} \cdots \sum_{S_N \in \{-1, 1\}} S_i e^{-\beta H[S_1, \ldots, S_N]}. \tag{6.32}$$

This expression computes the average value of S_i for an arbitrary site i by summing over all the possible ways of assigning the plus and minus values to the sites of the lattice: although S_i can only assume two values itself, each of these have to be combined with all the ways the spins on the remaining sites can be allocated the values ± 1 and each configuration contributes to the average with its probabilistic weight $p_{\text{state}} = \exp(-\beta E_{\text{state}})/Z$ where the energy E_{state} is obtained by computing the sums in the Hamiltonian in Eq. (6.29). The first term in the Hamiltonian in Eq. (6.29) makes the computation of the partition function and the average in Eq. (6.32) difficult to carry out for arbitrary dimension d. For $d = 1$ the calculation is straightforward though a bit involved, see Sec. 6.6.1. For $d = 2$ an exact result can be obtained but the computation is significantly more involved than for $d = 1$. This was first done by Lars Onsager in 1944, 24 years after the introduction of the model. Onsager's achievement is considered a major contribution to statistical mechanics. No exact solution is known in higher dimensions. But we can obtain a qualitative understanding of the behaviour of the magnetisation by use of a mean-field approach.

To do so we rewrite the Hamiltonian in the following way:

$$\begin{aligned}
H[S_1, \ldots, S_N] &= -\frac{J}{2} \sum_{i=1}^{N} \sum_{j \in \text{nb}(i)} S_i S_j - B \sum_{i=1}^{N} S_i \\
&= \sum_{i=1}^{N} -\left(\frac{J}{2} \sum_{j \in \text{nb}(i)} S_j + B \right) S_i \\
&:= -\sum_{i=1}^{N} \tilde{B}_i S_i,
\end{aligned} \tag{6.33}$$

where in the last equation we introduced the 'effective' field acting on S_i as

$$\tilde{B}_i = \frac{J}{2} \sum_{j \in \text{nb}(i)} S_j + B. \tag{6.34}$$

We can obtain a significant simplification by neglecting the dependence of $\tilde{B}_i$ on i. The rationale behind this goes as follows. Let $q = |\text{nb}(i)|$ denote the number of neighbours of each spin. We write

$$\sum_{j \in \text{nb}(i)} S_j = q \frac{\sum_{j \in \text{nb}(i)} S_j}{q} = q \bar{S}_i, \tag{6.35}$$

where $\bar{S}_i$ is the average value of the spins in the neighbourhood of spin number i. Since this is obtained by summing over all the sites connected to site i, the value of $\bar{S}_i$ may not vary too much from one location i to another. That is, we may expect that the typically on-average strength of interactions applied to the spin at, say, site $i = 7$ does not vary much from the on-average strength of interactions applied to the spin on site $i = 117$, for example. Under this assumption we can write

$$\bar{S}_i = \bar{S}. \tag{6.36}$$

To highlight that we choose to ignore the site dependence, we have left out the index i on the right-hand side. When we do this, the dependence on i also disappears from $\tilde{B}_i$ in Eq. (6.34). This approximation enables us to replace the Hamiltonian in Eq. (6.29) by

$$H_{MF}[S_1, \ldots, S_N] = -\sum_{k=1}^{N} \tilde{B} S_k. \tag{6.37}$$

This is called the mean-field Hamiltonian for the Ising model to stress that we have made use of on-average considerations. Using H_{MF} makes the mathematics much simpler because we can factorise[4]

$$\exp\{-\beta H[S_1, \ldots, S_N]\} = \exp\left\{-\beta \left(-\sum_{k=1}^{N} \tilde{B} S_k\right)\right\} = \Pi_{k=1}^{N} e^{\beta \tilde{B} S_k}, \tag{6.38}$$

and we can express Eq. (6.32) as

$$\langle S_i \rangle = \frac{1}{Z} \left(\sum_{S_i = \pm 1} S_i e^{\beta \tilde{B} S_i} \right) \Pi_{k \neq i} \left(\sum_{S_k = \pm 1} e^{\beta \tilde{B} S_k} \right). \tag{6.39}$$

The partition function is this mean-field approximation given by

$$Z = \Pi_{k=1}^{N} \left(\sum_{S_k = \pm 1} e^{\beta \tilde{B} S_k} \right). \tag{6.40}$$

---

[4] We use standard notation for the product:

$$\Pi_{k=1}^{N} A_k = A_1 A_2 \cdots A_N.$$

Hence, the factors in the partition function can cancel all the factors in the product term in Eq. (6.39), leading to

$$\langle S_i \rangle = \frac{\sum_{S_i = \pm 1} S_i e^{\beta \tilde{B} S_i}}{\sum_{S_i = \pm 1} e^{\beta \tilde{B} S_i}} = \frac{e^{\beta \tilde{B}} - e^{-\beta \tilde{B}}}{e^{\beta \tilde{B}} + e^{-\beta \tilde{B}}} = \tanh(\beta \tilde{B}). \tag{6.41}$$

We are not finished yet, since

$$\tilde{B} = \frac{J}{2} \sum_{j \in \mathrm{nb}(i)} S_j + B = \frac{J}{2} q \bar{S} + B \tag{6.42}$$

contains the so far indetermined $\bar{S}$. To close the equation we finally make the assumption that the average $\bar{S}$ is well approximated by the average over the Boltzmann weights $\langle S \rangle$, so we simply use $\bar{S} = \langle S \rangle$, which turns Eq. (6.41) into the following closed equation for $\langle S \rangle$:

$$\langle S \rangle = \tanh \left[\beta \left(\frac{Jq}{2} \langle S \rangle + B \right) \right]. \tag{6.43}$$

This equation determines $\langle S \rangle$ as a function of the temperature through $\beta = 1/k_B T$, the connectivity of the lattice given by the number of neighbours q and the external field B. We are not able to solve this equation by moving $\langle S \rangle$ to one side to express it in terms of the other variables. However, it is easy graphically to understand how the solution for $\langle S \rangle$ behaves. In Fig. 6.3 we have drawn the functions $f_1(x) = x$ and $f_2(x) = \tanh(ax + b)$, which are the functional dependencies of $\langle S \rangle$ on the left-hand side and the right-hand side of Eq. (6.43), respectively. Here $f_1(x)$ is just the identity line $y = x$ and has no parameter dependence. Whereas $f_2(x)$ depends on a and b. If $b > 0$ the curve representing $f_2(x)$ will always cross $f_1(x)$ and the x value that satisfies $f_1(x) = f_2(x)$ will give us the solution to Eq. (6.43) we are looking for when we use $a = Jq\beta/2 > 0$ and $b = B$.

If $b = 0$ the function $f_2(x)$ starts from 0 at $x = 0$ and there are now two possibilities. If the slope of $f_2(x)$ at $x = 0$ is too shallow, the function will remain below $f_1(x) = x$ for

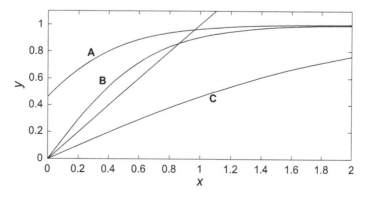

Figure 6.3 The functions in Eq. (6.43). The straight line represents $y = x$. The three curves are plots of the function $f_2(x) = \tanh(ax + b)$ with curve **A** corresponding to $a = 1$ and $b = 0.5$, curve **B** corresponds to $a = 1.5$ and $b = 0$, and curve **C** corresponding to $a = 0.5$ and $b = 0$.

all values of $x > 0$. But if the slope of $f_2(x)$ at $x = 0$ is sufficiently steep, the function will start out above $f_1(x)$ for small values of x. But since $\tanh(x) \to 1$ as $x \to \infty$ we know that for large values of x we will have $f_2(x) < f_1(x) = x$ and hence for some value of $x > 0$ we will have a solution x^* which satisfies $f_1(x^*) = f_2(x^*)$. These observations tells us immediately that when the slope of $f_2'(0) < 1$ the only solution to $f_1(x) = f_2(x)$ is $x = 0$, whereas for $f_2'(0) > 1$ an additional positive solution exists. Since $f_2'(x) = a[1 + \tanh^2(ax + b)]$ we see that $f_2'(0) = a$, so the criterion for the existence of a positive solution is $a > 1$ or $Jq\beta/2 > 1$. This means, in terms of the temperature, that when $T < T_c = \frac{Jq}{2k_B}$, a positive solution for $\langle S \rangle$ exist. Since $x = 0$ is always a solution to the equation $f_1(x) = f_2(x)$, one may ask how we know whether the solution $x = 0$ or the positive solution should be used when both exist. To answer this, one needs to look at what the solution represents in terms of the system being investigated. For the magnet, considerations concerning the thermodynamic stability show that for $T < T_c$ the stable solution is the positive one, see e.g. Sec. 6.5.2 in [392]. Figure 6.4 contains a summary of our analysis. The magnetisation $\langle S \rangle$ is an example of an **order parameter** since its value distinguishes the ordered region, or phase, $\langle S \rangle > 0$ from the disordered region $\langle S \rangle = 0$.

Although, as we will look at in the next section, the mean-field analysis presented possesses limitations, it does capture a number of important aspects of how the collective systemic behaviour is determined by the competition between the ordering effect caused by the interaction between the components and the disordering effect of the thermal fluctuations. First consider $J = 0$, i.e. no coupling between the degrees of freedom, in this case $\langle S \rangle = \tanh[\beta B]$ and the magnetic moment is directly induced by the applied field. If we think of the model as relating to opinions, $J = 0$ would correspond to the absence of any mutual influence, say through discussion, amongst the agents who simply respond to some external influence represented by B.

For $J > 0$ and $B > 0$ our graphic analysis tells us that only one solution $\langle S \rangle > 0$ exists, this means that even the smallest external field will induce a preferred direction or opinion.

Figure 6.4 Sketch of the behaviour of the order parameter near a critical point.

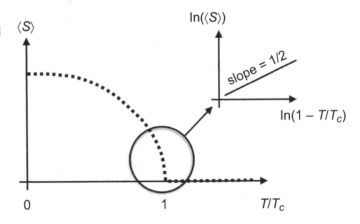

Finally, for $J > 0$ and $B = 0$ we found graphically that a non-zero solution appears when the temperature becomes lower than $T_c = \frac{Jq}{2k_B}$. The fact that a preferred direction occurs at low temperature is, as mentioned in Secs. 2.5 and 4.1, called symmetry breaking. The symmetry between up $S_i = 1$ and down $S_i = -1$, satisfied by the Hamiltonian in Eq. (6.29) when $B = 0$, is broken by the net effect of the mutual interaction. The mean-field mathematical discussion is just a first step towards a more refined and detailed understanding of how the collective emergent features can be completely different from the properties of the individual.

Here it is worthwhile to recall the discussion of the ideal gas in Sec. 1.1. The pressure of the ideal gas is also a sum of the effect of the individual particles in the same way as the magnetisation in Eq. (6.30) is. So why do we consider the interacting Ising model as an example of emergent behaviour but stress that the ideal gas does not support any emergent properties? The reason is that for the ideal gas the total effect of the N particles is obtained by adding together the effect of each of the non-interacting particles. This is indicated by the N on the right-hand side of the ideal gas equation in Eq. (1.1). The temperature is proportional to the average kinetic energy of each of the individual non-interacting gas particles and kinetic energy enters the ideal gas equation through the momentum transfer to the walls of the container of the independent individual gas particles.

The situation is very different for the Ising model. When the temperature $T < T_c$ we have $m > 0$, indicating a breaking of the up–down symmetry. But the average of a single isolated spin variable $\langle S \rangle$ is zero for all temperatures and a single spin is never able to break the up–down symmetry. Hence $m > 0$, the total magnetisation per spin, cannot be obtained by adding together contributions from individual non-interacting spins. It is the interaction included in the Hamiltonian in Eq. (6.29) between the spins that makes the breaking of symmetry possible. And since, per definition, ideal gas particles do not interact, they are not able to establish a quantity that breaks symmetry.

6.3 The Peculiar Nature of the Critical Point

We saw that at low temperature the interaction between the spins is able to induce a preferred direction which makes $\langle S \rangle > 0$. At high temperature, random thermal fluctuations are sufficiently strong to dominate over the interaction and it is equally likely for a spin to assume the value $+1$ as the value -1; this leaves $\langle S \rangle = 0$. We also saw that the transition between these two regimes happens at a specific temperature T_c, called the **critical temperature**. The behaviour and structure precisely at the temperature T_c is very unique and it is said that the system is critical for $T = T_c$. This is of particular interest to us because the phenomenology near and at the critical temperature is found to be of very broad general importance to many other types of systems far beyond the restricted class consisting of physical systems undergoing equilibrium phase transitions. So as an introduction to the behaviour of systems near or in a critical state, we will in some detail discuss the critical behaviour of the Ising model.

Table 6.1 Critical properties of the hypercube Ising model. The actual critical temperature is lower than the mean-field estimate because thermal fluctuations, neglected in the mean field, are able to destroy the order at a lower temperature.

d	MF T_c [J/k_B]	Actual T_c [J/k_B]	MF β	Actual β
1	1	0	1/2	—
2	2	1.13	1/2	1/8
3	3	2.26	1/2	0.326
4	4	3.34	1/2	1/2

Near a critical point the dependence of observables on control parameters in general takes the form of power laws. An example of this is obtained from Eq. (6.43). When we expand the hyperbolic tangent, $\tanh(x) \simeq x - x^3/3$, we obtain the following expression valid for T in the vicinity below T_c:

$$\langle S \rangle \simeq \sqrt{3}\left(1 - \frac{T}{T_c}\right)^{1/2} = \sqrt{3}\left(1 - \frac{T}{T_c}\right)^{\beta}. \tag{6.44}$$

This behaviour is sketched in Fig. 6.4. The exponent[5] $\beta = 1/2$ is an example of a critical exponent. These are often difficult to compute and mean field gives the exact result only when the dimension of the system is sufficiently high. See Table 6.1. An intuitive way to understand why the mean-field estimate might become correct in high dimensions is to note that the number of neighbours increases with dimension and accordingly the replacement of $\bar{S}_i$ by $\bar{S}$ and then by $\langle S \rangle$ is likely to be more accurate as the dimension of the system is increased. The sum over neighbours in $\bar{S}_i$ will contain more sites in higher dimensions and can therefore become more representative of the behaviour across the system. This may make neglecting the site dependence, which we did when replacing $\bar{S}_i$ by $\bar{S}$, less inaccurate. The difference between the mean-field estimate and the exact behaviour is due to fluctuations away from the average behaviour. It appears reasonable that these fluctuations are stronger in lower dimensions where the sum in Eq. (6.35) which determines the average spin value in the neighbourhood contains fewer terms. It turns out that these fluctuations can be so strong that they prevent the ordering, $\langle S \rangle > 0$, from being established at any $T > 0$. This is what happens for the Ising model in one dimension.

In general, one defines a lower (d_L) and an upper (d_U) critical dimension. The lower critical dimension is the dimension at and below which no phase transition occurs. And the upper critical dimension is the dimension at which mean-field theory starts to predict the correct values of the critical exponents. For the Ising model we have $d_L = 1$ and $d_U = 4$. The discussion about behaviour in different dimensions may seem somewhat academic if one has physical systems in mind, which of course tend always

---

[5] It is, of course, unfortunate to denote this so-called order parameter exponent by β when the same Greek letter is used in relation to the inverse temperature $\beta = 1/k_B T$. But this notation is used universally throughout the literature.

to exist in three space dimensions. But when we model biological or social systems, the neighbourhood of other agents interacting with a given agent may very well be described by a dimensionality different from the physical space dimension and even in physical systems, materials can be anisotropic and form one and two-dimensional structures within three-dimensional compounds like the two-dimensional graphene oxides, see the 2010 Nobel Physics Prize [2].

The deviation from the average behaviour captured by mean-field considerations is caused by the competition between ordering and disordering tendencies, which come to a head at T_c. Although these deviations from the average are strong enough to make the mean-field estimates inaccurate in dimensions below d_U and entirely misleading at, and below, d_L, we can still use mean-field analysis to get some idea about how power laws become very relevant in the vicinity of, and at, the critical state.

In the next section we will look at how, for temperatures close to T_c, the system's sensitivity and response to external forces can become singular and divergent at T_c. Right at T_c the system lacks characteristic scales; as we discussed conceptually in Sec. 3.1, mathematically this shows up as power-law dependence of spatial and temporal correlation functions. Correlation functions are related to the effect of external perturbations and therefore the long range of correlations between components in the critical state leads to divergences of sensitivity. It is worth keeping in mind that although the mathematical formalism is most straightforward and extensively developed for physical systems in equilibrium, the scenario is found to be of broad general relevance as e.g. indicated by the title of a recent review volume: *Criticality as a Signature of Healthy Neural Systems: Multi-scale Experimental and Computational Studies* [281].

6.4 Fluctuations, Response and Correlations

The internal state of a system can be probed by investigating how it responds to some externally applied changes. Although here we will look at the energy and magnetisation of a physical system, the relationship between response and internal state is very general. Anecdotally one may think of our mood and how touchy we are. Feeling self-confident and secure makes it more likely we can respond to external challenges in a constructive, creative way than if we for some reason feel worthless or depressed, in which case we might tend to react aggressively and destructively.

As a simple example, let us look at how a system responds to changes in the external temperature or changes in the external magnetic field. This is described by the heat capacity

$$C = \frac{\partial \langle E \rangle}{\partial T} \tag{6.45}$$

and the magnetic susceptibility

$$\chi = \frac{\partial \langle M \rangle}{\partial B} = N \frac{\partial \langle S \rangle}{\partial B}. \tag{6.46}$$

Here, $\langle E \rangle$ is the thermodynamic average of the total energy and $\langle M \rangle$ the thermodynamic average of the total magnetisation $M = \sum_i S_i = N\langle m \rangle$. The heat capacity C relates to how the internal energy changes if the temperature is changed a small amount and the magnetic susceptibility relates to how much the magnetisation changes if the applied magnetic field is changed by a small amount.

For the canonical ensemble we have Eq. (6.10) for $\langle E \rangle$, so

$$\langle E \rangle = -\frac{\partial \ln(Z)}{\partial \beta} \tag{6.47}$$

and, according to the Hamiltonian in Eq. (6.29), we have

$$\langle M \rangle = \frac{1}{\beta} \frac{\partial \ln(Z)}{\partial B}. \tag{6.48}$$

From these expressions it is straightforward to derive the following very similar expressions for the two **response functions**:

$$C = k_B \beta^2 (\langle E^2 \rangle - \langle E \rangle^2) \tag{6.49}$$

and

$$\chi = \beta(\langle M^2 \rangle - \langle M \rangle^2). \tag{6.50}$$

These two equations are examples of a general connection between the fluctuations in equilibrium, the right-hand side of the equations, and the response of the system given by the left-hand side of the equations. They are known as **fluctuation–dissipation relations** in the literature and attempts to establish such relations outside the field of equilibrium statistical mechanics exist e.g. in ecology [378] and neuroscience [377].

We notice that in both cases, the response is proportional to the variance, i.e. the size of fluctuations, of the relevant observable. As T_c is approached, both C and χ exhibit singular behaviour and are experimentally commonly found to diverge according to a power law, see e.g. [392] Chap. 7:

$$C \propto |T - T_c|^{-\alpha} \quad \text{and} \quad \chi \propto |T - T_c|^{-\gamma}. \tag{6.51}$$

It is simple to compute the divergence of the magnetic susceptibility, Eq. (6.46), in the mean-field approximation, by differentiation of Eq. (6.43). By implicit differentiation of Eq. (6.43) we obtain

$$\chi = N \frac{\partial \langle S \rangle}{\partial B} = N \left(\frac{\beta J q}{2} \frac{\partial \langle S \rangle}{\partial B} + \beta \right) \cosh^{-2} \left[\beta \left(\frac{J q}{2} \langle S \rangle + B \right) \right]. \tag{6.52}$$

We are interested in the limit $B \to 0$ and after moving factors around, we obtain

$$\chi = \frac{N\beta}{\cosh^2 \left(\frac{J q \beta}{2} \langle S \rangle \right) - \frac{J q \beta}{2}}. \tag{6.53}$$

Recall that $T_c = \frac{Jq}{2k_B}$. We will expand about $T = T_c$ and make use of Eq. (6.44) together with $\cosh(x) = 1 + x^2/2! + \cdots$ to derive that

$$\chi \simeq \frac{N}{k_B}(T - T_c)^{-1} \text{ for } T \to T_c^+. \tag{6.54}$$

The divergent sensitivity, as $T \to T_c$, captured by susceptibilities like C and χ originates in an increase in the range over which components are correlated significantly. To understand how this can be, we look at correlations.

To measure the correlation between events at position r_0 at time t_0 and events at another position $r_0 + r$ at another time $t_0 + t$, we look at the difference between the average of the product and the product of the average:[6]

$$\begin{aligned} C(r, t) &= \langle S(\mathbf{r_0}, t_0)S(\mathbf{r_0} + \mathbf{r}, t_0 + t) \rangle - \langle S(\mathbf{r_0}, t_0) \rangle \langle S(\mathbf{r_0} + \mathbf{r}, t_0 + t) \rangle \\ &= \langle S(\mathbf{r_0}, t_0)S(\mathbf{r_0} + \mathbf{r}, t_0 + t) \rangle - \langle S(\mathbf{r_0}, t_0) \rangle^2. \end{aligned} \tag{6.55}$$

The statistical-mechanics literature denotes $C(r, t)$ as the correlation function, though more correctly it is the variance. The important point is that if $S(\mathbf{r_0}, t_0)$ and $S(\mathbf{r_0} + r, t_0 + t)$ are independent then $C(r, t) = 0$. This is because, if two variables A and B are independent, the joint probability for A and B is the product of the probabilities for A and for B, or

$$P_{AB}(a, b) = P_A(a)P_B(b), \tag{6.56}$$

where $P_{AB}(a, b)$ is the probability that the variable A assumes the value a at the same time as variable B assumes the value b. For independent variables we have

$$\begin{aligned} \langle AB \rangle &= \sum_a \sum_b ab P_{AB}(a, b) = \sum_a \sum_b ab P_A(a)P_B(b) \\ &= \sum_a a P_A(a) \sum_b b P_B(b) = \langle A \rangle \langle B \rangle. \end{aligned} \tag{6.57}$$

Think of $S(\mathbf{r_0}, t_0)$ as A and $S(\mathbf{r_0} + \mathbf{r}, t_0 + t)$ as B and we see that if what happens at $\mathbf{r_0}$ at time t_0 is independent of what happens at $\mathbf{r_0} + \mathbf{r}$ at time $t_0 + t$, we must have $C(\mathbf{r}, t) = 0$. So $C(\mathbf{r}, t) \neq 0$ indicates that there is a dependence between $S(\mathbf{r_0}, t_0)$ and $S(\mathbf{r_0} + \mathbf{r}, t_0 + t)$. We will naturally assume that the larger the distance r between the two positions the less the correlation, or the longer the time interval t the less the correlation. Hence we can characterise how strong the interdependence between different positions at different times is by how fast the function $C(r, t)$ decays towards zero as r or t are increased. The functional form of $C(r, t)$ depends on whether the system is critical or not.

Away from the critical point, typically the leading functional dependence of $C(r, t)$ is of the form

$$C(r, t) \propto \exp(-r/\xi)\exp(-t/\tau), \tag{6.58}$$

---

[6] We have, for simplicity, assumed that statistically all directions are equivalent and that we are looking at a stationary system where the statistics do not explicitly depend on t_0. Then $C(r, t)$ will only depend on the distance r between the two positions $\mathbf{r_0}$ and $\mathbf{r_0} + \mathbf{r}$ and the separation in time t. And the average behaviour at $(\mathbf{r_0}, t_0)$ will be identical to the average behaviour at $(\mathbf{r_0} + \mathbf{r}, t_0 + t)$, which explains the second equality sign.

meaning exponential decay of correlations in space and time. The two parameters ξ and τ are discussed below. At the critical point T_c the exponential decay is replaced by much slower power-law decay in space and time. In general, at the critical point, for large values of r and t, the correlations decay as

$$C(r,t) \propto \frac{1}{r^{2-d+\eta}} \frac{1}{t^\alpha}, \quad \text{for } r \gg 0 \text{ and } t \gg 0. \tag{6.59}$$

Here d denotes the space dimension of the system and η and α are so-called critical exponents. The change from exponential decaying correlations to power-law decay comes about through the divergence of the characteristic length ξ, called the correlation length, and of the characteristic time scale τ, called the correlation time, namely

$$\xi \propto |T - T_c|^{-\nu} \text{ and } \tau \propto \xi^z. \tag{6.60}$$

We are now able to see how the divergent sensitivity, indicated by the divergence of susceptibilities like C and χ, is related to an increase in the range over which components of the system remain strongly correlated. To illustrate this, we look at χ in Eq. (6.50) and deal with the two terms separately. First we note

$$\langle M^2 \rangle = \left\langle \left(\sum_i S_i \right)^2 \right\rangle = \left\langle \sum_i S_i \sum_j S_j \right\rangle$$

$$= \left\langle \sum_i \sum_j S_i S_j \right\rangle = \sum_i \sum_j \langle S_i S_j \rangle. \tag{6.61}$$

Let us denote the spin at position i by $\boldsymbol{r}_0$ and the one at position j by $\boldsymbol{r}_0 + \boldsymbol{r}$, and since in M all spins are assessed at the same instance in time, say t_0, we can write

$$S_i S_j = S(\boldsymbol{r}_0, t_0) S(\boldsymbol{r}_0 + \boldsymbol{r}, t_0).$$

Now we have

$$\langle M^2 \rangle = \sum_{\boldsymbol{r}_0} \sum_{\boldsymbol{r}} \langle S(\boldsymbol{r}_0, t_0) S(\boldsymbol{r}_0 + \boldsymbol{r}, t_0) \rangle. \tag{6.62}$$

The second term in Eq. (6.50) can be rewritten in the following way:

$$\langle M \rangle^2 = \left\langle \sum_i S_i \right\rangle^2 = \left\langle \sum_i S_i \right\rangle \left\langle \sum_j S_j \right\rangle = \sum_i \sum_j \langle S_i \rangle \langle S_j \rangle$$

$$= \sum_i \sum_j \langle S \rangle^2. \tag{6.63}$$

The last equality makes use of the system being homogeneous, so the average $\langle S_i \rangle$ is the same for all positions i, i.e.

$$\langle S_i \rangle = \langle S_j \rangle \mapsto \langle S \rangle. \tag{6.64}$$

We substitute Eqs. (6.62) and (6.63) into Eq. (6.50) to get

$$\frac{1}{\beta}\chi = \sum_{r_0}\sum_{r}(\langle S(r_0, t_0)S(r_0 + r, t_0)\rangle - \langle S\rangle^2) = \sum_{r_0}\sum_{r}C(r, 0)$$
$$= N\sum_{r}C(r, 0). \tag{6.65}$$

For the last equality we used the fact that the sum over r_0 runs through all the N positions. The susceptibility per component, in this case per spin, is therefore determined by a sum over the correlation function

$$\frac{\chi}{\beta N} = \sum_{r}C(r, 0), \tag{6.66}$$

where r runs over all the possible distances between pairs of the components.

Note the factor $1/N$ on the left-hand side. The presence of this factor means that the sum involving the correlation function on the right-hand side determines the sensitivity *per component* and not the total sensitivity. The latter may always become large or diverge with increasing N, like the mass of a system will increase with its size, but the mass density will not. The important observation is that Eq. (6.66) demonstrates that the response per component, χ/N, to an external change can become very large if $C(r, 0)$ decreases slowly as r increases. Indeed, for systems from physics like magnets that contain huge numbers of atoms – of order Avogadro's number, which is about 6×10^{23} – one normally think of the limit $N \to \infty$ as the relevant size. Other systems typically encountered in complexity science are much smaller, though still substantial. For example the population of the entire earth is about 7×10^9, the number of cells in the human brain is about 10^{11} and the total number of cells of the human body is about 4×10^{13}. This is just to say that in such cases it makes good sense to think of the sum in Eq. (6.66) as running over values of r that cover the entire range from zero to infinity. This means that if $C(r, 0)$ does not decay very fast with increasing r, the sensitivity of the system per component given by χ/N will essentially diverge. This is exactly what happens in the critical state established at the critical point $T = T_c$. In the critical state the dependence of $C(r, 0)$ is given by (6.59), which decays algebraically as r increases[7] and the sum will diverge with N, except if the critical exponent $2 - d + \eta$ in Eq. (6.59) happens to be sufficiently large to make the sum convergent. Away from the critical state, the exponential decay of $C(r, 0)$, see Eq. (6.58), will always ensure that the sum in Eq. (6.66) remains finite even as $N \to \infty$.

The conclusion is that even when the direct interaction between components only involved the immediate neighbourhood, as is the case e.g. in the Ising model, this short-range interaction can generate very long-range *correlations* that make the system hypersensitive in the sense that what happens at one position may correlate in an

---

[7] A word of clarification: it looks like $C(r, 0)$ in Eq. (6.59) will become infinite if we put $t = 0$. This is not so. The expression in Eq. (6.59) refers to large values of r and t. Going back to the definition in Eq. (6.55) we see that $C(r, 0) = \langle S(r_0, t_0)^2\rangle - \langle S(r_0, t_0)\rangle^2$. This is simply the variance of the variable S at the position r_0 at time t_0, which is finite and, in fact in a homogeneous and stationary state, the variance will be the same for all positions r_0 and times t_0.

essential way with other far-away components. It is this long-range sensitivity described by the slow algebraic decay of $C(r, 0)$ with separation r that combines to give the diverging systemic sensitivity (per component) represented by χ/N.

Let us emphasise that the long-range correlations generating the diverging sensitivity are not the same as the long-range ordering observed below T_c, see Fig. 6.4. The correlation function in Eq. (6.55) decays rapidly, i.e. as an exponential function of separation in space and time, both below and above the critical point. Perhaps this at first seems counter-intuitive, since the configurations below the critical point, in the ordered phase, tend to align with a preferred value for each of the variables S_i, whereas above the critical point no ordering exists and the configurations are essentially random.

The reason that correlations decay exponentially both in the ordered, i.e. $m = \sum_i S_i/N > 0$, and in the disordered region, i.e. where $m = \sum_i S_i/N = 0$, is as follows. In contrast to the order parameter, the correlation function is concerned with deviations away from the average behaviour and measures how these are correlated. The deviation away from the average at position r_0 at time t_0 is given by $\Delta(r_0, t_0) = S(r_0, t_0) - \langle S(r_0, t_0) \rangle$ and the deviation at position $r_0 + r$ at time $t_0 + t$ is given by $\Delta(r_0 + r, t_0 + t) = S(r_0 + r, t_0 + t) - \langle S(r_0 + r, t_0 + t) \rangle$. It is straightforward to see that the correlation in Eq. (6.55) can also be written as

$$C(r, t) = \langle \Delta(r_0, t_0) \Delta(r_0 + r, t_0 + t) \rangle, \tag{6.67}$$

one just needs to substitute the expressions for the two Δs and make use of e.g. that

$$\langle S(r_0, t_0) \langle S(r_0 + r, t_0 + t) \rangle \rangle = \langle S(r_0, t_0) \rangle \langle S(r_0 + r, t_0 + t) \rangle = \langle S(r_0, t_0) \rangle^2.$$

Since $\langle \Delta(r, t) \rangle = \langle S(r_0, t_0) - \langle S(r_0, t_0) \rangle \rangle = 0$, the average of the product of deviations in Eq. (6.67) is the correlation function of the variable $\Delta(r, t)$, no need for any subtraction of the product of the average like in Eq. (6.55). Hence, $C(r, t)$ describes the correlations between deviations away from the average at positions and times separated by r and t. In the ordered phase, the deviations are deviating from the average $\langle m \rangle > 0$ and in the disordered phase the deviations are away from $\langle m \rangle = 0$.

We may ask ourselves what all these observations concerning correlation and critical states are good for, and if it is of any relevance to people not particularly interested in the physics of magnets or other systems from physics. The answer is that there are strong indications from observations on may different types of systems that states with very long-range correlations occur frequently. We will look at examples in the next section.

6.5   Examples of Correlation Functions: Brain, Flocks of Birds, Finance

Let us mention a few examples from outside physics where correlation functions have been measured and are indicative of critical behaviour in the sense of correlations extending far in space and time.

There exist many indications that the human brain operates in, or near, a critical state. It has even been argued that the vicinity of a critical state is essential for the health of the brain, see e.g. the collection of papers in [281]. Direct evidence in terms of the space and time dependence of the correlation function has also been measured in the case of fMRI scans in [136]. The correlation function of the signal measured from the many small brain regions monitored by the fMIR scanner, the so-called BOLD signal, does indeed decay as an algebraic function like in Eq. (6.59). The time behaviour of the same data set was also analysed and found to be consistent with a very slow, essentially logarithmic, decay with the separation in time.

The correlations of the velocity fluctuations of the individual birds in flocks of starlings were considered in [79] and the range of the correlations found to scale (i.e. grow) with the spatial extent of the flock. This is consistent with an algebraic decay of the correlation function, or perhaps an exponential decay with a correlation length larger than the size of the flock.

Financial time series are found to exhibit different types of correlations at different time scales. See e.g. the study in [399], which considers the correlations of return volatility and found evidence of correlations extending over many days.

An algebraic decay of the correlations between earthquakes was found in [266]. The magnitude distribution of energy released in earthquakes, the Gutenberg–Richter law, is in itself indicative of earthquakes being scale-free and consistent with a scaling analysis of the distribution of waiting times between aftershocks [35].

The field of self-organised criticality [34, 214, 354] tries to explain the many cases in nature where signs of criticality are found. In most cases this is done by an analysis of event distributions such as earthquake sizes, sizes of rain showers, the size of bursts of brain activity, the size of financial crashes, etc. Power laws of such distributions may indeed point towards criticality, but not necessarily so, since power laws of distributions can arise due to other mechanisms than long-range correlations and critical behaviour. See e.g. Sornette's catalogue of mechanisms that can lead to power laws in Chap. 14 of [422]. So the critical behaviour relating to diverging correlation lengths is best identified by use of correlation functions [325, 327]. For more detail see Sec. 12.1.

6.6 Diverging Range of Correlations

To develop an intuition for how one can compute the equal-time correlation function $C(r,0)$ in Eq. (6.55) when the probability distribution of the individual micro configurations is known, we consider the one-dimensional version of the Ising model using two approaches. First we will calculate exactly the correlation function by use of the so-called transfer matrix method. This mathematical technique can be generalised to different models, though in higher dimensions it can be a rather involved procedure. We will try to explain the mathematics in detail. Readers with mathematical experience can just skim through and readers who want to gain some mathematical experience should be able to use the discussion as an illuminating exercise.

The second approach we will consider is not exact or entirely rigorous, but it highlights the role of the interplay between a mechanism that favours order, namely the energy cost of placing anti-parallel spins next to each other, and a mechanism, the thermal fluctuations, that induce disorder.

6.6.1 Correlation Function – Exact Approach

In the one-dimensional case the energy given by the Hamiltonian in Eq. (6.29) can be written in the simple form[8]

$$H[S_1, \ldots, S_N] = -J \sum_{i=1}^{N} S_i S_{i+1} - B \sum_{i=1}^{N} S_i = \sum_{i=1}^{N} H[S_i, S_{i+1}]. \tag{6.68}$$

The factor $1/2$ disappears because we replace the sum over the neighbours to the left and to the right $(S_i S_{i-1} + S_i S_{i+1})/2$ in Eq. (6.29) by the single term $S_i S_{i+1}$ representing the interaction to the right. The last equality is just a way to simplify the notation with the definition

$$H[S_i, S_{i+1}] = -J S_i S_{i+1} - \frac{B}{2}(S_i + S_{i+1}).$$

The factor 2 dividing B is to compensate for including both S_i and S_{i+1} to make the expression more symmetric.

To simplify the mathematics, we use periodic boundary conditions, namely we assume the spins are placed on a ring instead of a line, which means that the spin to the right of S_N is spin S_1 or that $S_{N+1} = S_1$.

To calculate the correlation function in Eq. (6.55) we need to compute the following two terms $\langle S_i \rangle$ and $\langle S_i S_{i+k} \rangle$.

The following set of equations may at first look involved, but if one fetches pencil and paper and calmly follows each step, it will, hopefully, turn out to be perhaps a bit lengthy but otherwise uncomplicated. First we notice that

$$
\begin{aligned}
Z &= \sum_{S_1, S_2, \ldots, S_N} \exp(-\beta H[S_1, \ldots, S_N]) \\
&= \sum_{S_1, S_2, \ldots, S_N} \exp\left(-\beta \sum_{i=1}^{N} H[S_i, S_{i+1}]\right) \\
&= \sum_{S_1, S_2, \ldots, S_N} e^{-\beta H[S_1, S_2]} e^{-\beta H[S_2, S_3]} \ldots e^{-\beta H[S_{N-1}, S_N]} e^{-\beta H[S_N, S_1]} \\
&= \sum_{S_1} \left\{ \sum_{S_2} e^{-\beta H[S_1, S_2]} \sum_{S_3} e^{-\beta H[S_2, S_3]} \sum_{S_4} e^{-\beta H[S_3, S_4]} \ldots \right. \\
&\qquad\qquad \left. \ldots \sum_{S_{N-1}} e^{-\beta H[S_{N-2}, S_{N-1}]} \sum_{S_N} e^{-\beta H[S_N, S_1]} \right\}.
\end{aligned}
\tag{6.69}
$$

---

[8] The one-dimensional Ising model is a standard subject in many statistical mechanics-related textbooks. The two books [84, 392] contain very clear explanations of material closely related to the discussion given here.

This bulky expression is nothing more than a set of multiplications of a matrix with itself. We see this by introducing the *transfer matrix*

$$M = \begin{pmatrix} e^{\tilde{J}+\tilde{B}} & e^{-\tilde{J}} \\ e^{-\tilde{J}} & e^{\tilde{J}-\tilde{B}} \end{pmatrix}. \tag{6.70}$$

A number of comments are needed. Firstly, we introduced the notation $\tilde{J} = \beta J$ and $\tilde{B} = \beta B$. Next we observe that if we alter slightly the way we enumerate the elements of the matrix M and denote the four matrix elements in the following way[9] as $M_{1,1}, M_{1,-1}, M_{-1,1}$ and $M_{-1,-1}$, we can write the elements M_{ij} in Eq. (6.70) as

$$M_{S_i, S_{i+1}} = \exp\left[\tilde{J}S_i S_{i+1} + \frac{\tilde{B}}{2}(S_i + S_{i+1})\right]. \tag{6.71}$$

Our notation now allows us to write the partition function Z in Eq. (6.69) as

$$\begin{aligned} Z &= \sum_{S_1}\left\{\sum_{S_2} M_{S_1, S_2} \sum_{S_3} M_{S_2, S_3} \sum_{S_4} M_{S_3, S_4} \cdots\right. \\ &\qquad \left.\cdots \sum_{S_{N-1}} M_{S_{N-2}, S_{N-1}} \sum_{S_N} M_{S_N, S_1}\right\} \\ &= \sum_{S_1}\left\{M^N\right\}_{S_1 S_1} \\ &= \mathrm{Tr}(M^N). \end{aligned} \tag{6.72}$$

The last equality makes use of the definition of the *trace* of a matrix from linear algebra, namely that the trace is the sum of the diagonal elements of a matrix, i.e. for a general $n \times n$ matrix A we have per definition

$$\mathrm{Tr}\, A = \sum_{i=1}^{n} a_{ii},$$

the sum of the diagonal elements. Equation (6.72) tells us that the partition function is given by multiplying the matrix M by itself N times and then computing the trace. This is not very helpful in the first round, but if we make use of a couple of results from linear algebra it becomes easy to complete the computation of Z. First we use that any symmetric matrix can be diagonalised; this means we can write the matrix as a product (details can be found in Box 8.1 in Chap. 8):

$$M = U D U^T \tag{6.73}$$

of some orthogonal matrix[10] U, which we do not need to know explicitly just yet, and the diagonal matrix

---

[9] Usually we use (1,1), (1,2), (2,1) and (2,2). But this is of course irrelevant since the indices of the matrix elements simply label their position in the array.

[10] That is, a matrix which satisfies $U^{-1} = U^T$.

$$D = \begin{pmatrix} \lambda_1 & 0 \\ 0 & \lambda_2 \end{pmatrix}, \tag{6.74}$$

where the diagonal elements are given by the two roots in the equation $\det(M - \lambda I) = 0$. Here I is the identity matrix and since we are dealing with 2×2 matrices, the equation for λ is a square equation and its roots are

$$\lambda_\pm = e^{\tilde{J}} \left(\cosh \tilde{B} \pm \sqrt{e^{-4\tilde{J}} + \sinh^2 \tilde{B}} \right). \tag{6.75}$$

We can now complete the calculation of Z:

$$Z = \text{Tr}(M^N) = \text{Tr}(UDU^T)^N = \text{Tr}(UD^NU^T) = \text{Tr}(D^N). \tag{6.76}$$

The last equality follows because the trace satisfies

$$\text{Tr}(AB) = \text{Tr}(BA) \tag{6.77}$$

and $U^T U = I$. Therefore

$$Z = \lambda_+^N + \lambda_-^N = \lambda_+^N \left[1 + \left(\frac{\lambda_-}{\lambda_+} \right)^N \right]$$

$$\Downarrow \text{ for } N \to \infty$$

$$Z = \lambda_+^N. \tag{6.78}$$

The last equation follows since $\lambda_-/\lambda_+ < 1$.

Having determined Z, we are ready to move on to the calculation of $\langle S_i \rangle$. We go back to Eq. (6.29) and notice that

$$\langle S_i \rangle = \sum_{S_1, S_2, \ldots, S_N} S_i \frac{\exp(-\beta H[S_1, \ldots, S_N])}{Z}$$

$$= \frac{1}{N} \sum_{S_1, S_2, \ldots, S_N} \left(\sum_{q=1}^N S_q \right) \frac{\exp(-\beta H[S_1, \ldots, S_N])}{Z}$$

$$= \frac{1}{N} \frac{\partial}{\partial B} \sum_{S_1, S_2, \ldots, S_N} \frac{\exp(-\beta H[S_1, \ldots, S_N])}{Z} \tag{6.79}$$

$$= \frac{1}{N} \frac{\partial}{\partial B} \ln Z.$$

It is straightforward, with pencil in hand and recalling that

$$\frac{d}{dx} \cosh(x) = \sinh(x)$$

and

$$\sinh(x) \to 0 \text{ when } x \to 0,$$

to see from Eq. (6.79) that for $B = 0$ the average vanishes, i.e. $\langle S_i \rangle = 0$. Therefore, to calculate the correlation function in Eq. (6.55), we only need to care about the first term $\langle S_i S_{i+k} \rangle$ and can again make use of the transfer matrix formalism to write

$$\langle S_i S_{i+k} \rangle = \frac{1}{Z} \sum_{S_1, S_2, \ldots, S_N} S_i S_{i+k} \exp(-\beta H[S_1, \ldots, S_N])$$

$$= \frac{1}{Z} \sum_{S_1, S_2, \ldots, S_N} M_{S_1, S_2} \cdots M_{S_{i-1}, S_i} S_i M_{S_i, S_{i+1}} \cdots \qquad (6.80)$$

$$M_{S_{i+k-1}, S_{i+k}} S_{i+k} M_{S_{i+k}, S_{i+k+1}} \cdots M_{S_{N-1}, S_1}.$$

This expression can be rewritten in terms of a product of matrices when we make use of what is know as the third Pauli matrix

$$\sigma_3 = \begin{pmatrix} 1 & 0 \\ 0 & -1 \end{pmatrix}. \qquad (6.81)$$

We get

$$\langle S_i S_{i+k} \rangle = \frac{1}{Z} \mathrm{Tr}\left(M^i \sigma_3 M^k \sigma_3 M^{N-(i+k)} \right)$$

$$= \frac{1}{Z} \mathrm{Tr}\left(\sigma_3 M^k \sigma_3 M^{N-k} \right). \qquad (6.82)$$

The last equality follows from Eq. (6.77) and the fact that M is symmetric. We proceed by use of the formula Eq. (6.73) to introduce the diagonal form of M. Again we transform M to diagonal form by use of Eq. (6.73). This time we need to know the matrix U, which linear algebra tells us is determined by the **eigenvectors** v_+ and v_- of the matrix M. These are given by the equations $M v_\pm = \lambda_\pm v$. We are right now interested in how the multi-component system organises itself and for this reason we focus on the case where no external field is applied, i.e. $B = 0$. In this case one finds, after some linear algebra:[11]

$$U = \frac{1}{\sqrt{2}} \begin{pmatrix} 1 & 1 \\ -1 & 1 \end{pmatrix}. \qquad (6.83)$$

This matrix satisfies

$$U \sigma_3 U^T = -\begin{pmatrix} 0 & 1 \\ 1 & 1 \end{pmatrix} = -\sigma_1, \qquad (6.84)$$

where the last equality introduces the matrix σ_1, known as the first Pauli matrix. These couple of equations enable us to reduce the expression in Eq. (6.82):

$$\langle S_i S_{i+k} \rangle = \frac{1}{Z} \mathrm{Tr}\left(\sigma_3 M^k \sigma_3 M^{N-k} \right)$$

$$= \frac{1}{Z} \mathrm{Tr}\left(\sigma_3 U D^k U^T \sigma_3 U D^{N-k} U^T \right)$$

$$= \frac{1}{Z} \mathrm{Tr}\left(D^k U^T \sigma_3 U M^{N-k} U^T \sigma_3 U \right) \qquad (6.85)$$

$$= \frac{1}{Z} \mathrm{Tr}\left(D^k \sigma_1 D^{N-k} \sigma_1 \right).$$

---

[11] If this is unfamiliar to the reader, a Google search for matrix diagonalisation will provide immediate help.

The third equality follows by use of Eq. (6.77) and the last equality follows from Eq. (6.84). Now all that is left to do is to multiply these matrices and sum the diagonal elements of the product to obtain the trace. The result is

$$\langle S_i S_{i+k} \rangle = \frac{\lambda_-^k \lambda_+^{N-k} + \lambda_+^k \lambda_-^{N-k}}{\lambda_+^N + \lambda_-^N}$$

$$= \left(\frac{\lambda_-}{\lambda_+}\right)^k \frac{1 + (\frac{\lambda_-}{\lambda_+})^{N-2k}}{1 + (\frac{\lambda_-}{\lambda_+})^N} \tag{6.86}$$

$$\simeq \left(\frac{\lambda_-}{\lambda_+}\right)^k \text{ for } N \to \infty.$$

We used the fact that $\lambda_+ > \lambda_-$ for the last approximate equality. We have found that the correlation function decays exponentially. To highlight this we write

$$C(r) = \langle S_i S_{i+k} \rangle = \exp(-k/\xi), \tag{6.87}$$

where for $B = 0$

$$-\frac{1}{\xi} = \ln\left(\frac{\lambda_+}{\lambda_-}\right) = \ln\left(\frac{e^{\tilde{J}} - e^{-\tilde{J}}}{e^{\tilde{J}} + e^{-\tilde{J}}}\right) = \ln\left(\frac{1 - e^{-2\tilde{J}}}{1 + e^{-2\tilde{J}}}\right). \tag{6.88}$$

Next we make use of the fact that $\exp(-2\tilde{J}) \ll 1$ when we consider low-temperature $T \ll 1$ since $\tilde{J} = J/(k_B T)$. This allows us to make use of the approximations

$$\frac{1-x}{1+x} \approx 1 - 2x \text{ and } \ln(1+x) \approx x,$$

for x close to 0, to conclude that

$$\xi = \frac{1}{2} e^{\frac{2J}{k_B T}} \text{ for } T \to 0. \tag{6.89}$$

Let us summarise. We have computed the correlation function exactly in the limit of a large number of components N and found that the correlations decay exponentially for all temperatures $T > 0$. Furthermore, we find that the length scale ξ that sets the scale for the decay of correlations, diverges as the temperature goes to zero. The divergence of the correlation length happens typically when parameters are tuned towards a critical point, see Eq. (6.60) above. In this sense the one-dimensional Ising model has a phase transition to a critical state at $T = 0$. At zero temperature, the ordering, which favours the spins assuming the same sign, wins completely since the disordering effect caused by thermal fluctuations disappears entirely for $T = 0$. The exponential divergence of ξ is a bit unusual, though it is also seen in other models such as the two-dimensional XY model discussed in Sec. 6.7. The more usual behaviour is that ξ diverges as a power law [as pointed out in Eq. (6.60)].

The mathematical transfer matrix technique discussed here can be generalised and used in other situations such as for the two-dimensional Ising model, see e.g. Baxter's book on exact solutions [54]. But it is nevertheless somewhat specialised and it is not very easy from the mathematical discussion to identify what sort of mechanisms are responsible for the destruction of the correlations. We will therefore, in the next subsection, analyse the correlation function in a more intuitive way with a focus on the configurational structures that are responsible for the decay of the correlation function.

6.6.2 Correlation Function – Intuitive Discussion

In contrast to the exact mathematical treatment in the previous section, in this section we will discuss how 'excitations' can act to decorrelate events at different positions. By excitations we simply mean configurations that have higher energy than the configuration of lowest possible energy, referred to as the ground state. To focus on the effect of the interaction between the spins, we choose $B = 0$, no external field. The lowest energy is achieved when all the spins assume the same value, either all $S_i = 1$ or all $S_i = -1$. A configuration which deviates from this pattern is an excitation and has some higher energy. The excitations can be considered as building blocks capturing specific examples of collective configurations. This aspect makes it easier to generalise the philosophy of the approach. Perhaps even to non-physical systems. In an abstract, somewhat loose sense, one can consider concepts like species in biology or neurosis in physiology as examples of collective excitations. Within physics, the pair excitations we will look at in this section can be seen as simpler versions of the vortex excitations to be considered in Sec. 6.7 below.

Our starting point is the Hamiltonian in Eq. (6.68) and we note that a pair of spins can store the energy $E_0 = -J$ if they assume the values $+1, +1$ or $-1, -1$. And the higher energy $E_1 = J$ if they assume the values $+1, -1$ or $-1, +1$. If we think of the spins as arrows that either point up $S_i = +1$ or down $S_i = -1$, then the lowest energy E_0 corresponds to the arrows being parallel: ↑↑ or ↓↓. And it costs an amount of $\Delta E = E_1 - E_0 = 2J$ to turn two parallel spins into an antiparallel pair, see Fig. 6.5.

We can measure energies from the lowest energy state, i.e. use the ground-state energy E_{gr} as our base line. In the Boltzmann weights in Eq. (6.7) this corresponds to dividing both $\exp(-\beta E_s)$ and Z by the factor $\exp(-\beta E_{gr})$. The probability that the system is in a state of energy E_s can then be expressed as

$$p_s = \frac{\exp(-\beta(E_s - E_{gr}))}{\sum_{s'} \exp(-\beta(E_{s'} - E_{gr}))} = \frac{\exp(-\beta \Delta E_s)}{\sum_{s'} \exp(-\beta \Delta E_{s'})}. \tag{6.90}$$

Therefore the probability of an antiparallel pair at position i to $i+1$ can be expressed as

$$p_{\text{pair}} = \omega e^{-\beta \Delta E}, \tag{6.91}$$

where ω is a constant.

Let us now imagine that we initially placed the chain of spins at zero temperature, so no thermal energy is available and hence the system will be in the ground state with all spins parallel, so either all ↑↑ ... ↑↑ or all ↓↓ ⋯ ↓↓. As we increase the temperature from zero, thermal energy will become available and can create the antiparallel configurations ↑↓ or ↓↑. We also notice that the energy of an antiparallel pair is independent

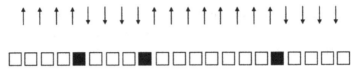

Figure 6.5 A section of the Ising spin chain. At zero temperature the entire chain is either pointing up or pointing down. As thermal energy becomes available for $T > 0$, pairs of antiparallel spins will start to occur. The black square indicates which of the interaction bonds have an energy ΔE above the lowest – ground state – energy.

of its position. Though of course they need to appear in the right order, so a $\uparrow\downarrow$ pair is followed by a $\downarrow\uparrow$ pair, see Fig. 6.5.

To compute the correlation function $C(k) = \langle S_i S_{i+k} \rangle$ we need to compute the number of times the spin direction has swapped as one goes from position i to position $i + k$, which is the same as how many pairs of antiparallel spins one encounters between the two positions. Let p_q denote the probability that q pairs are present between i and $i+k$. We then average over the two possible values of S_i and the number of possible swaps between position i and position $i + k$:

$$
\begin{aligned}
C(k) &= \langle S_i S_{i+k} \rangle \\
&= +1 \times \text{Prob}\{S_i = 1\}\{p_0 - p_1 + p_2 - \cdots\} \\
&\quad -1 \times \text{Prob}\{S_i = -1\}\{-p_0 + p_1 - p_2 - \cdots\} \\
&= p_0 - p_1 + p_2 - \cdots .
\end{aligned}
\tag{6.92}
$$

We used the fact that $\text{Prob}\{S_i = 1\} = \text{Prob}\{S_i = -1\} = 1/2$. To estimate the probabilities p_q we first notice, see Fig. 6.5, that a pair of antiparallel spins at i and $i+1$ can be considered equivalent to a black particle occupying the space between i and $i+1$. Equivalently, a pair of parallel spins corresponds to no particle occupying the space between the two members of the pair. At low temperature there will be few antiparallel pairs and we can estimate the probability that a square in Fig. 6.5 is occupied $p_{\blacksquare}$ or empty $p_{\square}$ in the following way with an eye on Eq. (6.91):

$$
\begin{aligned}
p_{\square} &= \frac{\omega}{\omega + \omega e^{-\beta \Delta E}} \simeq 1 - e^{-\beta \Delta E} \\
p_{\blacksquare} &= \frac{\omega e^{-\beta \Delta E}}{\omega + \omega e^{-\beta \Delta E}} \simeq e^{-\beta \Delta E}.
\end{aligned}
\tag{6.93}
$$

The last estimate for $p_{\square}$ and $p_{\blacksquare}$ assumes the temperature to be low and hence β large.

So p_q can be thought of as the probability that q black particles are found between i and $i + k$. An easy way to estimate this probability consists in replacing the discrete variables i and k by continuous variables along the x-axis. This approach allows us to make contact with **Poisson processes** and gives us an occasion to recall how these important processes are treated mathematically.

We will demonstrate that the probability, q, that particles are located on an interval of length x is given by

$$p_q(x) = \frac{(vx)^q}{q!}e^{-vx}, \tag{6.94}$$

where vdx is the constant probability that a particle is located in an interval of infinitesimal length dx. The expression for $p_q(x)$ is obtained in the following way by use of proof by induction.

Case $q = 0$

The probability that no black particle is found on the interval $[0, x + dx]$ is equal to the probability $p_0(x)$ that no particle is found on the interval $[0, x]$ times the probability $1 - vdx$ that no particle is found in the infinitesimal interval $[x, x + dx]$, i.e. we have

$$p_0(x + dx) = p_0(x)(1 - vdx)$$

$$\Downarrow$$

$$\frac{p_0(x + dx) - p_0(x)}{dx} = -vp_0(x)$$

$$\Downarrow \tag{6.95}$$

$$\frac{dp_0}{dx} = -vp_0(x)$$

$$\Downarrow$$

$$p_0(x) = e^{-vx}.$$

In the last step, when solving the first-order differential equation for $p_0(x)$, we made use of the fact that $p_0(0) = 1$, namely the probability for no particle being present on an interval of length 0 must be 1.

To obtain the formula in Eq. (6.94) for general values of q, we do the induction step, that is we assume the formula is correct for some arbitrary value of q and then show that if this is the case, then the formula will also hold for $q + 1$.

Induction step

We need to show that assuming $p_q(x)$ has the form in Eq. (6.94), then $p_{q+1}(x)$ is given by the same expression with q replaced by $q + 1$. Notice that $p_{q+1}(x)$ will be equal to the sum (or integral) of the product of the following three probabilities:

(1) The probability $p_q(x')$ that q particles are present on $[0, x']$.
(2) The probability vdx' that a particle is present on the interval $[x', x' + dx']$.
(3) The probability $p_0(x - x')$ that no particle is present on the interval $[x' + dx', x]$. Since dx' is infinitesimal we use the following approximation for the length of this interval: $x' + dx' - x \simeq x - x'$.

By use of the induction assumption that $p_q(x)$ is given by Eq. (6.94), we can summarise items (1), (2) and (3) as the following equation:

$$p_{q+1}(x) = \int_0^x p_q(x')v\,dx'\,p_0(x-x')$$

$$= \int_0^x \frac{(vx')^q}{q!}e^{-vx'}v\,dx'\,e^{-v(x-x')}$$

$$= \frac{v^q}{q!}e^{-vx}\int_0^x x'^q\,dx' \tag{6.96}$$

$$= \frac{v^{q+1}}{(q+1)q!}x^{q+1}e^{-vx}.$$

The last expression is of the form in Eq. (6.94) for $q+1$ particles on the interval $[0, x]$. And since we have demonstrated the formula for $q = 0$, the induction step allows us to conclude that the formula is correct for any value of $q = 0, 1, 2, \ldots$.

We can now return to Eq. (6.92) and substitute the expressions from Eq. (6.94) for the p_q to get

$$C(x) = e^{-vx} - \frac{vx}{1!}e^{-vx} + \frac{(vx)^2}{2!}e^{-vx} - \cdots$$

$$= e^{-vx}\left(1 - \frac{vx}{1!} + \frac{(vx)^2}{2!} - \cdots\right) \tag{6.97}$$

$$= e^{-2vx}.$$

For the last equality we made use of the formula $e^t = 1 + t/1! + t^2/2! + \cdots$.

The conclusion is that this more elucidatory and qualitative discussion of the correlations produces the same exponential decay as found by the exact mathematical analysis in Eq. (6.87). It is natural to assume $v = p_\blacksquare$, the correlation length of $C(r)$ in Eq. (6.97) is then given by

$$\xi = \frac{1}{2v} = \frac{e^{\beta\Delta E}}{2} = \frac{1}{2}e^{\frac{2J}{k_BT}}, \tag{6.98}$$

which is the same as the result in Eq. (6.89) derived from the exact calculation.

The appeal of this non-rigorous discussion is that it immediately lends itself to an interpretation in terms of collective emergent excitations, namely the particles of Fig. 6.5 consisting of the pairs of antiparallel spins. This is a successful example, though admittedly a very simple one, of identifying the relevant emergent building blocks needed to understand the systemic-level behaviour. We discussed in Chap. 2 that it is an ultimate aim of complexity science to identify such collective robust structures, which will allow a description of the emergent systemic phenomenology. We mention three examples. In biology we are, for example, in need of a proper consistent identification of the species concept valid across the entire taxonomy. In neurology we would like to be able to identify the building blocks of the dynamics of the mind, similar to e.g. Freud's neurosis concept. In sociology we are still looking for a robust identification of the building blocks of society, concepts able to expand and develop something similar to Marx's social classes. Of course the antiparallel pairs of the Ising model are in a way ridiculously far from such a goal, but it is after all a start and in the next section we will give our last example from physics of such an analysis which has had far-reaching consequences for the understanding of very different systems.

6.7 The Two-Dimensional XY Model

The name **2d XY model** is too technical and terse to fully convey its conceptual splendour and significance. The 2d refers to the model being defined on a two-dimensional lattice and the XY refers to the component of the model being rotors in the XY plane. None of this indicates that the model is of relevance to many different phenomena including melting of crystals, magnetism and superconductivity (see e.g. Chap. 9 in [80] and Thouless's Chap. 7 in [96]), growing surfaces (see Chap. 18 in [44]) and flocking of birds [33, 158].

Probably the easiest way to grasp what the model is about is by looking at a figure of a configuration captured in a computer simulation, see Fig. 6.6.

Similar to the Ising model, the individual components of the model are very simple, though they possess slightly more structure than do the Ising spins. An arrow of length one, S_i, is placed at each grid point of a two-dimensional square lattice. The arrow is like a compass needle and can rotate around its axis; we can specify its state by use of the angle θ_i between the arrow and the direction of the x-axis, namely $S_i = (\cos\theta_i, \sin\theta_i)$. These arrows are referred to in the literature as spins due to the connection to magnetism and quantum-mechanical spins, or we can call them rotors, since this is what they do: rotate about their midpoint.

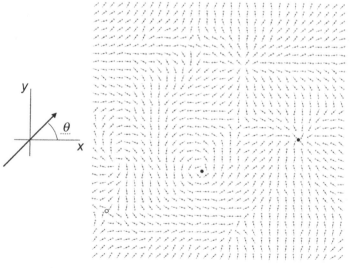

Figure 6.6 Rotor configuration of the XY model. Vortices (black circle) and anti-vortices (white circle) are clearly seen. The configuration is neutral, i.e. there is an equal number of vortices and anti-vortices; but only one circle is indicated. The reader may find it amusing to try to locate additional vortices and anti-vortex. Figure courtesy of Dr Hans Weber.

The rotors are assumed to interact with each other according to the energy given by the Hamiltonian:

$$H = -J \sum_{\langle i,j \rangle} S_i \cdot S_j = -J \sum_{\langle i,j \rangle} \cos(\theta_i - \theta_j). \tag{6.99}$$

Here J is the coupling constant between the rotors and $\langle i,j \rangle$ denotes summation over all nearest-neighbour sites in the lattice. The energy of the interaction between a pair i,j is lowest when the two arrows point in the same direction, i.e. when $\theta_i = \theta_j$, for which $-J \cos(\theta_i - \theta_j) = -J$. The energy is largest when the arrows point in opposite directions $\theta_i = \theta_j + \pi$, leading to $-J \cos(\theta_i - \theta_j) = -J \cos \pi = J$. This means that if no thermal energy is available, i.e. temperature $T = 0$, all the rotors will point in the same direction $\theta_i = \theta_j = \theta$ for all i and j. The energy of this fully aligned state is independent of the value of the common direction θ, it is therefore possible without injecting energy into the system to rotate all the rotors as long as we rotate all simultaneously. Because all these configurations have the same energy, they will contribute with the same probability given by Eq. (6.7) to the thermal average over states. Thus, even at $T = 0$ there is no preferred direction and therefore the average direction of the rotors is zero at all temperatures $\langle S_i \rangle = 0$. For this reason the 2d XY model was for a while considered to be not very interesting, since it appears that in contrast to the Ising model, see Sec. 6.2, the model does not exhibit a dramatic phase transition. This turned out to be wrong [240] in a very stimulating way. A phase transition does exist in the 2d XY model and the reason this was not initially realised, and how the misconception was rectified, makes the 2d XY model an ideal example of emergence. This model is a very informative example of the importance of understanding the emergent collective structure in a many-component system.

We start our discussion by considering the formation of **spin waves** and **vortices** in the sea of two-dimensional rotors. As the temperature is increased above zero, thermal energy becomes available. This allows configurations with energies above the lowest, or ground state, energy to start to appear because their Boltzmann weights in Eq. (6.7) increase when the factor $E_s/k_B T$ decreases.

The first kind of 'excited' configurations consist of gentle fluctuations in the directions of the nearly aligned rotors. In the literature, these configurations are said to exhibit spin waves because they take the form of wave-like perturbations of the directions of the rotors. In Fig. 6.6 one can discern spin waves perturbing the directions of the rotors in the top three rows.

As the temperature is increased further, it becomes energetically possible to excite a new type of configuration. These are the vortices and consist of the rotors turning around in patterns reminiscent of the flow pattern of a vortex in water. In Fig. 6.6, three of the vortices present are marked: two by black dots and one by a white dot. Try to locate all the vortices in Fig. 6.6. As one follows the direction of the rotors along an anticlockwise path around the black dot, the direction of rotors makes one entire 360-degree rotation. If we make the same tour around the white dot near the bottom-left corner, the rotors also circulate 360 degrees, but this time they rotate clockwise, so opposite to our anticlockwise path around the white dot. These vortex disturbances in the field of the rotor direction play a role similar to the 'particles' of pairs of antiparallel spins in the one-dimensional Ising model, see Fig. 6.5, though the vortices are substantially more subtle objects and have multiple effects on the properties of the systems in which they are found.

In summary, we noticed that two types of vortices exist: one of clockwise and one of anticlockwise rotation of the rotors around their centre. This is somewhat similar to the two types of antiparallel excitations we encountered for the Ising chain. A single vortex will influence the direction of the rotors throughout the entire system and will, for this reason, have an energy that grows with the size of the system and therefore for large systems be energetically unreachable. But a pair of clockwise–anticlockwise vortices will only significantly disrupt the rotor configuration over an area of linear size comparable to the separation between the vortices, see Fig. 6.7. A pair of vortices will therefore cost an energy which is given by the separation between them and as we increase the temperature from zero, pairs of small separation will occur first. With increasing temperature more vortex–anti-vortex pairs will be excited and the separation between the members of the pairs will also be able to increase.

The effect of the vortices on the macroscopic systemic properties is dramatic. So why did it take so long to realise their importance. This question is of general interest to complexity science since it highlights the need, as well as the difficulty, of identifying emergent collective structures. As mentioned above, it was understood that a preferred direction cannot spontaneously emerge in the 2d XY model. A preferred direction is probed by a local, single-component average, namely $\langle S_i \rangle$. And since this average is equal to zero for all temperatures, it cannot be used to detect the effect of the vortices. But the vortices are able to change global properties and so to detect their presence we need to probe the system as a whole.

The appearance of vortices has different consequences in different systems. They are responsible for melting of two-dimensional crystals, they destroy superconductivity of thin films of superconducting material, they make frictionless **superfluidity** become viscous normal fluidity for thin films of superfluid liquids [240], in particular helium, and the vortices of the 2d XY model are even of relevance to surface growth [44]. In this way, the vortex phenomenology is an example of how emergent properties can be shared, or manifest, across completely different systems [80, 240].

Let us explain the effect of vortices in terms of the 2d model of rotors by analogy with the difference between a liquid and a solid. A solid can sustain an elastic deformation, we need force to bend a rod of metal. We can induce the equivalent of a bend to the rotors of the XY model. Look again at Fig. 6.6 but now think of zero temperature where all the rotors will be parallel, say they are all pointing up. Starting from this configuration, we now clamp the rotors along the right edge at, for example, $\theta = \frac{\pi}{2}$ and at the left edge at $\frac{\pi}{2} + \delta$, see Fig. 6.8. The non-alignment of the rotors will induce an energy cost and to minimise this energy, the twist of the rotors is gradually distributed across the system. The increase in the energy ΔE can, for small values of δ, be written as

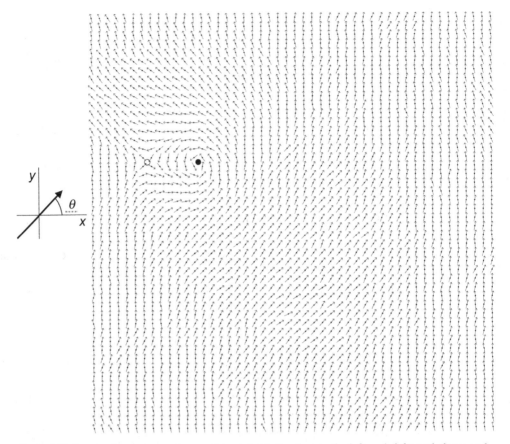

Figure 6.7 A pair of a vortex and an anti-vortex. Notice that to the left and right and above and below the vortex pair, the rotors are pointing in the same direction, namely upwards, and that along the line between the vortex and anti-vortex the rotors are pointing in the opposite direction, i.e. downwards. Figure courtesy of Dr Hans Weber.

Figure 6.8 A configuration of the rotors. A gradual twist of the directions of the rotors has been imposed by clamping the rotor direction at the right edge at $\theta = \frac{\pi}{2}$ and at the left edge at $\theta = \frac{\pi}{2} + \delta$. This induces a gradient in the direction of the rotors $\nabla\theta = \delta/L$.

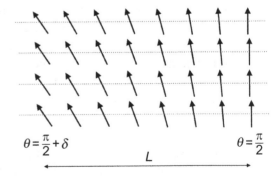

Figure 6.9 A sketch of the temperature dependence of helicity modulus Υ. The temperature T is along the x-axis and Υ along the y-axis. The size of the discontinuity at $T = T_{KT}$ is universal in the sense that by using relative variables, the same jump is seen in a variety of systems.

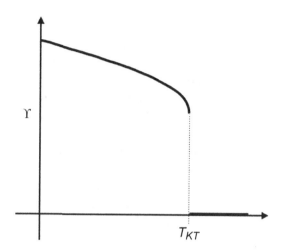

$$\Delta E = \frac{1}{2}\Upsilon\left(\frac{\delta}{L}\right)^2, \tag{6.100}$$

where Υ is a constant called the helicity modulus. This is the behaviour at zero temperature. As the temperature is increased, thermal fluctuations become important and one needs to consider the change induced by the gradient in the direction of the rotors of the *free* energy ΔF instead of ΔE. The free energy takes into account the effect of thermal fluctuations and is given by $F = E - TS$, where the entropy S accounts for the existence of a multitude of equivalent microscopic configurations at temperature $T > 0$, all with the same macroscopic properties. At non-zero temperatures, Eq. (6.100) is replaced by

$$\Delta F(T) = \frac{1}{2}\Upsilon(T)\left(\frac{\delta}{L}\right)^2. \tag{6.101}$$

The increase in the free energy depends on the temperature through the temperature dependence of $\Upsilon(T)$. It is found that $\Upsilon(T)$ first decreases gradually as T is increased for then suddenly to drop to zero discontinuously at the so-called Kosterlitz–Thouless temperature, denoted by T_{KT}, as sketched in Fig. 6.9 and e.g. explored by use of Monte Carlo simulations in [488].

The initial gradual decrease in Υ as T is increased from zero is due to thermally induced spin wave fluctuations in the rotor configuration. The more a configuration is disturbed by the thermal fluctuations, the smaller is the change $\Delta F(T)$ induced by the imposed twist. This is because, when the directions of the rotors in the bulk are somewhat randomised by the thermal fluctuations present at $T>0$, the additional twist distortion is less significant.

The discontinuous drop of Υ to zero at $T = T_{KT}$ is dramatic and signals a complete change in the systemic properties. At T_{KT} the density of vortex pairs and their sizes have become sufficiently large that the pairs disintegrate. The force between the members of a pair, which is proportional to Υ, vanishes and the vortices can move freely about. Since the direction of the rotors circles around the centre of a vortex, the motion of a vortex from e.g. the bottom to the top of the system will make the rotors on the left and right boundaries undergo a relative rotation of 360 degrees. When the vortices at T_{KT} unbind, they allow a relative rotation of the rotors on the boundaries to be done without any increase in the free energy. The system-spanning rigidity, or stiffness, which favours alignment of the rotors has disappeared, and this corresponds to melting. The description sketched here can be made mathematically precise [80, 240] and is of relevance to a large group of systems. For a system to be able to undergo the Kosterlitz–Thouless transition, it needs to support emergent collective excitations bound in pairs with an energy that increases logarithmically with the size of the pair.

Pairs of structural defects called dislocations and disclinations in two-dimensional crystals behave like vortices, as do so-called topological excitations in many other systems. Clearly this scenario is not immediately applicable to complex systems from outside physics, for which a description in terms of a Hamiltonian is not available. Nevertheless, it is a useful metaphor for how complex systems may exhibit hierarchies of emergent structures. In the 2d XY model the microscopic level consists of the individual rotors. At the next level the rotors collectively form the structures consisting of individual vortices. The vortices are able to form new structures by binding into pairs or unbinding to form a plasma of plus and minus charges.

This taxonomy of emergent creatures is robust in the sense that focusing on the macroscopic properties and the phase transition generated by the unbinding, one may forget about the rotors and simply concentrate on the positions of the cores of the vortices. By doing this, the phenomenology is reduced to the physics of point particles interacting through a potential energy depending logarithmically on separation. This representation of the 2d XY model is call the 2d electron gas [296].

Knowledge of the 2d XY phenomenology is also useful as a typical example of the difficulties in understanding the high-level behaviour of many-component systems without an identification of the emergent collective structures. The next section contains a brief schematic presentation of the mathematical analysis underpinning the phenomenology of the vortex physics. There exist many much more detailed discussions such as [80, 240], but we include for completeness an overview with a focus on how the vortices can be identified by use of the Boltzmann–Gibbs formalism of statistical mechanics.

6.7.1 2d XY: Some Mathematical Details

As repeatedly mentioned throughout this book, in complexity science we are interested in emergence. We like to understand how macroscopic structures can emerge from the interactions of the microscopic components. The 2d XY model is of particular interest in this respect because it is sufficiently realistic to relate to a number of real physical systems, as discussed in the sections above, but at the same time it is simple enough to allow for a detailed mathematical description of how the emergent structures, the vortices, are generated. We will now describe this.

In order to do the statistical mechanics of the 2d XY model we need to compute the energy corresponding to each specific configuration of rotors, such as those in Fig. 6.6 or 6.7. To do this we need to compute the sum in Eq. (6.99).

If we assume that the direction of the rotors varies smoothly from site to site, for a given pair i and j we can approximate $\cos(\theta_i - \theta_j)$ by the first two terms $1 - \frac{1}{2}(\theta_i - \theta_j)^2$ in the Taylor expansion of $\cos t = 1 - \frac{1}{2}t^2$. So we get

$$\cos(\theta_i - \theta_j) = 1 - \frac{1}{2}(\theta_i - \theta_j)^2. \tag{6.102}$$

To obtain the energy in Eq (6.99) we need, for each lattice site i, to sum over the four pairs connecting the neighbours to the left and right and above and below. Hence, for one specific i_0, we have

$$\sum_{\langle i_0, j \rangle} \cos(\theta_{i_0} - \theta_j) = 1 - \frac{1}{2}(\theta_{i_0} - \theta_\rightarrow)^2 + 1 - \frac{1}{2}(\theta_{i_0} - \theta_\leftarrow)^2$$
$$+ 1 - \frac{1}{2}(\theta_{i_0} - \theta_\uparrow)^2 + 1 - \frac{1}{2}(\theta_{i_0} - \theta_\downarrow)^2, \tag{6.103}$$

where $\theta_\rightarrow$, $\theta_\leftarrow$, $\theta_\uparrow$ and $\theta_\downarrow$ denote the rotors on the four sites that are neighbours of the rotor at location i_0.

It turns out that, instead of working with the rotors S_i and their angles θ_i located at the discrete positions of a square lattice labelled by i, it is convenient to think of a continuous field of rotors, one at each position r of the two-dimensional plane. At each lattice point this field of rotors is assumed to coincide with the rotors on the discrete lattice positions. We write $\theta(r)$ for all r in the plane and assume $\theta(r) = \theta_i$ when r is equal to the lattice position denoted by i. Using continuum as the version of the rotor field $\theta(r)$, we can write

$$\theta_{i_0} - \theta_\rightarrow = -a_0 \frac{\partial \theta}{\partial x}$$
$$\theta_{i_0} - \theta_\leftarrow = a_0 \frac{\partial \theta}{\partial x}$$
$$\theta_{i_0} - \theta_\uparrow = -a_0 \frac{\partial \theta}{\partial y}$$
$$\theta_{i_0} - \theta_\downarrow = a_0 \frac{\partial \theta}{\partial y}. \tag{6.104}$$

Here a_0 denotes the lattice constant for the square lattice on which the rotors are placed, i.e. the distance between neighbour sites. The expressions in the four equations above are the result of a Taylor expansion to first order of $\theta(x, y)$ about the site denoted i_0 with coordinates (x_0, y_0). For example:

$$\theta_{i_0} - \theta_{\rightarrow} = \theta(x_0, y_0) - \theta(x_0 + a_0, y_0)$$

$$= [i_0 - (i_0 + a_0)]\frac{\partial\theta}{\partial x} \tag{6.105}$$

$$= -a_0\frac{\partial\theta}{\partial x}.$$

When we substitute the expressions in Eq. (6.104) into Eq. (6.103), we get

$$-J\sum_{\langle i_0, j\rangle}\cos(\theta_{i_0} - \theta_j) = -4J + Ja_0^2\left[\left(\frac{\partial\theta}{\partial x}\right)^2 + \left(\frac{\partial\theta}{\partial y}\right)^2\right] = -4J + a_0^2 J(\nabla\theta)^2, \tag{6.106}$$

where the last equality makes use of the standard notation for the gradient vector

$$\nabla\theta = \left(\frac{\partial\theta}{\partial x}, \frac{\partial\theta}{\partial y}\right).$$

To get the energy of the system given by the Hamiltonian in Eq. (6.99), we need to sum the expression in Eq. (6.106) over all the lattice positions i_0. We replace this sum by an integral over all positions $\boldsymbol{r}$ and end up with the following continuum Hamiltonian:

$$H = E_0 + J\int d\boldsymbol{r}(\nabla\theta)^2. \tag{6.107}$$

Here $E_0 = -4JN$ is the energy of the completely aligned ground state of N rotors.

We recall that the statistical weight of a certain state of energy E_S in equilibrium is given by $\exp(-E_s/k_BT)/Z$, see Eq. (6.7). Hence, at a given temperature, the most likely states are those with the lowest energy. To find the configurations of rotors which contribute most to the behaviour of the 2d XY model, we will look for the local minima of the energy given by the Hamiltonian in Eq. (6.107). Similar to when we look for the local minima or maxima of a function $f(x)$ and solve $df/dx = 0$ to find x, we can look for the 'gradient' of H with respect to the configuration $\theta(\boldsymbol{r})$. Of course when we are dealing with a function of a single variable we simply differentiate the function directly to obtain $f'(x) = df/dx$, but we could also make use of the fact that $f(x + dx) - f(x) = f'(x)dx$. To make use of this approach to derive the gradient of H, we add a small position-dependent change $\delta(\boldsymbol{r})$ to $\theta(\boldsymbol{r})$ and write

$$H[\theta + \delta] - H[\theta] = J\int d\boldsymbol{r}(\nabla[\theta + \delta])^2 - J\int d\boldsymbol{r}(\nabla\theta)^2$$

$$= J\int d\boldsymbol{r}(\nabla\theta + \nabla\delta)^2 - J\int d\boldsymbol{r}(\nabla\theta)^2$$

$$= J\int d\boldsymbol{r}(\nabla\theta)^2 + J\int d\boldsymbol{r}(\nabla\delta)^2 + 2J\int d\boldsymbol{r}\nabla\theta\nabla\delta - J\int d\boldsymbol{r}(\nabla\theta)^2$$

$$= 2J\int d\boldsymbol{r}\nabla\theta\nabla\delta + J\int d\boldsymbol{r}(\nabla\delta)^2$$

$$= -2J\int d\boldsymbol{r}[\nabla^2\theta]\delta + O(\delta^2). \tag{6.108}$$

In the last line we have performed an integration by parts and assume that $\delta(\mathbf{r})$ vanishes on the boundary to obtain the first term.[12] We neglect the term that contains two factors of $\delta(\mathbf{r})$, which we indicate by the expression $O(\delta^2)$. We can do this since to obtain the first derivative, we only need to consider terms involving the first power of the small change δ. This is similar to when we write $f(x + dx) - f(x) = f'(x)dx$, we do not need to consider terms proportional to dx^2 or higher powers. We identify the derivative of $H[\theta]$ as the factor in front of δ and conclude

$$\frac{\delta H}{\delta\theta(\mathbf{r})} = -2J\nabla^2\theta(\mathbf{r}).$$

To determine the local minima of $H[\theta]$ we have to solve

$$\frac{\delta H}{\delta\theta(\mathbf{r})} = 0 \ \Rightarrow \ \nabla^2\theta(\mathbf{r}) = 0. \tag{6.109}$$

It turns out that there are two types of solutions to this equation. The first consists of the ground state $\theta(\mathbf{r}) = $ constant. The second consists of vortices in the rotator field, obtained by imposing the following set of boundary conditions on the circulation integral of $\theta(\mathbf{r})$:

(1) For all closed curves encircling the position $\mathbf{r}_0$ of the centre of the vortex

$$\oint \nabla\theta(\mathbf{r}) \cdot d\mathbf{l} = 2\pi n, \ \ n = 1, 2, \ldots . \tag{6.110}$$

(2) For all paths that don't encircle the vortex position $\mathbf{r}_0$

$$\oint \nabla\theta(\mathbf{r}) \cdot d\mathbf{l} = 0. \tag{6.111}$$

Condition (1) imposes a singularity in the rotator field. Note that the circulation integral *must* be equal to an integer times 2π since we circle a closed path and therefore $\theta(\mathbf{r})$ of the starting point and the end point of the closed path refers to the same rotor and must therefore represent the same direction.

We can estimate the energy of a single vortex in the following way. The dependence on r can be found from Eq. (6.110). We calculate the circulation integral along a circle of radius r centred at the position $\mathbf{r}_0$ of the vortex and make use of the spherical symmetry, hence the vortex field θ_{vor} must be of the form $\theta(\mathbf{r}) = \theta(r)$:

$$\begin{aligned}
2\pi n &= \oint \nabla\theta(\mathbf{r}) \cdot d\mathbf{l} \\
&= \int_0^{2\pi} d\theta|\nabla\theta|r d\theta = |\nabla\theta|r \int_0^{2\pi} d\theta \\
&= 2\pi r|\nabla\theta|.
\end{aligned} \tag{6.112}$$

Here we used that for a circle of radius r we have $dl = rd\theta$. And that the scalar product $\nabla\theta \cdot d\mathbf{l} = |\nabla\theta|r d\theta$.

---

[12] Recall $\int_a^b dx f(x)g'(x) = [f(x)g(x)]_a^b - \int_a^b dx f'(x)g(x)$.

We solve and obtain $|\nabla\theta(r)| = n/r$. Substitute this result into the Hamiltonian Eq. (6.107):

$$E_{vor} - E_0 = \frac{J}{2}\int dr [\nabla\theta(r)]^2 \tag{6.113}$$

$$= \frac{Jn^2}{2}\int_0^{2\pi} d\phi \int_a^L rdr\frac{1}{r^2} \tag{6.114}$$

$$= \pi n^2 J \ln\left(\frac{L}{a_0}\right). \tag{6.115}$$

Here a_0 denotes the lattice constant and L is the linear size of the considered lattice. The circulation condition Eq. (6.110) creates a distortion in the phase field $\theta(r)$ that persists away from the centre of the vortex. $|\nabla\theta|$ decays only as $1/r$, leading to a logarithmic divergence of the energy. Hence we need to take into account that the integral over r in Eq. (6.114) is cut off for large r-values by the finite system size L and for small r-values by the lattice spacing a_0. We recall that our continuum Hamiltonian is an approximation to the lattice Hamiltonian in Eq. (6.99). A vortex with the factor n in Eq. (6.110) larger than one is called multiple charged. We notice that the energy of the vortex is quadratic in the charge. In a macroscopically large system even the energy of a single charged vortex will be large, and therefore we do not expect individual vortices to be thermally induced.

Consider now a pair consisting of a single charged vortex and a single charged anti-vortex. When we encircle the vortex, we pick up $\oint dl \cdot \nabla\theta = 2\pi$ and when we encircle the anti-vortex, we pick up $\oint dl \cdot \nabla\theta = -2\pi$. Hence, if we choose a path large enough to enclose both vortices, we pick up a circulation of the phase equal to $2\pi + (-2\pi) = 0$. That is, the distortion of the phase field $\theta(r)$ from the vortex–anti-vortex pair is able to cancel out at distances from the centre of the two vortices that are large compared to the separation R between the vortex and the anti-vortex, see Fig. 6.7. This explains why the energy of the vortex pair is of the form [78, 297]

$$E_{2vor}(R) = 2E_c + E_1 \ln(R/a) \tag{6.116}$$

where E_c is the energy of a vortex core and E_1 is proportional to Υ in Eq. (6.101). In detail, the phase field $\theta_{2vor}(r)$ of a vortex located at $r = (-a, 0)$ and an anti-vortex located at $r = (a, 0)$ is given by [78]

$$\theta_{2vor}(r) = \text{arctg}\left(\frac{2ay}{a^2 - r^2}\right). \tag{6.117}$$

Significant aspects of the macroscopic behaviour of the XY model can be understood by treating the vortices as particles characterised by their position and their charge and ignoring the underlying sea of rotors. Indication of this follows from the expression for the energy of a pair of vortices in Eq. (6.116). This energy is given in terms of the relative position of the two vortices, no reference is needed to the microscopic rotor field given in Eq. (6.117). The pairs of vortices have dramatic effects on the macroscopic behaviour of the XY model. At low temperature, the vortices are organised in fairly small bound pairs. As the temperature is increased and more thermal energy is available, the separation between paired-up vortices grows, and at a certain temperature the size of pairs becomes comparable to the separation between the pairs and the pairs effectively break apart with the effect that the individual vortices can now move freely around as they are no longer kept in check by their partner of opposite charge. The result is the Kosterlitz–Thouless transition which manifests itself in various ways in different realisations of the XY model. We will return to this in Sec. 6.7.3 below.

6.7.2 Vortex Unbinding

An indication of the mechanisms behind the abrupt change with temperature brought about by the vortices can be obtained from the following simple and heuristic argument [240]. In thermodynamic equilibrium the macroscopic state of a system will minimise the Helmholtz free energy F given by $F = E - TS$, where E is the internal energy, T the temperature and S the thermodynamic entropy. The entropy can be written as $S = k_B \ln \Omega$, where Ω denotes the number of ways the state of the system can be realised. We can estimate the free energy when a single vortex is present in the following way. The energy is given by Eq. (6.115), and the entropy is given by the logarithm of the number of ways we can construct a state with one single vortex. This we estimate as the number of places where we can position the vortex centre on our lattice of size $L \times L$. The vortex can be placed on each of the $(L/a_0)^2$ plaquettes of the square lattice, i.e. $S = k_B \ln(L^2/a_0^2)$. Accordingly, the free energy is given by

$$F = E_0 + (\pi J - 2k_B T) \ln(L/a_0). \tag{6.118}$$

For $T < \pi J/2k_B$ the free energy of the state containing one vortex will diverge to plus infinity as $L \to \infty$. In contrast, for temperatures $T > \pi J/2k_B$ the system can lower its free energy by producing vortices $F \to -\infty$ as $L \to \infty$. This simple heuristic argument points to the fact that the logarithmic dependence on system size of the energy of the vortex combines with the logarithmic dependence of the entropy to produce the subtleties of the vortex unbinding transition. Assume a different dependence of the energy on system size and one will either have thermal activation of vortices at all temperatures (in case $E_{vor} \to$ constant $< \infty$) or vortices will not be activated at any temperature (in case $E_{vor} \sim (L/a)^b$ with $b > 0$). It is the logarithmic size dependence of the 2d vortex energy that allows the outcome of the competition between the entropy and the energy to change qualitatively at a certain finite temperature T_{KT}.

We recall from Sec. 6.7 that in reality it is not single vortices of the same sign that proliferate at a certain temperature. What happens is that the vortex pairs, which are bound together for temperatures below T_{KT}, unbind at T_{KT}. This is a collective effect that can be treated quantitatively by use of a special **renormalisation group** method design by Kosterlitz [239]. In the picture where we focus on the vortices and forget about the underlying rotors, the unbinding can be seen as driven by small bound pairs softening the interaction between the two vortices of larger pairs, eventually breaking the attraction between the partners of the larger pairs. More details can be found in the wonderful book by Chaikin and Lubensky [80].

6.7.3 The Vortex Unbinding Transition in Other Systems

We have mentioned above that not only the XY model exhibits the Kosterlitz–Thouless vortex unbinding transition [296]. Any two-dimensional system that supports thermally induced 'charges' or topological defects, which interact logarithmically, will undergo this transition. The symmetry of the phase field $\theta(r)$ of the XY model is also present in the Ginzburg–Landau free energy of superfluids and superconductors. The topological excitations in the case of a superfluid consist of vortices in the flow of the superfluid. Vortices like those observed when one empties a bath tub. In a thin layer of superfluid helium with increasing temperature, such vortices destroy the superfluid phase according to the scenario of the Kosterlitz–Thouless transition [13, 14].

The situation is slightly more complicated in superconductors. Because the superfluid in this case is charged (superconducting pairs of electrons), electric screening effects play a role and induce a cutoff in the interaction between vortices given by the so-called London penetration depth λ [80, 296, 444]. However, for thin superconductors of thickness δ the effective screening length is given by $\lambda_{\mathrm{eff}} = \lambda^2/\delta$, which can easily become a macroscopic length. In this case the loss of superconductivity is caused by the unbinding of vortex pairs according to the Kosterlitz–Thouless transition. The broken pairs can move freely when they respond to an applied electric current. As they move they cause phase slips in the superconducting order parameter equivalent to the change induced in the rotor field $\theta(\mathbf{r})$ in the 2d XY model. In the superconductor these phase slips induce a voltage drop according to the so-called Josephson relation. This relation is part of the achievement for which Josephson won the Nobel Prize in Physics 1973 [1]. The voltage drop implies that the superconductivity is lost [220, 444].

Dislocations in two-dimensional crystals interact through the strain field, which is the field of deformations in the lattice structure. Two edge dislocations of opposite sign correspond to an extra row of atoms inserted along the line connecting the location of the two dislocation cores. The extra line of atoms produces strain and leads to an increase in the energy which is logarithmic in the separation between the two dislocations. Thus the situation is very similar to the one encountered in the 2d XY model. When the dislocations unbind, free dislocations are produced. A shear applied to the system can now be accommodated by the mobile dislocations without an increase in the (free) energy. That is, the shear constant has dropped to zero and the material is melted. The 2d melting theory of Kosterlitz–Thouless–Halperin–Nelson–Young predicts that melting occurs in two stages. At the first stage, dislocations unbind and make the shear constant drop to zero. The dislocations are topological defects; their effects on the order of the lattice are, however, not very dramatic. Before the unbinding of dislocations, the translational and orientational order of the lattice are both described by correlation functions that depend algebraically on distance. When the dislocations unbind, the translational correlation function becomes exponential but the orientational correlations remain algebraic. At a somewhat higher temperature the topological defects called 'disclinations' unbind with the effect that the orientational order becomes exponential. Details can be found in [80].

There are many other cases where the logarithmic vortex interaction and the Kosterlitz–Thouless transition play a role. For instance, the shape of surfaces in three dimensions may undergo a transition from smooth to rough [44]. Assume that the surface energy of the two-dimensional surface is proportional to the area of the surface. And assume that the surface is defined in terms of its height $h(x,y)$ above the xy-plane, i.e. no overhangs. In other words, the points on the surface have the coordinates $(x, y, h(x, y))$. The Hamiltonian for the surface is then

$$H = \sigma \int dx \int dy \sqrt{1 + (\nabla h)^2}. \tag{6.119}$$

Here σ is a measure of the surface tension. If the height varies slowly as a function of (x, y) we can assume $|\nabla h| \ll 1$ and then expand the square root. In this approximation the Hamiltonian in Eq. (6.119) can be written as

$$H = \sigma L^2 + \frac{1}{2} \int dx \int dy (\nabla h)^2 \tag{6.120}$$

(L is the linear size of the system in the xy-plane), which is equivalent to Eq. (6.113), and we expect the same physical phenomenology to apply to the surface as we found for the XY model.

Summary: When each configuration of the components is associated with an energy, statistical mechanics can determine the probability with which a configuration contributes to the equilibrium behaviour. We discussed:

1. The dependence of fluctuations, systems response and correlations as a function of temperature.
2. At low temperature ordered regimes exist, which are commonly separated from high-temperature disordered regimes by a critical temperature T_c.
3. At T_c infinite correlations lead to power-law behaviour.
4. The 2d XY model offers an example of general relevance describing how emergent structures can be analysed by use of the partition function formalism.
5. Reduction of dimensionality: It is possible to go from a description in terms of zillions of individual rotors to the much fewer coordinates describing macroscopic vortices. This is similar to describing a game of billiards by following the billiard balls instead of all the molecules they are made of.

6.8   Further Reading

Introductory level:

An introduction to the basic ideas underpinning statistical mechanics, using only very basic mathematics, is given in the book by Glazer and Wark [162].

David Thouless explains, without the use of mathematics, critical phenomena, the correlation length, vortices and more in Chap. 7 of the book *The New Physics* [96]. The same book contains several highly recommendable chapters discussing emergence and collective phenomena in physics, such as the phenomenology of critical points and superconductivity.

Intermediate level:

Reif's classic book *Fundamentals of Statistical and Thermal Physics* [362] gives a comprehensive and very clear conceptual and mathematical introduction.

The book by Franz Schwabl [392] gives a mathematical based, very clearly written and impressively comprehensive introduction.

The statistical mechanics of collective topological objects such as vortices is discussed with care and mathematical clarity in Chap. 9 of the book *Principles of Condensed Matter Physics* [80] by Chaikin and Lubensky.

Analysis of complex systems by use of Tsallis's entropy is comprehensively discussed in the book *Introduction to Nonextensive Statistical Mechanics: Approaching a Complex World* [453] by Constantino Tsallis.

Advanced level:

A mathematically advanced explanation of how methods from theoretical particle physics can be used to analyse emergence in critical phenomena is given by Jean Zinn-Justin in his book *Quantum Field Theory and Critical Phenomena* [505].

6.9   Exercises and Projects

Exercise 1
Use the equations for the canonical ensemble to discuss a very simple system consisting of two Ising spins S_1 and S_2 with energy given by $H = -JS_1S_2$, where $J > 0$. Use the notation $x = \beta J$ and show:

(a) The partition function is given by $Z = 4\cosh(x)$.
(b) The average energy is given by $\langle E \rangle = -J\tanh(x)$.
(c) Determine $\langle E \rangle$ for $T \to 0$ and for $T \to \infty$.

Exercise 2

Show that in the canonical ensemble, the entropy given by Eq. (6.11) for the two-spin system in Exercise 1 is given by

$$S[p_1, p_2, p_3, p_4] = \ln[4\cosh(x)] - x\tanh(x), \tag{6.121}$$

again using the notation $x = \beta J$.

Exercise 3

Now consider a general system and show that the entropy given by Eq. (6.11) in the microcanonical ensemble is given by $S = \ln \Omega$.

Exercise 4

We consider again the system in Exercises 1 and 2 and make use of the same notation. We will study the limits $T \to 0$ and $T \to \infty$.

(a) In the limit $T \to 0$ compute the four probabilities given by Eq. (6.7) and compute directly the entropy in Eq. (6.11) in this limit.
(b) In the limit $T \to \infty$ compute the four probabilities given by Eq. (6.7) and compute directly the entropy in Eq. (6.11) in this limit.
(c) Now use the expression valid for all temperatures in Eq. (6.121) and from this expression derive the limting values of (a) and (b).

Exercise 5

Consider a particle moving on the lattice structure depicted in Fig. 6.10. We assume it moves to each of its nearest-neighbour sites with equal probability and that it moves in a way such that within each box it is equally likely to be found at any of the L_i^2 squares, where L_i is the linear size of box i, i.e. $L_i = 2^i$ with $i = 1, 2, 3, \ldots$.

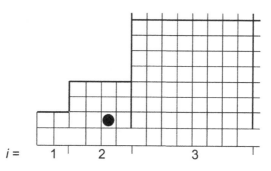

Figure 6.10 Particle moving in boxes. Imagine the figure extends to the right with ever more boxes $i = 1, 2, 3, 4, \ldots$ added. The particle is able to move between neighbour boxes through the hole at the bottom right corner of box i.

(a) Assume for a moment that the passage between boxes is blocked. What is the entropy of the particle when placed in box i?

Imagine we have an ensemble of systems like the one in Fig. 6.10. We can then think of what on average is happening in this set of equivalent systems. It is natural to estimate the average time between recurrent visits to a given site in box i as given by the inverse of the probability of visiting a site.

(b) Assume the particle is in box i. What is the probability that the particle is located next to an escape site, see Fig. 6.10?
(c) What is the probability per time that a particle in box i moves to box $i+1$?
(d) What is the probability per time that a particle in box $i+1$ moves to box i?

Now imagine one particle is in box i and one in box $i+1$.

(e) Show that the net probability (right move minus left move) for a particle moving from box i to $i+1$ is given by $2^{-2(i+1)}$.

We see that the particles will move towards higher entropy. Considerations like these were used in [28] to argue about Gaia, evolution and 'entropic pressure'.

Exercise 6
Derive the expression for the eigenvalues given in Eq. (6.75).

Exercise 7
Discuss the similarity and differences between a vortex–anti-vortex pair and a pair of $\uparrow\downarrow$ and $\downarrow\uparrow$. How does the energy of creating a single $\uparrow\downarrow$ perturbation scale with system size? How does the energy of a pair $\uparrow\downarrow$ and $\downarrow\uparrow$ scale with the distance between the two sets of antiparallel spins?

Exercise 8
Study of collective excitations in a one-dimensional system. Consider a set of rotors $S_n = (\cos(\phi_n), \sin(\phi_n))$ located at positions $x_n = na$ with $n = 1, 2, \ldots, N$. Here $\phi \in [0, 2\pi)$ is the phase of the rotor.
   The energy of the system is given by

$$H = \sum_{n=1}^{N} [-J\cos(\phi_n - \phi_{n+1}) - h\cos\phi_n].$$

Assume periodic boundary conditions, i.e. $S_{N+1} = S_1$. We confine our analysis to the limit where ϕ_n is slowly varying and consider the continuum limit

$$\cos(\phi_n - \phi_{n+1}) \simeq 1 - \frac{1}{2}(\phi_n - \phi_{n+1})^2 \mapsto 1 - \frac{a^2}{2}\left(\frac{d\phi}{dx}\right)^2.$$

In this approximation the Hamiltonian is given by

$$H = -\frac{J}{a} \int_{-\infty}^{\infty} dx \left[1 - \frac{a^2}{2} \left(\frac{d\phi}{dx} \right)^2 + \frac{h}{J} \cos \phi \right].$$

(a) Show that the extremum equation

$$\frac{\delta H}{\delta \phi} = 0$$

leads to the so-called sine-Gordon equation given by

$$\frac{d^2 \phi}{dx^2} = \mu^2 \sin \phi,$$

where $\mu^2 = \frac{h}{Ja^2}$.

(b) Demonstrate that the sine-Gordon equation is solved by the soliton

$$\phi(x) = 4 \operatorname{arctg}[\exp(\mu x)].$$

(c) The energy of the soliton and the fluctuation spectrum around the soliton is discussed in Secs. 2.1 and 2.2 of [144].

Exercise 9
Mean-field analysis of interacting vortices. Go through the phenomenological self-consistent calculation presented in [221].

Project 1 – Microcanonical ensemble
Consider a collection of N non-interacting particles. Each particle can be in one of R states. The microcanonical ensemble assumption states that, left to itself, the system will visit all allowed microstates with equal probability. The allowed microstates correspond to all the different ways one can allocate the R states to the N particles under the given constraints. Let n_i denote the number of particles in state i. If there were no constraints there would be a total of

$$\Omega(\mathbf{n}) = \frac{N!}{n_1! n_2! \cdots n_R!}$$

microstates corresponding to the configuration $\mathbf{n} = (n_1, n_2, \ldots, n_R)$. $\Omega(\mathbf{n})$ varies greatly depending on the configuration. For example, there is only one microstate for $\mathbf{n} = (N, 0, \ldots, 0)$, but many with $\mathbf{n} = (N/R, \ldots, N/R)$ (assuming N to be divisible by R). It is therefore sensible to assume that the configurations $\mathbf{n}$ in which one typically finds the system correspond to the largest number of allowed microstates. That is, the configuration that maximises $\Omega(\mathbf{n})$ under the constraints given by the particle number

$$\sum_{i=1}^{R} n_i = N \qquad (\dagger)$$

and total energy

$$\sum_{i=1}^{R} \epsilon_i n_i = U. \qquad (\dagger\dagger)$$

The typical configuration n is then computed by maximising $\Omega(n)$ under the constraints (†) and (††), which is done by introducing Lagrange multipliers and maximising

$$J(n) = \ln \Omega(n) + \alpha \left(N - \sum_{i=1}^{R} n_i \right) + \beta \left(U - \sum_{i=1}^{R} \epsilon_i n_i \right).$$

The solution n that solves this maximisation problem is taken to be the configuration of occupancies n_i for the system in equilibrium.

(a) Use Stirling's approximation for $N!$ and $n_i!$ and derive the equilibrium expression for n_i in terms of ϵ_i, α and β.
(b) Eliminate α by use of the normalisation $\sum_i n_i = N$.
(c) Write down the expression which determines β.

Now consider the case where $R = 2$ and the energies $\epsilon_1 = 0$ and $\epsilon_2 = \epsilon$.

(d) Use the result in (c) to determine β.
(e) Plot $1/\beta$ as a function U/N.
(f) How does the behaviour of $1/\beta$ compare with the thermodynamic identification of $1/\beta$ being proportional to the temperature?
(g) Relate your results to the discussion of negative temperature, e.g. given at http://en.wikipedia.org/wiki/Negative_temperature and in Sec. 73 of [253].

Project 2 – One-dimensional Ising model
Consider the one-dimensional Ising model given by the following Hamiltonian:

$$H = -J \sum_{i=1}^{N} S_i S_{i+1} + NJ,$$

with $S_i = \pm 1$ for $i = 1, \ldots, N$ and $S_{N+1} = S_1$.

(a) Calculate the energy E_1 of the first excited state of the system in which the spin configuration can be described in terms of a single kink, i.e. one single pair of antiparallel spins.
(b) Assume that the chain in a higher excited state can be described by a gas of non-interacting kinks distributed along the chain. Show that the partition function is given by

$$Z = \left[1 + e^{-\beta E_1} \right]^N.$$

(c) Show that the average number of kinks $\langle n \rangle$ is given by

$$\langle n \rangle = -\frac{1}{E_1} \frac{\partial}{\partial \beta} \ln Z.$$

(d) Consider $\langle l \rangle = N/\langle n \rangle$ to be a measure of the correlation length and derive its temperature dependence. In particular, study the limits $T \to 0$ and $T \to \infty$.

Project 3 – Computational study of fluctuations as indicator of critical transition

Consider a set of particles on a two-dimensional square lattice of linear size L. The particles interact through repulsive central unit forces with their nearest neighbours. Double occupancy is not permitted. Hence a square is either empty or contains one particle. All particles are updated simultaneously by moving the particles to neighbour sites according to the vector sum of the forces they are subject to.

One can define the dynamics as completely deterministic in the following way. If two particles want to move to the same site, the particle subject to the strongest force is moved, while in the case of equal forces no particle is displaced.

Or one can allow for a stochastic element by choosing at random one particle amongst the particles that all want to move on to the same square.

Assume periodic boundary conditions.

(a) Write a code able to simulate the dynamics of the particles.

Define the dissipating squares as the square involved in the motion from one time step to the next. That is, say a particle is at position $\mathbf{r}$ at time t and then moves on to position $\mathbf{r}'$ at time $t + 1$. Both squares $\mathbf{r}$ and $\mathbf{r}'$ belong to the set of dissipating squares at time t.

Now study the statistical properties of the dissipation by sampling over configurations separated by, say, L^2 time steps.

(b) Write a code to study the distribution of cluster sizes.

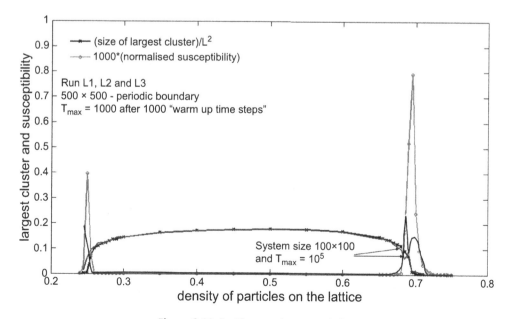

Figure 6.11 Lattice gas cluster statistics.

(c) Focus on the largest cluster for each generated configuration. Study the average and the standard deviation of the size of the largest cluster. The standard deviation is equivalent to the susceptibility.

(d) For affordable system sizes, reproduce Fig. 6.11.

(e) Discuss the significance of the peaks at low and high particle density in the standard deviation of the largest dissipation cluster.

Reference: A. Giometto, H. J. Jensen. Connecting the micro-dynamics to the emergent macro-variables: Self-organized criticality and absorbing phase transitions in the deterministic lattice gas. *Physics Reviews E*, 85: 011128, 2012.

7 Synchronisation

Synopsis: When interactions between individual dynamical components are sufficiently strong, coordinated dynamics at the systemic level can emerge. This is called synchronisation.

We mentioned briefly in Sec. 2.3 that synchronisation can be considered the emergence of structure in time. We observe around us multitudes of different types of synchronisation involving very different types of components. It can be neurones in the brain [466] or musicians in an orchestra [187], synchronisation is crucial as temporal coordination is linked intimately to functionality. Despite the difference between the components, we will see that synchronous behaviour across varieties of situations shares many similarities and that simple mathematical models can help to identify essential general mechanisms involved in synchronisation.

Since synchronisation can occur between a very broad range of dynamical components and the character of synchronised dynamics also varies greatly, the literature is accordingly extensive. For a good non-mathematical introduction to the topic, Strogtaz's book [429] is well written and comprehensive. An excellent general introduction with only a gentle use of mathematics can be found in [345]. A mathematically more detailed presentation based on a dynamical systems approach is given in [38] and a recent broad discussion with some mathematical details can be found in [62].

Here we will concentrate on two aspects of synchronisation, namely the transition from incoherent to synchronised dynamics and the structure of the synchronisation. To do this we will focus on the Kuramoto model [245], which is yet another example, like the Ising model or the branching process, of modelling that attempts to distil out the bare minimum needed for an analysis of the considered phenomena.

The elegant model of Kuramoto considers how synchronisation can be established as the outcome of a competition between incoherent dynamics of the individuals and some interaction which induces coordination amongst the individual components. After having looked at how synchronisation amongst many components may be established, we will turn to the structure of the synchronous dynamics, which can be much more lively than simply all components ticking along at the same rate. To describe this, we will look at the formation of chimera states [9].

7.1   The Kuramoto Model: The Onset of Synchronisation

A prototypical model of synchronisation was introduced in the mid-1970s by Kuramoto [245] in which he studied rotors each characterised by a phase, or angle, θ_k for $k = 1, 2, \ldots, N$. These can be thought of as the rotors of the 2d XY model, see Sec. 6.7. However, for the 2d XY model we looked at equilibrium thermodynamics. We studied the behaviour of the system in equilibrium by use of the statistical mechanics of Boltzmann and Gibbs. Now we look at the rotors as dynamical variables subject to the following equations of motion:

$$\frac{d\theta_k}{dt} = \omega_k + \frac{\epsilon}{N} \sum_{j=1}^{N} \sin(\theta_j - \theta_k) \text{ for } k = 1, \ldots, N. \tag{7.1}$$

First notice that if $\epsilon = 0$, hence when the second term in the equation is absent, each rotor θ_k will progress at speed ω_k. In the literature, ω_k is called a frequency or an angular frequency.

Now let us ignore the effect of the first term to focus on the effect of the second term, i.e. let us for a moment assume $\omega_k = 0$. The term $\sin(\theta_j - \theta_k)$ will clearly tend to make $\theta_j(t)$ and $\theta_k(t)$ equal since if, for example, θ_k is lagging behind θ_j, the term will be positive and therefore θ_k will speed up. And similarly, when θ_k is ahead of θ_j, the sin-term will make θ_k slow down.

However, when the first and second terms are taken together and if the frequencies ω_k are not all equal, the adjustment towards synchronisation effected by the second term will have to compete against the individual speeds imposed by the first term ω_k of Eq. (7.1).

We see that when the ω_k assumes different values for different rotors, in order to achieve synchronisation, the rotors have to find a compromise between the speed favoured by their ω_k term and some collectively 'agreed' speed, perhaps one would guess that the common speed would be given by $\overline{\theta_k} = (\sum_k \theta_k)/N$. But then this appears not always to be possible. Think of situations where the coupling ϵ is very weak and the distribution of the ω_k frequencies very broad, then for some k we will have that ω_k differs significantly from $\overline{\theta_k}$ and since the first term in Eq. (7.1) dominates when ϵ is small, it is unlikely that θ_k can move with a velocity that significantly differs from ω_k.

From these considerations we expect that the degree of synchronisation, i.e. how many rotors are moving with velocities that are equal, or nearly equal, will depend on the relationship between the width σ_g of the distribution $g(\omega)$ from which the intrinsic frequencies of the rotors are drawn and the strength ϵ of the interrelation between the rotors.

To analyse this behaviour in more mathematical detail, we introduce

$$Ke^{i\theta} = \frac{1}{N} \sum_{k=1}^{N} e^{i\theta_k}. \tag{7.2}$$

We will see below that the modulus K will act as an order parameter in the sense discussed in Sec. 6.3. Namely, K allows us to distinguish between no synchronisation $K = 0$ and some degree of synchronisation $K > 0$. The strength of the coupling ϵ acts as a control parameter. A critical value ϵ_c exists, such that for $\epsilon < \epsilon_c$ no synchronisation takes place so $K = 0$ and for $\epsilon > \epsilon_c$ synchronisation sets in and $K > 0$.

To analyse the degree of synchronisation amongst the rotors, we introduce the order parameter into Eq. (7.1) in the following way:[1]

$$
\begin{aligned}
\frac{d\theta_k}{dt} &= \omega_k + \frac{\epsilon}{N} \sum_{j=1}^{N} \sin(\theta_j - \theta_k) \\
&= \omega_k + \frac{\epsilon}{N} \mathrm{Im} \sum_{j=1}^{N} e^{i(\theta_j - \theta_k)} \\
&= \omega_k + \frac{\epsilon}{N} \mathrm{Im} \left\{ \left(\sum_{j=1}^{N} e^{i\theta_j} \right) e^{-i\theta_k} \right\} \\
&= \omega_k + \epsilon \mathrm{Im} \left\{ K e^{i\theta} e^{-i\theta_k} \right\} \\
&= \omega_k + \epsilon K \sin(\theta - \theta_k).
\end{aligned}
\tag{7.3}
$$

Now we introduce the average of the individual rotor velocities:

$$
\bar{\omega} = \frac{1}{N} \sum_{k=1}^{N} \omega_k \simeq \int_0^\infty \omega g(\omega) d\omega,
\tag{7.4}
$$

here the last approximative equality expresses the empirical average for the actual realisation of the set of ω_k values in terms of the statistical average of the probability density $g(\omega)$ describing the ω_k.

Consider the deviation ψ_k of rotor k from value $\bar{\omega}t$ the rotor would have at time t if it moved with the average speed, i.e.

$$
\theta_k(t) = \bar{\omega}t + \psi_k(t).
\tag{7.5}
$$

Next we assume that the phase θ introduced in Eq. (7.2) moves with the average speed, hence that $\theta(t) = \bar{\omega}t$. This assumption will have to be checked for consistency later, but for now we observe that substituting Eq. (7.5) into Eq. (7.3) we have

$$
\frac{d\psi_k}{dt} = \omega_k - \bar{\omega} - \epsilon K \sin(\psi_k).
\tag{7.6}
$$

This equation allows us to study the strength of the synchronisation and if any occurs at all. Equation (7.6) supports two types of solutions:

(1) If $\psi_k(t) \to$ constant as $t \to \infty$, this corresponds to entrainment, or synchronisation, of rotor k to θ.

(2) If $\psi_k(t)$ monotonously increases (or decreases), ψ_k does not synchronise to θ.

---

[1] Recall that $e^{ix} = \cos x + i \sin x$, so we have the imaginary $\mathrm{Im}(e^{ix}) = \sin x$.

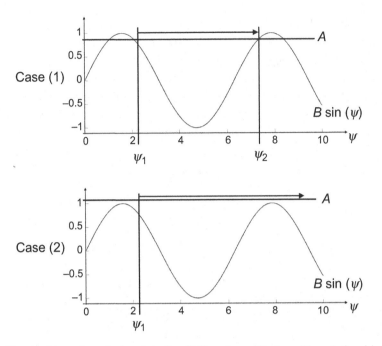

Figure 7.1 Sketch of how the solution ψ to Eq. (7.7) evolves with time. The relationship between A and B determines whether case (1) or case (2) applies. In the top panel ψ moves from ψ_1 to ψ_2 where it stops because the right-hand side of Eq. (7.6) becomes equal to zero at ψ_2. In the bottom panel $A > B$ and therefore $d\psi/dt$ remains positive for all ψ, allowing $\psi(t)$ to monotonously increase forever.

To reach this conclusion we analyse the equation

$$\frac{d\psi}{dt} = A - B \sin \psi, \tag{7.7}$$

which is of the same form as each Eq. (7.6). The graphical analysis in Fig. 7.1 shows that those rotors for which $A = |\omega_k - \bar{\omega}| < \epsilon K = B$ belong to case (1) and those with intrinsic frequencies farther away from $\bar{\omega}$, i.e. $A = |\omega_k - \bar{\omega}| > \epsilon K = B$, belong to case (2).

We make the assumption $\theta = \bar{\omega}t$ and substitute this ansatz[2] on the left-hand side of Eq. (7.2) and substitute on the right-hand side of this equation the expression for θ_k given in Eq. (7.5). This way we obtain

$$K e^{i\bar{\omega}t} = \frac{1}{N} \sum_{k=1}^{N} e^{i(\bar{\omega}t + \psi_k)}$$

$$\Downarrow \tag{7.8}$$

$$K = \frac{1}{N} \sum_{k=1}^{N} e^{i\psi_k}.$$

[2] The term 'anzats' is used in mathematics to denote an initial assumption about the properties of a function.

The next step is to approximate the sum over k in Eq. (7.8) by an integral over the probability density of the phases ψ_k, i.e. we write

$$K = \int_{-\infty}^{\infty} d\psi\, n(\psi) e^{i\psi}. \tag{7.9}$$

One arrives at this approximation by grouping the terms in the sum over k in Eq. (7.8) into bundles according to the value of ψ_k. Thus $n(\psi)d\psi$ denotes the fraction of rotors that have their ψ_k value in the interval $[\psi, \psi + d\psi]$.

To compute K from the integral in Eq. (7.9) we need to estimate the distribution of the phases ψ. To do this we distinguish between the rotors k that participate in synchronisation and those that do not. That is, we need to classify the rotors according to the cases (1) and (2) above. We write

$$n(\psi) = n_s(\psi) + n_{as}(\psi), \tag{7.10}$$

where n_s denotes the density of rotors that synchronise, i.e. case (1), and n_{as} denotes the density of those that move asynchronously, i.e. case (2).

For those that synchronise we have that $d\psi/dt = 0$ after some transient time. According to Eq. (7.6) these rotors satisfy

$$\omega = \bar{\omega} + \epsilon K \sin(\psi). \tag{7.11}$$

For a specific rotor this equation enables us to consider ω as a function of ψ while we keep $\bar{\omega}$ and ϵK fixed. We connect the distribution of ψ to the distribution of ω by use of $n_s(\psi)|d\psi| = g(\omega)|d\omega|$ to write

$$n_s(\psi) = g(\omega)\left|\frac{d\omega}{d\psi}\right| = g(\bar{\omega} + \epsilon K \sin\psi)\epsilon K \cos\psi. \tag{7.12}$$

Next we turn to the rotors that belong to case (2), i.e. the asynchronous rotors. For these the phase ψ increases monotonously with time and given the periodicity of Eq. (7.6) it follows[3] that $n_{as}(\psi + \pi) = n_{as}(\psi)$. This periodicity ensures that the asynchronous rotors do not contribute to K in Eq. (7.9) since

$$n_{as}(\psi + \pi)e^{i(\psi+\pi)} = -n_{as}(\psi + \pi)e^{i\psi} = -n_{as}(\psi)e^{i\psi}.$$

We can therefore concentrate on the rotors that belong to case (1) and write

$$\begin{aligned} K &= \int d\psi\, n_s(\psi)e^{i\psi} \\ &= \int_{-\pi/2}^{\pi/2} d\psi\, g(\bar{\omega} + \epsilon K \sin\psi)\epsilon K \cos\psi\, e^{i\psi}. \end{aligned} \tag{7.13}$$

---

[3] Note that in this case $\psi(t)$ is a solution to the equation $\frac{d\psi}{dt} = A - B\sin(\psi)$ with $A > B$. Integrate the equation to get

$$\begin{aligned} t &= \int \frac{d\psi}{A - B\sin\psi} \\ &= \frac{2}{\sqrt{A^2 - B^2}} \arctan \frac{A\tan\frac{\psi}{2} - B}{\sqrt{A^2 - B^2}}. \end{aligned}$$

We made use of the π periodicity to restrict the support of $n_s(\psi)$ to the interval $[-\pi/2, \pi/2]$. To proceed we need to make assumptions about how the intrinsic frequencies of the individual rotors are distributed. We will assume that the ω_k are drawn from a peaked distribution $g(\omega)$, which is symmetric about its mean:

$$g(\bar{\omega} + \omega) = g(\bar{\omega} - \omega), \; \omega \in \mathbb{R}. \tag{7.14}$$

This symmetry makes the integral of the imaginary part of Eq. (7.13) vanish and, at the same time, makes θ equal to the average of the time derivative of θ_k. Namely, we assumed $\theta(t) = \bar{\omega}t$ and from Eq. (7.5) we have $\theta_k = \bar{\omega}t + \psi_k$. So to ensure $\theta(t) = \langle \theta_k \rangle$ we must have

$$\frac{d\langle \psi_k \rangle}{dt} = 0.$$

The symmetry assumption on $g(\omega)$ makes this happen, since by use of Eq. (7.6) we have

$$\begin{aligned}
\left\langle \frac{d\psi_k}{dt} \right\rangle &= \frac{1}{N} \sum_{k=1}^{N} \frac{d\psi_k}{dt} \\
&= \frac{1}{N} \sum_k (\omega_k - \bar{\omega} - \epsilon K \sin \psi_k) \\
&= -\epsilon K \frac{1}{N} \sum_k \sin \psi_k = -\epsilon K \int_{-\pi/2}^{\pi/2} d\psi \, n(\psi) \sin \psi \\
&= \int_{-\pi/2}^{\pi/2} d\psi \, g(\bar{\omega} + \epsilon K \sin \psi) \epsilon K \cos \psi \sin \psi = 0.
\end{aligned}$$

The penultimate equality follows by use of Eq. (7.12) and the last equality from the fact that the assumed symmetry of $g(\omega)$ makes the integrand an odd function and therefore the integral vanishes.

Hence from Eq. (7.13) we have the following equation, which determines K in terms of $\bar{\omega}$ and ϵ:

$$1 = \epsilon \int_{-\pi/2}^{\pi/2} d\psi \, g(\bar{\omega} + \epsilon K \sin \psi) \cos^2 \psi. \tag{7.15}$$

We will first derive a general expression for K by expanding $g(\omega)$ to second order about $\bar{\omega}$, after that we will consider a specific functional form for $g(\omega)$.

First expand

$$g(\bar{\omega} + \epsilon K \sin \psi) \simeq g(\bar{\omega}) + \frac{g''(\bar{\omega})}{2}(\epsilon K \sin \psi)^2. \tag{7.16}$$

The symmetry assumed in Eq. (7.14) implies $g'(\bar{\omega}) = 0$, meaning that $g(\omega)$ has a peak at $\omega = \bar{\omega}$. Next, since

$$\int_{-\pi/2}^{\pi/2} d\psi \, \cos^2 \psi = \frac{\pi}{2} \quad \text{and} \quad \int_{-\pi/2}^{\pi/2} d\psi \, \sin^2 \psi \cos^2 \psi = \frac{\pi}{8},$$

we get

$$\epsilon g(\bar{\omega}) \frac{\pi}{2} - \frac{\epsilon^3 |g''(\bar{\omega})|}{2} K^2 \frac{\pi}{8} = 1, \tag{7.17}$$

where we made the assumption $g''(\bar{\omega}) < 0$ explicit by introducing the factor $-|g''(\bar{\omega})|$. Solving this equation we get

$$\epsilon_c = \frac{2}{\pi g(\bar{\omega})} \qquad (7.18)$$

and furthermore

$$K = \begin{cases} 0 & \text{for } \epsilon < \epsilon_c \\ \sqrt{\frac{8g(\bar{\omega})}{|g''(\bar{\omega})|\epsilon^3}}(\epsilon - \epsilon_c)^{\frac{1}{2}} & \text{for } \epsilon > \epsilon_c \end{cases}. \qquad (7.19)$$

The behaviour of K in this equation is depicted in Fig. 7.2. The dependence of the order parameter K on the strength of the coupling between the different rotors is strikingly similar to the temperature dependence of the order parameter for the Ising model, i.e. the average of the individual spin $\langle S \rangle$, see Fig. 6.4. In both cases a region of zero order parameter is found when the disordering tendencies are too strong. In the case of the rotors, the disorder is encouraged by the spread in the individual frequencies ω_k in Eq. (7.1). Only when the coupling to the other rotors becomes sufficiently strong at $\epsilon = \epsilon_c$ can alignment of the rotors begin to dominate the tendency for incoherent motion caused by the variance in the frequencies ω_k. Large values of the coupling strength ϵ are equivalent in thermodynamics to low temperature, in the sense that lower temperature corresponds to a lesser amount of erratic thermal motion and hence lesser randomness amongst the Ising spins. This explains why Fig. 6.4 is an inverted version of Fig. 7.2.

Let us finish with a concrete example for the distribution of frequencies $g(\omega)$. Consider a peak about some value $\bar{\omega}$ given by the so-called Cauchy–Lorentz distribution.

$$g(\omega) = \frac{\gamma}{\pi[(\omega - \bar{\omega})^2 + \gamma^2]}.$$

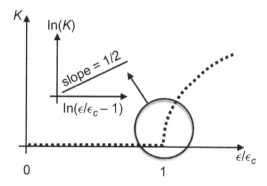

Figure 7.2 Kuramoto order parameter as a function of the relative coupling strength. No synchronous order is possible for couplings weaker than a critical value ϵ_c. At $\epsilon = \epsilon_c$ the coupling between rotors is able to overcome the disordering due to the variation in the ω_k. Gradually more rotors synchronise as ϵ is increased above ϵ_c, making $K > 0$. In the vicinity above ϵ_c, the order parameter K behaves as a power indicated by the log–log plot in the insert magnifying the behaviour just above ϵ_c

We have $g(\bar{\omega}) = \frac{1}{\pi\gamma}$ and differentiate twice to obtain

$$g''(\omega) = -\frac{2\gamma}{\pi}\frac{[(\omega - \bar{\omega})^2 + \gamma^2]^2 - 4(\omega - \bar{\omega})^2[(\omega - \bar{\omega})^2 + \gamma^2]}{[(\omega - \bar{\omega})^2 + \gamma^2]^4},$$

from which we see that $|g''(\bar{\omega})| = \frac{2}{\pi\gamma^3}$. The critical value of ϵ_c given in Eq. (7.18) becomes

$$\epsilon_c = 2\gamma$$

and for $\epsilon > \epsilon_c$ the order parameter in Eq. (7.19) now becomes

$$K^2 = \frac{8}{\pi\gamma}\frac{\pi\gamma^3}{2}\frac{1}{\epsilon^3}(\epsilon - \epsilon_c).$$

Introduce $\Delta = \epsilon - \epsilon_c$. Close to ϵ_c we have $\Delta \ll 1$ and therefore the leading dependence of K on the deviation Δ from the critical value is of the form $K \propto \Delta^\beta$ with $\beta = 1/2$, the same value of exponent as in Eq. (7.19) and the same value of the order parameter exponent we found in the mean-field treatment of the Ising model, see Eq. (6.44).

7.2   Chimera States

Synchronisation can involve very different types of coordinated dynamics. So far we have considered synchronisation, which after a transient period establishes a persistent homogeneous coherence amongst a subset n_s of components, with all components eventually participating in the synchronisation when the coupling ϵ becomes sufficiently strong. This type of synchronisation is of relevance to, for example, the millennium bridge problem mentioned earlier in Sec. 3.5 or, to some extent, to audience applause [308], though the degree of synchronisation of clapping does vary during applause.

In other situations much more elaborate patterns of synchronisation are at play. Obviously an orchestra performing is one such example, another is epileptic seizure. EEG measurements reveal that the healthy brain exhibits elaborate synchronisation structures both spatially and temporally [466]. During seizure, the synchronisation between brain regions changes. EEG recorded during seizure exhibits very intricate synchronisation dynamics, which involves instances of very strong rigid synchronisation and periods of desynchronisation [23, 222, 258].

Mathematically it has been realised that when the interaction between components varies across space, the dynamics may involve partial synchronisation in which different groups of the components go in and out of synchronisation and the level of synchronisation may vary spatially and with time. This behaviour was described by Kuramoto and Battogtokh in 2002 [246] and given the name 'chimera states' by Abrams and Strogatz in 2004 [9]. The term chimera is borrowed from the Greek mythological monster composed as a hybrid of parts from different animals such as a lion, a goat and a snake. The mathematical chimera state is composed of coexistence between groups of components

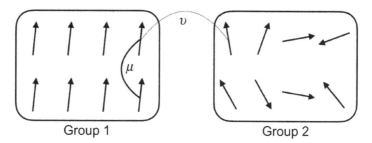

Figure 7.3 Oscillators arranged in two groups. Within a group, all oscillators are coupled together with strength μ. Across the groups, the all-to-all coupling strength is v. Although all oscillators are coupled, see Eq. (7.20), for transparency only one within-group and one across-group coupling are indicated in the figure. The oscillators in group 1 are depicted to move in synch, all pointing essentially in the same direction as they rotate. Group 2 behaves differently, these oscillators point in different directions and their degree of alignment may vary in time. See also Fig. 7.4.

that exhibit different forms of coherent synchronisation, while the dynamics of other groups remain incoherent. During the evolution of the dynamics, the coherence and incoherence may move between groups.

There is by now an extensive theoretical literature on the chimera states, see e.g. [62, 323]. To indicate the mathematical mechanisms, we will discuss a simple configuration of oscillators consisting of two groups, θ_k^1 and θ_k^2 of equal size, $k = 1, 2, \ldots, N$, with a stronger coupling within a group μ than between the groups v. Following [328] (see also [8, 323]) we write

$$\frac{d\theta_k^1}{dt} = \omega - \frac{\mu}{N} \sum_{j=1}^{N} \sin\left(\theta_j^1 - \theta_k^1 + \alpha\right) - \frac{v}{N} \sum_{j=1}^{N} \sin\left(\theta_j^2 - \theta_k^1 + \alpha\right)$$

$$\frac{d\theta_k^2}{dt} = \omega - \frac{\mu}{N} \sum_{j=1}^{N} \sin\left(\theta_j^2 - \theta_k^2 + \alpha\right) - \frac{v}{N} \sum_{j=1}^{N} \sin\left(\theta_j^1 - \theta_k^2 + \alpha\right)$$

(7.20)

and assume $\mu > v > 0$, see Fig. 7.3. One can easily check by simple substitution that the fully synchronised configuration $\theta_k^i = \Omega t$ for all $k = 1, \ldots, N$ and $i = 1, 2$ is a solution if $\Omega = \omega - (\mu + v) \sin \alpha$. The question is whether solutions corresponding to non-homogeneous chimera states also exist.

It has been shown that if the difference in coupling strength $A = \mu - v$ is large or the phase lag parameter α is very different from $\pi/2$, full synchronisation occurs. Detailed analysis, see [8], concludes that for appropriate initial conditions[4] and when α differs less than about 15% from $\pi/2$, persistent chimera states exist for small values of A. Figure 7.4 indicates the situation. Outside the triangular-shaped region bordered by the dotted and dashed curves, no chimera states exist. For these values of $\mu - v$ and α,

---

[4] Starting from an arrangement of all the θ_k^1 and θ_k^2 close to the chimera state, i.e. all θ_k^1 more or less equal and all θ_k^2 spread over the interval $[0, 2\pi]$.

Figure 7.4 Different types of chimera states exists for small values of $A = \mu - \nu$ and for α not too different from $\pi/2$. The difference between the in-group coupling μ and the between-group coupling ν is varied along the y-axis while keeping $\mu + \nu = 1$ [328].

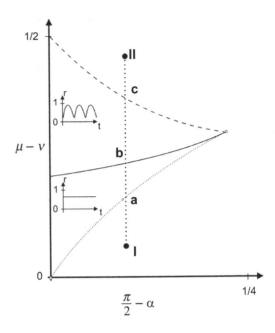

coherent homogeneous synchronisation will be established across the entire system and the two groups will move synchronously. This changes when α is tuned towards $\pi/2$. Let us describe how the behaviour of the system changes as we tune the parameters so as to follow the dotted line in Fig. 7.4 from point I to II. At I, we have full synchrony as we make the in-group coupling constant μ stronger compared to the between-group coupling ν, so A increases, we move up along the line and will cross the finely dotted line at **a**. In the region between point **a** and point **b** chimera states exist in which one group of oscillators, say group 1, synchronise completely while oscillators of group 2 do not.

Two types of chimera synchronisation are possible and differ by the time dependence of the degree of coordinated motion in the group that does not synchronise. In order to quantify this difference we make use of the order parameter introduced in Eq. (7.2), but in this case we only include the oscillators in group 2, the ones that do not fully synchronise. We define

$$r(t) = |\frac{1}{N} \sum_{k=1}^{N} e^{i\theta_k^2}|. \tag{7.21}$$

The time dependence of $r(t)$ is indicated in Fig. 7.4 by the small inserts. In the region between **a** and **b**, $r(t)$ is constant in time, indicating that the mismatch of θ_k^2 across the population in group 2 does not vary with time. As the in-group coupling μ is increased and we move into the region between **b** and **c**, $r(t)$ exhibits a periodic variation with time. The maximum of $r(t)$ increases towards 1 and the period goes to infinity at **c**, i.e. at the boundary indicated by the dashed curve. Above the dashed curve the two groups yet again fully synchronise.

We have several times discussed how emergent collective behaviour is frequently controlled by a competition between ordering and disordering tendencies. Recall e.g. the behaviour at phase transitions, see Sec. 3.1. A similar mechanism is also, though in a subtle way, involved in determining the different regions where chimera states are possible in Fig. 7.4. We notice that for $\alpha = 0$, Eq. (7.20) encourages full synchronisation in the same manner as in the Kuramoto model in Eq. (7.1). This effect is sufficiently strong to prevent the existence of chimera states when α is small, i.e. $\pi/2 - \alpha$ is large. For α close to $\pi/2$, chimera states are possible only if the ordering effect caused by the couplings to the surrounding oscillators is sufficiently weak. Below the dotted curve the in-group and the between-group coupling μ and ν are not too different, making the effect of an oscillator from the other group more or less equal to the effect from oscillators within the same group. That is, the system is too homogeneous to prevent chimera states and full synchronisation is established. On the other hand, in the region above the dotted and dashed curves, the difference in strength between the within-group and the between-group coupling is big enough to let the two groups maintain different dynamical behaviour as a consequence of their different initial starting conditions. The within-group coupling μ will favour synchronisation but the effect of the between-group coupling ν is able to prevent total synchronisation. Above the dashed line the within-group coupling is so strong that each group fully synchronises. Global homogeneous synchronisation between the two groups is then established by the between-group coupling.

Here we have only tried to indicate that synchronisation can exhibit very complicated spatial and temporal structure. Much more detail and applications can be found in the very accessible, mathematically transparent exposition in Secs. 4.3 and 4.5.3 of [62].

The realisation that the emergent dynamics of collections of oscillators can exhibit such a varied mix of coherent and incoherent synchronous and non-synchronous structures in space and time has been met with great enthusiasm and in particular the possible relevance for the dynamics of the neuronal networks in brains is under investigation. Kang et al. [227] studied a more involved two-group system in which the subgroups have internal structure. This arrangement of oscillators produces a great variety of intricate chimera dynamics, which the authors suggest may be related to brain dynamics. In attempts to substantiate the connection to brain dynamics, neuronal oscillators were placed on the well-known network of the 302 neurones of the nematode *C. elegans* and evidence was found of dynamics that resemble chimera states [192, 350]. Probably one of the most daring and optimistic suggestions along this line of research is the work by Bansal and co-workers [43]. It is suggested that time-dependent patterns of synchronisation seen in chimera states may inform us about how large-scale brain networks relate to cognition. A review of the relevance of chimera states and brain dynamics can be found in [274].

We notice that the dynamical equations supporting chimera dynamics are of a general structure which one may find of very broad relevance. It is therefore to be expected that this kind of mix of synchrony and incoherent spatiotemporal dynamics will be found across very different types of systems, an expectation born out e.g. by the observation of chimeras in semiconductor systems [400].

Summary: The Kuramoto model was discussed as a paradigm for how the dynamics of diverse individual components may synchronise when the interaction becomes sufficiently strong.

- The interaction strength has to overcome the spread in the distribution of the intrinsic speed of the components.
- This phase transition is sharp and takes place at a critical value of the interaction strength.
- The partly synchronised chimera states of the inhomogeneous Kuramoto model are relevant to a broad range of real systems.

7.3 Further Reading

General audience:

A non-mathematical introduction aimed at the general reader but scientifically accurate can be found in the book *Sync: The Emerging Science of Spontaneous Order* [429] by Steven Strogatz.

Intermediate level:

The book *Synchronization: From Coupled Systems to Complex Networks* [62] by Stefano Boccaletti, Alexander N. Pisarchik and Charo I. del Genio contains a thorough discussion of different types of synchronisation starting from dynamical systems theory.

The mathematical description of a range of general mechanisms of synchronisation is discussed in the book *Synchronisation: From Simple to Complex* [38] by Alexander Balanov, Natalia Janson, Dmitry Postnov and Olga Sosnovtseva.

Advanced level:

The important engineering problem of *Synchronization Control for Large-Scale Network Systems* [497] is discussed at advanced mathematical level by Yuanqing Wu, Renquan Lu, Hongye Su, Peng Shi and Zheng-Guang Wu.

7.4   Exercises and Projects

Exercise 1

We consider two rotors given by the angles θ_1 and θ_2.

(a) Assume their time evolution is given by

$$\frac{d\theta_1}{dt} = \omega_1 + \epsilon(\theta_2 - \theta_1)$$

$$\frac{d\theta_2}{dt} = \omega_2 + \epsilon(\theta_1 - \theta_2).$$ (7.22)

Write down the equation for the time evolution of the difference $\Delta = \theta_1 - \theta_2$. Solve this equation and show that for any value of $\epsilon > 0$, synchronisation will happen as $t \to \infty$ with a constant so-called phase lag where $\theta_1 = \theta_2 - (\omega_2 - \omega_1)/(2\epsilon)$.

(b) Now consider the following somewhat unusual time evolution:

$$\frac{d\theta_1}{dt} = \omega\theta_1 + \epsilon(\theta_2 - \theta_1)$$

$$\frac{d\theta_2}{dt} = \omega\theta_2 + \epsilon(\theta_1 - \theta_2).$$ (7.23)

Show that for $\epsilon > \omega/2$ synchronisation will happens, whereas for $\epsilon < \omega/2$ no synchronisation happens.

Exercise 2

Go through the details of the calculation leading to Eq. (7.19).

Exercise 3

Use Eq. (7.7) and Fig. 7.1 to establish the details of how the separation of the rotors into the two classes in Eq. (7.10) works.

Project 1 – Chain of rotors

Consider an infinite linear chain of rotors each described by a phase variable θ_n, with $n \in \mathbb{Z}$. The rotors are coupled to their neighbours and their equation of motion is given by

$$\frac{d\theta_n}{dt} = \Omega_n + \gamma(\theta_{n+1} - \theta_n) + \gamma(\theta_{n-1} - \theta_n).$$

Make use of the continuum approximation given by

$$\partial_t\theta(x, t) = \Omega(x) + \gamma\partial_x^2\theta(x, t).$$

Assume that the initial configuration of the chain is given by $\theta(x, 0) = \phi(x)$.

(a) Fourier transform in space and solve the continuum equation for $\hat{\theta}(k, t)$.
(b) Estimate the long-time limit of $\Delta = \int_{-\infty}^{\infty} dx |\partial_x \theta(x, t)|^2$.
(c) How is synchronisation related to the behaviour of Δ?
(d) Determine the small-k behaviour needed of the Fourier transform $\hat{\Omega}(k)$ of $\Omega(x)$ to ensure Δ remains finite.

8 Network Theory

Synopsis: Emergent phenomena require some interdependence between components. Networks of nodes and connecting links are therefore a very natural and powerful language for the analysis and characterisation of complex systems.

The mathematics of networks has a long history, dating back to 1736 when Euler invented network theory, usually called graph theory in the mathematical literature. Euler showed that it was impossible to visit all the parts of the city of Königsberg by crossing each of the city's seven bridges exactly once. He demonstrated the power of extracting the essential topological structure and applying mathematical analysis at the abstract level. This kind of simplification is particularly important when dealing with complex many-component systems.

Unfortunately, the science and mathematics literature on networks uses different terms for the same objects. Sometimes networks are called graphs, sometimes they are considered to consist of vertices and edges, sometimes of nodes and links. There is perhaps a tendency to use the term 'network' in relation to applications and the term 'graph' when focusing on mathematics. But this is far from always the case and the difference in terminology has much to do with taste, style and the background of the authors. Physicists often prefer the term 'network', whereas graph theory is a long-established discipline in mathematics. This variation in terminology is of course not at all uncommon in the scientific literature and not much of a problem, but it is convenient to keep this in mind when, for instance, one performs literature searches.

We will call the components nodes and links and will often think of the nodes as representing some components of a complex system and the links as representing relationships between these components. This is the reason we prefer to talk about links rather than the often used term 'edges'. Namely, the links link, through some relationship, the nodes together.

However, how the network relates to a concrete complex system will not be our concern in this chapter. Here we will simply present the basic mathematical language and concepts used when describing the structure of, or processes on, networks. The hope is that reading this section will prepare the reader for the existing literature. The last 20 years have seen a dramatic increase in the number of publications on network, or graph, theory, both in terms of research papers, textbooks and popular books. Simultaneously, a number of new scientific journals have been created to accommodate the explosive production of research. New books and journal papers appear continuously, so a web search is needed every now and then to keep up. We will just mention a few books here, which are very well written and enjoyable to read. Trudeau's *Introduction to Graph Theory* [452] does not assume any mathematical background and enthusiastically points out that although pure mathematics may sound like a remote intellectual activity, it has for millennia shaped our world. Biggs's *Algebraic Graph Theory* [58] is also introductory, but does assume some familiarity with mathematics. Bollobás's *Modern Graph Theory* [63] is a very comprehensive mathematics book with an eye on applications. Newman's *Networks: An Introduction* [312] is an excellent survey of the developments since around 2000 when an explosion in the interest in networks occurred. Dynamics and synchronisation on networks are transparently discussed in *Synchronization: From Coupled Systems to Complex Networks* [62] by Boccaletti, Pisarchik and del Genio.

In this chapter we will introduce network theory as a toolbox for the complexity scientist who needs to handle emergent structure and phenomena in fields such as ecology, economics, sociology, linguistics or neuroscience, just to mention a few. We will discuss examples of how structures of networks can be mathematically characterised and some examples of dynamics *of* networks, concerned with the evolution in time of the structure of the network. We will also look at dynamics *on* networks, relating to how information travels across a network.

8.1   Basic Concepts

To characterise the structure of a network, one first looks at how the nodes and the links are put together. A first step is to look at the geometry or topology of the network. When we deal with ordinary three-dimensional objects, we describe the object by use of concepts from geometry such as lines, angles, circles, triangles and so on. To characterise the shape and topology of networks, we have to describe how different nodes are related to each other in terms of the links between them. We want to have ways to quantify how central, or important, a node is. Of course, depending on exactly what the node represents, the importance of a node to the overall functioning of the network may not be captured entirely just by describing the node's position in the network. Not everything of the same shape has the same function. The sun is circular and so is the steering wheel of a car. Nonetheless, if the links between the nodes are expressions of some kind of influence between the nodes, think for example of a network of friends or a network of airports, then the significance of the node to the overall functioning of the network is likely to be related to how the node is placed in the network.

We can characterise a node's relative location in many different ways, each emphasising different aspects of what might make a node important. We can look at the number of links attached to the node, or how 'important' its neighbours are. Or we can determine how easy it is to get to our node starting from some other arbitrary node, or how likely it is that we will pass this specific node when we move via the links of the network from one arbitrary node to another on the network. Each of these measures of importance are called centralities and we will now introduce them in more detail.

8.2 Measures of the Importance of Nodes

At the most coarse-grained level we may characterise a network by its number of nodes N, the total number of links L and use these two numbers to give us a measure, the connectance C, of how dense the network is. The connectance is defined in the following way. For a network of N nodes there are $\binom{N}{2}$ pairs of nodes, which each may be linked together. We define the connectance as the ratio between the actual number of links and the maximal possible number of links, namely

$$C = \frac{L}{\binom{N}{2}} = \frac{2L}{N(N-1)}.$$ (8.1)

The number of nodes, links and the connectance are global overall properties of the network, in the following subsections we will look at measures which indicate various aspects of the importance of individual nodes.

8.2.1 Degree Centrality

The **degree**, or degree **centrality**, of a node is equal to the number of links connected to that node. In Fig. 8.1 the nodes labelled α and β each have degree 4, so in this respect they are of equal centrality to the network. The probability that a randomly picked node has a certain degree k is called the degree distribution $P_{\deg}(k)$ and can, for a concrete realisation of a network, be obtained by making the histogram of the number of nodes of a certain degree. This is spelled out for the network in Fig. 8.1 in Table 8.1.

To be able to analyse mathematically the properties of networks, we need a convenient way of representing how nodes are connected. This is done by introducing the **adjacency matrix A**. First we consider networks where the links all count with equal strength and have no direction, meaning a link from node α to β is the same as a link from node β to α.

Figure 8.1 A very simple example of a small network.

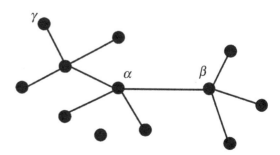

Table 8.1 Imagine we, with equal probability, select one of the 12 nodes in Fig. 8.1. The probability, $P_{\deg}(k)$, that the selected node is of degree k is given by the right column.

Degree k	# Nodes with degree k	$P_{\deg}(k)$
0	1	1/12
1	8	2/3
2	0	0
3	0	0
4	3	1/4

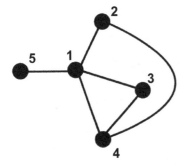

Figure 8.2 Small network with an arbitrary enumeration of the nodes.

In this case the elements a_{ij} of the adjacency matrix $\mathbf{A}$ assume the values $a_{ij} = 0$ for no link and $a_{ij} = 1$ when a link connects i to j. Since we assume the links do not carry a direction, $\mathbf{A}$ is a symmetric matrix, hence $a_{ij} = a_{ji}$. To illustrate how this works, take a look at Fig. 8.2.

We have enumerated the nodes from 1 to 5, which node is given which number does not matter, but we need the labels to be able to represent the network in terms of the adjacency matrix, which for this specific enumeration is given by

$$A = \left\{ \begin{array}{ccccc} 0 & 1 & 1 & 1 & 1 \\ 1 & 0 & 0 & 1 & 0 \\ 1 & 0 & 0 & 1 & 0 \\ 1 & 1 & 1 & 0 & 0 \\ 1 & 0 & 0 & 0 & 0 \end{array} \right\}. \tag{8.2}$$

The adjacency matrix can, in multiple ways, be used to understand the structure of the network. The degree k_i of a node i in a network of N nodes is, for instance, given by

$$k_i = \sum_{j=1}^{N} a_{ij}. \tag{8.3}$$

And the degree distribution $P_{\deg}(k)$, like the one given in Table 8.1, can be obtained from the adjacency matrix in the following way. First we note

$$P_{\deg}(k) = \frac{\sum_{i=1}^{N} \delta_{\mathrm{K}}[k_i, k]}{N}. \tag{8.4}$$

We have introduced a simple function, the Kronecker delta, which allows us to count how many times the degrees k_i are found to be equal to k. The way this works is by defining

$$\delta_K[x,y] = \begin{cases} 1 & \text{if } x = y \\ 0 & \text{if } x \neq y \end{cases}. \tag{8.5}$$

By use of this function we notice that every time we, in the sum over i in Eq. (8.4), find a degree with $k_i, = k$, the term in the sum is equal to one, otherwise the term is zero. So the numerator simply adds up the number of nodes with degree k and the denominator normalises by the total number of nodes. To obtain an expression for the degree distribution in terms of the adjacency matrix, we just needed to substitute the expression for k_i in Eq. (8.3) into Eq. (8.4) and obtain the slightly clumsy-looking expression

$$P_{\text{deg}}(k) = \frac{\sum_{i=1}^{N} \delta_K[\sum_{j=1}^{N} a_{ij}, k]}{N}. \tag{8.6}$$

This expression may look a bit heavy but the important point is that it reduces the computation of the degree distribution for a given network to manipulations on the adjacency matrix. The latter could, for example, be obtained from data indicating relations between the nodes, or from a model of some kind of processes performed by the nodes.

It is natural to assume that the importance of a node in a network will in some way relate to its degree and a network with a very broad degree distribution will correspond to a very heterogeneous network where the effect of the individual nodes differs greatly. Of course, how exactly the degree of a node is connected to the node's importance will depend on what the network represents. But in any case a node of large degree is in direct contact with a larger part of the network than is a low-degree node, meaning that even from this simple geometrical or structural observation we may expect that different nodes differ significantly in their roles. Say the network represents how a social community is linked through acquaintanceships, we will then expect that high-degree nodes represent influential people. In Sec. 8.5.1 below we will describe different mathematical models that lead to different degree distributions and discuss how they relate to observed networks. But to have a reference point, let us here return to the degree distribution, first considered in Sec. 3.2, of a network where links between pairs of nodes are allocated at random with a fixed probability, this is called a Gilbert[1] graph [157]. We consider a set of N nodes and to each of the $N(N-1)$ pairs of nodes we allocate a link with probability p and no link with probability $1 - p$, or expressed in terms of the adjacency matrix for the network we have $a_{ii} = 0$ and for the non-diagonal elements $a_{ij} = a_{ji}$ we have

$$\begin{aligned} \text{Prob}\{a_{ij} = 1\} &= p \\ \text{Prob}\{a_{ij} = 0\} &= 1 - p, \end{aligned} \tag{8.7}$$

---

[1] In the literature this network is often called an Erdös–Rényi network, although that network is slightly different. Namely, it is defined as a network of N nodes with L links placed at random [130].

for $i > j$. We can immediately realise that the degree distribution for this network is a binomial distribution, since a node will have degree k with the probability that we pick k nodes among the possible $N - 1$ nodes which can be paired with a given node and allocate a link to these k nodes. This gives us the following degree distribution for the Gilbert network:

$$p(k) = \binom{N-1}{k} p^k (1-p)^{N-1-k} \simeq \frac{(Np)^k}{k!} \exp(-Np), \tag{8.8}$$

where the approximate expression to the right is accurate in the limit of N and fixed Np, i.e. fixed average degree. This is of course just the usual Poisson limit of the binomial distribution.

In the following sections we will look at some of the other centrality measures used to describe the structure of a network and to distinguish the role played, and contribution made, by the individual nodes.

Box 8.1 Matrices and eigenvectors

For a good introduction to linear algebra see e.g. [287]. We recall that when we multiply a vector

$$v = \begin{pmatrix} v_1 \\ \vdots \\ v_N \end{pmatrix} \tag{8.9}$$

by a matrix

$$M = \begin{pmatrix} m_{11} & m_{12} & \cdot & \cdot & \cdot & m_{1N} \\ m_{21} & m_{12} & \cdot & \cdot & \cdot & m_{2N} \\ \cdot & \cdot & \cdot & \cdot & & \cdot \\ \cdot & \cdot & & \cdot & \cdot & \cdot \\ \cdot & \cdot & & & \cdot & \cdot \\ m_{N1} & m_{12} & \cdot & \cdot & \cdot & m_{NN} \end{pmatrix} \tag{8.10}$$

we obtain a new vector given by

$$Mv = \begin{pmatrix} \sum_{j=1}^{N} m_{1,j} v_j \\ \vdots \\ \sum_{j=1}^{N} m_{N,j} v_j \end{pmatrix}. \tag{8.11}$$

If the multiplication of v by the matrix M amounts to multiplying v by a number λ, i.e.

$$Mv = \lambda v, \tag{8.12}$$

we say that v is an eigenvector of the matrix M with eigenvalue λ. We can determine the possible eigenvalues by the following observation:

$$Mv = \lambda v$$
$$\Downarrow$$
$$Mv - \lambda v = 0 \tag{8.13}$$
$$\Downarrow$$
$$(M - \lambda I)v = 0.$$

In the last equation I denotes the identity matrix, with 1 on all the diagonal elements and 0 everywhere else. We are looking for eigenvectors $v \neq \mathbf{0}$. If the matrix $M - \lambda I$ on the left-hand side of the last equation can be inverted, we can multiply by its inverse $(M - \lambda I)^{-1}$ on both sides of the equation and we will find $v = \mathbf{0}$ so to ensure $v \neq \mathbf{0}$, we need that $M - \lambda I$ cannot be inverted. This is the case if

$$\det[M - \lambda I] = 0. \tag{8.14}$$

By use of the definition of the determinant, see e.g. [287], we notice that Eq. (8.14) is a polynomial of degree N in λ. We find the N roots and have N eigenvectors which fulfil

$$Mv^q = \lambda_q v^q, \tag{8.15}$$

for $q = 1, \ldots, N$. The vectors v^q corresponding to the eigenvalues λ_q are determined by solving Eq. (8.15) for each of the N vectors v^q in terms of λ_q and the $N \times N$ coefficients m_{ij}.

It is known from linear algebra [287] that the set of eigenvectors can be used as a basis, by which we mean that any N-dimensional vector $\mathbf{a}$ can by written as a sum over the vectors v^q, i.e.

$$a = \sum_{q=1}^{N} c_q v^q, \tag{8.16}$$

for one unique set of coefficients c_q.

For a symmetric matrix, i.e. M for which the matrix elements are symmetric across the diagonal, namely $m_{ij} = m_{ji}$, it is possible to use the eigenvectors to transform M into an equivalent diagonal matrix D. The diagonal matrix has all elements away from the diagonal equal to zero, i.e. $d_{ij} = 0$ if $i \neq j$.

The diagonal matrix D corresponding to M is related to M in the following way:

$$M = UDU^{-1}, \tag{8.17}$$

where the columns of the matrix U consist of the coordinates of the eigenvectors v^q:

$$U = \begin{pmatrix} v_1^1 & v_1^2 & . & . & . & v_1^N \\ v_2^1 & v_2^2 & . & . & . & v_2^N \\ . & . & . & . & . & . \\ . & . & . & . & . & . \\ v_n^1 & v_N^2 & . & . & . & v_N^N \end{pmatrix} \tag{8.18}$$

and the diagonal elements of D equal the eigenvalues:

$$D = \begin{pmatrix} \lambda_1 & 0 & . & . & . & 0 \\ 0 & \lambda_2 & . & . & . & 0 \\ . & . & . & . & . & . \\ . & . & . & . & . & . \\ 0 & 0 & . & . & . & \lambda_N \end{pmatrix}. \tag{8.19}$$

8.2.2 Eigenvector Centrality

The degree centrality simply counts the number of nodes linked to a given node. Each of the neighbours is counted with the same weight but there can of course be situations where the effect of one node is very different from another. Think of a network of scientific colleagues, where having a relationship to a well-connected scientist, say the President of the Royal Society, may help you to promote your ideas more than do relationships with less influential scientists. A natural way to try to include such effects is to improve on the degree centrality by including a measure of how influential each neighbour is. We want a measure that makes a node more 'central' if it is connected to other nodes of high centrality.

This means that we are looking for a measure of the importance of nodes that, in a self-consistent manner, incorporates the importance of other nodes. This will make the measure more collective and in itself an emergent property of the network.

The so-called eigenvector centrality [66, 67] is an example of such a measure and since it is an example of a collective emergent measure that can be discussed in some detail using analytical mathematical arguments, we will here elaborate on this specific measure more extensively than we do on other measures of node importance.

To define eigenvector centrality we need to make use of the adjacency matrix and some linear algebra, see Box. 8.1. Recall that the adjacency matrix A for a network consisting of N nodes is defined as the $N \times N$ matrix of elements a_{ij} for which $a_{ij} = 1 = a_{ji}$ if node number i is connected to node number j. Since A is a symmetric matrix, linear algebra tells us that there exist a set of eigenvectors v_q with $q = 1, \dots, N$ such that the result of multiplying v_q by the matrix A is to multiply v_q by a real number λ_q, i.e.

$$A v_q = \lambda_q v_q. \tag{8.20}$$

We also know from linear algebra that any N-dimensional vector x can be written as a linear combination of the eigenvectors. Accordingly we know a set of number c_q with $q = 1, \ldots, N$ exists such that

$$x = c_1 v_1 + \cdots + c_N v_N = \sum_{q=1}^{N} c_q v_q. \tag{8.21}$$

Can we make use of this to find weights $x = (x_1, \ldots, x_N)$ that can be used as our self-consistent importance measure? Yes, and it is done in the following way. At first we allocate equal weight to all nodes $x(0) = (x_1, \ldots, x_N) = (1, \ldots, 1)$. We then imagine we find the set of coefficients $(c_1, c_2, \ldots, c_N)$ in Eq. (8.21). And then we keep multiplying the vector x by our adjacency matrix A. After n multiplications by A we have

$$
\begin{aligned}
x(n) &= A^n x(0) \\
&= A^n \sum_{q=1}^{N} c_q v_q \\
&= \sum_{q=1}^{N} c_q A^n v_q \\
&= \sum_{q=1}^{N} c_q \lambda_q^n v_q.
\end{aligned} \tag{8.22}
$$

Now comes a manipulation that allows us to establish the measure we are looking for by letting n go to infinity. Let λ_{max} denote the largest of the eigenvalues. We pull that eigenvalue out in front of the summation and obtain

$$x(n) = \lambda_{max}^n \sum_{q=1}^{N} c_q \left[\frac{\lambda_q}{\lambda_{max}}\right]^n v_q. \tag{8.23}$$

All the fractions $\lambda_q / \lambda_{max} < 1$ except when q denotes the largest eigenvalue for which the fraction is identical to 1. So

$$\left[\frac{\lambda_q}{\lambda_{max}}\right]^n \to 0 \text{ as } n \to \infty, \tag{8.24}$$

and we have, as n is increased, that the only term in the sum that does not vanish is the term corresponding to the eigenvector with the largest eigenvalue and therefore

$$x(n) \approx \lambda_{max}^n c_{max} v_{max}. \tag{8.25}$$

If we multiply one more time by A we get

$$
\begin{aligned}
x(n+1) &\approx A\lambda_{max}^{n} c_{max} v_{max} \\
&= \lambda_{max}^{n} c_{max} A v_{max} \\
&= \lambda_{max}^{n} c_{max} \lambda_{max} v_{max} \\
&= \lambda_{max} c_{max} \lambda_{max}^{n} v_{max} \\
&= \lambda_{max} x(n)
\end{aligned}
\tag{8.26}
$$

or since $x(n+1) = Ax(n)$ we have, at least approximately, for large n

$$
Ax(n) = \lambda_{max} x(n).
\tag{8.27}
$$

Next we move the eigenvalue to the other side of the equation and write out the multiplication between the matrix A and the vector $x(n)$. Finally we substitute the index n by the superscript ∞ to indicate that we are looking at the limit of large n, and end up with the relation

$$
x_i^{\infty} = \frac{1}{\lambda_{max}} \sum_{j=1}^{N} a_{ij} x_j^{\infty}
$$

$$
\Downarrow
$$

$$
x^{\infty} = \frac{1}{\lambda_{max}} A x^{\infty}.
\tag{8.28}
$$

We notice that, independent of the specific choice of the initial vector $x(0)$, after multiplying by the adjacency matrix A many times we end up with a vector x^{∞} of weights for each node, which according to Eq. (8.28) are given as a combination of how many neighbours a node i has, namely how many non-zero a_{ij} terms there are on the right-hand side of Eq. (8.28), and the weights x_j^{∞} of these neighbours.

From the discussion above we see that the eigenvector centralities of the individual nodes x_i^{∞} can be determined in two ways: either by solving the eigenvalue problem in Eq. (8.28), which is possible for small networks but for networks with many nodes, the $n \times n$ adjacency matrix A makes it difficult to compute the eigenvalues and eigenvectors directly, in which case we can make use of the iterative procedure indicated by the step from Eq. (8.23) to Eq. (8.25). One simply starts with an arbitrary set of weights $x(0)$ and keeps multiplying $x(0)$ by A until $x(n+1) = x(n)$ with a desired accuracy. Clearly, if x^{∞} is a solution to Eq. (8.28), then x^{∞} multiplied by a constant is also a solution, so we can normalise the iterated solution found in any way that is convenient. One will mainly be interested in the relative importance, i.e. is node number i more central than node number j? An absolute measure of centrality is not needed.

Applications of eigenvector centrality are widespread. Google's PageRank, which is used to order the pages in the World Wide Web to enable effective searching, is an elaboration on eigenvector centrality. Eigenvector centrality is frequently used to characterise the structure of networks. For instance, Wink [496] used consecutive time windows of fMRI scans to derive a connectivity matrix in the form of correlations between different voxels of the fMRI scan. This connectivity matrix is then used as an adjacency matrix and by use of the procedure in Eq. (8.26), the eigenvalue centralities are computed and maps across the brain can be established and their time evolution studied. The behaviour was reported to depend on gender and age.

Despite these successful applications of eigenvector centrality, it is worth being aware that recently it has been realised that the measure can sometimes allocate questionable centrality strengths to some nodes. Sharkey [396] found that nodes in identical or similar subgraphs may end up with very different values for their eigenvalue centrality.

8.2.3 Closeness Centrality

The degree of a node is a local property in the sense that it only tells us something about how a specific node is connected to its neighbours. If we are interested in flow, or transport, of some sort across a connected component of a network, we will like to know the distance between nodes in terms of the (smallest) number of links one needs to traverse in order to go from one node to another. In Fig. 8.1, nodes α and β are separated by one step, while nodes β and γ are separated by a distance of three links. All very straightforward. The importance of a node's position in a network can be characterised by various measures. The closeness centrality C_i of a node i is given in terms of the inverse of the average of the shortest distance l_i of node i to all other $N-1$ nodes in that network component:

$$C_i = \frac{1}{l_i} = \frac{1}{\frac{1}{N} \sum_{j=1}^{N} d_{ij}}, \tag{8.29}$$

where d_{ij} denotes the shortest distance form node i to another node j. Note that we include the distance $d_{ii} = 0$ just for simplicity of notation when writing down the expression in Eq. (8.29). If the node i is located a short distance from all the other nodes, C_i will be large.

Goldstein and Vitevitch used closeness centrality of words to analyse the speed with which we are able to make contact between different concepts [165].

8.2.4 Betweenness Centrality

Besides being well connected to other important nodes, or being close to other nodes, a node can also occupy a special position in the network by being placed such that many routes on the network pass through the node. Think of the network of international airports. It is clear that when going from one part of the world to another, it is more likely that one passes though London, New York or Istanbul than through the airport in Nuuk, Greenland. In fact, according to Song and Yeo [419], it is most likely if starting from one arbitrary city going to another arbitrary city that one passes through Istanbul.

The betweenness centrality of a node is introduced to measure this property. The betweenness of a certain node, say α (see Fig. 8.1), is the number of shortest paths from somewhere, for instance γ, to somewhere else, say β passing through node α. In Fig. 8.1

the betweenness of node α is high whereas the betweenness of node β is much lower, since only paths leading to the three nodes to the right of β have to pass through β. When one defines betweenness mathematically, one needs to decide how to weight a path passing though a node if it is one of multiple shortest paths between two other nodes. Or should one include paths starting and ending on a node in the betweenness of that node? All this does not alter the essential idea behind the betweenness measure [312].

8.2.5 How Well Does it Work?

The betweenness measure has found broad use. For example, Kirkeley and collaborators demonstrated the measure's relevance to the analysis of congestion patterns in cities [236] and Hafner-Burton and Montgomery [177] have used the three centrality measures, degree, betweenness and closeness, to illuminate the distribution of power of individuals, organisations or states.

In particular, when analysing networks in biological cells, how well the importance of a specific node is represented by the theoretic network measures does depend on context [208]. For gene regulatory networks, it has been found that the entire plethora of centrality measures – degree, betweenness, closeness, PageRank and eigenvalue – are informative about a gene's regulatory importance, see e.g. [206]. When analysing microbial metabolic networks, it has been noticed [234] that the degree and betweenness centrality are less useful to identify the importance of nodes (metabolites). The importance of a node in the reaction graph appears to be captured more reliably by centrality measures (bridging centrality and cascade numbers), which more directly focus on the flow of information.

8.3   Community Detection

So far we have characterised the structure of a network by allocating different measures of importance to the individual nodes. Often the heterogeneous way links are organised between nodes will suggest that the nodes belong to different clusters. Here we will explain some of the ideas behind the identification of communities. The aim is to divide the nodes into subsets, such that the nodes of a given subset are more related to each other than they are to nodes in the other subsets.

In Fig. 8.3 the group of nodes to the left of node α form a close-knit structure with many links starting from and ending on nodes within this cluster and only a single link, the one between α and β, reaching out of the cluster. In the same way the nodes to the right of node β form a well-connected cluster. Such a collection of well-connected nodes is called a community and identifying and characterising these communities of nodes can help to understand the overall structure of the network and may help to understand the functioning of the network or the processes taking place on the network.

Although the relation between community structure and network function is likely, at least to some extent, to depend on the nature of the relation between nodes – represented

Figure 8.3 Network with two communities of nodes, one to the left of node α and another to the right of node β.

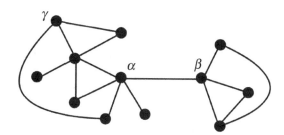

by the links – and also on what the nodes represent and their dynamics, seen from the perspective of complexity science, network analysis is of particular interest in so far as it is able to summarise the relationships between different components and how emergent structure and dynamics are generated from the web of these relations. The identification of the nodes and the nature of their relationship for a given complex system, say the brain, will be discussed in Sec. 9.3, after we have considered how information theory can help to derive networks from time series. In the present chapter we will assume that a network is given in terms of a collection of nodes and the links between them. Here we will look at the structure of clusters, or communities, from a purely geometrical or topological point of view.

There are many ways one may identify the community structure of a network [312], and the literature is huge and expanding. Fortunately, software packages are made available. Fortunato and Hric [145] includes a list of approaches and links to software and a multidisciplinary review was recently published by Javed et al. [210]. Details are best studied in the specialised literature, such as the recently published books *Community Search over Big Graphs* [203], *From Security to Community Detection in Social Networking Platforms* [229] or *Advances in Network Clustering and Blockmodeling* [111].

But even without specific systems in mind, we can anticipate that the identification of communities may depend on how we treat the links between the nodes. One approach consists of partitioning the nodes of the network into groups in such a way that one optimises the number of links *within* groups and minimises the number of links *between* different groups. This is the procedure our eyes make use of when we look at Fig. 8.3 and identify the two communities.

Another approach is to place dynamical processes on the network. This could be diffusion [103], where random walkers perform walks from node to node along the existing links; we look at this approach in Secs. 8.4 and 8.5.2. This is directly of relevance to spreading of diseases in case the network represents physical contact relations between people, or to rumour spreading if the links represent who can communicate with whom. Or one can study synchronisation of dynamic processes taking place on the nodes and coupled together through the links [26]; this, of course, is related to our study in Chap. 7. We discuss synchronisation on networks in Sec. 8.5.3.

In the following we will first discuss the geometric partitioning approach to clustering. Before we can develop a mathematical procedure able to partition a network into

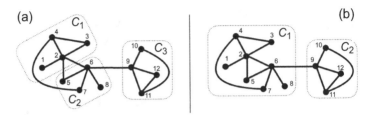

Figure 8.4 Two attempts at partitioning the network into clusters. Clearly the division into the two clusters indicated in panel (b) appears to be most natural.

clusters, we need to have a mathematical measure of clustering. Such a measure should take a large value when a certain division of the nodes of the network into a set of clusters captures the geometry of the network and a small value if our grouping of nodes does not reflect the way the links between the nodes are distributed.

Motivated by network structures like the one depicted in Fig. 8.4, we want a measure which assumes a high value on the partitioning sketched to the right in the figure and a low value when computed on the partitioning sketched to the left in the figure. The measure of modularity should assume a high value if an imposed partitioning sorts nodes into groups which have a high number of links within the group and fewer between the groups. But we also want our measure to be able to distinguish clustering which signifies some relations between the nodes in a cluster beyond accidental random clustering. Let us assume we have divided the network into a set of clusters $C = \{C_1, C_2, \ldots, C_K\}$, like in Fig. 8.4, where $C = \{C_1, C_2, C_3\}$ in the left panel.

We want to compare the actual number of links, L_{C_q}, within a proposed cluster C_q to the expected number of links, $\langle L_{C_q} \rangle$, inside C_q when links are placed at random with uniform probability as in a Gilbert network.

We can compute L_{C_q} as[2]

$$L_{C_q} = \frac{1}{2} \sum_{i \in C_q} \sum_{j \in C_q} a_{ij}. \qquad (8.30)$$

Next we compute the expected number of links $\langle L_{C_q} \rangle$. We denote by $|C_q|$ the number of nodes in the cluster C_q. For each of these nodes $i = 1, 2, \ldots, |C_q|$, we look at the k_i links emerging from the node. In case of uniform random connection of links, the probability that a link starting from node $i \in C_q$ connects to another node $j \in C_q$ can be estimated in the following way.

We denote the total number of links in the entire network by L. The link we are focusing on starting from node i can, with equal probability, attach to any of the $L - k_i$

---

[2] In this equation, the sum over i and j runs through the nodes belonging to cluster C_q. For example, for cluster C_2 in the left partitioning we would have i and j each running through the nodes 5, 6, 7 and 8.

links in the network not emerging from node i. We are interested in the link from i connecting within the cluster C_q to a node j of degree k_j; the probability for this is

$$\frac{k_j}{L - k_i} \simeq \frac{k_j}{L}.$$

Since k_i links propagate from node i, the average number of links connecting i to a node $j \in C_q$ will be $k_i k_j / L$. To obtain the total average number of links internal to C_q, we sum over i and since we will visit each node twice, both as originator and receiver of links, we need to divide by a factor of 2 to avoid double counting. This leads to

$$\langle L_{C_q} \rangle = \sum_{i \in C_q} \sum_{j \in C_q} \frac{1}{2} \frac{k_i k_j}{L}. \tag{8.31}$$

The measure of modularity introduced by Newman [312] consists of adding across all the clusters of a given partitioning the discrepancy between the actual number L_{C_q} and the expected number $\langle L_{C_q} \rangle$ of links within C_q, and normalising by an extra factor $\frac{1}{2L}$. Recall that the total degree $\sum_i k_i = 2L$. This leads us to the following definition of modularity $Q[C]$ of a given partitioning C:

$$Q[C] = \frac{1}{2L} \sum_{q=1}^{K} \sum_{i \in C_q} \sum_{j \in C_q} \left(a_{ij} - \frac{k_i k_j}{2L} \right). \tag{8.32}$$

We need to consider the meaning of $i = j$ in these sums. Including $i = j$ corresponds to including links starting from and ending at the same node, so-called self-loops. In simple networks we do not include such loops, but there can be situations where a node connecting to itself makes sense. For example, Park et al. [332] looked at the network of people editing Wikipedia. Repeated work by one person on the same Wikipedia page is represented as a self-loop in the network relating people who have all edited a given page. The number of self-loops is represented in the adjacency matrix by the diagonal terms a_{ii}. The estimate of self-loops appearing at random is given by the probability that a link starting from node i connects to this node, i.e. by $(k_i - 1)/2L \simeq k_i/2L$. Hence, the two cases of existence of self-loops and no self-loops are covered by our discussion of L_{C_q} and $\langle L_{C_q} \rangle$ above.

However, when considering networks where self-loops are excluded, the sum over the $i = j$ terms in the computation of $\langle L_{C_q} \rangle$ should not be included. Fortunately, there is no problem when computing L_{C_q}, since summing over $i = j$ in this case corresponds to including the zero diagonal terms $a_{ii} = 0$. Conveniently, this is not of any concern when we make use of $Q[C]$ in Eq. (8.32) to find the partitioning C which leads to the largest vale of $Q[C]$. The reason for this is that the diagonal terms will contribute the same amount to $Q[C]$ for all possible partitionings, which can be seen by separating out the contribution from the $i = j$ terms to the $Q[C]$ for two different ways of partitioning the graph C and C'. We have

$$Q[C] - Q[C'] = (\text{terms from } i \neq j)$$

$$+ \frac{1}{2L} \sum_{q=1}^{K} \sum_{i \in C_q} \left(a_{ii} - \frac{k_i k_i}{2L} \right) - \frac{1}{2L} \sum_{q=1}^{K'} \sum_{i \in C'_q} \left(a_{ii} - \frac{k_i k_i}{2L} \right). \qquad (8.33)$$

The sums

$$\sum_{q=1}^{K} \sum_{i \in C_q} = \sum_{q=1}^{K'} \sum_{i \in C'_q} = \sum_{i=1}^{N}$$

and therefore the two $i = j$ terms in Eq. (8.33) will cancel.

Before we turn to the practicalities of how one divides a network up into the clusters that correspond to a high modularity, i.e. how to devise an algorithm that identifies the clusters C_1 and C_2 of the right panel in Fig. 8.4 rather than the clusters C_1, C_2 and C_3 of the left panel of that figure, we will rewrite $Q[C]$ in Eq. (8.32) to develop a feeling for which cluster structures will make $Q[C]$ assume high values. For a careful mathematical discussion of the properties of $Q[C]$, see [69]. Here we notice

$$Q[C] = \frac{1}{2L} \sum_{q=1}^{K} \sum_{i \in C_q} \sum_{j \in C_q} \left(a_{ij} - \frac{k_i k_j}{2L} \right)$$

$$= \frac{1}{2L} \sum_{q=1}^{K} \sum_{i \in C_q} \sum_{j \in C_q} a_{ij} - \frac{1}{2L} \sum_{q=1}^{K} \sum_{i \in C_q} \sum_{j \in C_q} \frac{k_i k_j}{2L} \qquad (8.34)$$

$$= \frac{1}{L} \sum_{q=1}^{K} L_{C_q} - \frac{1}{(2L)^2} \sum_{q=1}^{K} \left(\sum_{i \in C_q} k_i \right)^2,$$

where the last equality makes use of Eq. (8.31). Written this way, we see that a large value of the modularity can be achieved if the nodes of the clusters are well connected within the clusters, leading to large values of L_{C_q} terms in the first sum. The second term in the expression above needs to be made small to make $Q[C]$ large, and since $(\sum_{i \in q} k_i)^2$ is the square of the number of all the links connecting to nodes in the cluster C_q, arrangement of the nodes into clusters with small total number of links leads to higher values of the modularity measure $Q[C]$.

Now we have a measure $Q[C]$ of how well a specific partitioning of the network into clusters of nodes represents the modular structure of the network. We are interested in the grouping, i.e. a given realisation of $C = C_1, \ldots, C_K$, that maximises $Q[C]$ and expect that this configuration of clusters will be a useful representation of the network's modular structure.

For large graphs, implementing this procedure by brute force, i.e. going through all possible ways of partitioning the graph, is not computationally feasible, so approximate and optimised algorithms are needed. Luckily, established procedures and software are available, see e.g. [145, 272]. We will describe the efficient and intuitive hierarchical procedure called the **Louvain algorithm** invented by Blondel, Guillaume, Lambiotte and

Lefebvre in 2008 [61]. The algorithm works both for networks with no weight on the links, i.e. networks for which the elements of the adjacency matrix a_{ij} assume the values 0 or 1, and for networks where the links carry a strength, or weight. In the latter case a_{ij} can have values different from 1 between two connected nodes and the actual value of a_{ij} determines how strongly or weakly the nodes i and j are coupled together.

The algorithm consists of two stages: a cluster assignment stage and a network reduction stage in which new nodes are formed by aggregating nodes within the same cluster to form new nodes. The combination of these two stages is called a pass.

(1) During the cluster assignment stage, one keeps visiting the nodes at random and tentatively tests if by assigning a node to the cluster of one of its neighbours, the value of $Q[C]$ can be increased. One does this for each of the neighbours of the node and computes the change in $Q[C]$ each time. The largest change in $Q[C]$ determines which cluster the node is assigned to. Of course it may be that there is no increase in $Q[C]$ at all, in which case one leaves the node in the cluster where it is at present. Or it can occur that the largest change in $Q[C]$ is not unique to one of the neighbours. In that case, one chooses at random between the neighbours leading to the same maximum increase in $Q[C]$. During this stage each node may be tested more than once and may be reassigned to different clusters as these keep changing during the testing and cluster reassignment of the nodes. One continues this process of reassigning cluster membership to nodes until $Q[C]$ doesn't increase any more.

(2) After having assigned a configuration of clusters which corresponds to an increased value of $Q[C]$, one redefines the nodes of the network. The cluster identified in the previous step is now defined as the nodes of the new network. All the links within the clusters are kept as a self-loop in this new network. And links between the nodes of this aggregated network are given a strength by adding up the strengths of the links reaching from one cluster to another. The number of nodes in this new network will be equal to the number of clusters identified during stage (1).

By repeating (1) and (2), the algorithm produces a hierarchy of clustering of the original network. To clarify the details, let us go through how the network in Fig. 8.5 is handled

Figure 8.5 Iteratively grouping nodes into clusters.

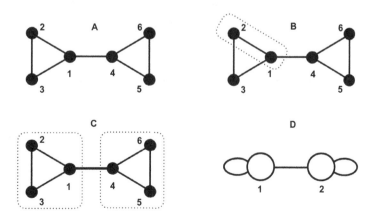

by the Louvain algorithm. To begin with, each node defines one cluster. Let us start by testing node number 1. We can try to combine it with node 2, 3 or 4 to form a new cluster. The changes to $Q[C]$ are as follows (see Exercise 5 at the end of this chapter): $\delta Q_{1\to2} = \delta Q_{1\to3} = 18/196$ and $\delta Q_{1\to4} = 10/196$. So we should assign node 1 to the same cluster as either node 2 or 3, and choose at random between the two. In panel B of Fig. 8.5 we indicate that node 1 is assigned to the cluster of node 2. We continue to compute the changes to $Q[C]$ as we test assignments of the nodes to clusters. As we go along, we need to keep in mind that the sums over the clusters in the definition of $Q[C]$, see Eq. (8.31), keep changing. One reason for the algorithm's efficiency is, however, that when we compute the changes to $Q[C]$, we only need to include terms in the sums for $Q[C]$ involving links connected to the node and cluster being considered.

One may wonder about the structure of the communities constructed by the Louvain algorithm's optimisation of the modularity measure $Q[C]$ in Eq. (8.32). Will the algorithm identify the 'real' communities in the network. In simple cases like the networks depicted in Figs. 8.4 and 8.5, simply looking at the network immediately lets us identify what we intuitively may think of as the well-connected substructures that can play the role of communities, and in such cases the Louvain algorithm will identify these. However, it has been pointed out that for more complicated networks, the communities identified by the Louvain algorithm may not have the properties we would expect. For example, it is possible for nodes placed in the same community to form disconnected subclusters within this community [451].

These subtleties remind us that identifying communities in networks can be an intricate matter.[3] Besides the difficulties of identifying communities solely from the way nodes are linked up, we also need to keep in mind that the functionality and properties of the nodes may be of importance to community organisation. From the perspective of complexity science, we are interested in how components work together to generate the systemic properties. The modularity measure defined in Eq. (8.32) quantifies how much the density of links within a community deviates from what we should expect if the nodes were linked together at random without any particular mechanisms responsible for adding a link between node i and node j.

The modularity in Eq. (8.32) was established by comparing the actual connectivity between node i and j, given by the adjacency matrix element a_{ij}, to the probability that a link exists between i and j in the case of random uniform attachment:

$$P_{ij} = \frac{k_i k_j}{2L}. \tag{8.35}$$

This comparison allows us to conclude that if community structures can be found which make $Q[C]$ significantly different from zero, the links between nodes of the network are likely to originate from mechanisms beyond the purely random. The question is whether this deviation from the purely random helps us to understand the complex system we are investigating. Let us assume the nodes possess some properties which make some

---

[3] We stress that a large and growing specialised literature exists, e.g. [111, 203, 229].

nodes more likely to connect than others. As an example, we will consider people living in different geographical positions and speaking different languages [135]. In a case like this we would not assume that people living far apart or speaking different languages connect with the same probability as people of the same language living near each other. This means that it is unreasonable to compare the actual connectivity a_{ij} to P_{ij} given by Eq. (8.35); we should instead compare the actual connectivity to how likely it is that two nodes are connected given what we know about their properties. Let us focus on distance. We recall that in the simplest case the elements of the adjacency matrix a_{ij} assume the value either 0 or 1, corresponding to either no link or a link between i and j, but a_{ij} can also be taken to be a positive real number, a weight, that indicates a strength of the connection between the two nodes. Along this line of thought we can replace the estimate in Eq. (8.35) by an estimate that depends on the distance of the two nodes and some measure of their respective importance:

$$P_{ij}^{Gr} = m_i m_j f(d_{ij}). \tag{8.36}$$

Here, m_i and m_j are relevant measures of the importance of the two nodes; this could be the total population in the case of a network of cities, or it might be the total flow through a node in the case of a transport network, or it could simply be the degree of the node. The function $f(d_{ij})$ describes how the distance d_{ij} between node i and node j influences the expected strength of the connection between the two nodes. Often the functional form $f(x) = x^{-\alpha}$ is found to capture well the distance dependence, which makes Eq. (8.36) look similar to Newton's law of gravitation between two bodies of masses m_i and m_j. For this reason, the network literature calls Eq. (8.36) the gravity law.

When we replace the expression in Eq. (8.35) by the estimate in Eq. (8.36) into Eq. (8.32) to obtain

$$Q^{Gr}[C] = \frac{1}{2L} \sum_{q=1}^{K} \sum_{i \in C_q} \sum_{j \in C_q} (a_{ij} - m_i m_j f(d_{ij})), \tag{8.37}$$

the new modularity measure will compare the actual clustering to the clustering expected when links are allocated at random while still respecting spatial restrictions.

The authors of [135] analysed a mobile communication network from Belgium. They compared the communities identified by maximising the purely random reference of $Q[C]$ in Eq. (8.32) to the communities that maximise the modularity measure $Q^{Gr}[C]$. The communities obtained from $Q[C]$ are dominated by the spatial location of the nodes and do not relate to the important fact that Belgium is a bilingual country with Flemish and French-speaking subpopulations. In contrast, using $Q^{Gr}[C]$ to adjust the expected clustering for geographical aspects identifies communities that follow the language divide. Given that we already know that Belgium has two language communities, this is not a too surprising discovery, but the fact that the community structure identified by the analysis depends crucially on what we take as our reference for comparing the actual network connectivity is of significance.

The important lesson is that when we use network analysis to identify emergent structures in a complex system, in particular how components may fall into separate groups, we need to keep in mind that whether a link exists between two nodes may depend on multiple factors and the insight we are able to gain from the network analysis increases the more we are able to incorporate from the set of known structural restrictions. These restrictions can, like in the analysis of the mobile phone network in [135], relate to real physical space. The importance of this approach to the analysis of the brain is discussed in [51]. But 'spatial restrictions' may also have their origin in a more abstract space. For example, in the analysis of the transcriptional regulatory network of *E. coli* [238], space refers to the one-dimensional *E. coli* chromosome.

8.4   Spreading on Networks – Giant Cluster

The formalism of network theory is of particular relevance to the analyse of spreading phenomena such as epidemics, fires or rumours, where spreading from one component to another happens through some kind of network linking the parts together. Let us assume we are dealing with a phenomenon like a highly infectious virus and that as soon as one component is contaminated, the contamination will spread to all social neighbours. This means that the size of the outbreaks of contamination will be determined by the size of the connected clusters of the network. To discuss how large a fraction of the network is that will be affected if a random component suffers infection, we will use the mathematics of generator formalism, which finds widespread applicability across mathematics and in particular when dealing with applications of probability theory; see also Box 5.1 and Sec. 5.1.

We will assume that we have access to the degree distribution of the network on which the contamination is spreading. Clearly, it may sometimes be difficult in practice to obtain this information. When a disease is spreading through a population, one of the difficult problems is to determine the relevant network representation of the population. The network relevant to sexually transmitted diseases may resemble the network representing the social friendship relations of the population, whereas an airborne viral infection can spread between people who have no social relation to each other at all, but simply happen to pass close enough for the virus to jump. Although these complications are important and often one of the major obstacles to accurate epidemiological modelling, we will ask more general questions that can be studied on the basis of the degree distribution.

We will combine network theory with probability theory to discuss how the existence of a giant cluster is related to the degree distribution in the case of random networks. By a **giant cluster**, or component, we mean a connected part of the network that contains an appreciable fraction of all the nodes in the network. Moreover, the giant cluster should also increase in size when we increase the size of the network. The giant cluster is, for example, of importance when estimating how serious the impact will be of a disease

Figure 8.6 Sketch of a network. The excess degree of node β is 3, since the link back to node α isn't included in the excess degree.

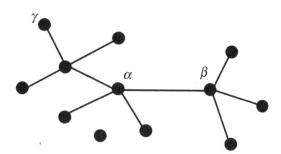

spreading in a population. If we can be sure the disease will remain confined to a cluster of some size $N_0 \ll N$, and N_0 does not increase with population size N, the situation is much more manageable than if we find that the disease will spread to a cluster of size N_0 with $N_0 = \gamma N$, where γ is a significant fraction so N_0 increases as larger populations are considered.

First we need some definitions. Box 5.1 briefly introduces a powerful way of analysing probability distributions for discrete variables. This is the formalism we will use to describe how the existence of a giant cluster is related to the degree distribution of a random network.

Consider an ensemble of networks for which the probability that a node selected with uniform probability (i.e. the same probability for all nodes independent of whatever attributes they may have) has degree k, i.e. k links attached, is given by p_k for $k = 0, 1, 2, \ldots$. This is the degree distribution introduced in Eq. (8.6). The generator for this degree distribution is given by

$$g_0(s) = \sum_{k=0}^{\infty} p_k s^k. \tag{8.38}$$

We next introduce the concept of **excess degree**. Select a node, call it α, of degree $k \geq 1$. Follow one of the links attached to node α in order to reach a neighbour node, call it β. Focus on the links emerging from β *different* from the link that took us from α to β. The number of such links is called the excess degree of β. In Fig. 8.6 the degree of α is 4 and the degree of β is also 4, but its excess degree is 3 since the link to α does not count for the excess degree when we arrive at β from α.

Let q_k be the probability that the neighbour has excess degree k. The generator function for this distribution is

$$g_1(s) = \sum_{k=0}^{\infty} q_k s^k. \tag{8.39}$$

We will now derive a relation between $g_0(s)$ and $g_1(s)$. Recall that we consider random uncorrelated networks for which the degree distributions are given.[4] Assume that the total number of links is L and the total number of nodes is N. The average degree is

---

[4] The procedure for constructing a random network with a given degree distribution is called a **configuration model**, see Sec. 8.10 for details.

then given by $\langle k \rangle = 2L/N$. The expected number of nodes of degree k is Np_k, hence we have kNp_k links sticking out from nodes of degree k. We call these half-links. In total there are $2L$ half-links. Hence the probability that a link we follow lands on a node of degree k is

$$\frac{kNp_k}{2L} = \frac{kp_k}{\langle k \rangle}. \tag{8.40}$$

The probability, q_k, that the node at the end of the link has excess degree k, and therefore total degree $k + 1$, is

$$q_k = \frac{(k+1)p_{k+1}}{\langle k \rangle},$$

and we can conclude that the generator for the excess degree distribution is given by

$$\begin{aligned} g_1(s) &= \sum_{k=1}^{\infty} q_k s^k = \frac{1}{\langle k \rangle} \sum_{k=0}^{\infty} (k+1)p_{k+1} s^k \\ &= \frac{1}{\langle k \rangle} \sum_{k=1}^{\infty} k p_k s^{k-1} = \frac{1}{\langle k \rangle} \frac{s}{ds} \sum_{k=0}^{\infty} p_k s^k \\ &= \frac{1}{\langle k \rangle} \frac{d}{ds} g_0(s), \end{aligned} \tag{8.41}$$

and since $\langle k \rangle = g_0'(1)$ we conclude that

$$g_1(s) = \frac{g_0'(s)}{g_0'(1)}. \tag{8.42}$$

Next we turn to the giant cluster and define the following probability:

$$\pi_Q := \text{Prob}\{\text{a randomly chosen node belongs to a (non-giant) cluster of size } Q\}. \tag{8.43}$$

The related generator function is

$$h_0(s) = \sum_{Q=1}^{\infty} \pi_Q s^Q. \tag{8.44}$$

We allow for the possible existence of giant clusters by assuming $\sum_Q \pi_Q = 1 - S$, where S is the strength of the giants. By strength S we mean the probability that a node chosen at random with uniform probability belongs to a giant. The idea is that the sum $\sum_Q \pi_Q$ is the total probability that a random node belongs to a cluster which is *not* a giant cluster. So, if no giant cluster exists then $\sum_Q \pi_Q = 1$, but if a fraction $S > 0$ of the nodes of the network belong to some giant cluster, then $\sum_Q \pi_Q = 1 - S$.

Below we will show that

$$h_0(1) = g_0(h_1(1)) = 1 - S, \tag{8.45}$$

with the generator function $h_1(s)$ to be defined in due course. First we demonstrate that to a good approximation, small clusters can be considered to be of tree structure, i.e. we can ignore chains of links forming loops within the cluster. To see this we compare the number of ways we can add a new link to an existing cluster of size $Q \ll N$ nodes out of a total of N nodes. Assume the cluster is of tree structure. We can then compare the number of ways the new link can be added such that it produces a loop. We attach one end of a link to one of the Q nodes in the cluster. In order for this new link to produce a loop, the other end has to be connected to one of the other $Q-1$ nodes in this cluster. This can be done in $Q-1$ ways, which, for large N, is much smaller than the number of ways the new link can be connected to a node that doesn't belong to the cluster, of which there are $N-Q$.

This suggests that it is a reasonable approximation to neglect loops and concentrate on tree clusters. We analyse the distribution of cluster sizes by making use of their self-similar structure. This is similar to how we treated the branching process in Chap. 5.

We want to relate the size of the cluster containing a specific node α, see Fig. 8.7, to the size of the clusters originating from the neighbours of α. It is possible to use an approach similar to the one we used when discussing Z_n in Chap. 5, see in particular Fig. 5.2. To do this we introduce the probability ρ_Q given by

$$\rho_Q = \text{Prob\{node at end of link belongs to a cluster of size } Q\} \tag{8.46}$$

and denote its generator by

$$h_1(s) = \sum_{Q=1}^{\infty} \rho_Q s^Q. \tag{8.47}$$

Let

$$P(Q|k) = \text{Prob\{the } k \text{ neighbours of node } \alpha \text{ belong to a cluster} \tag{8.48}$$
$$\text{with a total of } Q \text{ nodes\}}.$$

Figure 8.7 The size of the cluster containing node α is given by the sum of the sizes of the clusters originating from the neighbours of α.

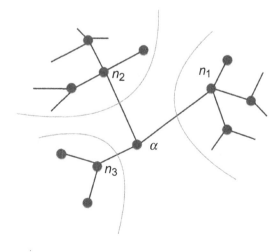

We therefore have

$$\text{Prob}\{\text{node } \alpha \text{ belongs to a cluster of size } Q\} = P(Q-1|k). \tag{8.49}$$

After the introduction of this rather cumbersome set of definitions, we are now able to observe that the probability that an arbitrarily chosen node α belongs to a cluster of size Q is given by

$$\pi_Q = \sum_{k=0}^{\infty} P(Q-1|k)p_k, \tag{8.50}$$

where the factor p_k is the probability that the chosen node α happens to have degree k. We make use of the above to derive the following expression for the generator $h_0(s)$ for the cluster size distribution:

$$
\begin{aligned}
h_0(s) &= \sum_{Q=1}^{\infty}\sum_{k=0}^{\infty} P(Q-1|k)p_k s^Q \\
&= s\sum_{Q=1}^{\infty}\sum_{k=0}^{\infty} P(Q-1|k)p_k s^{Q-1} \\
&= s\sum_{k=0}^{\infty} p_k \sum_{Q=0}^{\infty} P(Q|k)s^Q.
\end{aligned}
\tag{8.51}
$$

We observe that

$$\sum_{Q=0}^{\infty} P(Q|k)s^Q$$

is the generator for the stochastic variable

$$Q = \sum_{n=1}^{k} Q_n,$$

which is the sum of the k subclusters connected to the neighbours of a node of degree k. Hence Q is a variable of the form already considered in Eqs. (5.6) and (5.8), see Box 5.1. In the present case, the upper limit (the number of neighbours) is fixed at k; this corresponds to applying Eq. (5.8) to a case where the upper limit $N = N_0$ with probability 1. The generator for such a variable is $g_N(s) = s^{N_0}$. So for a fixed upper limit, Eq. (5.8) reduces to $g_{S_N}(s) = g_X(s)^{N_0}$. Now apply this to $Q = \sum_{n=1}^{k} Q_n$. Keeping in mind that the generator for the subcluster sizes Q_i of the neighbour clusters is $h_1(s)$, we obtain

$$\sum_{Q=0}^{\infty} P(Q|k)s^Q = [h_1]^k. \tag{8.52}$$

We substitute this result into Eq. (8.51) and get, for the generator $h_0(s)$ of the probability that a randomly chosen node belongs to a cluster of a certain size:

$$h_0(s) = s \sum_{Q=0}^{\infty} [h_1(s)]^Q = sg_0(h_1(s)), \tag{8.53}$$

where we use that the generator for the degree distribution is denoted by $g_0(s)$.

Let us summarise: Eq. (8.53) relates the generator $h_0(s)$ for the probability that a randomly chosen node belongs to a cluster of size Q and the generator $g_0(s)$ for the degree distribution and the generator $h_1(s)$ for the probability that the size of the cluster at the end of a link has size Q_i. We are going to use this relation to discuss the existence of a giant cluster.

The established formalism enables us to investigate under what conditions a giant cluster forms. Since $h_0(s)$ is the generator for the probability that a randomly chosen node belongs to a cluster of some specific finite size $Q = 1, 2, 3, \ldots < \infty$, we have that $h_0(1)$ is the total probability that a random node belongs to some finite cluster. The complementarity to belonging to some finite cluster is to belong to the giant cluster. The strength of the giant cluster, S, will according to Eq. (8.53) satisfy

$$1 - S = \text{Prob\{belonging to some finite cluster\}} = h_0(1) = g_0(h_1(1)), \tag{8.54}$$

i.e.

$$S = 1 - g_0(h_1(1)), \tag{8.55}$$

and we have established Eq. (8.45). See Fig. 8.8 for an example to be discussed below, after Eq. (8.58).

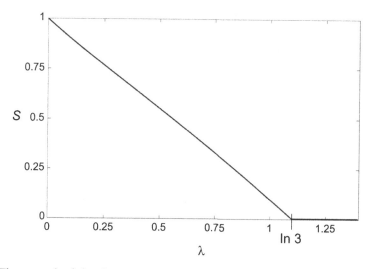

Figure 8.8 The strength of the giant component as a function of λ. There is no giant component for $\lambda > \ln 3$: the degree distribution decays too fast. For $0 < \lambda < \ln 3$, a giant component exists and for $\lambda = 0$ all nodes belong to the giant.

Next we have to determine $h_1(s)$. We recall that $h_1(s)$ is the generator for the probability that the node at the end of the link is of a specific *finite* size Q_i. This means that

$$h_1(1) = \text{Prob\{node not connected to giant through any neighbour\}}$$
$$= \text{Prob\{each neighbour is not connected to giant\}}. \tag{8.56}$$

Assume that the specific neighbour has excess degree k, which happens with probably q_k. To compute the generator for the probability that each of the neighbours *are not connected*[5] to the giant, we have to sum over all possible excess degrees, i.e.

$$h_1(1) = \sum_{k=0}^{\infty} q_k [h_1(1)]^k = g_1(h_1(1)). \tag{8.57}$$

Here we used the fact that $g_1(s)$ is the generator function for the excess degree. Equation (8.57) determines $h_1(1)$ as the smallest root[6] in the equation

$$x = g_1(x). \tag{8.58}$$

Example

Consider an exponential degree distribution

$$p_k = (1 - e^{-\lambda})e^{-\lambda k}, \text{ with } \lambda > 0. \tag{8.59}$$

The generator for this degree distribution is

$$g_0(s) = \sum_{k=0}^{\infty} p_k s^k = (1 - e^{-\lambda}) \sum_{k=0}^{\infty} e^{-\lambda k} s^k$$
$$= (1 - e^{-\lambda}) \sum_{k=0}^{\infty} \left[\frac{s}{e^{\lambda}}\right]^k = \frac{1 - e^{-\lambda}}{1 - se^{-\lambda}} = \frac{e^{\lambda} - 1}{e^{\lambda} - s}. \tag{8.60}$$

The generator for the excess degree is, according to Eq. (8.42):

$$g_1(s) = \frac{g_0'(s)}{g_0'(1)} = \left(\frac{e^{\lambda} - 1}{e^{\lambda} - s}\right)^2. \tag{8.61}$$

According to Eq. (8.58), we need to find the root of the following equation:

$$x = \left(\frac{e^{\lambda} - 1}{e^{\lambda} - x}\right)^2. \tag{8.62}$$

Note that $x = 1$ is a trivial solution, and factorise the equation to obtain

$$x^2 - (2e^{\lambda} - 1)x + (e^{\lambda} - 1)^2 = 0. \tag{8.63}$$

The root gives us $h_1(1)$ and from Eq. (8.55) we obtain

---

[5] Or, in other words, that no neighbour is connected to the giant.
[6] Note that $x = 1$ is always a root.

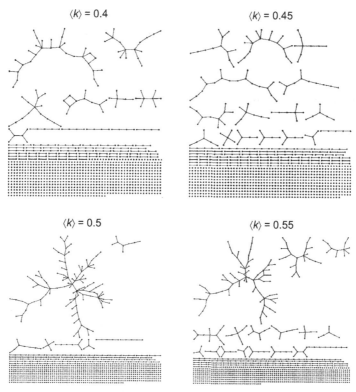

Figure 8.9 Simulations of configuration model (see Sec. 8.10) networks consisting of $N = 1000$ nodes, all with degree distribution given by Eq. (8.59) and different values of λ. To confirm by simulations that the large component starting to appear at $\langle k \rangle = 0.5$ is a giant, one will need to study the dependence of the size S of this large component on N to ensure S increases when N is increased. Simulation and figure courtesy of Dr Tim Evans.

$$S = 1 - g_0(h_1(1)) = 3/2 - \sqrt{e^\lambda - 3/4}. \tag{8.64}$$

When $\lambda = \ln 3$, the expression in Eq. (8.64) becomes equal to zero and becomes negative for $\lambda > \ln 3$, which corresponds to the strength of the giant becoming equal to zero, i.e. there is no giant for $\lambda > \ln 3$. See Fig. 8.8. In the region $\lambda > 3$, the degree distribution in Eq. (8.59) decays so fast that according to our analysis, only finite clusters can exist. The appearance of a giant component at $\lambda = 3 \Rightarrow \langle k \rangle = 1/2$ can be perceived in Fig. 8.9. See also Exercise 8 at the end of this chapter.

8.5 Analysis of Dynamics of and on Networks

We now turn to dynamics. First we will look at the emergence of networks as a consequence of various algorithmic procedures for generating structure. In this sense the network emerges as a result of the dynamical mechanisms.

Next we will assume a given static network structure and look at how dynamics unfolding on this hardwired frame can inform us about the network's topological properties.

8.5.1 Generating Networks

First we will consider the configurational model, which is a purely algorithmic method to produce a network with a given degree distribution. The procedure is not intended to represent any particular real mechanism behind the emergence of observed networks. It is simply a mathematical method to obtain a network with some desired properties.

Next we will look at the construction of networks through dynamical procedures that are intended to focus on mechanisms expected to be relevant to some real situations. We will look at random allocation of links, the Gilbert and Erdös–Rényi random networks already encountered in Sec. 3.2. Then we will consider growth of networks where new nodes are brought in one by one and their links attached preferentially to nodes with certain properties. Finally we will look at a steady-state model where nodes are removed and duplicated in a way inspired by evolutionary dynamics. Except for the configurational model, the philosophy behind these models is to consider an observed degree distribution as emerging from the interactions and dynamics of the nodes and then try to classify which mechanisms may be responsible for which class of degree distributions. We will consider four different algorithms for network generation.

The Configurational Model

This is not really a model, at least not in the sense that it proposes a mechanism suggested to represent some dynamical aspect of network creation relevant to some real system. It is rather a procedure to construct an N-node network which will have nodes with degrees according to a desired distribution $P_{\deg}(k)$.

We start out with N nodes without any links attached. Next, we draw with probability $P_{\deg}(k_i)$ numbers $k_1, k_2, \ldots, k_N$. We attach k_i **half-links** to each of the N nodes, see Fig. 8.10. If $\sum_i k_i$ is even, we are able to produce L links, since the sum over the half-links will add up to twice the number of links between pairs of nodes. If $\sum_i k_i$ is odd, we simply add one more half-link to a randomly selected node; for N large this will have a negligible effect on the degree distribution of the constructed network.

In the next step we pick pairs of half-links at random with uniform probability and join them together to form links, see Fig. 8.10.

The nodes of the resulting network will have degrees according to the desired degree distribution. But the network may contain self-loops and more than one link between two nodes, an example of the latter is seen between node 3 and node N in Fig. 8.10. As N is increased, self-loops and double links will become of relatively less importance. This is because the larger the number of nodes the more likely it is that a half-link starting

Figure 8.10 Configuration model. (a) Allocation of half-links. (b) Connecting pairs of half-links. (c) The resulting network.

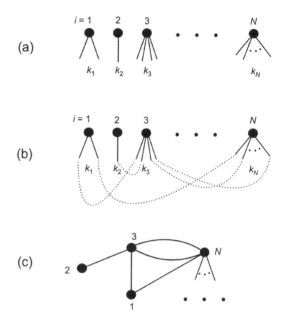

from node i is paired with a half-link from another node $j \neq i$. Similarly, when selecting half-links for pairing, the more nodes are available, the less likely it is to pair half-links from the same pair of nodes more than once. For a detailed analysis of the configuration model, see [312].

Dynamically Generated Networks

There are many procedures for assembling networks according to dynamical rules which appear to be relevant in different contexts. We mentioned the Gilbert and the Erdös–Rényi random networks before, see Secs. 3.2 and 8.2.1, where links are allocated to nodes with uniform probability, leading to a binomial degree distribution

$$p(k) = \binom{N-1}{k} p^k (1-p)^{N-1-k} \simeq \frac{(Np)^k}{k!} \exp(-Np). \tag{8.65}$$

This process is relevant if the links between nodes represent a relationship which is unrelated to the properties of the individual nodes. In Sec. 2.6 we suggested as an example a situation where people are milling about in a crowded space such as a shopping street or railway station, and links are added between people who accidentally bump into each other. Totally random allocation of links will obviously not always be relevant to networks representing real systems.

In real networks, nodes will correspond to some entities such as persons, airports, neurones, species, etc. and the links between the nodes will represent some relationship between the nodes. It is therefore likely that as a network develops, links are connected according to some kind of preference, as we touched upon in Sec. 2.6. For example, consider the network of airports and flight routes connecting these; intuitively it seems reasonable that well-connected airports are attractive destinations for new routes.[7] This suggests that links from new nodes may preferentially be attached to nodes which already have a high degree. Recall the discussion in Sec. 3.2 and see [45, 352, 411, 499] for different versions of preferential growth.

We will now sketch mathematically how preferential growth can produce networks with degree distributions which follow a power law. We look at the simple procedure known as the Barabási–Albert model [45]. Start with n_0 isolated nodes, i.e. all of degree zero. The network will be grown by successively adding new nodes. Each new node has m links attached. These links will be attached to existing nodes with a preference for attaching links to nodes that already possess a large number of links. So when, at a given stage of the growth process, we introduce a new node, each of its m links will be attached to one of the existing nodes with probability

$$\pi_i(k_i) = \text{Prob}\{\text{attach to node } i \text{ of degree } k_i\} = \frac{k_i}{\sum_j k_j}, \tag{8.66}$$

where the sum over j in the denominator runs over the n_0 initial nodes plus each of the nodes added up to the present stage. Figure 8.11 contains a sketch of a realisation of this stochastic growth process.

The degree distribution $P_{\text{deg}}(j)$ of the resulting network will asymptotically, i.e. after adding many nodes and for large degree k, behave as a power law $P_{\text{deg}}(j) \sim 1/k^3$-which can be established by the following simple mean-field-like considerations. As nodes are added, the degree of a node may change. Consider a node i and let $k_i(t)$ denote its degree at time t, after t new nodes have been added to the initial n_0 nodes. A new node is added carrying m links; the probability that a link attaches to node i is $\pi_i(k_i)$, so on average the degree of node i will increase by $m\pi_i$ as a result of the new node added, i.e. we have

$$k_i(t+1) = k_i(t) + m\pi_i(k_i). \tag{8.67}$$

Since each link contributes to the degree of two nodes, the total degree and the number of links L satisfies $\sum_i k_i = 2L$. After t nodes are added, each contributing m links, the total number of links will be $L = mt$, so at time step t we have $\sum_i k_i = 2mt$ and therefore $\pi_i = k_i/(2mt)$.

Putting this together, we can write Eq. (8.67) as

$$k_i(t+1) = k_i(t) + \frac{k_i}{2t}. \tag{8.68}$$

---

[7] Though of course, besides connectedness, other considerations such as cost may play a role when airlines choose airports. Presumably, the cost of landing and airport services is the reason Ryanair often serves minor airports not too far from bigger cities and their bigger airports.

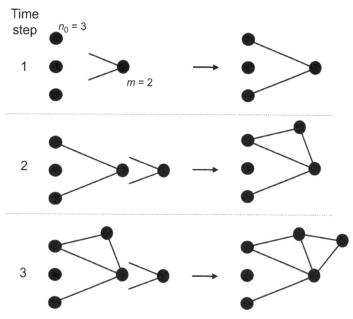

Figure 8.11 Network constructed through three preferential attachment steps. Starting from $n_0 = 3$, three nodes each equipped with $m = 2$ links are added. The two new links arriving in each time step are more likely to be attached to existing nodes of high degree.

We solve this equation by turning it into a differential equation. Move $k_i(t)$ onto the left-hand side and assume $k_i/2t$ to be small, so we can use the approximation $k_i(t+1) - k_i(t) \simeq dk_i/dt$ to write

$$k_i(t+1) - k_i(t) = \frac{k_i}{2t}$$

$$\Downarrow$$

$$\frac{dk_i}{dt} = \frac{k_i}{2t}$$

$$\Downarrow$$

$$\int \frac{dk_i}{k_i} = \int \frac{dt}{2t} \tag{8.69}$$

$$\Downarrow$$

$$\ln k_i(t) = \frac{1}{2} \ln t + c$$

$$k_i = \tilde{c}\sqrt{t}.$$

Here c is a constant of integration and $\tilde{c} = \exp c$. We can determine $\tilde{c}$ from the initial condition of this differential equation. Let t_i denote the time at which node i is added to the network. Since each node arrives with m links attached, we have $k_i(t_i) = m$, which means $\tilde{c} = m/\sqrt{t_i}$ and therefore

$$k_i(t) = m\sqrt{\frac{t}{t_i}} \quad \text{for} \ t \geq t_i. \tag{8.70}$$

We can use this result to determine the degree distribution. The probability, $P(k_i < k)$, that a randomly chosen node has degree less than a certain value is simply a matter of when the node was introduced into the network. We have

$$P(k_i < k) = \text{Prob}\left\{m\sqrt{\frac{t}{t_i}} < k\right\} = \text{Prob}\left\{t_i > \frac{m^2 t}{k^2}\right\}. \tag{8.71}$$

Since we introduce one node at each time step $1, 2, \ldots, t$, the entrance time t_i for a node chosen at random assumes one of the values $1, 2, \ldots, t$ with equal probability, i.e. $\text{Prob}\{t_i = \tau\} = 1/t$ for $\tau = 1, 2, \ldots, t$. This tells us that

$$\text{Prob}\left\{t_i > \frac{m^2 t}{k^2}\right\} = \frac{t - \frac{m^2 t}{k^2}}{t} = 1 - \frac{m^2}{k^2}, \tag{8.72}$$

or since $k_i = m\sqrt{t/t_i}$ we have

$$P(k_i < k) = 1 - \frac{m^2}{k^2}$$

$$\Downarrow$$

$$\text{Prob}\{k_i = k\} = P(k_i < k+1) - P(k_i < k) = m^2\left(\frac{1}{k^2} - \frac{1}{(k+1)^2}\right) \tag{8.73}$$

$$\Downarrow$$

$$\text{Prob}\{k_i = k\} = m^2 \frac{2k+1}{k^2(k+1)^2} \simeq 2\frac{m^2}{k^3},$$

where the last approximate equality holds for large values of k.

We have demonstrated that growing a network according to the simple preferential attachment probability in Eq. (8.66) generates a network with a power-law degree distribution $P_{\text{deg}}(k) \sim 1/k^\alpha$ and that the exponent $\alpha = 3$ is independent of the values of n_0 and m. However, the value of the exponent α does depend on how the growth of the network is implemented. The growth procedure can be modified, for example, by allowing new nodes to arrive with a number of links determined by a probability distribution, or changing the probability with which links are attached to existing nodes corresponding to different implementations of preferential growth. The effect on the power law is to make α assume values between 2 and 3 [352, 411, 499]. The attachment of links to preferred nodes can also be done, for example, by assigning a so-called 'fitness' to each node which determines the likelihood of attaching a new link. A careful and comprehensive discussion can be found in Chap. 13 in [312].

We will probably not expect real networks to exhibit pure power-law behaviour for the entire range of the degree. The exact relevance of power-law degree distributions to real networks has been debated; most will agree that the important point is how the tail behaves for large values of the degree [197]. A network with a degree distribution that follows a power law for large values of the degree will behave very differently from one which decays exponentially fast with the degree, because the power-law tail corresponds to a relatively high density of nodes with very large degree of connections. These nodes will influence the properties, such as transport on the network, in a very significant way. See Fig. 3.6 for examples.

Steady-State Evolutionary Dynamics

Not all dynamical networks will necessarily grow in size. We will finish this section with a discussion of a birth–death process inspired by evolutionary dynamics in which nodes can give rise to offspring by duplication and can die and vanish from the network [249, 250]. Think of agents undergoing reproduction with mutations and death. In general terms, the agents could represent reproducing organisms or even species undergoing speciation events and extinction. This kind of model is also of relevance to protein–protein interaction networks and a large literature exists, see e.g. [137, 334] and references therein.

We start from a network consisting of N nodes linked together according to some arbitrary degree distribution. After a transient period, the dynamics will converge towards a steady state characterised by a degree distribution which is independent of the starting point.

The dynamics is indicated in Fig. 8.12. A time step consists of two events. First, a node is chosen at random with uniform probability. With probability P_k this node is removed (hence subscript k for killing) from the network together with all the links attached to the node, as indicated in the top row of Fig. 8.12. Next we choose with uniform probability a 'parent' node amongst the remaining $N - 1$ nodes. This node is duplicated to produce an 'offspring'. Each of the links attached to the parent is inherited by the offspring with probability P_e (subscript e for existing link). Since parent and offspring are of the same kind, it appears natural that they may influence each other and therefore a link between parent and offspring is added with probability P_p (subscript p for link to parent). Finally, to represent that offspring and parent are not identical, for example due to the occurrence of mutations, the offspring is also attached to nodes that are not attached to the parent. Each of these new links are attached to the offspring with probability P_n (subscript n for new link).

Because four stochastic processes are involved in the dynamics, full analytical treatment of the model is challenging, though some insights can be obtained by neglecting stochastic fluctuations to establish equations for average behaviour of, for example, the degree distribution [137, 215, 249, 334] or the connectance [249]. Here we will limit ourselves to an illuminating mathematical argument indicating how the dynamics for any initial configuration converges towards an attractive state. After that we will describe the results of combined analytic and numerical studies concerning the behaviour of the degree distribution.

Recall that a time step from t to $t + 1$ involves two events: removal of a randomly chosen node followed by duplication of another randomly chosen node. To make the bookkeeping simple, we break the time step down into two half steps. The removal of

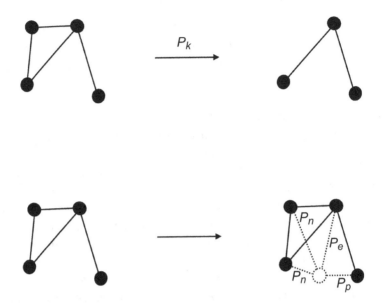

Figure 8.12 Network undergoing evolutionary dynamics. The top row depicts the removal of the top-left node and all links attached to this node. The probability that a node is removed is P_k. The bottom row represents the duplication event of the bottom-right node. The probability that the offspring, indicated by the dotted circle, inherits links attached to the parent is P_e. The probability that the offspring establishes new links to nodes not connected to the parent node is P_n and the probability that the offspring connects to the parent is P_p.

a node takes us from t to $t + \frac{1}{2}$ and changes the number of nodes from N to $N - 1$. The following duplication of a node completes the time step from $t + \frac{1}{2}$ to $t + 1$ and takes the number of nodes from $N - 1$ back to N.

We denote by L_t the total number of links in the network at time t and by $\bar{k}_t$ the average degree at time t. The removal of a node will, on average, make the link number change according to

$$L_{t+\frac{1}{2}} = L_t - \bar{k}_t, \tag{8.74}$$

due to the loss of all the links connected to the removed node. The duplication of a node is more involved and leads to the following change:

$$L_{t+1} = L_{t+\frac{1}{2}} + P_e \bar{k}_{t+\frac{1}{2}} + P_n \left(N - 2 - \bar{k}_{t+\frac{1}{2}} \right) + P_p. \tag{8.75}$$

The term on the right-hand side proportional to P_e represents the average number of links inherited by the offspring. On average, the duplicated node has $\bar{k}_{t+\frac{1}{2}}$ links and each will be attached to the offspring with probability P_e. The term proportional to P_n represents the on-average number of new links attached to the offspring. Since at time $t + \frac{1}{2}$, the network has $N - 1$ nodes, there are $N - 2 - \bar{k}_{t+\frac{1}{2}}$ *new* nodes the offspring may attach to, namely the $N - 2$ nodes different from itself minus the on-average $\bar{k}_{t+\frac{1}{2}}$ nodes already attached to the parent node. These nodes are already accounted for by the P_e term. Finally, the last term in the equation represents the average effect of placing with probability P_p a link between offspring and parent.

To close the equations, we can express the average degrees at time t and time $t + \frac{1}{2}$ as

$$\bar{k}_t = \frac{2L_t}{N}$$

$$\bar{k}_{t+\frac{1}{2}} = \frac{2L_{t+\frac{1}{2}}}{N-1} = 2\frac{L_t - \bar{k}_t}{N-1}. \tag{8.76}$$

Substituting these expressions for the average degrees and adding the changes over the two half time steps we end up with the following equation for the time evolution of the number of links:

$$L_{t+1} = L_t\left[1 - \frac{2}{N} + 2\frac{P_e - P_N}{N(N-1)}(N-2)\right] + P_n(N-2) + P_n. \tag{8.77}$$

We write the equation in the form

$$L_{t+1} = (1+a)L_t + b, \tag{8.78}$$

where

$$a = -\frac{2}{N} + 2\frac{P_e - P_N}{N(N-1)}(N-2)$$

$$b = P_n(N-2) + P_n. \tag{8.79}$$

This iterative equation has the solution

$$L_t = A(1+a)^t - \frac{b}{a}, \tag{8.80}$$

where A is a constant. It is easy by substitution to check that this expression for L_t solves Eq. (8.78), but one may of course ask where did this solution come from? Iterating Eq. (8.78) is very suggestive. We have

$$\begin{aligned}
L_{t+1} &= (1+a)L_t + b \\
&= (1+a)[(1+a)L_{t-1} + b] + b \\
&= (1+a)^2 x_{t-1} + (2+a)b,
\end{aligned} \tag{8.81}$$

so we notice that a factor $(1+a)$ will be multiplied on x_t at each time step and an additive term is generated as well. This inspires us to look for a solution of the form

$$L_t = A(1+a)^t + B. \tag{8.82}$$

We need to check if we can determine the constants A and B to balance Eq. (8.78). By substituting Eq. (8.82) into Eq. (8.78) we get

$$\begin{aligned}
\text{left-hand side } \quad & L_{t+1} = A(1+a)^{t+1} + B \\
\text{right-hand side } \quad & (1+a)L_t + b = (1+a)[A(1+a)^t + B] + b \\
& = A(1+b)^{t+1} + (1+a)B] + b.
\end{aligned} \tag{8.83}$$

To match the right-hand and the left-hand side we need $(1 + a)B + b$ or $B = -\frac{b}{a}$, which gives us the solution in Eq. (8.80).

If $|1 + a| < 1$, the first term of the solution in Eq. (8.80) will go to zero as $t \to \infty$ and we can conclude that in the long-time limit the connectance, defined in Eq. (8.1), of the network will converge towards

$$C_\infty = \frac{L_\infty}{\binom{N}{2}} = \frac{P_n(N - 2) + P_p}{N - 1 + (P_n - P_e)(N - 2)}. \tag{8.84}$$

Here we substituted the expressions for a and b from Eq. (8.79) into $L_\infty = -b/a$. This expression for the value of the connectance in the limit of many removal and duplication events is in agreement with available simulation results, see [137].

The degree distribution $P_{\text{deg}}(k)$ is more difficult to analyse analytically, though combined simulation and analytic analyses have established that the functional dependence of $P_{\text{deg}}(k)$ on k depends on the fidelity of the duplication process [249]. Reproduction with a large amount of stochasticity generates a binomial degree distribution which changes through exponential to exponentially cutoff power-law behaviour as the reproduction process becomes more accurate, i.e. as $P_e \to 1$ and $P_n \to 0$. The exponent of the power law is found to behave as $1 - 2P_p$ [137, 249].

This functional dependence appears to be of relevance to biological systems such as food webs, which are sometimes, though not always, reported to exhibit exponential degree distributions [116]. The change towards power-law behaviour as the duplication process becomes more accurate has been related to protein interaction networks [137], since such networks have been reported to exhibit power-law-like degree distributions with exponents not too far above one, see e.g. [160]. How relevant and in which way power-law-like degree distributions are to real networks continues to be a matter of some controversy [197]. Analysis of a large number of biological, social and transport networks suggested that when extrapolated to the limit of infinite size, the degree distribution frequently exhibit a power law [393].

8.5.2 Random Walk on Networks

After having looked at how different types of dynamics can assemble networks with different properties, we will take the structure of the network to be a static frame on which different types of dynamics can unfold. This is a way to isolate how the topology of the network may influence the dynamical processes unfolding on the network. As an illustration, we will consider random walkers moving from node to node. We discussed in Sec. 2.4 the ordinary random walk and noticed that a local peak in the concentration of walkers will gradually disappear and in the limit of infinite time become a homogeneous flat distribution. When the walkers are confined to the nodes of a network the situation is different, in the long-time limit any initial distribution of walkers on the nodes will be heterogeneous and the number of walkers resident on the different nodes will, as time goes to infinity, be given by the degree of the nodes.

This can be seen in the following way. We imagine that a population of agents have been placed on the nodes of a network and denote by $q_i(t)$ the number of agents resident on node number i at time step $t = 1, 2, \ldots$. At each time step an agent on node i with degree k_i chooses at random with equal probability one of the links attached to the node and moves to the node at the end of the link. That is, with probability $1/k_i$ the agent moves to node j if $a_{ij} = 1$. This means that the probability π_{ij} that a walker on node i makes a move to node j is given by

$$\pi_{ij} = \frac{a_{ij}}{k_i}. \tag{8.85}$$

The arrival of agents from the k_i neighbours to node i and the leaving of agents from node i to the k_i neighbours will, in a mean-field approximation v, satisfy the following balance equation:

$$
\begin{aligned}
q_i(t+1) &= q_i(t) + \sum_{j=1}^{k_i} \frac{1}{k_j} q_j(t) - \sum_{j=1}^{k_i} \frac{1}{k_i} q_i(t) \\
&= q_i(t) + \sum_{j=1}^{N} q_j(t) \pi_{ji} - q_i(t) \\
&= q_i(t) + \sum_{j=1}^{N} q_j(t)[\pi_{ji} - \delta_{ji}].
\end{aligned}
\tag{8.86}
$$

We can rewrite this equation in vector and matrix form by introducing the vector $\boldsymbol{Q}$ and its transpose $\boldsymbol{Q}^{\mathrm{T}}$:

$$\boldsymbol{Q} = \begin{pmatrix} q_1 \\ q_2 \\ \vdots \\ q_N \end{pmatrix}, \quad \boldsymbol{Q}^{\mathrm{T}} = (q_1, q_2, \ldots, q_N), \tag{8.87}$$

together with the transition matrix $\boldsymbol{\Pi}$ with matrix elements $\boldsymbol{\Pi}_{ij} = \pi_{ij}$ and the identity matrix $\boldsymbol{I}$ with matrix $\boldsymbol{I}_{ij} = \delta_{ij}$:

$$\boldsymbol{Q}(t+1) = \boldsymbol{Q}(t) + \boldsymbol{Q}^{\mathrm{T}}(t)[\boldsymbol{\Pi} - \boldsymbol{I}]. \tag{8.88}$$

Next we assume that a stationary configuration has been reached so $\boldsymbol{Q}(t+1) = \boldsymbol{Q}(t)$, which implies

$$
\begin{aligned}
&\boldsymbol{Q}^{\mathrm{T}}(t)[\boldsymbol{\Pi} - \boldsymbol{I}] = \boldsymbol{0} \\
&\Downarrow \\
&\sum_{l=1}^{N} q_l \frac{a_{li}}{k_l} = q_i.
\end{aligned}
\tag{8.89}
$$

Now recall that, for the adjacency matrix, we have $\sum_l a_{li} = k_i$, from which we see that Eq. (8.89) is solved by

$$Q = c \begin{pmatrix} k_1 \\ k_2 \\ \vdots \\ k_N \end{pmatrix}. \tag{8.90}$$

We can choose c to obtain a convenient normalisation. We choose $c = 1/2L$, then Eq. (8.90) tells us that the probability, ρ_i, that a randomly chosen agent is at node i when the stationary state has been reached is given by

$$\rho_i = \frac{k_i}{2L}. \tag{8.91}$$

This tells us that if we let a population of random walkers loose on a network, they will as time passes eventually move around so nodes with high degree have a correspondingly large number of walkers visiting at any given time. This seems to be intuitively reasonable, since many routes lead to and from high-degree nodes.

Next, we consider the average time a random walker spends travelling from node i to another node j. We use the approach in Sec. 11.5 in [173]. The walker may arrive at node j multiple times after leaving node i. We are interested in the *first* time the walker from node i arrives at node j and we define m_{ij} to be the average of this first passage time and by convention and convenience define $m_{ii} = 0$. We can derive an equation for m_{ij} by observing that the walker can get from node i to node j in two ways: the first step from i may take the walker directly to j, or the first step may take the walker to a node $k \neq j$ and then the walker will have to make their way to j. See Fig. 8.13.

Adding these contributions together, we establish the equation

$$m_{ij} = \pi_{ij} \times 1 + \sum_{k \neq j} \pi_{ik}(1 + m_{kj}). \tag{8.92}$$

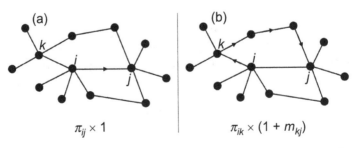

$$\pi_{ij} \times 1 \qquad\qquad \pi_{ik} \times (1 + m_{kj})$$

Figure 8.13 Random walk on a lattice. In panel (a), the walker moves in one step from node i to node j. In panel (b), the walker makes first a step to node k and will then have to travel from this node to node j. The contribution to the average first passage time is indicated below each panel.

We have $\sum_k \pi_{ik} = 1$ and therefore $\sum_{k \neq j} \pi_{ik} = 1 - \pi_{ij}$ and Eq. (8.92) can be written as

$$m_{ij} = 1 + \sum_{k \neq j} \Pi_{ik} m_{kj}. \tag{8.93}$$

Now consider the case where $j = i$. This corresponds to the average first return time of the walker and is denoted by r_i. The walker needs to make a move away from node i to some node k and then return from there. The average of the $1 + m_{ki}$ moves involved, summed over all the possible nodes k, gives us the equation

$$r_i = \sum_{k=1}^{N} \pi_{ik}(1 + m_{ki}). \tag{8.94}$$

We used $\pi_{ii} = 0$ to allow the sum over k to include node i. We want to solve Eqs. (8.93) and (8.94) in order to find r_i in the stationary state given by Eq. (8.90). To do this we combine the two equations into one equation by introducing three matrices: the average first passage time matrix M, the average first return time matrix R and a matrix with all elements equal to 1. That is, we define

$$M = \begin{pmatrix} 0 & m_{12} & \dots & m_{1N} \\ m_{21} & 0 & \dots & m_{2N} \\ \vdots & \vdots & \ddots & \vdots \\ m_{N1} & m_{N2} & \dots & 0 \end{pmatrix}, \quad R = \begin{pmatrix} r_1 & 0 & \dots & 0 \\ 0 & r_2 & \dots & 0 \\ \vdots & \vdots & \ddots & \vdots \\ 0 & 0 & \dots & r_N \end{pmatrix}, \quad K = \begin{pmatrix} 1 & 1 & \dots & 1 \\ 1 & 1 & \dots & 1 \\ \vdots & \vdots & \ddots & \vdots \\ 1 & 1 & \dots & 1 \end{pmatrix}. \tag{8.95}$$

Equipped with this notation and making use of the transition matrix in Eq. (8.85), we can now write Eqs. (8.93) and (8.94) together in the following compact form:

$$M = K + \Pi M - R. \tag{8.96}$$

The off-diagonal elements in this matrix equation are equivalent to Eq. (8.93) and the diagonal elements correspond to Eq. (8.94). We rearrange to be able to make contact with Eq. (8.89) and write

$$(I - \Pi)M = K - R. \tag{8.97}$$

We use Eq. (8.89) to obtain

$$0 = Q^T(I - \Pi)M = Q^T(K - R). \tag{8.98}$$

The elements in the vector $Q^T K$ are all equal to $c \sum_{i=1}^{N} k_i = 2Lc$ and the elements in the vector $Q^T R$ are equal to $ck_i r_i$ for $i = 1, 2, \dots, N$, which allows us to conclude that

$$r_i = \frac{2L}{k_i}. \tag{8.99}$$

It is easier for the walker to find their way back to a node of high degree, since more routes lead back to where they came from, cutting down on the return time.

Now that we have determined the matrix $\boldsymbol{R}$, we can in principle solve Eq. (8.97) to express the average first passage times as $\boldsymbol{M} = (\boldsymbol{I} - \boldsymbol{\Pi})^{-1}(\boldsymbol{K} - \boldsymbol{R})$, though the right-hand side is not in general easy to express in a simple way. The intimate relationship between the occupancy of the walkers on the nodes of the network and the times for the walkers to travel around on the network has inspired studies that relate random walk dynamics and its unfolding in time to the structural properties of networks, see e.g. [68, 282].

Networks encountered in the real world will change over time. Social networks change because people's circumstances are ever-changing, food webs change because of changes to habitat, and functional brain network keep changing depending on tasks. This makes it interesting to consider dynamics, such as diffusion or random walks, on networks with time-dependent topology. For such a study see e.g. [103].

8.5.3 Synchronisation on Networks

In Chap. 7 we studied how interacting rotors may synchronise and found that the stronger the interaction, the easier is the synchronisation. This finding suggests that the time dependence of the onset of synchronisation of rotors connected through a network of interactions may be influenced by the structure of the network of interactions. We expect that communities in networks consisting of nodes which are mutually well connected will synchronise faster than regions of nodes which share few connections.

Let $\nu(t)$ denote the number of non-synchronised components of the network at time t. If we start the rotors in some initially non-synchronised state, we will have $\nu(0) = N$. Each time a community of nodes synchronise, $\nu(t)$ will drop by the number of nodes belonging to this community. Simulations of Kuramoto rotors on hierarchical networks exhibit this behaviour [26]. So, the way in which the synchronised state emerges will be shaped by the topological structure of the network.

Box 8.2 Laplacian on a Network

When we study the dynamics of, for example, liquids flowing through space, we encounter equations which involve derivatives with respect to time and space variables. When the dynamics takes place on a network the derivatives we are used to from calculus are replaced by similar discrete expressions.

Let us recapture the definition of the first and second derivative of a function $f(x)$ of a single variable. We have for the first derivative

$$f'(x) = \frac{df}{dx} = \lim_{\delta \to 0} \frac{1}{\delta}[f(x+\delta) - f(x)] \simeq \frac{1}{\delta}[f(x+\delta) - f(x)]. \quad (8.100)$$

Here, the last approximate relation holds when δ is small. For the second derivative we have

$$f''(x) = \frac{d^2 f}{dx^2} = \frac{df'}{dx}$$

$$= \lim_{\delta \to 0} \frac{1}{\delta}[f'(x+\delta) - f'(x)]$$

$$= \lim_{\delta \to 0} \frac{1}{\delta^2}[f(x+2\delta) - f(x+\delta) - (f(x+\delta) - f(x)]$$

$$= \lim_{\delta \to 0} \frac{1}{\delta^2}[f(x+2\delta) + f(x) - 2f(x+\delta)]. \tag{8.101}$$

To obtain a more symmetric expression we replace x by $x - \delta$ which in the limit $\delta \to 0$ makes no difference[a] and obtain

$$f''(x) = \lim_{\delta \to 0} \frac{f(x+\delta) + f(x-\delta) - 2f(x)}{\delta}$$

$$\simeq \frac{1}{\delta^2}[f(x+\delta) + f(x-\delta) - 2f(x)]. \tag{8.102}$$

When we have functions of more than one variable $f(x_1, \ldots, x_N)$, we recall that we work with partial derivatives for each of the variables, which in the discrete case takes the form

$$\frac{\partial f}{\partial x_i} = \simeq \frac{1}{\delta}[f(x_1, \ldots, x_i + \delta, \ldots, x_N) - f(x_1, \ldots, x_i, \ldots, x_N)] \tag{8.103}$$

and the second derivative along x_i becomes

$$\frac{\partial^2 f}{\partial x_i^2} \simeq \frac{1}{\delta^2}[f(x_1, \ldots, x_i + \delta, \ldots, x_N) + f(x_1, \ldots, x_i - \delta, \ldots, x_N)$$

$$- 2f(x_1, \ldots, x_i, \ldots, x_N)]. \tag{8.104}$$

From the gradient vector, usually called the gradient operator or nabla, given by

$$\nabla = \left(\frac{\partial}{\partial x_1}, \ldots, \frac{\partial}{\partial x_N} \right)$$

we can form the scalar product of the ∇ with itself, which is called the Laplacian and we have in the discrete case

---

[a] When we are dealing with continuous and differentiable functions.

$$\Delta f := \nabla \cdot \nabla f$$

$$= \sum_{i=1}^{N} \frac{1}{\delta^2} [f(x_1, \ldots, x_i + \delta, \ldots, x_N) + f(x_1, \ldots, x_i - \delta, \ldots, x_N)$$

$$- 2f(x_1, \ldots, x_i, \ldots, x_N)] \tag{8.105}$$

$$= \frac{1}{\delta^2} \sum_{x_n} f(\boldsymbol{x}_n) - 2Nf(\boldsymbol{x}).$$

Here we introduce the vector notation $\boldsymbol{x} = (x_1, \ldots, x_N)$ and use $\mathbf{x}_n$ with $n = 1, \ldots, N$ to denote the neighbours obtained by adding and subtracting δ from the coordinates of $\boldsymbol{x}$, see Fig. 8.14.

Let us now consider a network and a function f. We denote the value of f on node i by f_i. This means that we can collect the values of the function evaluated on the network in a vector

$$f = \begin{pmatrix} f_1 \\ f_2 \\ \vdots \\ f_N \end{pmatrix}. \tag{8.106}$$

In analogy with Eq. (8.105), the network Laplacian is denoted by L and when evaluated for function f on node i is defined as

$$(Lf)_i = k_i f_i - \sum_{j=1}^{N} a_{ij} f_j = \sum_{j=1}^{N} (k_i \delta_{ij} - a_{ij}) f_j. \tag{8.107}$$

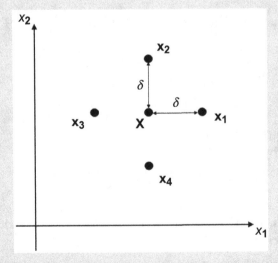

Figure 8.14 Laplacian neighbourhood in two dimensions. The Laplacian evaluated at a point x involves x and the four neighbours $\boldsymbol{x}_1, \ldots, \boldsymbol{x}_4$.

Since the degree k_i of the node is equal to its neighbours and the sum in the second term evaluates f on the nodes connected to i we see that this definition is essentially the same as the expression in Eq. (8.105), except that the tradition in the network literature is to use the opposite sign of the one in Eq. (8.105).

We can write Eq. (8.107) in a more compact form by use of matrix notation. We denote the effect of the Laplacian on the function f by the vector Lf, which has the components $(Lf)_i$ given by Eq. (8.107). In this way we can write

$$Lf = [D - A]f, \qquad (8.108)$$

where the matrix D has zero off-diagonal elements and the elements along the diagonal are equal to the degree of the nodes, i.e. $D_{ij} = k_i \delta_{ij}$, and A is the adjacency matrix.

The eigenvectors and eigenvalues of the Laplacian, given by

$$Lv_q = \lambda_q v, \qquad (8.109)$$

contain useful information concerning the structure of the network and properties of dynamics such as diffusion, random walks and synchronisation unfolding on the network, see e.g. [25, 60] and [312] Sec. 6.14.

Here we will just point out two facts. First, the number of disconnected clusters is equal to the number of eigenvectors with eigenvalue equal to zero and second, the eigenvalues are all non-negative, i.e.

$$0 = \lambda_0 \leq \lambda_1 \leq \lambda_2 \leq \cdots \leq \lambda_{N-1}. \qquad (8.110)$$

That the number of zero eigenvalues is equal to the number of disconnected components of the network can be seen in the following way. First consider a network consisting of a single connected cluster. The vector containing a 1 for each of the nodes, $e_0^T = (1, 1, 1, \ldots, 1)$, is an eigenvector for L with eigenvalue zero. This is easily checked by computing the components of Le_0, these are $k_i - \sum_j a_{ij} = 0$. If the network has multiple disconnected clusters we will have a zero eigenvector for each cluster. To see this we take a closer look at the adjacency matrix for such a network.

Namely, consider a network composed of K disconnected groups of nodes, each containing N_i nodes for $i = 1, \ldots, K$ with $\sum_{i=1}^{K} N_i = N$. The adjacency matrix of the network has no non-zero elements connecting nodes in one component with the nodes in another. This implies that the adjacency matrix for the combined network A is constructed from the K adjacency matrices for the individual components $A^{(i)}$ with $i = 1, \ldots, K$ by placing the $A^{(i)}$ down along the diagonal of A and zeros everywhere else, namely

$$A = \begin{pmatrix} A^{(1)} & & & 0 \\ & A^{(2)} & & \\ & & \ddots & \\ 0 & & & A^{(K)} \end{pmatrix}. \tag{8.111}$$

The Laplacian for this network will have a similar block-diagonal structure, since $L = D - A$, i.e.

$$L = \begin{pmatrix} L^{(1)} & & & 0 \\ & L^{(2)} & & \\ & & \ddots & \\ 0 & & & L^{(K)} \end{pmatrix}. \tag{8.112}$$

The block-diagonal structure of the Laplacian allows us to construct K eigenvectors v_i all with eigenvalue zero by placing a string of 1s at the elements in v_i corresponding to the block $L^{(i)}$ in L and zeros everywhere else. So a vector of the form

$$v_i = (0, 0, \ldots, 0, \underbrace{1, 1, \ldots, 1}_{N_i}, 0, 0, \ldots, 0).$$

The relationship between topological structure and the Laplacian eigenvalues and eigenvectors is illustrated in Fig. 8.15.

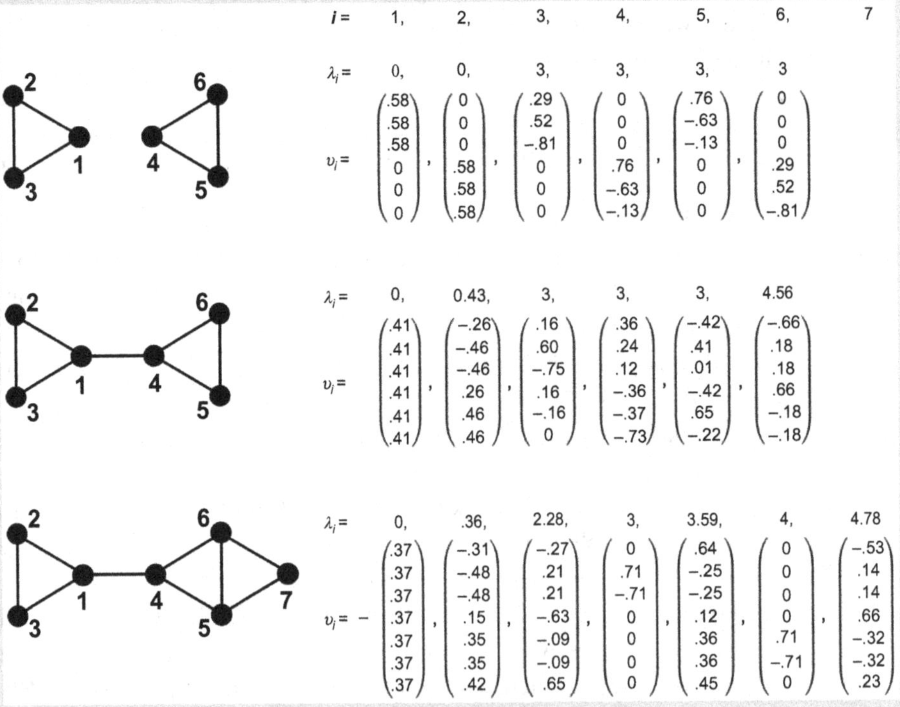

Network 1

$i =$	1,	2,	3,	4,	5,	6,	7
$\lambda_i =$	0,	0,	3,	3,	3,	3	

$$v_i = \begin{pmatrix} .58 \\ .58 \\ .58 \\ 0 \\ 0 \\ 0 \end{pmatrix}, \begin{pmatrix} 0 \\ 0 \\ 0 \\ .58 \\ .58 \\ .58 \end{pmatrix}, \begin{pmatrix} .29 \\ .52 \\ -.81 \\ 0 \\ 0 \\ 0 \end{pmatrix}, \begin{pmatrix} 0 \\ 0 \\ 0 \\ .76 \\ -.63 \\ -.13 \end{pmatrix}, \begin{pmatrix} .76 \\ -.63 \\ -.13 \\ 0 \\ 0 \\ 0 \end{pmatrix}, \begin{pmatrix} 0 \\ 0 \\ 0 \\ .29 \\ .52 \\ -.81 \end{pmatrix}$$

Network 2

$\lambda_i =$	0,	0.43,	3,	3,	3,	4.56

$$v_i = \begin{pmatrix} .41 \\ .41 \\ .41 \\ .41 \\ .41 \\ .41 \end{pmatrix}, \begin{pmatrix} -.26 \\ -.46 \\ -.46 \\ .26 \\ .46 \\ .46 \end{pmatrix}, \begin{pmatrix} .16 \\ .60 \\ -.75 \\ .16 \\ -.16 \\ 0 \end{pmatrix}, \begin{pmatrix} .36 \\ .24 \\ .12 \\ -.36 \\ -.37 \\ -.73 \end{pmatrix}, \begin{pmatrix} -.42 \\ .41 \\ .01 \\ -.42 \\ .65 \\ -.22 \end{pmatrix}, \begin{pmatrix} -.66 \\ .18 \\ .18 \\ .66 \\ -.18 \\ -.18 \end{pmatrix}$$

Network 3

$\lambda_i =$	0,	.36,	2.28,	3,	3.59,	4,	4.78

$$v_i = - \begin{pmatrix} .37 \\ .37 \\ .37 \\ .37 \\ .37 \\ .37 \\ .37 \end{pmatrix}, \begin{pmatrix} -.31 \\ -.48 \\ -.48 \\ .15 \\ .35 \\ .35 \\ .42 \end{pmatrix}, \begin{pmatrix} -.27 \\ .21 \\ .21 \\ -.63 \\ -.09 \\ -.09 \\ .65 \end{pmatrix}, \begin{pmatrix} 0 \\ .71 \\ -.71 \\ 0 \\ 0 \\ 0 \\ 0 \end{pmatrix}, \begin{pmatrix} .64 \\ -.25 \\ -.25 \\ .12 \\ .36 \\ .36 \\ .45 \end{pmatrix}, \begin{pmatrix} 0 \\ 0 \\ 0 \\ 0 \\ .71 \\ -.71 \\ 0 \end{pmatrix}, \begin{pmatrix} -.53 \\ .14 \\ .14 \\ .66 \\ -.32 \\ -.32 \\ .23 \end{pmatrix}$$

Figure 8.15 Networks and corresponding eigenvalues and eigenvectors of the Laplacian.

To study mathematically how network structure can influence the onset of synchronisation we will consider the Kuramoto model (see Sec. 7.1) on a network, i.e. we place a rotor variable $\theta_i(t)$ on each node and couple these together using the adjacency matrix. In order to isolate the effect of the network topology, we assume that all rotors are characterised by the same inherent speed ω and consider the following dynamical equations for the phase variables θ_i of each of the N rotors:

$$\frac{d\theta_i}{dt} = \omega + \epsilon \sum_{j=1}^{N} a_{ij} \sin(\theta_j - \theta_i). \tag{8.113}$$

If we choose $\epsilon = 0$, all rotors will move according to $\theta_i = \omega t$. Similar to the analysis in Sec. 7.1, we introduce the variable ψ_i as the deviation from ωt, so we have

$$\theta_i(t) = \omega t + \psi_i(t). \tag{8.114}$$

In the fully synchronised state we will have $\psi_i = $ constant for all $i = 1, 2, \ldots, N$. By substitution into Eq. (8.113) we obtain the following equations for the $\psi_i(t)$ variables:

$$\frac{d\psi_i}{dt} = \epsilon \sum_{j=1}^{N} a_{ij} \sin(\psi_j - \psi_i). \tag{8.115}$$

Let us now assume that the rotors have moved close to the synchronised state, so $\psi_i - \psi_j \simeq 0$, and make use of the first term in the Taylor expansion $\sin x \simeq x - \frac{1}{3!}x^3 + \cdots$ to write

$$
\begin{aligned}
\frac{d\psi_i}{dt} &= \epsilon \sum_{j=1}^{N} a_{ij}(\psi_j - \psi_i) \\
&= \epsilon \left(\sum_{j=1}^{N} a_{ij}\psi_j - \sum_{j=1}^{N} a_{ij}\psi_i \right) \\
&= \epsilon \left(\sum_{j=1}^{N} a_{ij}\psi_j - k_i\psi_i \right) \\
&= -\epsilon \sum_{j=1}^{N} (k_i\delta_{ij} - a_{ij})\psi_j.
\end{aligned}
\tag{8.116}
$$

By use of the vector and matrix notation and the **Laplacian** introduced in Box 8.2, we can write this equation as

$$\frac{d\boldsymbol{\psi}}{dt} = -\epsilon \boldsymbol{L}\boldsymbol{\psi}, \tag{8.117}$$

where $\boldsymbol{\psi}$ denotes the vector with components ψ_i.

Let e_q denote the eigenvectors of the Laplacian, i.e. $Le_q = \lambda_a e_q$. We expand the vector containing the rotor phases ψ on these eigenvectors:

$$\psi = \sum_{q=1}^{N} c_q(t) e_q. \tag{8.118}$$

Next we substitute this expression for ψ into Eq. (8.117) to obtain

$$\frac{d}{dt} \sum_{q=1}^{N} c_q(t) e_q = -\epsilon L \sum_{q=1}^{N} c_q(t) e_q$$

$$\Downarrow$$

$$\sum_{q=1}^{N} \frac{dc_q(t)}{dt} e_q = -\epsilon \sum_{q=1}^{N} c_q(t) L e_q \tag{8.119}$$

$$\Downarrow$$

$$\sum_{q=1}^{N} \frac{dc_q(t)}{dt} e_q = -\epsilon \sum_{q=1}^{N} c_q(t) \lambda_q e_q.$$

For each value of q, we equate the term on the left-hand side with the corresponding term on the right-hand side and conclude

$$\frac{dc_q(t)}{dt} = -\epsilon \lambda_q c_q(t)$$

$$\Downarrow \tag{8.120}$$

$$c_q(t) = c_q(0) \exp(-\epsilon \lambda_q t) \text{ for } q = 1, 2, \ldots, N.$$

These expressions for the coefficients, together with Eq. (8.118), lead to

$$\psi = \sum_{q=1}^{N} c_q(0) \exp(-\epsilon \lambda_q t) e_q. \tag{8.121}$$

From Eq. (8.114) we recall that the coefficients ψ_i of ψ represent how much the rotor on node i deviates from the synchronised state. Equation (8.121) corresponds to finding the components of, for example, a three-dimensional velocity vector v along the different coordinate axes x, y and z. From the expression $v = x e_x + y e_y + z e_z$ we can tell how much the motion is aligned with the various directions e_x, e_y and e_z. Motion along the x-axis has only $x \neq 0$, whereas motion along a line forming a 45° angle with each of the coordinate axes will have $x = y = z \neq 0$.

Similarly, the expansion in Eq. (8.121) separates the contributions to the deviation vector ψ into the components aligned with the different directions in the N-dimensional vector space of the phases of the N rotors.

We are interested in the temporal behaviour of the approach to the synchronised state for which $\psi_1 = \psi_2 = \cdots = \psi_N$. We can see from Eq. (8.121) that all the coordinates of ψ decay exponentially, each with a rate set by λ_q. Large values of λ_q correspond to a rapid decrease and small values to a slow vanishing of the coordinate along that eigenvector.

The coordinates of ψ along the eigenvectors with $\lambda_q = 0$ are time-independent. These vectors represent phase deviations ψ_i which are constant across the nodes i corresponding to the cluster associated with this eigenvector, see Box 8.2 for details. This means that if we align the deviation vector ψ with one of the eigenvectors which has $\lambda_q = 0$, the phases across this cluster will be given by

$$\theta_i(t) = \omega t + \psi_i(t) = \omega t + \text{constant},$$

corresponding to full synchronisation. The approach towards the synchronised state is controlled by the components of ψ along the eigenvectors with positive eigenvalues. The coefficient $c_q(0) \exp(-\epsilon \lambda_q t)$ becomes negligible when $\lambda_q t > 1$. Hence we expect that the deviation ψ will experience a rapid decay each time t passes through one of the values

$$t = \frac{1}{\lambda_N}, \frac{1}{\lambda_{N-1}}, \ldots, \frac{1}{\lambda_2}, \tag{8.122}$$

where we ordered the eigenvalues according to $0 < \lambda_2 < \lambda_3 < \cdots < \lambda_N$. Full synchronisation sets in on time scales longer than $1/\lambda_2$. This behaviour was observed for a highly symmetric and hierarchical network in [26]; the situation is more involved for more realistic and less symmetric networks [12, 25, 273, 314, 339, 395]. Nevertheless, the analysis of the eigenvalues and eigenvectors of the Laplacian has been used to analyse functional brain activity [6, 201].

Finally, just for completeness, we mention that expansion of ψ in Eq. (8.121) is an example of a general approach called the **graph fourier transform** or the **temporal graph signal transform**. These are generalisations of the very powerful Fourier transform used extensively when solving differential equations or analysing the frequency content in a time signal. See Box 10.1.

The usual Fourier transform [446] takes a function $f(x)$ from the real numbers and writes it as a sum of the eigenvectors of the one-dimensional Laplacian, namely in somewhat schematic notation

$$f(x) = \sum_q c_q e_q(x), \tag{8.123}$$

where the eigenvectors in this case are functions which solve

$$-\frac{d^2}{dx^2} e_q(x) = \lambda_q e_q(x). \tag{8.124}$$

This leads to complex exponential functions, also called plane waves:

$$e_q(x) = e^{\sqrt{\lambda_q} i x}. \tag{8.125}$$

When we consider a function $F_i(t)$, defined on the nodes of a network, the nodes $i = 1, 2, \ldots, N$ correspond to the x variable in Eqs. (8.123), (8.124) and (8.125). So by analogy we expand for each value of the time variable t the $F_i(t)$ on the set of eigenvectors for the network Laplacian v_q, see Eq. (8.109). We write the function $F_i(t)$ as the vector

$$F(t) = \begin{pmatrix} F_1(t) \\ F_2(t) \\ \vdots \\ F_N(t) \end{pmatrix}, \tag{8.126}$$

and arrive at the expansion

$$F = \sum_q c(t) v_q. \tag{8.127}$$

This was what we did above for the phase deviation $\psi_i(t)$ of the rotor variable $\theta_i(t)$.

This kind of Fourier analysis on networks has been used to analyse temporally varying functions on networks such as optical or brain scan signals [207, 402] or the dynamical states of rotors placed on the nodes of a network [134].

Summary: We discussed network theory as the mathematical language of sets of inter-linked nodes and focused on:

- How the importance of a given node can be characterised in various ways according to the role played by the node.
- That eigenvector centrality is an example of a self-consistent measure of importance. Google's PageRank is another.
- The local structure of networks is captured by measures such as the degree distribution.
- How more global structures can be characterised by measures such as betweenness or community detection.
- The analysis of the relation between dynamics on, or of, networks and the topological structure of the network.
- That the analysis of dynamical aspects is relevant to the analysis of real complex systems such as ecosystems, brain dynamics and epidemiology, for example.

8.6   Further Reading

General audience:

The very entertaining and broad-reaching book *Six Degrees: The New Science of Networks* [486] gives a very readable introduction to the hectic development of this field of research.

The popular science book *Linked: How Everything is Connected to Everything Else and what it Means for Business, Science, and Everyday Life* [46] by Albert-László Barabási and Jennifer Frangos gives a lively introduction.

Intermediate level:

The introductory text *A First Course in Network Science* [290] by Filippo Menczer, Santo Fortunato and Clayton A. Davis offers a non-technical but still scientifically sophisticated introduction which may be of particular interest to people who like to simulate networks.

A very instructive introduction to the analysis and simulation of networks and dynamics of and on networks can be found in Chaps. 15, 16, 17 and 18 in the book *Introduction to the Modeling and Analysis of Complex Systems* [379] by Hiroki Sayama.

An exceptionally comprehensible and well-explained introduction including mathematics is given in the book *Networks: An Introduction* [312] by Mark Newman.

The lucid book *Synchronization: From Coupled Systems to Complex Networks* [62] contains a very broad range of examples of dynamics supporting synchronisation.

Advanced level:

Dynamics on networks is addressed with clarity and rigour in the book *Random Graph Dynamics* [120] by Rick Durrett.

Network theory from the perspective of pure mathematics is discussed in two classic volumes:

- At an introductory level in *Introduction to Graph Theory* [452] by Richard J. Trudeau.
- At a more advanced level in *Algebraic Graph Theory* [58] by Norman Biggs.

8.7 Exercises and Projects

Exercise 1
Check that node α in Fig. 8.1 has $l_\alpha = 14/11$ and therefore $C_\alpha \approx 0.79$ and that node γ has $l_\gamma = 26/11$ and accordingly $C_\gamma \approx 0.42$.

Exercise 2
Study the relation between network structure and the dynamics of of an infectious disease. Simulate spreading on different network structures such as random Gilbert and a network with a power-law distributed degree distribution. See also [92, 231].

Exercise 3

Derive an expression for the change in $Q[C]$ corresponding to reassigning a node from one existing cluster to another cluster, see [61].

Exercise 4

Calculate Q in Eq. (8.32) and Eq. (8.34) for each of the two partitionings into clusters given in Fig. 8.4.

Exercise 5

Check the numerical calculation of $Q[C]$ in Eqs. (8.32) and (8.34) for the network in Fig. 8.5.

Exercise 6

Apply the Louvain algorithm to the network in Fig. 8.4.

Exercise 7

A study of the paper [451].

(a) Explain the advantages of the modern view of communities.

Assume you have access to the social network structure of all the students at a certain university. You want to identify communities among the students.

(b) Which algorithm would you apply?
(c) How would you go about assessing the validity of the identified communities?

Exercise 8

The criterion of Eq. (8.57) for the existence of a giant component is equivalent to the so-called Molloy and Reed [312] condition

$$\langle k^2 \rangle - 2\langle k \rangle > 0. \tag{8.128}$$

Use generator formalism for the exponential degree distribution in Eq. (8.59) to check that the result in Eq. (8.64) is consistent with Eq. (8.128) and that $\langle k \rangle = 1/2$ at the onset of the giant component. *Remark:* This result is a reminder that the result that the giant component appears when $\langle k \rangle = 1$ given in Figs. 11.1 and 11.4 in [312] is specific to the Gilbert or Erdös–Rényi network.

Exercise 9

We will look at walks on a network. Consider a network of N nodes. Assume that each pair of nodes are connected by a link with probability p. In the following we neglect loops and work in the limit of large N. Start from a randomly chosen node and perform a random walk from node to node. You are not allowed to walk back along the link you arrived along. See Fig. 8.16. The walk in panel A is allowed but the walk in panel B is not.

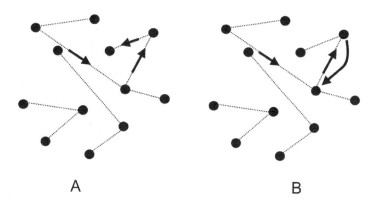

Figure 8.16 Walk on a directed network. A walker can move from one node to another if these are connected by a link. It is not permitted to turn around and move back along the link one has just arrived via. So the walk in panel B is illegal but the walk in panel A is permitted.

Think of increasing p from zero and determine the first value of p at which, with probability larger than zero, the walk may be able to reach any of the $N - 1$ other nodes.

Project 1 – Mean-field theory of percolation: giant cluster
Note that this project is of direct relevance to the concepts of a giant component in Sec. 8.4.

The process consisting of **percolation** on a lattice is another paradigmatic model, which conceptually is of great relevance to complex systems [84, 263, 424]. Consider a d-dimensional hypercubic lattice of size $N = L^d$. Assume that a site is occupied (by whatever you fancy: a coin, an elephant, a tree, ...) with probability p. On average, the number of occupied sites is $N_{occ} = pN$. We define clusters as sets of occupied sites that are connected through nearest-neighbour adjacency. See Fig. 8.17.

This is a purely geometric process. No interaction is assumed between the sites, so in this respect the situation is simpler than in the Ising model in Sec. 6.2.

In the usual spirit of mean-field theory we have the following identity for the probability P_∞ that a randomly selected site belongs to the system-spanning giant cluster:

$$1 - P_\infty = (1 - pP_\infty)^{q_c}. \qquad (\star)$$

Here, p = density of occupied site.

- First satisfy yourself that the expression above follows from the fact that:
  (1) $1 - P_\infty$ = probability site doesn't belong to the system spanning cluster.
  (2) pP_∞ = probability that a neighbour site is occupied and belongs to the giant cluster.
  (3) For a site not to belong to the giant cluster only, none of its neighbours should be part of the giant cluster.

Figure 8.17 Cluster of occupied sites. We see two clusters of size 1, one of size 3, one of size 5 and one systems-spanning **giant cluster**.

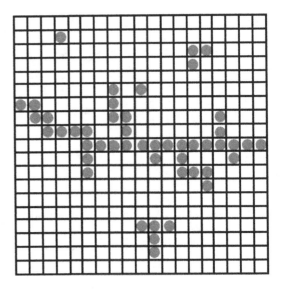

- Next use the equation to verify that $p_c = 1/q_c$ is a critical value for p, in the sense that for $p < p_c$ we have $P_\infty = 0$ and for $p > p_c$ we have $P_\infty > 0$.
- Consider the critical exponent β given by

$$P_\infty \propto (p - p_c)^\beta$$

for p in the vicinity above p_c. By expanding Eq. ($\star$) about p_c show that $\beta = 1$.

Project 2 – Configurational models and centrality measures

This project uses computer simulations to study networks with different degree distribution and various centrality measures.

(a) Use the configurational mode described in Sec. 8.5.1 to numerically generate networks which have exponential, algebraic or binomial degree distributions. Choose a reasonable number of nodes, not too small but neither so large that your computer struggles to run the simulations.

(b) For each of the above network categories, determine the eigenvector centrality for the nodes and inspect visually how the nodes with high rank are situated in the network.

(c) Use available library routines to compute the PageRank for the nodes of the networks considered in (b) and compare with the eigenvector centrality.

(d) Compute the closeness and betweenness centralities for the nodes of the networks in (a) and compare the behaviour across the different functional forms of the degree distribution.

(e) Use community detection algorithms available online, e.g. the Louvain algorithm, to determine the community structure of the networks in (a) and compare the behaviour across the different functional forms of the degree distribution.

Project 3 – Return times on networks

Consider a network given by the adjacency matrix $a_{ij} = a_{ji} = 1$ with probability p and $a_{ij} = a_{ji} = 0$ with probability $1 - p$.

(a) In the limit of large N and fixed pN, compute the degree distribution and calculate the connectivity.

(b) Now exclude the nodes of degree zero and neglect degree correlations. Use the distribution derived in (a) in the limit of large N and Np fixed to express the average return time $\langle r \rangle$ for the nodes of non-zero degree in terms of $Ei(pN)$, where $Ei(pN)$ denotes the exponential integral.

(c) Compare the expression in (b) with the estimate one obtains by the definition of r_i in Eq. (8.99), replacing k_i by the average degree and L by the average number of links. Consider the limit $Np \to 0$ and $N \gg 1$.

9 Information Theory and Entropy

Synopsis: Assume we are able to obtain the joint probability for a set of time series representing a complex system. Based on the joint probabilities, information theory can help to analyse the nature of the interdependence in the system. It is particularly important to be able to distinguish between different types of emergent behaviour, such as synergy or redundancy.

When we try to develop an understanding of the aggregated behaviour of multitudes of interacting components, we need formalisms that enable us to identify interactions and dependencies operating at different levels of aggregation. We may have available measured time series of various kinds of activity. In economics we could hope to have data on transactions between individual companies, or economical data for different geographical regions, or exchange rates between different currencies. In neuroscience we might have access to recordings of electric activity, firing of individual neurones, or fMRI data representing activity within some millimetre-size volumes or EEG data picking up activities from regions of centimetre size. Upon inspection, these time series will not immediately show us causal connections or even inform us about how one part of the system relates to another, or how activity at one level influences activity at other levels.

The large number of components and their many ways of influencing each other may produce time evolution that at first glance appears erratic and random. To penetrate what may look like random unrelated fluctuations, we can look for statistical trends and use probability theory in the form of information theory to establish correlations and even try to obtain measures of possible causal relations. Our focus in this chapter is to develop an understanding of how information theory can help us to identify emergent behaviour hidden in data. For a good basic introduction to information theory see [427] and for a more comprehensive and mathematically advanced text see [87].

We will assume that a set of N time series $X_i(t)$, with $i = 1, 2, \ldots, N$, are available, so we have a data vector

$$\mathbf{X}(t) = (X_1(t), X_2(t), \ldots, X_N(t)). \tag{9.1}$$

We call each of the X_i an observable. This could, for example, be the recording of 64 electrodes from an EEG recording. From the observations we can form histograms which, if we have sufficient data, will make it possible to estimate the probability distribution[1]

---

[1] For simplicity we consider observables X_i, which assume discrete values, x_i, and therefore consider probabilities for discrete variables.

$$P_{\mathbf{X}}(x_1, x_2, \ldots, x_N) = \text{Prob}\{X_1 = x_1, X_2 = x_2, \ldots, X_n = x_n\}. \tag{9.2}$$

If we have access to this joint distribution we will of course also have access to the marginal probabilities $P_{X_i}(x_i)$ for the individual observables and to any combination of joint distributions such as, say, $P_{X_i, X_j}(x_i, x_j)$ by summing over the other variables. For example

$$P_{X_i}(x_i) = \sum_{x_1} \cdots \sum_{x_{i-1}} \sum_{x_{i+1}} \cdots \sum_{x_N} P_{\mathbf{X}}(x_1, x_2, \ldots, x_N). \tag{9.3}$$

In the following sections we will look at how, by use of information theory, we can use knowledge about $P_{\mathbf{X}}(x_1, x_2, \ldots, x_N)$ to identify emergent behaviour. At the most basic level we would say that whenever

$$P_{X_i, X_j}(x_i, x_j) \neq P_{X_i}(x_i) P_{X_j}(x_j)$$

some degree of emergence, though not necessarily one of great interest, may be at play in the sense that the behaviour of X_i and X_j together encapsulated by the joint distribution $P_{X_i, X_j}(x_i, x_j)$ is different from what we would expect if each behaved as if in isolation. In the latter case of two entirely independent observables, we of course have $P_{X_i, X_j}(x_i, x_j) = P_{X_i}(x_i) P_{X_j}(x_j)$.

We need methods that can distinguish between correlations and causal relations and between indirect and direct relations. Figure 9.1 shows relations between three observables X_1, X_2 and X_3. In the left panel we assume that at any moment in time t the value $X_3(t)$ at time step t of observable X_3 is influenced directly by the value $X_2(t - 1)$ of observable X_2, which in turn is influenced by the value $X_1(t - 2)$ of observable X_1 at time step $t - 2$. A mathematical formalism should be able to identify the *direct* causal relations from X_1 to X_2 and from X_2 to X_3 and distinguish this kind of relation from the correlations, which the $X_1 \to X_2 \to X_3$ will generate between X_1 and X_3.

In the right panel of Fig. 9.1 the value of the two observables X_2 and X_3 is directly influenced by the value of the observable X_1 in the previous time step. This will lead to X_2 and X_3 being correlated, although there is no direct causal relation between X_2 and X_3. A good mathematical formalism should also be able to distinguish the correlations between X_2 and X_3 from a causal relation between the two observables.

In the next section we will look at information theory based on the entropy used by Shannon. After that we will briefly touch on the current interest in alternative

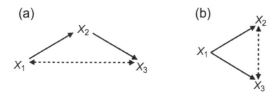

Figure 9.1 Causal relations (solid directed arrows) and correlations (dashed double-headed arrows) between three observables. The solid-line causal structure in (a) is called a chain and the one in (b) is called a fork [337].

entropies and how they may be helpful when dealing with complex systems with strong interdependencies amongst the components.

9.1   Information Theory and Interdependence

To simplify the notation we consider two stochastic variables X and Y assuming values x and y, respectively. In concrete applications X and Y can be two times series of two measured observables. We will discuss how approaches from information theory seek to quantify the interrelatedness between X and Y through an analysis of $P_{XY}(x,y)$.

Probably the simplest way to check for relations between two observables X and Y is to compute the Pearson **correlation coefficient**, given by

$$C_{XY} = \frac{\langle (X - \langle X \rangle)(Y - \langle Y \rangle) \rangle}{\sigma_X \sigma_Y} = \frac{E[(X - \mu_X)(Y - \mu_Y)]}{\sigma_X \sigma_Y}. \tag{9.4}$$

Here σ_X and σ_Y denote the standard deviations, given by

$$\sigma_X = \sqrt{E[(X - \mu_X)^2]}. \tag{9.5}$$

On the right-hand side of Eq. (9.4) we include two frequently used notations. The angular brackets $\langle X \rangle$ we have used for example in Chap. 6 to denote the average of X and the $E[X]$ for average, or expectation, of X is routinely used in probability theory and information theory. Also note that by multiplying out the parentheses we get the alternative expression

$$C_{XY} = \frac{\langle XY \rangle - \langle X \rangle \langle Y \rangle}{\sigma_X \sigma_Y}. \tag{9.6}$$

The correlation coefficient assumes values between -1 and 1 and if $P_{X,Y}(x,y) = P_X(x)P_Y(y)$ we have $C_{XY} = 0$, but the reverse it not necessarily true. That is, one may have $C_{XY} = 0$ but still have $P_{X,Y}(x,y) \neq P_X(x)P_Y(y)$. This means that we cannot be sure that X and Y are independent even if the correlation coefficient between them is zero. The reason for this is that the factor $(x - \mu_X)(y - \mu_Y)$ can assume both positive and negative values, which can make

$$C_{XY} = \frac{1}{\sigma_X \sigma_Y} \sum_x \sum_y P_{X,Y}(x,y)(x - \mu_X)(y - \mu_Y)$$

add up to zero even if $P_{X,Y}(x,y) \neq P_X(x)P_Y(y)$. On the other hand, if $P_{X,Y}(x,y) = P_X(x)P_Y(y)$ we always have

$$C_{XY} = \frac{1}{\sigma_X \sigma_Y} \sum_x \sum_y P_{X,Y}(x,y)(x - \mu_X)(y - \mu_Y)$$

$$= \frac{1}{\sigma_X \sigma_Y} \sum_x \sum_y P_X(x)P_Y(y)(x - \mu_X)(y - \mu_Y)$$

$$= \frac{1}{\sigma_X \sigma_Y} \sum_x P_X(x)(x - \mu_X) \sum_y P_Y(y)(y - \mu_Y)$$

$$= \frac{1}{\sigma_X \sigma_Y} \left[\sum_x P_X(x)x - \mu_X \sum_x P_X(x) \right]\left[\sum_y P_Y(y)y - \mu_Y \sum_y P_Y(y) \right]$$

$$= \frac{1}{\sigma_X \sigma_Y} \left[\sum_x P_X(x)x - \mu_X \right]\left[\sum_y P_Y(y)y - \mu_Y \right]$$

$$= 0 \cdot 0.$$

The correlation coefficient can be used as a first probe of possible interdependencies between observables. It is given in terms of averages that can be estimated from the data by use of empirical estimates such as $\mu_X \simeq \sum_{i=1}^{N} x_i/N$, even in situations where there are insufficient data to be able to estimate the probability distributions.

But if one is able to obtain estimates of $P(x,y)$, a more fundamental measure of interdependence consists of the **mutual information** given by

$$I(X;Y) = \sum_x \sum_y P_{X,Y}(x,y) \log \frac{P_{X,Y}(x,y)}{P_X(x)P_Y(y)}, \tag{9.7}$$

which assumes values $I(X;Y) \geq 0$; and $I(X;Y) = 0$ only if $P_{X,Y}(x,y) = P_X(x)P_Y(y)$.

As measures of the interdependence between two observables X and Y, the correlation coefficient and the mutual information have different advantages and drawbacks. To obtain an estimate of C_{XY} we only need to compute averages of the data available. This requires access to less data than needed to determine $I(X;Y)$, for which we need to estimate the joint probability $P_{X,Y}(x,y)$ and the marginal probabilities $P_X(x)$ and $P_Y(y)$, which can be done from histograms of the available data sets but is statistically more demanding than the empirical estimates of averages.

On the other hand, if enough data are available so $I(X;Y)$ can be estimated, more basic insights can be obtained from the mutual information. Firstly, $I(X;Y) = 0$ ensures X and Y are independent in contrast to $C_{XY} = 0$. Further, as we will see in a moment below, $I(X;Y)$ can be used as a building block for information-theoretic estimates of possible causal relations between X and Y when data are available in the form of time series.

Both the correlation coefficient in Eq. (9.4) and the mutual information in Eq. (9.7) are symmetric in X and Y. They can be used to quantify the degree of interdependence between X and Y, but their symmetry means that they cannot be used to determine a direction of the flow of any potential dependencies. That is, they do not allow us to measure if X might in some way have a causal influence on Y or if the opposite is the case. If X and Y are represented by time series it is possible to use the mutual information as a basis for a measure that can probe possible causal relations.

The economist Clive Granger suggested in 1969 that a causal relation may exist between two time series X and Y if knowledge of the past of X *and* Y makes us able to predict the future of X with higher success rate than if we only predict the future of X based on the past of X itself [170]. This strategy can of course be implemented in many different ways. Granger's original approach made use of fitting the times series to autoregression series and then studied the correlations of the autoregression series. This is done by finding the best-fit coefficients $C_{a,b}(i)$ with $a, b \in \{x, y\}$ and $i \in \{1, 2, \ldots, p\}$ for some integer p, which minimises the fluctuating part $\eta_X(t)$ and $\eta_y(t)$ in the following set of equations:

$$
\begin{aligned}
X(t) &= \sum_{i=1}^{p} C_{x,x}(i) X_{t-i} + \sum_{i=1}^{p} C_{x,y} Y_{t-i} + \eta_X(t) \\
Y(t) &= \sum_{i=1}^{p} C_{y,x}(i) X_{t-i} + \sum_{i=1}^{p} C_{y,y} Y_{t-i} + \eta_Y(t).
\end{aligned}
\tag{9.8}
$$

If the variance of $\eta_X(t)$ can be reduced by including terms $C_{xy}(i) \neq 0$, we say that knowledge of Y helps us to predict X and that there is Granger causality from Y to X. For more details see e.g. [421]. We notice that this approach combines Granger's definition of a causal relation (improved forecasting of X by including knowledge of the past of Y) with the assumption of the linear representation in Eq. (9.8). Commonly we will not be able to justify the linear model assumption and in this sense it is preferable to avoid the assumption about linearity and instead test in a more direct way how much the knowledge of the past of one time series improves our ability to forecast the future of another.

A very simple implementation of Granger's statistical measure of causal relation, which works directly with the data without any model assumption, consists in time-delayed correlation coefficients. We look at two observables X and Y measured at discrete time steps denoted by t and we denote the values assumed by X and Y at these time steps by $x(t)$ and $y(t)$. To obtain a measure of how one observable may 'drive' or 'causally' influence the other, we can compare the two correlation coefficients across a time delay. Let us denote by X^- the observable that assumes the value $x(t-1)$ of the previous time step, i.e. $X_t^- = x(t-1)$ and similar for Y^-. We use Eq. (9.6) and define the two correlation coefficients $C_{X \to Y}$ and $C_{Y \to X}$ by the following expression:

$$
\begin{aligned}
C_{X \to Y} &= \frac{\langle X^- Y \rangle - \langle X^- \rangle \langle Y \rangle}{\sigma_X \sigma_Y} \\
C_{Y \to X} &= \frac{\langle Y^- X \rangle - \langle Y^- \rangle \langle X \rangle}{\sigma_X \sigma_Y}.
\end{aligned}
\tag{9.9}
$$

A strong imbalance like $|C_{X \to Y}| \gg |C_{Y \to X}|$ might indicate a causal direction from X to Y. A simple example illustrates this.

We consider examples of observables denoted by capitals X, Y, etc. and their corresponding realisations in terms of time series $x(t)$, $y(t)$, etc.

Times Series Example 1: Two observables

$$x(t) = \xi_1(t)$$
$$y(t) = \alpha x(t-1) + \xi_2(t). \tag{9.10}$$

Here α is a constant while ξ_1 and ξ_2 denote two independent stochastic variables and $\xi_1(t)$ and $\xi_2(t)$ the values they assume at time step t. Neglecting for the moment the factor $\sigma_X \sigma_Y$, we have

$$C_{X \to Y} = \langle X(t-1)Y(t) \rangle - \langle X(t-1) \rangle \langle Y(t) \rangle \tag{9.11}$$

where we assume an average over the distribution of ξ_1 and ξ_2 and over the time variable t. From the definitions in Eq. (9.10) we get

$$
\begin{aligned}
C_{X \to Y} &= \alpha \langle \xi_1^2(t-1) \rangle + \langle \xi_1(t-1)\xi_2(t) \rangle - \alpha \langle \xi_1(t)\xi_1(t-1) \rangle - \langle \xi_1(t)\xi_2(t) \rangle \\
&= \alpha \langle \xi_1^2 \rangle - \alpha \langle \xi_1 \rangle^2 + \langle \xi_1 \rangle \langle \xi_2 \rangle - \langle \xi_1 \rangle \langle \xi_2 \rangle \\
&= \alpha \sigma_{\xi_1}^2.
\end{aligned}
\tag{9.12}
$$

We made use of the assumption that ξ_1 and ξ_2 are independent, which implies that the values of $x(t)$ and $x(t')$, and $y(t)$ and $y(t')$, at different time steps t and t' are independent.

The time-delayed correlations from Y to x are given by

$$
\begin{aligned}
C_{Y \to X} &= \alpha \langle \xi_1(t-2)\xi_1(t) \rangle + \langle \xi_2(t-1)\xi_1(t) \rangle - \alpha \langle \xi_1(t-2)\xi_1(t) \rangle - \langle \xi_2(t-1)\xi_1(t) \rangle \\
&= \alpha \langle \xi_1 \rangle^2 + \langle \xi_2 \rangle \langle \xi_1 \rangle - \alpha \langle \xi_1 \rangle^2 - \langle \xi_2 \rangle \langle \xi_1 \rangle \\
&= 0.
\end{aligned}
\tag{9.13}
$$

Time-delayed correlation coefficients were used e.g. to study the network structure of earthquakes in Greece, see [83]. The Granger causality considers in principle the dependence of one time series on the entire past of the other time series and not necessarily just the previous time step as we did in this example. To illustrate the effect of including longer time lags than just one time step we look at three time series as given in the following example, where we use the same notation as in Example 1.

Times Series Example 2: Three observables

$$x(t) = \xi_1(t)$$
$$y(t) = \alpha x(t-1) + \xi_2(t) \tag{9.14}$$
$$z(t) = \beta y(t-1) + \xi_3(t).$$

The correlation coefficients delayed by one time step between Z and Y, and Y and X, are non-zero and furthermore

$$C_{X \to Z} = \langle X(t-1)Z(t) \rangle - \langle X(t-1) \rangle \langle Z(t) \rangle = 0.$$

This correctly indicates the drive from X to Y and from Y to Z. However, when we look at correlation coefficients with a time lag of two we find e.g.

$$C_{X \to Z}^{(-2)} = \langle X(t-2)Z(t) \rangle - \langle X(t-2) \rangle \langle Z(t) \rangle \neq 0$$

because Z depends on the value of X two time steps back through the dependence of Z on Y. However, the causal effect of X on Z is indirect in nature since it is mediated through Y and we would like measures that can distinguish direct causal influences from indirect ones. This can be achieved in various ways by looking beyond correlations, for instance by involving the mutual information. This will also address the limitations mentioned above after Eq. (9.6) concerning the relation between interdependence and the correlation coefficient.

Hence there are several good reasons to use the mutual information as an indicator of directed causal relationships, should sufficient data be available. One way to do this was suggested by Thomas Schreiber [388] and is called **transfer entropy**. In the next section we will return to the concept of entropy in more detail, in the context of information theory. Right now we simply accept it as part of the name of the expression defined by Schreiber. In its simplest form the transfer entropy $TE_{Y \to Y}$ from Y to X is given by

$$TE_{Y \to X} = \left\langle \log \frac{\text{Prob}\{X(t) = x(t)| Y(t-1), X(t-1)\}}{\text{Prob}\{X(t) = x(t)|X(t-1)\}} \right\rangle$$

$$= \sum_{x(t)} \sum_{x(t-1)} \sum_{y(t-1)} \text{Prob}\{X(t) = x(t), X(t-1) = x(t-1), Y(t-1) = y(t-1)\}$$

$$\times \log \frac{\text{Prob}\{X(t) = x(t)| Y(t-1), X(t-1)\}}{\text{Prob}\{X(t) = x(t)|X(t-1)\}}. \tag{9.15}$$

The transfer entropy is equivalent to Granger's original use of autoregression series when the signals X and Y are joint Gaussian processes [48]. By this is meant that the probability densities $\text{Prob}\{X(t)=x\}$ and $\text{Prob}\{Y(t)=y\}$ are Gaussian and the autocorrelations and cross correlations of X and Y are fully specified by the covariance matrix $\text{cov} X(t_1) Y(t_2)$. See e.g. [465] for details. But not all stochastic process are Gaussian processes, see [465], and it can be difficult to check to what extent a measured time series satisfies the Gaussian assumptions. In this sense the transfer entropy may be a safer choice since it offers a consistent estimate of possible statistical causal relations without assuming Gaussian behaviour or relying on the model-dependent estimation via autoregression.

Let us decipher this expression in Eq. (9.15). When we are dealing with stationary processes the probabilities will be independent of time and the actual value of time step t and $t+1$ will not matter. The important point is that Eq. (9.15) relates the observables across time steps. The numerator in the logarithm is the probability that the observable X at time step t assumes a certain value $x(t)$ conditioned on the values assumed by X itself and the other observable Y in the previous time step $t-1$. If there is no dependence of the value X at time step t on the value of Y at time step $t-1$ we have $\text{Prob}\{X(t) = x(t)| Y(t-1), X(t-1)\} = \text{Prob}\{X(t) = x(t)|X(t-1)\}$ and the argument of the logarithm is equal to one and therefore the logarithm is equal to zero. The transfer entropy is a measure of how much, on average, the behaviour of $X(t)$ is related to Y in the previous time step. We note that the expression is not symmetric in X and Y but explicitly looks for how the future value of X depends on the Y value in the past. This means that we can relate a direction to the transfer entropy and think of it as a measure of how much Y influences X. Of course we can also look at the influence in the opposite direction by swapping X and Y in Eq. (9.15) to obtain $T_{X \to Y}$. If, for example, we find that $T_{Y \to X} \gg T_{X \to Y}$ we can take this as indicative that a causal mechanism may be at play that allows the observable Y to drive the observable X.

For simplicity in Eq. (9.15), we only represent the past by the time step $t - 1$ just before time step t. From Examples 1 and 2 above, we know this can be too limited and in reality one should condition on as much of the past of the time series as is *necessary*. A practical problem is that the more of the past one includes, the more difficult it is to estimate the histograms needed to compute the mutual information. To figure out how much of the past to include and to distinguish direct from indirect causation can be a difficult problem and we will return to this discussion, see Sec. 9.3, after we have looked in the next section at the concept of entropy from the perspective of information theory.

Let us finish this section by pointing out that the concept of Granger causality is used to find potential causal relations between observables from an analysis of their time series. Judea Pearl [337] has developed this approach into what he calls a calculus of causality. Pearl's formalism is nicely visualised by use of diagrams of the form depicted in Fig. 9.1.

9.2 Entropy and Estimates of Causal Relations

Entropy can mean different things to different people and in different situations. Here we will look at the entropy studied by Claude Shannon in [394].

At first it may not be clear in what sense Shannon's entropy $H(X)$ of a stochastic variable X, defined as

$$H(X) = \sum_x P_X(x) \log \frac{1}{P_X(x)} = -\sum_x P_X(x) \log P_X(x), \qquad (9.16)$$

has to do with information.[2] When we use the term 'information' in daily life we think of some piece of guidance or knowledge such as the fact that in the UK and Japan you have to drive on the left side of the road, whereas in Germany and Canada you drive on the right. Seen from this perspective, Eq. (9.16) does not seem to have much to do with information. But if we, like Shannon, think of the probability $P_X(x)$ as being related to how

---

[2] Science often makes use of bewildering names. One is certainly justified in asking why Shannon's expression is called entropy. This goes back to thermodynamics. In 1865 Clausius introduced the word 'entropy' in his study of transformation between heat and work. The word is derived from the greek *tropos*, meaning transformation, and also turn and change. The expression used by Shannon is, in statistical mechanics, related to Clausius's thermodynamic entropy. Apparently van Neumann pointed out to Shannon that the expression was known in statistical mechanics under the name entropy.

big our surprise is when the event $X = x$ actually happens, then it makes sense to consider $\log \frac{1}{P_X(x)}$ as a measure. If $P(x)$ is very small, corresponding to the event $X = x$ being unlikely and hence not something we really expect to happen, then our surprise when it happens, like throwing 20 heads with a coin in a row, is the greater, and so is $\log \frac{1}{P_X(x)}$ and therefore it does make sense to say that $\log \frac{1}{P_X(x)}$ can be considered as a measure of the surprise related to the event $X = x$. From this perspective the expression in Eq. (9.16) is the average surprise we are expected to experience when dealing with the observable X.

In our daily life, the terminology of surprise and information is not exactly the same. The *Oxford English Dictionary* defines information as 'facts provided or learned about something or someone' and surprise is defined as 'an unexpected or astonishing event, fact, etc.'. Nevertheless, the surprise measured by $H(X)$ can be seen as providing information about X. Namely, $H(X)$ assumes small values when the distribution $P(X)$ is narrow and large values if $P(x)$ is broad. The ultimate case is $P(x) = 1$ for some value $x = X_0$ and $P(x) = 0$ for all other values of x. In this case $H(X) = 0$. The opposite case is $P(x) = 1/N$ for each of the N values $X_1, X_2, \ldots, X_N$ the observable X may assume. In this case $N(X) = \log N$. This means that the value of $H(X)$ does tell us how much information one observation of a specific value $X = x$ is able to give us about the observable X. If the Shannon entropy of the observable is small, we know that a specific observation may be more representative of what is typical to expect, i.e. more informative, about how X behaves than if the entropy is large.

The fact that $H(X)$ is related to the shape of the distribution suggests that one may be able to determine the distribution of X from $H(X)$. This is exactly what the maximum entropy principle does. We will discuss this principle in some detail below in Sec. 9.4; here we focus on how we can extract information about dependencies, including causal relations, from a knowledge of joint probability distributions.

Conditioning is the essential tool employed by information theory to extract knowledge stored in probability distributions. The idea is to check for relations by investigating if the likelihood of observing some event depends on certain conditions or not. Think of the connection between lung cancer and smoking. What we now know as a causal link driven by the effect of certain chemicals in the smoke on the lung tissue was first discovered as a statistical possibility by looking at the frequency of lung cancer amongst smokers and non-smokers, so the event of lung cancer conditioned on either not smoking or smoking. Had no relation existed, the frequency of lung cancer amongst smokers and non-smokers would be the same.

Such considerations are captured mathematically in the following way. Consider two observables X and Y (e.g. $X =$ lung cancer, $Y =$ smoker). We imagine we have access to the joint distribution $P_{XY}(x, y)$ and we want to understand if X and Y are related by analysing $P_{XY}(x, y)$. To do this we look at the conditioned probability given by

$$P_{X|Y}(x, y) = \frac{P_{X,Y}(x, y)}{P_Y(y)}. \tag{9.17}$$

If X and Y are completely unrelated, i.e. independent, we have $P_{X,Y}(x, y) = P_X(x)P_Y(y)$ and therefore $P(X|Y) = \frac{P_{X,Y}(x,y)}{P_Y(y)} = P_X(x)$. We can combine these considerations with Shannon's entropy and introduce the conditioned entropy. In accordance with the definition of the entropy of the average surprise of a distribution, the conditioned entropy is defined as the average surprise of X conditioned on Y, averaged over the probability that the pair of values (x, y) is observed, i.e.

$$
\begin{aligned}
H(X|Y) &= \sum_x \sum_y P_{X,Y}(x,y) \log \frac{1}{P_{X|Y}(x,y)} \\
&= -\sum_x \sum_y P_{X,Y}(x,y) \log \frac{P_{X,Y}(x,y)}{P_Y(y)}.
\end{aligned}
\tag{9.18}
$$

If X and Y are independent, $P_{X,Y}(x,y) = P_X(x)P_Y(y)$, we have

$$
\begin{aligned}
H(X|Y) &= \sum_x \sum_y P_{X,Y}(x,y) \log \frac{1}{P_{X|Y}(x,y)} \\
&= -\sum_x \sum_y P_X(x)P_Y(y) \log P_X(x) \\
&= -\sum_x P_X(x) \log P_X(x) \sum_y P_Y(y) \\
&= H(X).
\end{aligned}
\tag{9.19}
$$

For independent events it makes no difference to condition, and the surprise contained in the distribution $P_{X,Y}(x,y)$ is equal to the one contained in $P_X(x)$.

We can now express the mutual information in Eq. (9.7) in terms of entropies in a number of ways:

$$
\begin{aligned}
I(X;Y) &= H(X) + H(Y) - H(X,Y) \\
&= H(X,Y) - H(X|Y) - H(Y|X) \\
&= H(X) - H(X|Y) \\
&= H(Y) - H(Y|X).
\end{aligned}
\tag{9.20}
$$

Here $H(X,Y)$ denotes the entropy of the joint distribution $P_{X,Y}(x,y)$. Each expression for $I(X;Y)$ in terms of different entropies can be considered as different ways of exposing the interdependence between X and Y captured by $I(X;Y)$.

The first equality in Eq. (9.20) looks at how much the total entropy, or total average surprise, contained in each observable considered individually differs from simultaneously observed. The second equation compares the simultaneous observation to the sum of looking at each of the two observations conditioned on the other. The last two equations consider how much of a difference it makes to the entropy if we observe one variable, say X on its own, or we confine the observation of X to those cases which also satisfy Y.

Obviously, this kind of information-theoretic Granger causality does not tell us what might be the causal mechanism in terms of specific processes, but it can be used to identify the possibility of some underlying causality relation.

The transfer entropy is also related to differences in the conditioned entropies introduced in Eq. (9.18), and hence mutual information. To make this connection we introduce the notation X^+ to denote the future of the time series X starting from an arbitrary time step t, i.e. X^+ includes all the events $\{X_{t+1}, X_{t+2}, \ldots\}$. And similarly we let X^- and Y^- denote the past of X and Y counting from time step t. So e.g. $X^- = \{X_t, X_{t-1}, X_{t-2}, \ldots\}$. We can then write

$$
\begin{aligned}
TE_{Y \to X} &= H(X^+|X^-) - H(X^+|X^-, Y^-) \\
&= I(X^+; Y^-|X^-) \\
&= I(X^+; (X^-, Y^-)) - I(X^+; X^-).
\end{aligned}
\tag{9.21}
$$

It is straightforward and worthwhile to check this identity, to get a feeling for how the information-theoretic expressions work. The first equality follows directly from the definition of $TE_{Y \to X}$ in Eq. (9.15) and the definition for the conditioned entropy in Eq. (9.18). The second equality follows by use of $I(X; Y) = H(Y) - H(Y|X)$ in Eq. (9.20). The last equality in Eq. (9.21) follows by adding and subtracting $H(X^+)$ in the right-hand side of the first equality.

Equation (9.21) indicates an illuminating interpretation of the transfer entropy. Recall that the mutual information $I(X; Y)$ can be thought of as a measure of how much information, in the sense we discussed after Eq. (9.16), is shared between the two observables X and Y. From this perspective, the second equality in Eq. (9.21) tells us that the transfer entropy from Y to X is related to the amount of information the past, given by X^-, Y^-, shares with the future X^+ compared with what the past of X, namely X^-, shares with its own future X^+.

From Eq. (9.19) we know that if X^+ does not depend on the behaviour of Y in the previous time step we have $H(X^+|X^-, Y^-) = H(X^+|X^-)$, in which case the right-hand side of the top identity in Eq. (9.21) is zero. We recall that the more narrow a distribution is, the smaller its entropy. If conditioning on Y narrows down the uncertainty in X, i.e. if knowledge about the past of Y helps to determine X, then the entropy conditioned on Y given by $H(X^+|X^-, Y^-)$ is smaller than the entropy $H(X^+|X^-)$ not conditioned on Y. The transfer entropy from Y to X is a measure of how much the uncertainty of the value of X in the next time step is lowered by conditioning on, or in other words incorporating, the past of the other observable Y.

The last equality in Eq. (9.21) tells us that the transfer entropy from Y to X is also a measure of how much 'information' the future of X and the past of Y 'share' when the past of X is taken into account.

A study of the transfer entropy in the context of identifying correlations and the phase transition in the Ising model can be found in [7].

In the next section we will discuss how transfer entropy and related measures can help to analyse possible causal relations amongst different parts of a complex system, in particular how to distinguish between indirect and direct causal influences. We will also mention the practical problem of estimating joint and conditioned probabilities needed e.g. to estimate the mutual information when handling the data sets consisting of many observables.

9.3   From Time Series to Networks

In this section we look at the practical problems one faces when applying the information-theoretic measures of causal relations discussed in the previous section. We want to connect the analysis of how the time series are interrelated to the theory of networks. Each time series will correspond to a node in the network, the question is then how to add the links between the nodes.

Let us imagine we have access to a number of time series measuring certain observables of our complex system. It could be 64 time series from the electrodes of an EEG helmet, or some thousands of time series from the individual voxels measured in an fMRI scan. Or it could be population sizes of different types in an ecosystem or data from a stock exchange. In any such cases we have a multivariate data set; at each time step[3] we record a vector $X(t) = X_1(t), X_2(t), \ldots, X_N(t)$ and we want to construct a network from $X(t)$. We will associate a node i to each time series $X_i(t)$. The next question is how we establish links between the nodes.

The simplest thing we can do is consider the correlation coefficients C_{X_i, X_j} and decide that nodes i and j are linked if C_{X_i, X_j} is larger than a chosen threshold t_{thr}. This may be a good first step, but we notice that the network may depend strongly on t_{thr}. Since $C_{X_i, X_j} \in [-1, 1]$ we will get a fully connected network if we choose $t_{thr} = -1$ and expect very few links if we choose $t_{thr} = 1$. Moreover, the links will be undirected because the correlation coefficient C_{X_i, X_j} is symmetric in i and j. So this kind of network will not be able to indicate possible causal structures.

Let us now discuss how we can construct links, which relate to possible causal influences between the N time series. Consider the probability that the variables assume a certain set of values $X_1, X_2, \ldots, X_N$ at a certain set of times $t_1, t_2, \ldots, t_N$, i.e. the probability

$$P_{X_1 X_2 \ldots X_N}(x_1, t_1; x_2, t_2; \ldots; x_N, t_N) = \text{Prob}\{X_1(t_1) = x_1, \ldots, X_N(t_N) = x_N\}. \quad (9.22)$$

If the probabilities depend on the absolute values of the times $t_1, t_2, \ldots, t_N$ we will not be able to estimate the joint probabilities in Eq. (9.22) from histograms sampled from the observed time series. But if we can assume that on the time scale we are interested in, the system can be considered to be stationary, then only the relative time differences $t_i - t_j$ are important and histograms can be computed by adding up occurrences separated by the same time interval. For example, we record how many times we observe that when X_i assumes a certain value $x_i(t)$ at a specific time t, the observable X_j assumes the value $x_j(t + \Delta)$ at time $t + \Delta$, where Δ is some time interval. To what extent any data set obtained from real systems can be considered to be stationary will of course depend strongly on the system and the time window we are looking at. Clearly EEG from the brain will not be stationary over hours across different activities like exercise, eating and

---

[3] Although time is a continuous variable, we imagine that data are collected at a fixed frequency, e.g. every 10 ms for an EEG recorder, every few seconds for an fMRI scanner and perhaps daily data for a stock exchange.

sleep. But it may be reasonable to treat the data as stationary over the some short time span during one type of activity.

Assume we are dealing with data that can be considered stationary. We now want to make use of information-theoretic methods to study the structure of the data. We can think of estimations of causal connections such as the transfer entropy given in Eq. (9.15). Ideally we want to check for all possible kinds of causal relationships between all the X_i observables. The causal relation between one specific observable X_{i_0} and the remaining X_j can, in the spirit of Granger, in principle be extracted by conditioning the future of X_{i_0} on the past of all the other X_j. To distinguish direct and indirect causal relations between the variables X_{i_0} and X_{j_0}, like in Eq. (9.14) of Example 2 in Sec. 9.1, we can isolate the direct from indirect contributions by factoring out the contribution obtained when conditioning on all the other observables *except* X_{j_0}.

We will illustrate this method in the context of transfer entropy and afterwards mention different ways of handling the difficulties arising in practice from the fact that we face exponential growth of the sample spaces every time we add another time instant to the past (or to the future) or another observable to the joint or conditioned probabilities. Recall that we have to estimate the probabilities by sampling histograms and to obtain good estimates for histograms of multivariate variables like in Eq. (9.22) over, say, the l_p last times in the past and the next l_f in the future we will need to construct histograms over a data set of $N \times l_p \times l_f$ dimensions. One will seldom have sufficient statistics to be able to do this without some careful optimisation procedure. Before we return to this important practical problem, we first look at how comparing different conditionings can help to separate direct from indirect causal influences.

Consider three observables X, Y and Z, represented by three time series $x(t)$, $y(t)$ and $z(t)$, which could be available in the form of data series. Here we will consider the following explicit example given by three equations:

$$
\begin{aligned}
x(t) &= \xi^X(t) \\
y(t) &= x(t-1) + \xi^Y(t) \\
z(t) &= y(t-1) + \xi^Z(t).
\end{aligned}
\tag{9.23}
$$

We assume ξ^X, ξ^Y and ξ^Z to be independent and identical stochastic variables. We also assume that the values of ξ^k at different time steps are independent, i.e. $P(\xi^k(t_1), \xi^k(t_2)) = P(\xi^k(t_1))P(\xi^k(t_2))$ when $t_1 \neq t_2$. When we are dealing with unknown processes or data series we cannot beforehand know how much of the past we need to include to capture possible causal relations. So in principle we ought to include the entire past as in Eq. (9.21). The transfer entropy in Eq. (9.23) between X and Z:

$$
TE_{X \to Z} = \left\langle \log \frac{\text{Prob}\{Z(t) = z(t) | Z(t-1), Z(t-2), \ldots, X(t-1), X(t-2), \ldots\}}{\text{Prob}\{Z_n = z_n | Z(t-1), Z(t-2), \ldots\}} \right\rangle
\tag{9.24}
$$

conditioned on the entire past will be non-zero because Z depends on the value assumed by x two time steps earlier, namely

$$z(t) = y(t-1) + \xi^Z(t) = x(t-2) + \xi^Y(t-1) = \xi^X(t-2) + \xi^Y(t-1) + \xi^X(t-2),$$
$$(9.25)$$

so $TE_{X \to Z}$ will pick up a non-zero contribution from the mutual information between $z(t)$ and $\xi^X(t-2)$. Of course this dependence does exist but since it is mediated through Y it is a more indirect dependence than the dependence of Y on X or Z on Y. If we only include one past time step, i.e. $X(t-1)$ and $Y(t-1)$ in Eq. (9.24), when estimating $T_{X \to Z}$ we would avoid the indirect causation between X and Z. This illustrates that the detection of causal structure will depend on the precise set of instances included in the past. We can vary the set of past time steps included in Eq. (9.24) and only include those that lead to significant values of the transfer of information from X to Z. For the example in Eq. (9.23), instead of including all $X(t-1), X(t-2), \ldots$ and $Z(t-1)$, $Z(t-2), \ldots$ we could just include $X(t-2)$.

In principle, when computing the transfer entropy in its general form, see Eq (9.21), we are supposed to include the entire set of future, and past, events. Of course in practice this is not possible. In Eq. (9.24) we only included the current time step Z_t. This is sufficient for the example considered in Eq. (9.23) but in general may not suffice. It is better to determine a subset $Z_{t+t_1}, Z_{t+t_2}, \ldots, Z_{t+t_l}$ amongst all future events $Z_t, Z_{t+1}, \ldots$ which leads to the largest possible transfer from X to Z. This strategy is called **mixed embedding** [475] and addresses in an efficient way the serious practical problem of handling the so-called **curse of dimensionlity** we face when including many past and future events in estimating the probabilities entering the information-theoretic expressions such as the mutual information or the transfer entropy.

The mixed embedding does not, however, help us to distinguish between direct and indirect causation, as for example the flow from X to Z through Y in Eq. (9.23) and situations where a direct causation exists from X to Z. This can be done in many ways. A conceptually simple method is to modify the transfer entropy by including further conditioning. For a set of N observables $\mathbf{X}(t) = X_1(t), X_2(t), \ldots, X_N(t)$ we can define a *direct* transfer entropy (also called partial transfer entropy) by the following expression [115, 329, 481]:

$$DTE_{X_i \to X_j} = \left\langle \log \frac{\text{Prob}\{X_j^+ | \mathbf{X}^-\}}{\text{Prob}\{X_j^+ | \mathbf{X}_{(-i)}^-\}} \right\rangle. \qquad (9.26)$$

Here X_j^+ denotes the future of the observable X_j and $\mathbf{X}^-$ the past of *all* the observables $X_1, X_2, \ldots, X_N$ while $\mathbf{X}_{(-i)}^-$ denotes the past of all these observables *except* the observable X_i. If, in the example in Eq. (9.23), we let X_i denote the variable X, we have $\mathbf{X}^- = (X^-, Y^-, Z^-)$ and $\mathbf{X}_{(-i)}^- = (Y, Z)$.

The following rewriting of Eq. (9.26) shows that the direct transfer entropy can be interpreted as separating out the causal influence of observable X_i on X_j from the contributions made by all the other observables, namely

$$DTE_{X_i \to X_j} = TE_{X \to X_j} - TE_{X_{(-i)} \to X_j}. \qquad (9.27)$$

The literature contains a large and growing number of methods to extract in a practical feasible way networks of information-theoretic Granger causality between a set of time series. Though these methods all share similarities with the description above in terms of transfer entropy and direct entropy, they differ in the detailed way they handle the practical computational problems one faces when dealing with many data series and investigating relations between multiple past and future events. We will just mention two prominent examples for which at the moment software packages are available.

The handling of the curse of dimensionality encountered when one estimates the joint probability distributions and needs to include a growing number of past and future events for a significant number of time series is addressed by the work of Dimitris Kugiumtzis and coworkers first by the bi-variant procedure call MIME (mutual information mixed embedding) and later for the conditioned multivariate procedure PMIME (partial mutual information mixed embedding). The latter makes use of conditioning in a way similar to the direct transfer entropy to distinguish direct causal relations from indirect, see [475, 244] for the original introduction of the methods and [408] for a review. PMIME is relatively fast because the mixed embedding helps to reduce the dimensionality even when a sizeable number of time series are considered. The procedure consists of iteratively including from the past and future only those events with the largest information transfer. Software packages can be obtained from Kugiumtzis's webpage http://users.auth.gr/dkugiu/. For applications of PMIME to analysis of EEG, see [241]. MIME was used to analyse the difference in structure of the network consisting of measured EEG signals from musicians and their audience during performances of classical music [482]. Difference in the causal structure could be identified between performances with a degree of improvisation and those using a strict rendition.

With a focus on conditioning as the tool to identify direct causal relations, Jacob Runge and collaborators have developed a packaged called Tigramite which makes use of three related but different ways of estimating causal relations: (1) linear time-delayed partial correlations; (2) linear Gaussian processes similar to Granger's original approach; and (3) conditioned mutual information. The method was used to analyse the world climate in [372] and an overview of the method is given in [371]. The conditioning – without mixed embedding – rapidly makes the full Tigramite procedure computationally heavy as the number of time series increases. Documentation and software is available, see e.g. https://jakobrunge.github.io/tigramite/.

We have described how joint probability distributions contain information about the relations between different parts of a system and how the information-theoretic entropy studied by Shannon is a central building block for methods which from probabilities derive a numerical estimate of causal relations.

In the next two sections we study how entropy can be used to derive probability distributions and use entropic ideas to develop measures that can be used to estimate the degree of complexity of a signal or an entire system.

9.4   From Entropy to Probability Distribution

In 1957 Edwin T. Jaynes pointed out in two papers [211, 213] that the probability distribution functions of Boltzmann and Gibbs, which are used in equilibrium statistical mechanics, can be viewed as the distributions that maximise the entropy expression in Eq. (9.16). See also the discussion in Chap. 6. When derived using the maximum entropy principles, the resulting probability distribution can be viewed as the least biased estimate given the restrictions we know about the system. When we throw dice, and we do not have any specific information about the given dice, we make the assumption that each face is equally likely to appear. Accordingly we expect one eye and six eyes each to occur with probability 1/6. Of course we are willing to change this assumption if we find out that the dice is loaded, say one half of the dice is heavier than the other. As soon as we have this information we will want to change the probabilities, e.g. to 1/3 for the face with one eye and 1/12 for the one with six eyes depending on the exact shape and weight distribution of the dice.

Jayens's maximum entropy principle works in the following way. We want to find the distribution $P(x)$ which maximises $H(X)$ in Eq. (9.16) subject to whatever restrictions we know about the observable X. Since $P(x)$ is a probability distribution, $P(x)$ inevitably has to fulfil the restriction that it is normalised, i.e. when we maximise $H(X)$ we have to ensure that

$$\sum_x P(x) = 1. \tag{9.28}$$

We can also imagine that the average $\langle x \rangle$ of X is fixed by some mechanics, in which case we only want to look at distributions that are normalised, maximise $H(X)$ and also ensure that

$$\sum_x x P(x) = \langle x \rangle. \tag{9.29}$$

To compute $P(x)$ from the maximisation of $H(X)$ under prescribed constraints we make use of the method of Lagrange multipliers, see Box 6.1. So we have to solve

$$\vec{\nabla} J = \vec{\nabla} [H(X) - \lambda_1 \sum_x P(x) - \lambda_2 \sum_x x P(x)] = 0. \tag{9.30}$$

We make use of the notation $p_i = P_X(x_i)$ with $i = 1, 2, \ldots, W$. To be consistent with the relevant literature, we denote the number of states by W instead of the notation Ω used on p. 114. For our present discussion it is convenient to write the entropy in Eq. (9.16) in the way we used when discussing statistical mechanics in Chap. 6, i.e.

$$H[p] = H(p_1, p_2, \ldots, p_W) = - \sum_{i=1}^{W} p_i \log p_i. \tag{9.31}$$

This is to emphasise that the entropy $H[p]$ is a function of the W variables which consist of the probabilities of the W 'events' or possible outcomes of the variable X.

We can now write

$$\frac{\partial J}{\partial p_k} = 0$$

$$\Downarrow$$

$$\frac{\partial}{\partial p_k} \sum_{i=1}^{W} \left[-p_k \log p_k - \lambda_1 p_k - \lambda_2 x_k p_k \right] = 0 \tag{9.32}$$

$$\Downarrow$$

$$-\log p_k - 1 - \lambda_1 - \lambda_2 x_k = 0$$

$$\Downarrow$$

$$p_k = \exp(1 + \lambda_1 + \lambda_2 x_k) = \exp(1 + \lambda_1) \exp(\lambda_2 x_k).$$

We notice that if we only included the constraint concerning normalisation in Eq. (9.28), corresponding to $\lambda_2 = 0$, we have that all p_k are equal. The Lagrange multiplier λ_1 will be determined from the condition $\sum_{k=1}^{W} p_k = 1$ to be given by $\exp(1 + \lambda_1) = 1/W$ and therefore $p_k = 1/W$. This is the microcanonical ensemble we discussed in Chap. 6. Including the constraint on the average leads to $p_k \propto \exp(\lambda_2 x_k)$, which is the canonical ensemble discussed in Chap. 6.

The maximum entropy principle has proven very successful [212]. Not only does it establish an information-theoretic perspective on statistical mechanics, the approach is remarkably useful in very many different situations. It is natural to consider what may happen if we maximise a different entropic expression than the one in Eq. (9.16) used by Shannon. The question is how to determine the expression for the entropy, i.e. the dependence on the probabilities p_i.

It was shown by Shannon [394] and Khinchin [233] that the functional dependence on the variables p_i in Eq. (9.31) is the only possible form if the entropy $H[p]$ satisfies four seemingly very natural properties. The entropy in the above equation is, as also discussed in Box 6.1, mathematically an ordinary function of a function of W variables, the probabilities p_i of the individual events. These four properties are called the Shannon–Khinchin axioms and consist of

(1) *Continuity.* The entropy is a continuous function of the variables $p_1, p_2, \ldots, p_W$. This makes sure that the entropy of a probability distribution does not change dramatically when one of the probabilities p_i is changed by a small amount.

(2) *Maximum on uniform distribution.* The entropy $H[p]$ is assumed to have its maximum value when evaluated for the uniform distribution $p_i = 1/W$. This property relates to the second law of thermodynamics. The thermodynamic entropy never decreases in a closed system and will increase till thermodynamic equilibrium is reached. In equilibrium, a closed system in statistical mechanics is described by the uniform distribution for which all microstates occur with equal probability. From

the information-theoretic perspective, the entropy is a measure of how uncertain we are about the state of the system. The most uncertain situation is when all states are equally likely.

(3) *Effect of zero probability events.* This is required to make the entropy a sound mathematical entity. It is required that the entropy of a distribution is unchanged if we add an additional state to the W possible states which occurs with zero probability, i.e. $p_{W+1} = 0$. Since a state with zero probability never occurs, it has no effect at all and hence we want for $p_{W+1} = 0$ that

$$H(p_1, p_2, \ldots, p_W) = H(p_1, p_2, \ldots, p_W, p_{W+1}).$$

(4) *Additivity.* This property is inspired by thermodynamics. The thermodynamic entropy of two non-interacting systems is additive. This is related to the fact that Clausius entropy is defined in terms of its change when heat, i.e. energy, is exchanged at a given temperature. Since energy is additive it seems reasonable that entropy also has to be, but since it is the *change* in entropy that is given in terms of heat exchange, some work is needed in order to show that additivity is demanded by thermodynamics, see [264].

The uniqueness tells us that whenever these four properties are satisfied, the only possible functional form for the entropy is the one given in Eq. (9.31). If we assume that all relevant probability distributions arise from the maximum entropy principle, we simply need to identify the appropriate constraints for a given problem and we can determine the probabilities p_i by use of the Lagrange multiplier technique. We know that for the two simplest constraints $\sum_i p_i = 1$ and $\sum_i E_i p_i = \langle E \rangle$, the probabilities p_i will either be uniform or exponential.

Of course, distributions different from the uniform and the exponential are often encountered. The question is whether these are related to more complicated constraints, or if the four Shannon–Khinchin axioms may not always be relevant. Which of the four axioms could possibly not be satisfied for a real system? The first three refer to purely mathematical properties that ensure a sensible analysis. Non-continuity of $H[p]$ will lead to erratic changes in the amount of entropy of a distribution and without property (2), we lose the relation between certainty versus uncertainty and the shape of the corresponding distribution. Property (3) appears totally obvious. If an event has zero probability it should not make a contribution.

Property (4) is different in nature from the first three. Additivity is not a mathematical requirement, but a property that is fulfilled for the thermodynamic entropy, though when we think of entropy purely as functions of probability distributions we do not have to impose additivity. Think of two independent systems A and B, see Fig. 9.2. Let us denote the probabilities of the W_A microstates of system A by p_i^A and similarly p_i^B for the W_B microstates of B. We can of course think of A and B together. Since A and B are independent, the microstates of the combined system will be all the possible states we can produce by combining a state s_i^A from A with a state s_j^B from B. We write $s_{(i,j)}^{A \times B} = (s_i^A, s_j^B)$. The probabilities for the combined microstates will be

Figure 9.2 The two systems A and B can either be considered as two separate systems or as one single system. Of course, in reality parts of A may interact with parts of B.

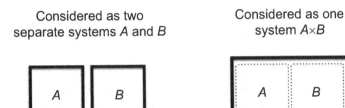

Considered as two separate systems A and B

Considered as one system $A{\times}B$

A B

A ⋮ B

States of $A = a_i$, States of $B = b_j$

States $= (a_i, b_j)$

given by the product of the individual probabilities, since we assume A and B are independent:

$$p_{(i,j)}^{A \times B} = p_i^A p_j^B. \tag{9.33}$$

The expression for the entropy in Eq. (9.31) is additive, which is seen by direct computation in the following way:

$$
\begin{aligned}
H(A \times B) &= -\sum_{(i,j)} p_{(i,j)}^{A \times B} \log p_{(i,j)}^{A \times B} \\
&= -\sum_{(i,j)} p_i^A p_j^B \log(p_i^A p_j^B) \\
&= -\sum_{(i,j)} p_i^A p_j^B [\log p_i^A + \log p_j^B] \\
&= -\sum_i p_i^A \log p_i^A \Big(\sum_j p_j^B\Big) + \Big(\sum_i p_i^A\Big) \sum_j p_j^B \log p_j^B \\
&= H(A) + H(B).
\end{aligned}
\tag{9.34}
$$

Hence the Boltzmann–Gibbs–Shannon entropy is additive when computed for combinations of independent systems.[4] Macroscopic subsystems considered in physics, such as when dealing with gasses and liquids and most materials, will in practice behave as if independent if the lengths of interactions between the molecules are short-ranged van der Waals forces for example. For short-range forces the influence of one subsystem on another will be limited to the interactions through the shared surfaces and these contributions to the physical properties will be negligible compared to contributions from the bulk when the system size is large. But for long-range interactions, such as gravitational forces or other kinds of far-reaching interactions, additivity does not apply.

Strong interactions in complex systems might, for example, originate from reactions that allow new microstates of the combined system which are different from those

---

[4] For systems that are not independent, i.e. when $p_{(i,j)}^{A \times B} \neq p_i^A p_j^B$, the entropy in Eq. (9.31) of $A \times B$ can be expressed as a sum of the entropy of A and the entropy of B conditioned on A, namely $H(A \times B) = H(A) + H(B|A)$, see Eq. (9.18) and e.g. Khinchin's discussion in the first chapter of [233].

obtained by matching a microstate from A with one from B, see e.g. [217]. As a simple illustrative example, think of hydrogen atoms H forming a new hydrogen molecule H_2. The states of the molecule are different from those that can be obtained by combining states of two isolated hydrogen atoms. The H atoms 'pair' to form a new entity, the hydrogen molecule, with properties different from simply matching one H atom with another. Or in biology, think of the fertilised state resulting from sperm pairing with an egg. When new states are produced, the state space, i.e. the set of all microstates, of the combined system AB, will be bigger than the state space $A \times B$ obtained from **Cartesian combinations** (s_i^A, s_j^B). The number of states in $A \times B$ is of course $W_A \times W_B$, so when new paired states can form the number of microstates, W_{AB}, of the fully interacting combined system AB will grow faster than exponentially. To see this, think of combining n identical subsystems A to get a system $AA \cdots A = A^n$; we will have that W_{A^n} grows faster than W_A^n because of the new paired states in A^n.

We can also imagine that strong interdependence amongst the components of a complex system may be able to exclude certain combinations of microstates of the subsystems. Think of something like a collection of prey, system A, and predators, system B. Assume the predators simply make the existence of the prey impossible if A and B are brought together in the same region. Hence when combined, the AB system will not allow any of the B states to be present. Or imagine a set of n coins which individually can show either heads or tails. But when brought together a very strong force between them ensures that the face of the first coin put down forces all following coins to show the same face. Instead of 2^n possible sequences of the n coins, the strong force eliminates all sequences except for the one where all show heads and the one where all show tails. Again we end up with a number of allowed microstates of the combined system which is very different from the exponential value obtained from the Cartesian combination.

When strong interdependence makes the number of available microstates, W_{AB}, for the fully interacting combined system significantly different from the product $W_A W_B$, the entropy in Eq. (9.31) will not be additive even for the simple uniform distribution obtained by maximising under the normalisation constraint $\sum_i p_i = 1$. For the uniform case we have for the subsystems before combining them that $p_i^A = 1/W_A$ for $i = 1, \ldots, W_A$ and $p_j^B = 1/W_B$ for $j = 1, \ldots, W_B$. For the fully interacting combined system AB, the uniform distribution is $p_k^{AB} = 1/W_{AB}$ for $k = 1, \ldots, W_{AB}$. The corresponding entropies are

$$H(A) = -\sum_{i=1}^{W_A} p_i^A \log p_i^A = -\sum_{i=1}^{W_A} \frac{1}{W_A} \log \frac{1}{W_A} = \log W_A, \qquad (9.35)$$

and similarly $H(B) = \log W_B$ and $H(AB) = \log W_{AB}$. But it is only when $W_{AB} = W_A W_B$ that $H(AB) = \log W_{AB} = \log W_A + \log W_B = H(A) + H(B)$.

These considerations suggest that it is of interest to investigate what happens if one considers entropic forms that are not additive, meaning that one needs to find a replacement for the fourth Shannon–Khinchin axiom. People have handled this in different ways. Probably the most famous example is the statistical mechanics of Constantino Tsallis derived from the entropic form

$$S_q = \frac{1 - \sum_{i=1}^{W} p_i^q}{q - 1},$$ (9.36)

suggested by Tsallis on the basis of great intuition [453, 454]. The parameter q is a real number. This entropy is not additive. The S_q entropy of the combination of two independent systems with probabilities given by Eq. (9.33) leads, instead of the additive expression in Eq. (9.34), to

$$S_a(A \times B) = S_q(A) + S_q(B) + (1 - q)S_q(A)S_q(B).$$ (9.37)

The hypothesis is that the S_q entropy relevant to a given system is obtained by choosing the parameter q such that S_q becomes extensive for the uniform distribution in the sense that S_q evaluated on the uniform distribution $p_i = 1/W$ is proportional to N, the number of components of the system. In Tsallis's statistics, additivity is replaced by the requirement that the entropy is extensive. Extensivity is an important property in thermodynamics since it ensures that as the limit of infinite systems is considered, we can still make sense of quantities such as energy computed per component. If the total internal energy of a system consisting of N parts is denoted by $U(N)$, we call $u = U(N)/N$ the energy density and similarly for the thermodynamic entropy, but a meaningful well-defined value of S_q/N will only exist if S_q is extensive, i.e. for large values of N is proportional to N. Hence it appears natural to choose q such that S_q is extensive.

In the limit $q \to 1$ the S_q entropy becomes equal to the entropy in Eq. (9.31) and additivity is restored. Maximising S_q under constraints leads to the so-called q-exponential functions

$$\exp_q(x) = [1 - (1 - q)x]^{\frac{1}{1-q}}.$$ (9.38)

This functional form has been found to fit well to distributions observed in a remarkably broad range of systems [453].

Classification and analysis of entropies in terms of the dependence on the number of available states $W(N)$ for increasing number of components N has been studied by Hanel and Thurner, see e.g. [181, 182, 442]. They generalise the framework of the Shannon–Khinchin axioms simply by leaving the fourth axiom out. They assume the validity of the first three axioms and classify entropies according to how they behave for large values of W. Working with only the first three Shannon–Khinchin axioms means that one does not specify a rule for how the entropy of composite systems relates to the subparts. The summation law imposed by the fourth Shannon–Khinchin axiom is, as discussed above, natural seen from the perspective of thermodynamics and Shannon also considered it because he thought of entropy in terms of information and found it useful to have an additive measure of information: double the length of a binary string and it may contain twice as much information.

When we turn to strongly interdependent complex systems where the parts significantly influence each other and thereby produce new emergent behaviour at the systemic level, it is not obvious that the entropic functional, needed to generate probability distributions through the maximum entropy principle, will be additive. On the other hand, we do need to be sure that our entropy can consistently deal with the situation depicted in Fig. 9.2 where we formally consider two independent systems A and B as a combined system $A \times B$. It should not matter if we first compute the entropy of $A \times B$ as one system and then afterwards realise that we are dealing with a system consisting of the two independent parts A and B. We need a rule for how the entropies of the two parts A and B are to be combined to equal the entropy of $A \times B$. In the Shannon–Khinchin case one simply adds, that is axiom four; the question is what we can replace the sum by.

Replacing the summation in axiom four by a composition rule inspired by the fact that combining systems follows a mathematical group structure has been studied by Piergiulio Tempesta [438, 439]. The summation is replaced by a composability axiom that makes use of the mathematics of formal group theory to establish a rule, potentially a non-additive rule, for how the entropy of the Cartesian combination of two independent subsystems relates to the entropy of the parts. We will call these entropies group entropies and to distinguish them from the entropy given in Eq. (9.31) we will denote the group entropy of a system A by $S(A)$. The entropy of the Cartesian combination $A \times B$ of two subsystems is given by

$$S(A \times B) = \Phi(S(A), S(B)). \tag{9.39}$$

Of course this equation is only useful if we have a prescription for how the composition function $\Phi(x, y)$ can be determined for a given system. The link to formal group theory is established by imposing on $\Phi(x, y)$ requirements that follow naturally from how we combine two systems. The order of combining A with B or B with A does not matter. If we have three systems A, B and C it should not matter if we first combine A and B and then combine with C or if we first combine B and C and then combine with A. Finally, if the entropy of one of the two systems, for example B, is zero, we want the entropy of the combined system $A \times B$ to be equal to the entropy of A. This is natural when we think of entropy as a measure of certainty. Entropy zero means that B for certain is in one specific state and therefore any uncertainty in the combined $A \times B$ will originate from A.

Mathematically the above considerations can be expressed as the following three conditions on $\Phi(x, y)$:

- *Symmetry* $\Phi(x, y) = \Phi(y, x)$.
- *Associativity* $\Phi(x, \Phi(y, z)) = \Phi(\Phi(x, y), z)$.
- *Null-composability* $\Phi(x, 0) = x$.

Of course it is not enough to know how to combine entropies, we need a procedure for how to determine which functional expression for the entropy is to be used for a given system. Should we use the form in Eq. (9.31) or in Eq. (9.36), or some different expression. The mathematical expression for the entropy can be determined by combining the three Shannon–Khinchin axioms with the rule for combining systems in Eq. (9.39) together with the requirement of extensivity.

This can be done by use of formal group theory, see [219, 439, 441], which ensures the existence of a function, the group generator, given by the power series $G(t) = t + A_2 t^2 + \cdots$ such that the function $\Phi(x, y)$ needed for combining systems, see Eq. (9.39), can be expressed as

$$\Phi(x, y) = G(G^{-1}(x) + G^{-1}(y)). \tag{9.40}$$

Within this approach the entropy is linked to the group generator $G(t)$ either through the so-called trace form

$$S[p] = \sum_{i=1}^{W} p_i G\left(\ln \frac{1}{p_i}\right) \tag{9.41}$$

or through the non-trace form

$$S[p] = \frac{G(\ln(\sum_{i=1}^{W} p_i^\alpha))}{1 - \alpha}, \tag{9.42}$$

where α is a real number. We notice that the trace form has a mathematical form similar to the entropy in Eq. (9.31) and that the non-trace form is related to the Rényi entropy defined as

$$S_\alpha = \frac{1}{1 - \alpha} \ln\left(\sum_{i=1}^{W} p_i^\alpha\right). \tag{9.43}$$

Detailed analysis shows that the non-trace form is the more general as it leads to entropic forms that satisfy the combination rule in Eq. (9.39) for all probability distributions p_i, whereas the trace form, except in the Boltzmann–Gibbs–Shannon case of Eq. (9.31), only satisfies the composition in the special case of uniform probabilities $p_i = 1/W$. Details can be found e.g. in [439, 440]. The reason the non-trace group entropy is based on a generalisation of Rényi is that this entropy is the simplest additive entropy that satisfies the composition rule for all probability distributions p_i. So this is a natural foundation to start from.

Let us recapitulate. Inspired by the success of the maximum entropy starting from the Boltzmann–Gibbs–Shannon entropy in Eq. (9.31), which is the unique entropy consistent with the four Shannon–Khinchin axioms, the group-theoretic approach looks for entropies for which a composition rule applies, though not necessarily in the form of summation. The combination rule is ensured by use of formal group theory. What is now left to be done is to determine the form of $G(t)$. We show one way of determining $G(t)$, namely by ensuring that the group entropy is extensive.

The requirement that the entropy of a system with N components is asymptotically extensive for the uniform distribution $p_i = 1/W$ is expressed as

$$S(1/W) = \lambda N \text{ for } N \to \infty. \tag{9.44}$$

Here λ is a proportionality constant and only the most important N dependence has been included. We can combine this expression with the expression for the entropy in either trace form in Eq. (9.41) or non-trace form in Eq. (9.42). Let us, as an example, consider the non-trace form. For $p_i = 1/W$ the entropy $S[p]$ in Eq. (9.42) becomes $S(1/W) = G(\ln(W^{1-\alpha}))$, which we substitute into Eq. (9.44) to get

$$G(t) = \lambda(1 - \alpha)W^{-1}\left[\exp\left(\frac{t}{1 - \alpha}\right)\right]. \tag{9.45}$$

We notice that this expression for $G(t)$ does not satisfy the requirement from formal group theory that $G(0) = 0$. That is natural since we have determined $G(t)$ from the behaviour at large arguments corresponding to the limit $N \to \infty$. This we rectify by subtracting the value of the right-hand side of Eq. (9.45) evaluated at $t = 0$ to obtain the final expression

$$G(t) = \lambda(1 - \alpha)\left\{W^{-1}\left[\exp\left(\frac{t}{1 - \alpha}\right)\right] - W^{-1}(1)\right\}. \tag{9.46}$$

We now need to consider the merit of this generalisation of the entropies. In the next section we will consider how the composition rule in Eq. (9.39) generates a new measure of how complex a given system is, but first, in the remainder of this section, we will see how the group entropic approach helps us to understand where and when we will expect to use the Boltzmann–Gibbs–Shannon entropy in Eq. (9.31) or Tsallis entropy in Eq. (9.36), and we will further see that the q-exponential distributions, so often encountered when analysing data can arise as a result of maximising entropies different from the Tsallis entropy.

Perhaps the most interesting point to emphasise is that the group-theoretic approach to entropies enables us to derive an expression for the entropy in terms of the asymptotic behaviour of $W(N)$, the number of microstates. So different functional behaviours of $W(N)$ will lead to different entropies. It looks like every system will now have its own entropy and, by maximising the entropy, we will be able to derive infinities of different functional behaviours of the probability distributions. This is not the case; what matters is only how $W(N)$ depends on N as N goes to infinity. Therefore we will expect some main classes of entropies corresponding to the three main classes of asymptotic behaviour of $W(N)$. Namely, algebraic $W(N) = N^a$, exponential $W(N) = k^N$ or super-exponential $W(N) = N^{\gamma N}$, see [219, 441].

Once the entropic form for a given $W(N)$ is given, we can maximise this entropy under the usual constraints and derive the corresponding distributions. Starting from the trace form in Eq. (9.41) or non-trace in Eq. (9.42) leads to different expressions, so it is possible at first to be slightly overwhelmed. But one does not need to look at all the details. In Box 9.1 we list the results so they are handy for reference. Luckily, when we look at the distributions obtained from maximising the entropies the situation becomes simpler.

Box 9.1 From $W(N)$ to entropy

Trace-form case

(I) Algebraic, $W(N) = N^a$:

$$S[p] = \lambda \sum_{i=1}^{W(N)} p_i \left[\left(\frac{1}{p_i} \right)^{\frac{1}{a}} - 1 \right] \tag{9.47}$$

$$= \frac{1}{q-1} \left(1 - \sum_{i=1}^{W(N)} p_i^q \right). \tag{9.48}$$

To emphasise the relation with the Tsallis q-entropy, we have introduced $q = 1 - 1/a$ and $\lambda = 1/(1-q)$. Note that the parameter q is determined by the exponent a, so it is controlled entirely by $W(N)$.

(II) Exponential, $W(N) = k^N$, $k > 0$:

$$S[p] = \frac{\lambda}{\ln k} \sum_{i=1}^{W(N)} p_i \ln \frac{1}{p_i}. \tag{9.49}$$

This is of course the Boltzmann–Gibbs case.

(III) Super-exponential, $W(N) = N^{\gamma N}$, $\gamma > 0$:

$$S[p] = \lambda \sum_{i=1}^{W(N)} p_i \left\{ \exp \left[L \left(-\frac{\ln p_i}{\gamma} \right) \right] - 1 \right\}. \tag{9.50}$$

Here $L(x)$ denotes the Lambert function.

Non-trace-form case

(I) Algebraic, $W(N) = N^a$:

$$S[p] = \lambda \left\{ \exp \left[\frac{\ln \left(\sum_{i=1}^{W(N)} p_i^\alpha \right)}{a(1-\alpha)} \right] - 1 \right\}. \tag{9.51}$$

(II) Exponential, $W(N) = k^N$:

$$S[p] = \frac{\lambda}{\ln k} \frac{\ln \left(\sum_{i=1}^{W(N)} p_i^\alpha \right)}{1 - \alpha}. \tag{9.52}$$

This is of course the Rényi entropy.

(III) Super-exponential, $W(N) = N^{\gamma N}$:

$$S[p] = \lambda \left\{ \exp \left[L \left(\frac{\ln \sum_{i=1}^{W(N)} p_i^\alpha}{\gamma(1-\alpha)} \right) \right] - 1 \right\}. \qquad (9.53)$$

A model where parts combine to produce new states leading to a super-exponential $W(N)$ is studied in [217]. This is also relevant to networks, see e.g. Sec. 15.2.3 in the first edition of the book [312] and [441].

We are particularly interested in how entropy can help to identify the relevant probability distributions by use of the maximum entropy principle, so we introduce Lagrange multipliers $J[p] = S[p] - \sum_{n=1}^M \lambda_n g_n[p]$ to impose constraints. Here we consider the normalisation constraint $g_1[p] = \sum_{i=1}^W p_i - 1$ and constraining the average $g_2[p] = \sum_{i=1}^W p_i(E_i - E_0) - \bar{E}$. We have introduced E_0 to denote the lowest possible energy and $\bar{E}$ is the average of $E_i - E_0$ imposed on the system. This corresponds to the canonical ensemble studied in statistical mechanics, see Chap. 6.

We can solve $\delta S/\delta p_i = 0$ for the different cases corresponding to the trace and non-trace forms for the different $W(N)$ behaviours. Luckily we end up with only two different types of probability distributions. The exponential case $W(N) = k^N$ leads to the well-known Boltzmann weights

$$p_i = \frac{\exp[-\beta(\Delta E_i - \bar{E}]}{Z}, \qquad (9.54)$$

where $\Delta E_i = E_i - E_0$, $\beta = \lambda_2/\lambda_1$ and $Z = \sum_i \exp[-\beta(\Delta E_i - \bar{E})]$. This is expected since, see [218], for $W(N) = k^N$ the composition rule in Eq. (9.39) corresponds to addition $\Phi(x, y) = x + y$ and the traditional statistical mechanics of Boltzmann and Gibbs corresponds to additive entropy.

For the algebraic $W(N) = N^a$ and super-exponential $W(N) = N^{\gamma N}$ cases the maximum entropy probability weights are equivalent to q-exponentials[5]

$$p_i = \frac{[1 + \beta(\Delta E_i - \bar{E})]^{\frac{1}{1-\alpha}}}{Z}, \qquad (9.55)$$

where formally $\beta = \lambda_2/\lambda_1$ and $Z = \sum_i [1 + \beta(\Delta E_i - \bar{E})]^{\frac{1}{1-\alpha}}$.

What did we learn? That if we start from the usual Shannon–Khinchin axioms, and modify the fourth axiom to ensure we have a systematic procedure for how to handle the composition of systems, we end up with only two functional forms for the probability weights: the Boltzmann weights or the Tsallis q-exponentials. That q-exponentials appear for both algebraic and super-exponential $W(N)$ behaviour is very interesting. The q-exponentials are of more general applicability than is the Tsallis entropy, which according to the group-theoretic analysis corresponds to algebraic N

---

[5] Here we neglected the trace form super-exponential case since it corresponds to entropies which only obey the composition rule on the uniform probabilities $p_i = 1/W$ and the expression for p_i derived from $\delta S/\delta p_i = 0$ cannot be expressed in closed form, see [219].

dependence of $W(N)$ only. This may help to explain why the q-exponential form is ubiquitously found to fit the data well [453, 455], seemingly far beyond algebraic $W(N)$ dependencies.

The q-exponential appears in this context of group entropies because the non-trace form entropies in Eq. (9.42) are based on the Rényi entropy and maximising this entropy leads to q-exponentials. The generalisation of the Rényi entropy through the expression (9.42) does not lead to different functional forms for the probability weights, but it ensures families of entropies which satisfy the composition rule Eq. (9.39) beyond the additive case. Rényi corresponds to $W(N) = k^N$ and $\Phi(x, y) = x + y$.

Although the different non-trace group entropies all lead to the same functional shape for the probability weights, the entropies can be used to distinguish the degree of complexity contained in different systems. This is done by use of the composition rule $\Phi(x, y)$ in Eq. (9.39), which allows us to compare the entropy of a fully interacting combination of subsystems and a Cartesian equivalent. In the next section (particularly Sec. 9.5.3), we will look at the details of this together with other information-theoretic approaches to the quantification of complexity.

9.5   Measures of Degrees of Complexity

In this section we will consider how information theory and entropy can be used to put scales on how complex or how much emergence a signal or a system exhibit. We will start with the well-established Lempel–Ziv (LZ) measure of the complexity of a signal and see how this is related to the entropy used by Shannon. The LZ measure quantifies the rate at which new patterns keep appearing in a time series and therefore relates to predictability.

After the LZ measure we will turn to emergence and complexity of multivariate signals or systems consisting of many components and include a discussion of three measures called integrated information, Ω information and Σ information. These make use of entropy and information-theoretic concepts to characterise the nature of the emergent interdependence. One example is inspired by attempts to capture the complexity of the brain and relate it to consciousness. Another example suggests an information-theoretic description of emergence. Finally we will describe how the group entropies introduced in the previous section offer a measure of emergence relating to how the interdependence of the components are able to make the available number of states $W(N)$ deviate from the exponential behaviour corresponding to Cartesian combination of the microstates of the individual components.

9.5.1   Lempel–Ziv Complexity Measure

Think of a time signal, e.g. the daily Dow Jones index or the output from one EEG electrode. It is reasonable to consider a signal that is repetitive, where the same variation of the signal repeats itself again and again, to be less *complex* than a signal where new variations keep showing up. The signal 1001001001001001 $\cdots$ is clearly less complex than the signal 1010001110010101 $\cdots$. The LZ measure is a measure of how much new variation keeps emerging as we travel down along the signal. For a good textbook discussion see Sec. 13.4 in [87]. The measure was developed by Jacob Ziv and Abraham Lempel [506, 507] to quantify how compressible a string of data is. The more the data contained in a file varies, the less it can be compressed. Assume we have a file containing the same number 100 times, then all we need to store is the number and 100. In contrast, if the file contains 100 different random numbers, we are unable to represent the data in a more efficient way than simply repeating all the numbers.

The simplest version of the LZ procedure first turns a time signal $f(t)$ into a binary sequence. This can for example be done by detecting whether the signal is above or below the average. That is, we can replace $f(t)$ by 1 if $f(t) \geq \langle f \rangle$ and replace $f(t)$ by 0 if $f(t) < \langle f \rangle$. This makes it easier to identify patterns in the sequence $f(1), f(2), f(3), \cdots$ which now becomes a string of 0 and 1. Patterns are identified by what is called factorisation, which consists of grouping the zeros and ones together, starting from the left. Each new group is to be different from any previous group, but at the same time we also want the group to be as small as possible. This is ensured by demanding that the new group without its rightmost element will be found somewhere among the sequences encountered so far.

Mathematically this can be expressed in the following way [133]. We consider the sequence $S(N) = s_1, s_2, s_3, \ldots, s_N$ where $s_i = 0$ or 1 and look for a factorisation of the sequence $S(N)$ in m factors of the form

$$S(N) = s(1, l_1)s(l_1 + 1, l_2)s(l_2 + 1, l_3), \ldots, s(l_{m-1} +, N) \tag{9.56}$$

where the factors need to satisfy the following two properties:

$$
\begin{aligned}
&\textbf{(1)} \;\; s(l_{k-1} + 1, l_k - 1) \subset s(1, l_k - 2) \\
&\textbf{(2)} \;\; s(l_{k-1} + 1, l_k) \not\subset s(1, l_k - 1).
\end{aligned}
\tag{9.57}
$$

This looks perhaps somewhat dense, but the first property is the way to ensure that the new group $s(l_{k-1} + 1, l_k)$ is only slightly different from what has been seen before by demanding that if we remove the rightmost bit from the new group, this shorter group is present somewhere along the sequence consisting of all the bits up to two places left of where we are now. The second property means that the new group itself, $s(l_{k-1}+1, l_k)$, is not already present somewhere to the left. This procedure leads to a unique factorisation of the sequence in terms of distinct subgroups.

Here is an example from [133]. Consider the sequence

$$
\begin{array}{cccccccccccccc}
1 & 1 & 0 & 1 & 1 & 1 & 0 & 1 & 0 & 0 & 0 & 0 & 1 & 1 \\
i=1 & 2 & 3 & 4 & 5 & 6 & 7 & 8 & 9 & 10 & 11 & 12 & 13 & 14
\end{array}
\tag{9.58}
$$

The factorisation for this sequence is $1 \times 10 \times 111 \times 010 \times 0001 \times 1$. As we start from the left, our first group is $s(1, 1) = 1$. We move one step to the right and find yet another 1. So we need to include one more bit to obtain a group $s(2, 3) = 10$ which is new.

We move again to the right and find the bits 11, but we have already encountered this pattern at the first two bits of the sequence. To make our new group different from any previously encountered one, we need to include one more bit and so the next factor becomes $s(4,6) = 111$. And so on. A few words about the fifth factor $s(10,13) = 0001$. Why not, after the fourth factor $s(7,9) = 010$, choose as fifth factor $s(10,11) = 00$. There is no pattern 00 in the preceding sequence $s(1,9)$. The point is that if we did choose $s(10,11) = 00$ as our fifth factor, we would have the pattern 00 repeated as $s(9,10)$ and $s(10,11)$. This is what property (2) in Eq. (9.57) makes sure should not happen.

The LZ measure $C_{LZ}(s)$ of the complexity of a sequence s is equal to the number of groups in the factorisation. For the sequence in Eq. (9.58) we have $C_{LZ}(s) = 6$. As the length of the string N becomes large, one can show that

$$C_{LZ}(S(N)) < \frac{N}{\log_2 N}, \tag{9.59}$$

where $\log_2 N$ is the base two logarithm of N. To get a measure relative to this upper bound and thereby a measure sensitive to the length of the considered string, one introduces the normalised measure

$$c_{LZ}(S(N)) = \frac{C_{LZ}(S(N))}{\frac{N}{\log_2 N}}. \tag{9.60}$$

For long sequences the LZ measure is related to the Shannon entropy. Denote by $p(s_1, s_2, s_3, \ldots, s_N)$ the probability that a certain specific realisation of $S(N)$ occurs. The Shannon entropy for this distribution is given by

$$H[S(N)] = -\sum_{s_1, s_2, \ldots, s_N} p(s_1, s_2, \ldots, s_N) \log p(s_1, s_2, \ldots, s_N)$$

and in the limit $N \to \infty$ the LZ measure converges to the so-called entropy rate [507], namely

$$c_{LZ}(s) \sim \frac{H[S(N)]}{N} \text{ for } N \to \infty. \tag{9.61}$$

For more detail see Chap. 13 in [507]. Even for finite N the relation in Eq. (9.61) can be useful, see [133].

Although this relation suggests that the highest LZ complexity will be a random string, since such a string has the highest entropy, for finite strings the situation is different. The measure c_{LZ} counts the rate with which new patterns arise in the string $s_1, s_2, \ldots, s_N$ as N is increased. For the consecutive s_{N+1}, s_{N+2} and so on to give rise to patterns that were not present somewhere in the past, the best choice of producing new patterns will depend on the exact configuration of the past sequence $s_1, s_2, \ldots, s_N$. Except in the limit $N \to \infty$, these correlations are strong enough to make the string that maximises $c_{LZ}(s)$ different from the purely random string [132].

This focus on the diversity of the structure of a signal has been found to make the LZ measure useful, for example when analysing brain scans. The LZ complexity of EEG signals is found to relate to the degree of awareness. It is observed that the unconscious brain has a lower c_{LZ} than the awake brain [381] and that EEG from the brain of a person on LSD has a larger c_{LZ} index than a brain on placebo [288].

Inspired by this ability of the LZ measure in a quantitative way to relate to the degree of awareness. The LZ measure was used to assess the EEG response of classical musicians and audience members listening to the same piece of music played either in a strict or in a more 'creative' let-go rendition including an element of improvisation. The c_{LZ} of the EEG time series from the let-go performance is higher than from the strict performance [109].

LZ has also found applications in economics recently, e.g. to analyse the time evolution of cryptocurrency [418].

9.5.2 Information-Theoretic Approach to Emergence

We have already considered in the previous sections of this chapter how entropic measures and information theory can be used to investigate possible causal relations (see Sec 9.2), degrees of complexity (see Sec. 9.5.1) and derive the probabilities that the system occupies a specific microscopic configuration (see Sec. 9.4). In this section we will describe approaches that allow for a differentiation between different types of emergent behaviour. Here we will focus on **synergy** and **redundancy**. Observables that are not independent can of course be related in many different ways. As soon as the joint probability does not factorise, we know that some kind of emergent behaviour is taking place in the sense that the observables taken together behave differently than on their own. This collective emergence may be more or less interesting.

Measures of synergy and redundancy can help to quantify the nature of the collective emergent behaviour. An extreme example of redundancy is when interactions force the observables to be identical. Imagine we throw coins and there is some kind of strong interaction between them which forces all the coins to land face up or all land face down. This means that if we know the value of one observable, we know immediately the value of all the others. Ultimate synergy, on the other hand, refers to a situation where one cannot know anything about the collective state without knowing the value of each of the observables. Below we will use the so-called xor operation to illustrate synergy.

Let us elaborate some more on these two concepts. We saw above how the LZ measure can classify some states of the mind as more or less complex. But surely even if LZ increases during improvisation of music in a way similar to how the measure increases when we wake up, all awake states of the mind are not identical to improvisatorial states of mind. We will expect the brain of a musician performing a creative improvisation to be in a more 'elevated' state than if the same musician tries to follow a score as accurately and mechanically as possible. We may expect the improvising brain to possess more synergy. Synergy is in general used synonymously with Aristotle's description of emergence as 'The whole is greater than the sum of its parts', see Sec. 1.2. For musicians, we expect that the set of signals of EEG electrodes attached to an improvising musician contains a higher degree of collective behaviour, more synergy, than is the case during mechanical performance.

As an example of redundancy, think of financial time series such as exchange rates; they may be highly interdependent but not necessarily in any synergetic way. For example, the Danish kroner is tied to the euro within 2.5%, so if one knows how the euro is moving, one knows with good accuracy the motion of the Danish kroner. In this sense, information about the Danish kroner is redundant if we know about the euro.

There exists a large information-theoretic literature developing measures that are supposed to quantify the degree of complexity in a many-component system such as the brain. One such example is the very influential line of research initiated by Tononi, Edelman, Sporns and collaborators. Two measures were suggested. First the so-called neural complexity measure, see [449, 450], and the later integrated information theory (IIT) measure, see [319, 448]. These measures were developed to be able to capture how the different parts of a complex system, in particular the brain, may perform distributed tasks in a differential way, each component being responsible for specific activities, while at the same time a coordination of the different processes is needed to integrate, or collect, the overall result. Think, for example, of trying to hit a target with a dart. A careful coordination of multiple types of brain processing is needed: at least visual processing to identify the target and motor processes is needed to be sure you throw the dart with the correct speed in the right direction, and perhaps also audible processing if you make use of sound to judge the wind speed. Over the years it has become clear that the integrated information measure is not always behaving as intended [289]. Here we will not discuss the subtleties of integrated information but instead illustrate the philosophy behind such measures by looking at a recently developed related measure, which appears to explain some unwanted features of the IIT measure.

Our starting point is to assume we have access to a data stream from a specific system from which we imagine that it is possible to obtain the joint probability distribution

$$P_{X_1 X_2 \ldots X_N}(x_1, t_1; x_2, t_2; \ldots; x_N, t_N) = \text{Prob}\{X_1(t_1) = x_1, \ldots, X_N(t_N) = x_N\}. \quad (9.62)$$

Now we want to look for ways to examine whether the observables X_i are related synergistically or redundantly. To have a frame of reference, we will compare our measures to a set of independent X_i.

To illustrate the effect of the restrictions on the configurations available to variables that are interdependent in one way or another, we will first look at a simple example of two coins. Each coin can, when thrown independently with equal probability, show a 0 or a 1, i.e. there are a total of four configurations available to the two coins: 00, 01, 10 and 11. The probability of finding a configuration is $P_{\text{indp}}(X_1, X_2) = 1/4$, where X_1 and X_2 denote the value of the face of each of the coins. To contrast this independent case we consider the extreme opposite example of interdependence between the two coins where we assume the second coin is linked to the first in a way that forces the second always to show the same face as the first. Now the coins can only occupy the two configurations 00 and 11. This gives us $P_{\text{copy}}(0,0) = P_{\text{copy}}(1,1) = 1/2$ and $P_{\text{copy}}(0,1) = P_{\text{copy}}(1,0) = 0$.

Finally we want to mix these two extremes. To do this we imagine that when we throw the coins they can either land according to the probability $P_{\text{indp}}(x_1, x_2)$ or according to the probability $P_{\text{copy}}(x_1, x_2)$, the former we call an independent throw and the latter a copy throw. Let us assume an independent throw occurs with probability α and a copy throw with probability $1 - \alpha$. This will result in the probability for the outcome (x_1, x_2) being given by the following linear combination:

$$P(x_1, x_2) = \alpha P_{\text{indp}}(x_1, x_2) + (1 - \alpha)P_{\text{copy}}(x_1, x_2). \qquad (9.63)$$

As the mixing parameter α is varied from 0 to 1 the amount of times the coins are found in a specific configuration changes, or in other words the probability $P(x_1, x_2)$ of observing the coins in configuration $X_1 = x_1$ and $X_2 = x_2$ changes. Decreasing α from 1 towards 0 corresponds to putting a restriction on the configurations.

We can use the entropy to compute the effective number of configurations visited by the coins as we keep throwing them according to the probabilities $P(x_1, x_2)$. We notice that the entropy in Eq. (6.11) or (9.16) computed for a uniform distribution $p_i = 1/W$ over W possible configurations becomes[6]

$$H[P] = -\sum_{i=1}^{W} p_i \log(p_i) = -\sum_{i=1}^{W} \frac{1}{W} \log\left(\frac{1}{W}\right) = \log(W). \qquad (9.64)$$

Since we are dealing with information theory we follow the sensible tradition and use the base 2 log, which means that if $y = \log(x)$ then $x = 2^y$. Only when the probabilities are uniform and all equal to $1/W$ will $2^{H[p]}$ be an integer, but since we may think of the probabilities p_i as how long the system spends in configuration i, we can still use $2^{H[p]}$ as an effective measure of how many of the total number of configurations are actually utilised.[7] We will call this number W_{Tot}; in a moment we will look at other effective measures describing the numbers of configurations visited.

For two coins we have to compute H from Eq. (9.63). In Fig. 9.3 we plot $W_{\text{Tot}} = 2^{H[P]}$ together with frequencies of occupation of the configurations $P(x_1, x_2)$. We notice that W_{Tot} nicely indicates how the number of configurations gradually decreases from 4 to 2 as the copying rule increases in importance and restricts the ability of the second coin to make full use of its states 0 and 1. Is a reduction from 4 to 2 big or small? We are interested in the effective number of configurations visited relative to the number of possible configurations when the coins are unrelated. In general, when we consider a system of N components where each component is able to occupy Q configurations when not interacting with anything, we can define a restriction ration γ_{Tot} as

$$\gamma_{\text{Tot}} = \frac{W_{\text{Tot}}}{Q^N}. \qquad (9.65)$$

We will have that $\gamma_{\text{Tot}} = 1$ for systems consisting of independent components where each component can visit the Q single-component configurations with equal probability. Otherwise $\gamma_{\text{Tot}} < 1$.

---

[6] We encountered this expression in Eq. (6.23).

[7] This intuitive argument for why 2^H is a measure of the effective number of states can be made more rigorous by use of large deviation theory, see Chap. 3 in [87].

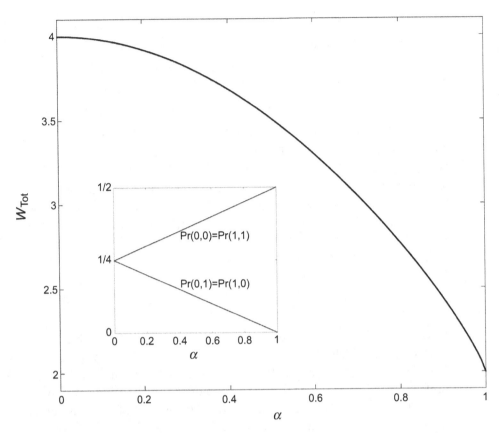

Figure 9.3 The main panel shows the effective number of configurations occupied by the two coins given by $W = 2^H$. The insert shows how the probability of finding the coins in one of the configurations changes as the mixing coefficient α varies.

We considered this initial warm up to introduce entropy as a way to measure how the interdependence between components restricts the number of configurations available to the interacting components. Now we want to go one step further and use entropy to probe degrees of redundancy and synergy amongst the observables.

First we need a precise mathematical definition of synergy and redundancy. In information theory, synergy means that knowing the value of one of the individual components X_i does not improve our ability to predict the value of the other X_j with $j \neq i$. In contrast, redundancy in information theory means that the different X_i observables influence each other in such a way that from knowing a subset of the X_i we improve our ability to predict the value of some of the other X_i observables. Below we will use **logical gates** from information theory as a simple way of quantifying synergy and redundancy. These logical gates form the basis of computing devices but this is not the reason we discuss them here. Rather, for us they offer minimalistic models of information-theoretic interrelationships much in the same way as the Ising model is a paradigmatic model for the statistical description of interaction amongst many

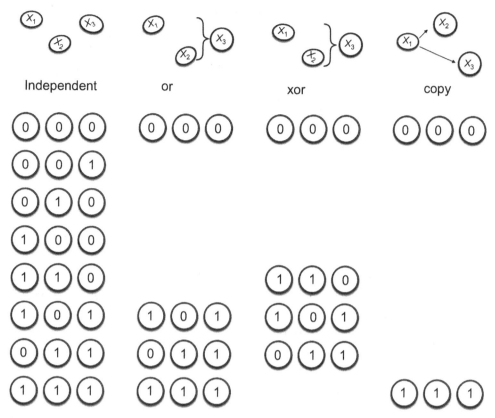

Figure 9.4 The outcome of throwing three coins. Each column corresponds to different relationships between the coins. The first column corresponds to independent coins which can occupy a total of eight different configurations. The next two columns show the allowed configurations when the face of the third coin is determined either as $X_3 = X_1 \text{or} X_2$ or $X_3 = X_1 \text{xor} X_2$. In the right column only the first coin is thrown at random and the other two coins copy the face of the first.

components. We consider three observables represented by variables X_1, X_2 and X_3 that each can assume the values 1 or 0. We will think of these as representing the faces of three coins and consider four types of relations between the coins, see Fig. 9.4.

Independent – First, as reference point, we consider X_1, X_2 and X_3 to be independent. This corresponds to no interaction between the coins; each coin can freely assume the values 0 or 1 and hence no emergence can occur. In this case eight different outcomes are possible as indicated by the first column in Fig. 9.4.

Or – Next, we imagine throwing the first two coins at random and let the value of X_3 be equal to the logic 'or' between X_1 and X_2, i.e. $X_3 = X_1 \text{or} X_2$. This reduces the number of possible configurations of the three coins to 4, as indicated in the second column of Fig. 9.4.

Table 9.1 The probability of each of the coin configurations for the four cases: independent, or, xor and copy.

X_1	X_2	X_3	P_{indp}	P_{or}	P_{xor}	P_{copy}
0	0	0	$\frac{1}{8}$	$\frac{1}{4}$	$\frac{1}{4}$	$\frac{1}{2}$
0	0	1	$\frac{1}{8}$	0	0	0
0	1	0	$\frac{1}{8}$	0	0	0
1	0	0	$\frac{1}{8}$	0	0	0
1	1	0	$\frac{1}{8}$	0	$\frac{1}{4}$	0
1	0	1	$\frac{1}{8}$	$\frac{1}{4}$	$\frac{1}{4}$	0
0	1	1	$\frac{1}{8}$	$\frac{1}{4}$	$\frac{1}{4}$	0
1	1	1	$\frac{1}{8}$	$\frac{1}{4}$	0	$\frac{1}{2}$

Xor – Then, we use the so-called logic 'xor' and let $X_3 = X_1 \mathbf{xor} X_2$. This is similar to the **or** operation but we will see that the **xor** generates a synergetic relationship between the variables X_1, X_2 and X_3. However, the number of possible configurations is again equal to 4. See the third column of Fig. 9.4.

Copy – Finally, we throw the first coin at random and let X_2 and X_3 copy the outcome, i.e. $X_2 = X_1$ and $X_3 = X_1$. In this case the three coins can only occupy two configurations. See the fourth column of Fig. 9.4.

We assume that the coins, which are thrown at random, have equal probability of showing face 0 or face 1. This implies that each of the *allowed* configurations will appear with equal non-zero probability and the configurations excluded by the interdependence between the coins will occur with probability zero. Table 9.1 lists these probabilities.

Let us now imagine a dynamical process consisting of throwing the coins in every time step. We could think of the three coins as the result of pulling the arm of a slot machine in every time step. The probabilities $P(x_1, x_2, x_3)$ listed in Table 9.1 then correspond to the relative time spent by the slot machine in the different configurations. This perspective is highlighted by the histogram in Fig. 9.5.

The difference in importance to us between the four different types of relationships consists in how much we are able to predict about the faces of one coin by observing the other coins. When the coins are independent, knowing the face of the first coin, or the face of both the first and the second coin, does not help us predict the face of the others since the coins are free to visit all possible configurations.

When the coins are linked together via the 'or' relationship, observing that face of the first coin shows a 1, $X_1 = 1$, lets us predict with probability 1 that the third coin shows a 1, $X_3 = 1$. Whereas observing $X_1 = 0$ leaves equal probability 1/2 for $X_3 = 0$ or $X_3 = 1$. This means the 'or' relationship shows some degree of redundancy, knowing about X_1 or about X_2 carries information about X_3. But there is also an element of synergy, since X_3 is not completely determined by knowing either X_1 or X_2.

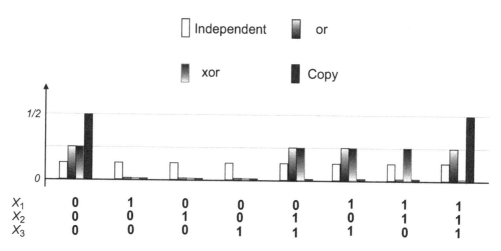

Figure 9.5 Histogram of the relative frequencies with which each configuration is occupied. Only the independent coins are able to visit all eight possible configurations. For the or, xor and copy the interdependence between the coins restricts the coins from visiting certain configurations.

This situation is different for the 'xor'. Observing $X_1 = 0$ or $X_1 = 1$ still leaves the probability 50:50 for X_3. The same is the case if we know the value of X_2, again no improvement on the prediction of X_3. In this sense the 'xor' relationship represents synergy. We can only improve our prediction of X_3 by knowing both X_1 and X_2, in which case X_3 is completely determined.

Finally, the 'copy' relation allow us to predict the faces of the two others by observing one of the coins. In this sense there is redundancy between the coins. The allowed number of configurations is equal to the two allowed for a single coin.

In order to have a broader range of relationships to analyse, we imagine building slot machines that let the three coins visit the eight possible configurations according to mixtures of the occupancy allowed by the four relationships: independent, or, xor and copy. We do this by defining the joint probability $P(x_1, x_2, x_3)$ as a linear mixing controlled by a parameter $\alpha \in [0, 2]$. For $\alpha \in [0, 1]$ we choose

$$P(x_1, x_2, x_3) = (1 - \alpha)P_{\text{copy}}(x_1, x_2, x_3) + \alpha P_{\text{indp}}(x_1, x_2, x_3). \tag{9.66}$$

When $\alpha = 0$ the relation between X_1, X_2, and X_3 is that of copying, and for $\alpha = 1$ the coins are independent. For values of α between 0 and 1 we have a mix of these two relationships.

For α between 1 and 2 we produce a mix of independent with the 'or' relationship or with the 'xor' relationship. To do this we study either

$$P(x_1, x_2, x_3) = (2 - \alpha)P_{\text{indp}}(x_1, x_2, x_3) + (\alpha - 1)P_{\text{or}}(x_1, x_2, x_3) \tag{9.67}$$

or

$$P(x_1, x_2, x_3) = (2 - \alpha)P_{\text{indp}}(x_1, x_2, x_3) + (\alpha - 1)P_{\text{xor}}(x_1, x_2, x_3). \tag{9.68}$$

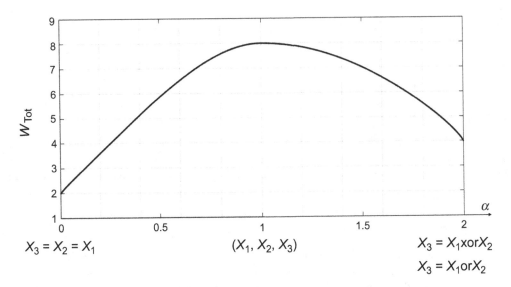

Figure 9.6 Change in the effective total number of configurations available to the three coins for different values of the mixing parameter α. When $\alpha = 0$ the coins follow pure copying. At $\alpha = 1$ the three coins are independent and at $\alpha = 2$ they are related through the logic 'or' or 'xor' operation.

As we did for the two coins above, we can look at the effective number of configurations visited by the three interrelated coins as our slot machine visits the eight possible configurations according to the probabilities $P(x_1, x_2, x_3)$. Figure 9.6 presents $W_{\text{Tot}} = 2^H$ for the three-coin system. We notice that as we change the mixing parameter, α, the effective allowed number of configurations W_{Tot} changes and only assumes the maximum number of configurations, namely 8, for the independent relationship corresponding to $\alpha = 1$. Tying the coins together by mixing an amount of any of the three relationships 'copy', 'or' and 'xor' prevents the coins from exploring the entire set of configurations. We notice that, as expected, $W_{\text{Tot}} = 2$ for pure coping and $W_{\text{Tot}} = 4$ for pure 'or' or 'xor'.

Our conclusion remains that for the two-coin system W_{Tot} is a good measure of how the interdependence between the coins restricts their ability to explore the entire set of configurations which would have been available if no interdependence between the coins was imposed. We observe that the stronger the restrictions, the smaller the value of W_{Tot}, with W_{Tot} smallest for pure copying, the most redundant relation. We also notice that W_{Tot} cannot distinguish whether we mix the 'or' or 'xor' relationship with the independent relationship.

We noticed above that 'or' and 'xor' differ in the way knowledge of the outcome of X_1 aids the prediction of X_3. This suggests that we need to look more closely at the restrictions imposed on one coin given the specific state of the other coins. That is, we need to construct the effective number of configurations subject to the condition that one or more coins assume specific configurations. We can do this by use of the marginal and conditioned probabilities and their corresponding entropies.

Let us first consider the marginal probability, for example

$$P_{X_1}(x_1) = \sum_{x_2} \sum_{x_3} P_{X_1 X_2 X_3}(x_1, x_2, x_3) \tag{9.69}$$

and

$$P_{X_1 X_2}(x_1, x_2) = \sum_{x_3} P_{X_1 X_2 X_3}(x_1, x_2, x_3), \tag{9.70}$$

and similarly for the other X_2 and X_3 and combinations of pairs.

First we consider the effective number of configurations available to the three coins if they are thrown independently according to the marginal probabilities. Similar to the discussion of W_{Tot} and γ_{Tot} we can define for each coin $i = 1, 2, 3$ the number $W_i = 2^{H_i}$, where H_i is the entropy of $P_{X_i}(x_i)$. The total effective number of available configurations when the three coins are thrown independently according to the marginal probabilities is given by the product $\Pi_i W_i$. We want to compare this number to the effective number of configurations when the coins follow the joint probability distribution in Eqs. (9.66), (9.67) and (9.68), so we define the relative change as

$$\gamma_{\text{mar}} = \frac{\Pi_i W_i}{W_{\text{Tot}}}. \tag{9.71}$$

Next we look at how each coin is restricted by the configurations assumed by the other coins. We do this by conditioning on the two other coins. Hence we consider, for example:

$$P(x_1 | x_2, x_3) = \frac{P_{X_1 X_2 X_3}(x_1, x_2, x_3)}{P_{X_2 X_3}(x_2, x_3)}. \tag{9.72}$$

From Eq. (9.18) we have that the corresponding conditioned entropy is defined as

$$H_{\neg 1} = -\sum_{x_2} \sum_{x_3} P_{X_1 X_2 X_3}(x_1, x_2, x_3) \log \left[\frac{P_{X_1 X_2 X_3}(x_1, x_2, x_3)}{P_{X_2 X_3}(x_2, x_3)} \right]. \tag{9.73}$$

Here we used the logic 'not' symbol $\neg 1$ as subscript to indicate that we condition on all the other observables different from X_1. We define three effective numbers of configurations visited when the statistics of the coins is conditioned on all other coins as $W_{\neg i} = 2^{H_{\neg i}}$ for $i = 1, 2, 3$ and a corresponding complement reduction coefficient as

$$\gamma_{\neg} = \frac{\Pi_i W_{\neg i}}{W_{\text{Tot}}}. \tag{9.74}$$

In the language of information theory, $\log \gamma_{\text{mar}}$ is called the total correlation and $-\log \gamma_{\neg}$ is called the total dual correlation [370].

As an overall measure of how the interdependence between the three observables puts restrictions on the exploration of configurations, we can use the product and ratio of the two coefficients γ_{mar} and $\gamma_{\neg}$.

Figure 9.7 shows how γ_{mar} and $\gamma_{\neg}$ change as the mixing between the different relationships changes. We see that $\gamma_{\text{mar}} \geq 1$, which means that the number of available

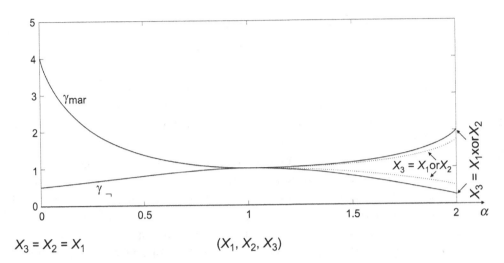

Figure 9.7 The coefficients of relative effective change for different types of mixing controlled by the parameter α. We notice that the 'xor' leads to larger changes compared to the independent case than does the 'or' relation.

configurations is larger when the coins follow the independent statistics given by marginals:

$$P_{\mathrm{mar}(X_1X_2X_3)}(x_1, x_2, x_3) = P_{X_1}(x_1)P_{X_2}(x_2)P_{X_3}(x_3), \tag{9.75}$$

than when the coins are linked together according to the joint distribution. This is to be expected since the joint distribution expresses restrictions on the collective set of configurations.

In contrast, we see that $\gamma_{\neg} < 1$, which corresponds to the conditioning of one coin on the other, restricts the effective number of available configurations compared to those allowed by the joint distribution. Since $\gamma_{\neg}$ refers to the reduction in the number of available configurations of a coin when we consider a coin i subject to specifying the value of the other coins $j \neq i$ and take the average over the possible configurations of these coins $j \neq i$, we can interpret $\gamma_{\neg} \leq 1$ as indicative of how the freedom of a coin i is restricted by the specification of the other coins. This is formulated in information-theoretic terminology as the randomness that remains in coin i after conditioning on coins $j \neq i$.

We also notice from Fig. 9.7 that the two coefficients γ_{mar} and $\gamma_{\neg}$ each assume the same kind of values as we vary the mixing parameter α from 0 to 1 and from 1 to 2. This near symmetry about $\alpha = 1$ means that γ_{mar} and $\gamma_{\neg}$ cannot be used directly to distinguish redundancy, $\alpha \in [0, 1]$, from synergy, $\alpha \in [1, 2]$.

Since γ_{mar} corresponds to an increase while $\gamma_{\neg}$ indicates a decrease, we can capture the net combined effective change by the product

$$\gamma_T = \gamma_{\mathrm{mar}}\gamma_{\neg} = \frac{\Pi_i W_i}{W_{\mathrm{Tot}}} \frac{\Pi_i W_{\neg i}}{W_{\mathrm{Tot}}}, \tag{9.76}$$

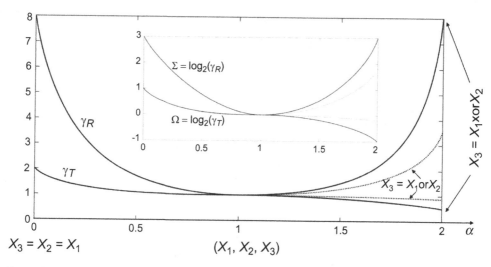

Figure 9.8 The total γ_T and the relative γ_R effective changes for different values of the mixing parameter α. We notice the monotonous change in γ_T as the degree of redundancy decreases on the interval $0 \leq \alpha \leq 1$ and that γ_T continues to decrease while the degree of synergy increases on the interval $1 \leq \alpha \leq 2$. In contrast, the ratio γ_R increases nearly symmetrically as we tune α away from the independence at $\alpha = 1$. The insert shows the behaviour of the information-theoretic quantities Ω and Σ corresponding to the two coefficients γ_T and γ_R.

and measure the relative increase compared to the decrease as the ratio

$$\gamma_R = \frac{\gamma_{mag}}{\gamma_\neg} = \frac{\Pi_i W_i}{\Pi_i W_{\neg i}}. \tag{9.77}$$

Figure 9.8 shows the behaviour of these coefficients. First we notice that they are both equal to 1 for the independent case. Moreover, the product given by γ_T varies monotonously as the mixing goes from pure redundancy via independence to pure synergy, while γ_R is equally large for pure redundancy and pure synergy. While γ_T is able to capture the degree of redundancy versus synergy, γ_R is a measure of structure amongst the observables in the sense that the larger values of γ_R correspond to a higher degree of interdependence between the observables, corresponding to systems where the components are farther away from behaving independently. This is because $\gamma_R > 1$ when the effective number of states available to the coins controlled by the marginal probabilities is larger than the corresponding number of states available when the coins follow the probabilities including the restrictions imposed by the other coins represented by the conditioned probabilities.

We can summarise these considerations in the following way. For a system consisting of independent components we have $\gamma_T = \gamma_R = 1$. The total combined reduction in available states represented by γ_T is larger than 1 when variables are dominated by redundancy. When synergy dominates $\gamma_T < 1$. Pure redundancy and pure synergy are similar in the way structure amongst the observables is imposed, which moves us away from the non-structured case of independent components.

Traditionally, information theory works directly with entropies rather than with the effective number of states, though a prominent exception is Einstein's 1910 paper [123].[8] The two entropies corresponding to γ_T and γ_R

$$\Omega = \log(\gamma_T) \quad \text{and} \quad \Sigma = \log(\gamma_R) \tag{9.78}$$

are discussed in information-theoretic terms in [370].

In terms of Ω and Σ we have, for independent components, $\Omega = \Sigma = 0$. When the organisation of the components leans towards redundancy, $\Omega > 0$ whereas when the components are overall synergistically organised, $\Omega < 0$. In this sense Ω captures *Organisation* and on the other hand we can think of Σ as capturing the *Structured* behaviour of the components. It was shown in [370] that Σ can be expressed as a sum of mutual information, and since mutual information is non-negative $\Sigma \geq 0$. So according to Eq (9.78) we have $\gamma_R \geq 1$, which according to Eq. (9.77) tells us that $\Pi_i W_i > \Pi_i W_{\neg i}$ or that the dependence on the surroundings can only make the effective number of configurations smaller than the number available when the components are independent of each other. However, the reduction in the number of available effective states induced by dependency on the other components can be due to redundancy or to synergy, see Fig. 9.9.

We can use Ω and Σ to construct a diagram of organisation in the sense of synergy–redundancy measured by Ω versus structure in the sense of distance from independent behaviour measured by Σ. See Fig. 9.9, where the Ω and Σ values for the three coins are plotted for different joint probability distributions. The curve connecting the point in the top-right corner denoted 'Pure copy' and the point denoted 'Independent' corresponds to the distributions given in Eq. (9.66) as α is changed from 0 to 1. And the curve connecting the point labelled 'Independent' with the point in the top-left corner denoted 'Pure xor' corresponds to the distributions given by Eq. (9.68) as α changes from 1 to 2.

In order to indicate how Ω and Σ behave for distributions with various degrees of redundancy and synergy, Fig. 9.9 contains two curves representing the linear superposition of the independent distribution in Table 9.1 with the weighted sum of the distributions P_{xor} and P_{copy} in the same table. The curves connecting the point 'Independent' to the point 'Pure 1/2 copy & 1/2 xor' and to the point 'Pure 4/3 copy & 1/4 xor' represents the Ω and Σ of the following distributions. First we mix the redundant 'copy' with the synergetic 'xor' by defining the distribution

$$P_{(a,b)}(x_1, x_2, x_3) = aP_{copy}(x_1, x_2, x_3) + bP_{xor}(x_1, x_2, x_3), \tag{9.79}$$

with $a + b = 1$.

Next we combine this composite distribution with the independent distribution in order to control the degree of dependence included among the three coins. This is done by defining the joint distribution as our usual linear combination

$$P(x_1, x_2, x_3) = (1 - \alpha)P_{indp}(x_1, x_2, x_3) + \alpha P_{(a,b)}(x_1, x_2, x_3). \tag{9.80}$$

---

[8] An English translation is available from Princeton University's Collected Papers of Albert Einstein at https://einsteinpapers.press.princeton.edu/vol3-trans/245

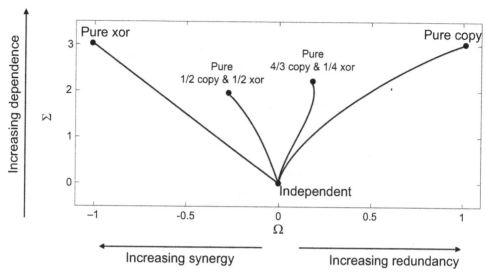

Figure 9.9 Plot of the variation in the Σ and Ω measures as redundancy and synergy are mixed in different proportions. Organisation along the x-axis and structure along the y-axis. See the text for a detailed description of the four curves.

Choosing $(a, b) = (\frac{1}{2}, \frac{1}{2})$ and tuning α from 0 to 1 generates the curve connecting the point 'Independent' to the point 'Pure 1/2 copy & 1/2 xor' and similarly choosing $(a, b) = (\frac{4}{3}, \frac{1}{4})$ leads from the point 'Independent' to the point 'Pure 4/3 copy & 1/4 xor'.

This exercise gives us some feeling for how Ω and Σ can be used to quantify the nature of the interrelation between a system's components. The absolute values of Ω and Σ are not meaningful right away, but the relative values between different situations or different systems are indicative of differences in correlation structure.

The measures can be generalised to N observables $\mathbf{X} = (X_1, X_2, \ldots, X_N)$; details can be found in [370]. The generalisation is straightforward and the final expressions are

$$\Omega = \sum_{i=1}^{N}(H_i + H_{\neg i}) - 2H(\mathbf{X})$$

$$\Sigma = \sum_{i=1}^{N}(H_i - H_{\neg i}), \tag{9.81}$$

where, along the lines of the notation above in this section, we introduced the entropy of the entire set of observables as

$$H(\mathbf{X}) = -\sum_{\mathbf{X}} P_{\mathbf{X}}(\mathbf{x}) \log P_{\mathbf{X}}(\mathbf{x}), \tag{9.82}$$

and the marginal entropy of observable X_i as

$$H_i = \sum_{x_i} P_{X_i}(x_i) \log P_{X_i}(x_i). \tag{9.83}$$

Finally, the entropy of the observable X_i conditioned on all the observables X_j with $j \neq i$ as

$$H_{\neg i} = - \sum_{\mathbf{X}} P_{\mathbf{X}}(\mathbf{x}) \log \left(\frac{P_{\mathbf{X}}(\mathbf{x})}{P_{\mathbf{X}_{\neg i}}(\mathbf{x}_{\neg i})} \right). \qquad (9.84)$$

Here $\mathbf{X}_{\neg i}$ and $\mathbf{x}_{\neg i}$ denote the vectors with the component number i removed, i.e.

$$\mathbf{x}_{\neg i} = (x_1, \ldots, x_{i-1}, x_{i+1}, \ldots, x_N). \qquad (9.85)$$

When we think of applications, looking at these general expressions for any number of components, we realise that to analyse large numbers of data streams it may become a practical problem to handle the joint distribution and various conditioned distributions in order to be able to compute the entropies. How severe a limitation this requirement is will depend on the available amount of data. If data are collected at high frequency, such as when dealing with data from algorithmic trading or brain data from EEG scanners, it may indeed be possible to estimate the joint probabilities from histograms of the data. But if we are dealing with sparse data like extinction data from the fossil record, it will be difficult to use measures like Ω and Σ to determine the relationship between the different components. And even when one has plenty of data, it will be computationally demanding to compute all the needed histograms, so one may have to compromise and select subsets of the available data streams.

In [370] an analysis of composers' synergistic use of voices was done by comparing Ω measured from scores by Johan Sebastian Bach and Arcangelo Corelli. Overall, Ω was found to be negative for Bach and positive for Corelli, indicating Bach using voices to create new emergent structures to a higher degree than Corelli made use of.

The way Ω and Σ depend on the level of redundancy, synergy and interdependence suggests why various implementations of the IIT measure did not behave as desired [289]. It turns out, see [370], that IIT is approximately proportional to Σ, which means that the IIT is unable to distinguish a highly interdependent set of components dominated by a redundant relation from an equally strongly interacting set of components exhibiting strong synergy. This is a limitation, since redundancy indicates that although strong interaction is needed to ensure that a component is a repeat of the others it does not imply, like synergy does, that the set of components supports collective emergent behaviour which is entirely different from that of the individual components.

9.5.3 Group Entropy Measure of Complexity

We will finish this chapter by briefly returning to the group-theoretic entropies discussed in Sec. 9.4. We want to consider how generalised entropies may offer a measure of complexity that does not utilise conditioning.

Imagine that we form the Cartesian product $A \times B$ of the microstates available to A on its own with the equivalent microstates for the isolated system B. According to the composition rule in Eq (9.39), the group-theoretic entropy of $A \times B$ is given by $S(A \times B) = \Phi(S(A), S(B))$. On the other hand, the entropy of the fully interacting combination AB, in which microstates may be frozen out or new states may appear due to interaction, is $S(AB)$. Imagine that we are able to obtain the individual entropies $S(A)$, $S(B)$ and $S(AB)$. This can perhaps be done by measuring histograms in simulations or from observational data to estimate the probabilities p_i^A, p_i^B and p_i^{AB}.

We can then construct the following measure of how much the Cartesian system $A \times B$ is different from the fully interacting system AB in the following way:

$$\Delta(AB) = S(A \times B) - S(AB) = \phi(S(A), S(B)) - S(AB). \tag{9.86}$$

To illustrate the behaviour of this complexity measure we evaluate it for different N dependences of $W(N)$ and simply consider the uniform case $p_i^A = 1/W_A$, $p_i^B = 1/W_B$ and $p_i^{AB} = 1/W_{AB}$. The composition rule for both trace and non-trace forms of the group-theoretic entropy be can expressed as

$$\phi(x, y) = \lambda \left\{ W^{-1} \left[W \left(\frac{x}{\lambda} + W^{-1}(1) \right) W \left(\frac{y}{\lambda} + W^{-1}(1) \right) \right] - W^{-1}(1) \right\}, \tag{9.87}$$

where $\lambda = 1/W^{-1}(1)$ and $W^{-1}(x)$ denotes the inverse function of $W(X)$, see [441]. We apply this procedure to compute the measure in Eq. (9.86) in the case where $B = A$, so combining a system of N components with itself. This leads to the following three expressions.

(I) Algebraic – $W(N) = N^a$:

$$\Delta(N) = \lambda \left(N + N + \frac{N^2}{\lambda} \right) - \lambda(N + N) = N^2. \tag{9.88}$$

We might call this the 'Tsallis case': the interdependence between particles strongly restricts the available state space. The entropy of the Cartesian combination $A \times A$ overshoots the entropy of the fully 'entangled' system AA by N^2. One may think of this as indicating that the reduction of state space involves a restrictive relation between each particle and the $N - 1$ other particles.

(II) Exponential – $W(N) = k^N$:

$$\Delta(N) = \lambda(N + N) - \lambda(N + N) = 0. \tag{9.89}$$

The Boltzmann–Gibbs case where the entire system effectively is composed of a non-interdependent set of subsystems.

(III) Super-exponential $W(N) = N^{\gamma N}$:

$$\Delta(N) = \lambda \left\{ \exp[L(2(1 + N) \ln(1 + N))] - 1 \right\} - \lambda(N + N)$$
$$\simeq \lambda(\exp[\ln(2N \ln N)] - 2N \simeq 2\lambda(N \ln N - N) \simeq 2\lambda \ln N!. \tag{9.90}$$

Here we assumed $N \gg 1$ and made use of the fact that the Lambert function $L(x) \simeq \ln x$ asymptotically.

The effective dependence of the complexity measure on the factorial suggests a relation to the super-exponential behaviour of $W(N)$ originating in the creation of new states by forming combinations of the components [218].

Let us finish this brief discussion of group entropies by mentioning that these entropies have been applied to time series analysis to produce a new measure of complexity able to handle time series, which contains numbers of perturbations of patterns that grow super-exponentially fast with the length of the signal, for details see [15, 16].

Summary: Information theory provides mathematical methods for the analysis of multiple aspects of interrelated components by analysing joint probbilites.

- Functions of probabilities called entropies are central to the formalism. The so-called Shannon entropy is of fundamental importance.
- Maximising an appropriate entropy can sometimes be used as a way to determine probabilities.
- Indicators of possible causal relations can be obtained by conditioning probabilities across time.
- Quantifying the degree of complexity is possible through information-theoretic methods such as the Lempel–Ziv measure.
- Different types of emergent interdependence, such as redundancy and synergy, can be quantified by entropy-based measures such as the Ω and Σ information.

9.6   Further Reading

General audience:

Information theory is engagingly explained in *The Information: A History, A Theory, A Flood* [163] by James Gleick without use of mathematics.

An ambitious and inspiring, though perhaps somewhat optimistic, attempt to explain life itself in terms of information is in the book *The Demon in the Machine: How Hidden Webs of Information Are Finally Solving the Mystery of Life* [97] by Paul Davies.

Intermediate level:

A careful, basic introduction to the mathematics of information theory can be found in *Information Theory: A Tutorial Introduction* [427] by James V. Stone.

Advanced level:

The book *Science and Information Theory* [70] by the physicist Leon Brillouin, first published in 1956, contains a wealth of interesting discussions of the meaning of the term 'information'.

The book *Mathematical Foundation of Information Theory* [233] by one of the pioneers of the field, A. Kinchin, is very readable and of great interest to anyone worrying about the cornerstones.

For a modern, very comprehensive and mathematically thorough introduction turn to *Elements of Information Theory* [87] by Thomas M. Cover and Joy A. Thomas.

The book *Information and the Nature of Reality: From Physics to Metaphysics* [98] edited by Paul Davies and Niels Henrik Gregersen contains 16 essays by scientists, philosophers and theologians discussing basic ontological questions.

9.7   Exercises and Projects

Exercise 1

Think of throwing a coin. Let $X = 1$ when the outcome is heads and $X = -1$ when the outcome is tails. We assume the coin is biased and that $P_X(1) = p$ and $P_X(-1) = 1 - p$. Compute the entropy of X and find the p for which $H(X)$ is maximal.

Exercise 2

An example of dependent variables with zero correlation coefficient. We consider the following two cases. Let X be uniformly distributed

(I) on the set $\{0, 1, 2, \ldots, N\}$;
(II) on the set $\{-N, -N+1, \ldots, -1, 0, 1, \ldots, N\}$.

Define another, clearly dependent, variable Y simply by assuming $Y = X^2$.

(a) Show that the correlation coefficient between X and Y is non-zero for case (I) and zero for case (II).
(b) Show that the mutual information between X and Y is non-zero both for case (I) and case (II).

Exercise 3

Consider two Ising spins in the canonical ensemble with a Hamiltonian given by $H = -S_1 S_2$. Compute the correlation coefficient $C_{X_1 X_2}$ and the mutual information $I(X_1; X_2)$ between the two variables X_1 and X_2 given by $X_1 = S_1$ and $X_2 = S_2$. Discuss the limits of the temperature going to zero and infinity, respectively.

Exercise 4

Derive the expressions for the mutual information given in Eq. (9.20).

Exercise 5

Use the relation between the transfer entropy and mutual information to derive Eq. (9.27).

Exercise 6

We generalise the set of Eqs. (9.14) to include an additional dependence between X and Z:

$$x_t = \xi_1(t)$$
$$y_t = \alpha x_{t-1} + \xi_2(t) \tag{9.91}$$
$$z_t = \beta y_{t-1} + \xi_3(t) + \gamma x_{t-1}.$$

Use simulations to compute $TE_{X \to Z}$ and $DTE_{X \to Z}$ for $\gamma = 0$ and $\gamma > 0$ and a unit time lag in Eqs. (9.24) and (9.25).

Exercise 7

Show that in the limit $q \to 1$ the Tsallis entropy in Eq. (9.36) becomes equal to the entropy in Eq. (9.31). And show that in the same limit the q-exponential in Eq. (9.38) becomes an ordinary exponential. *Hint:* l'Hôpital's rule may be helpful and recall that $\lim_{\kappa \to \infty}(1 + x/\kappa)^\kappa = \exp x$.

Exercise 8

Derive the different expressions for the entropies $S[p]$ corresponding to the different functional forms of $W(N)$ in Box 9.1.

Exercise 9

(a) Discuss what it can be that makes the Lempel–Ziv complexity of the EEG signal higher the more aware the brain.

(b) Do we expect our brain activity to produce more new 'thought patterns' if we are more aware? Study how Lempel–Ziv relates to different stages of sleep [22, 5].

Project 1

Consider three stochastic variables X_t, Y_t and Z_t which assume the values -1 or 1 for $t = 1, 2, 3, \ldots$. We assume that Z can swap freely between -1 and 1 with constant probability μ_Z at each time step.

We assume that when $Z(t-1) = 1$, at time t the two variables X and Y swap independently of each other between -1 and 1 with probability μ_X and μ_Y. If $Z(t-1) = -1$, the variables X and Y are unable to change at time t.

Assume that at time $t = 1$ X, Y and Y assume the values -1 and 1 with equal probability equal to $\frac{1}{2}$.

(a) Show that $\text{Prob}\{X_t = a\} = \frac{1}{2}$ for $a \in \{-1, 1\}$ and $t = 2, 3, \ldots$.

(b) Show that the covariance and the mutual information between X and Y are zero, i.e. $\langle XY \rangle - \langle X \rangle \langle Y \rangle = 0$ and $I(X, Y) = 0$.

(c) Compute the conditioned probabilities $P(\chi_t = a | \chi_{t-1} = b)$ for $\chi = X, Y, Z$ and $a, b \in \{-1, 1\}$ and show for example that

$$P(X_t = a | X_{t-1} = b) = \begin{cases} 1 - \mu_X P(Z_{t-1} = 1) = 1 - \frac{1}{2}\mu_X & \text{if } a = b \\ \mu_X P(Z_{t-1} = 1) = \frac{1}{2}\mu_X & \text{if } a \neq b \end{cases} \tag{9.92}$$

(d) Show that the transfer entropy between Z and X is given by

$$T_{Z \to X} = \log 2 + \frac{1}{2}[(1 - \mu_X)\log(1 - \mu_X) - (2 - \mu_X)\log(2 - \mu_X)]. \quad (9.93)$$

(e) Plot $T_{Z \to X}$. Can you think of an explanation of why $T_{Z \to X}$ is an increasing function of μ_X.

Now simulate the three processes X, Y and Z with the slight change that X_t and Y_t can swap only when $Z_{t-5} = 1$.

(f) Study the transfer entropy between $Z_{t-\tau}$ and X_t and Y_t and plot the transfer entropy as a function of the time lag τ.

(You can check your computations against Sec. 7 in [359].)

Project 2

We consider the so-called not-and and not-or logic gates denoted by nand and nor. Their truth tables are defined in the following way:

A	B	A **nand** B	A **nor** B
0	0	1	1
0	1	1	0
1	0	1	0
1	1	0	0

As in Sec. 9.5.2 we use these operations to define a relation between three stochastic variables X_1, X_2 and X_3, by letting $X_3 = X_1 \text{nand} X_2$ or $X_3 = X_1 \text{nor} X_2$.

(a) Similar to Table 9.1 construct the probabilities for P_{nand} and P_{nor}.

Now mix the processes with the copy process as in Eq. (9.67), replacing **or** by **nand** and **xor** by **nor**.

(b) Make a figure equivalent to Fig. 9.6.
(c) Make a figure equivalent to Fig. 9.7.
(d) Make a figure equivalent to Fig. 9.8.
(e) Make a figure equivalent to Fig. 9.9.
(f) Discuss the differences and similarities between the figures produced in (b), (c), (d) and (e) and Figs. 9.6, 9.7, 9.8 and 9.9.

Project 3

This project studies, in the case of stochastic dynamics, the relationship between the network deduced by use of various information-theoretic measures and the underling network of interactions given by a Hamiltonian.

First construct a random network with weighted directed links consisting of N nodes. You can, for example, generate a non-symmetric adjacency matrix where you draw independent weights J_{ij} from a normal distribution $N(\mu, \sigma^2)$ and let $a_{ij} = |J_{ij}|$

with probability p. We use p to control the overall connectivity and only use positive couplings to make the model behaviour less involved. We can choose $\mu = 1$ to define our energy scale. Now define a random asymmetric Ising Hamiltonian by the following expression:

$$H = -\sum_i \sum_j J_{ij} S_i S_j, \qquad (9.94)$$

where we have placed an Ising variable $S_i = \pm 1$ on each node of the network $i = 1, 2, \ldots, N$.

Next we use the usual Metropolis Monte Carlo (MC) algorithm to produce a time series for each node consisting of the value of spin $S_i(t)$ at time sweep t. The MC procedure is, for example, well explained in Appendix H of [84]. We imagine that these N time series constitute our observational knowledge about the system and want to see what we can learn about the interactions and interdependence of the network from an analysis of these. We will be interested in the behaviour at low and high MC temperature, $T = 1/\beta$, and for sparse, p well below $1/N$, and well connected networks, p well above $1/N$. It will also be interesting to study the behaviour for narrow and wide distributions of the weights J_{ij}. It is also of interest to know how results depend on the size of the network, but to keep the numerics affordable it may be a good idea to keep N fairly small, for example $N = 10$ and $N = 20$.

For different sets of parameters N, p, T and σ, simulate the time series $S_i(t)$ and compute:

(a) The matrix of correlation coefficients C_{ij} given in Eq. (9.6) between $S_i(t)$ and $S_j(t)$ for all $i, j = 1, 2, \ldots, N$.
(b) The matrix TE_{ij} of transfer entropies given by Eq. (9.15) between $S_i(t)$ and $S_j(t)$ for all $i, j = 1, 2, \ldots, N$.
(c) The matrix M_{ij} of MIME coefficients by use of the MIME algorithm obtainable from D. Kugiumtzis's webpage http://users.auth.gr/dkugiu/.
(d) Repeat (c), but this time for the matrix PM_{ij} of PMIME coefficients.

After having obtained these four measures of interdependence and information-theoretic causation, we want to plot networks using the four matrices C_{ij}, TE_{ij}, M_{ij} and PM_{ij} as weighted, and the three later directed, adjacency matrices.

(e) Compare the networks obtained from C_{ij}, TE_{ij}, M_{ij} and PM_{ij} with the underlying network of interactions given by J_{ij}.

Consider the networks obtained from C_{ij}, TE_{ij}, M_{ij} and PM_{ij} including links only with strength above a given threshold w_{thr}.

(f) Compare the networks with the J_{ij} for different choices of w_{thr}.

10 Stochastic Dynamics and Equations for the Probabilities

> **Synopsis:** We consider dynamics represented by successive stochastic moves. Assuming we know the transition probabilities for going from one configuration to the next, we will discuss ways to determine the probabilities of the individual configurations.

The time evolution of many interconnected components can exhibit erratic stochastic temporal dependence even if the individual components in principle are governed by deterministic processes. We are all used to estimating the likelihoods of the outcome of throwing dice by applying simple probability. We do this despite the dice following Newton's deterministic laws of mechanics. An accurate definite prediction of the result of a throw will require minute knowledge of the initial condition and the elastic properties of the dice and the walls of the cup in use and the surface of the table where the dice bounce around. Since this is not possible, probability is in practice the way to deal with the dice. The situation is the same when we, for example, consider the reproduction of a multitude of organisms prone to mutations or the behaviour of a group of people where each person's action results from an examination of a large, and often not very transparent, set of considerations.

To describe the time evolution of such systems, a range of mathematical tools have been developed and we will in the next few sections describe, by use of simple examples, some of the frequently used terminology and approaches. The intention is that after working through this chapter, the reader will find it easier to approach some of the many good books on stochastic dynamics. The list of these books is long and ever increasing, we just refer to a few classics. Durrett's *Essentials of Stochastic Process* [119] assumes very little mathematical background and is aimed at a general audience. Van Kampen's *Stochastic Processes in Physics and Chemistry* is a more advanced text, which has been very influential in mathematical natural science [465] and new editions have appeared repeatedly. Øksendal's *Stochastic Differential Equations* [320] and Gardiner's *Handbook of Stochastic Methods* [150] are careful mathematical texts, both assume some level of mathematical experience.

In what follows we will use an important and simple stochastic process to introduce ideas and approaches. Namely, we will consider the simplest version of a so-called random walk [205, 338] along the x-axis, which pictorially can be thought of a person, see Fig. 10.1, moving along a street in a confused manner making a step to the right or a step to the left at random. One may at first question how the dynamics of a single component can be relevant to emergence in complex systems when we have repeatedly

Figure 10.1 Random walker. Whether to make a step to the right or to the left is decided stochastically. The walker throws a coin, which show Right with probability p_+ and Left with probability p_-.

stressed the importance of interactions between multiple components. The connection consists in how the random motion is brought about: the random walk will typically itself be an emergent phenomenon that is the net effect of a multitude of interactions or influences amongst a set of underlying components. This was exactly what motivated the studies of this particular stochastic process by Louis Bachelier when he investigated the fluctuations of stockmarket prices [32] and when Albert Einstein analysed the erratic motion of small particles suspended in a liquid and subject to zillions of collisions by the molecules of the liquid [124], known as Brownian motion.

To know what to expect, we will first make use of the simplicity of the random walk to compute exactly the probability that the walker is at a certain position at a certain time. We will then show how the general approach of **master equations** leads to a difference equation for this probability distribution. To be able to make use of mathematical techniques developed for differential equations, we turn the difference equation into a **partial differential equation** which is known as the diffusion equation. This equation is an example of a **Fokker–Planck** equation, which are used ubiquitously to describe stochastic dynamics. Next we demonstrate how this formalism allows us to analyse distributions for so-called **first passage times** and **first return times**. Finally we turn the diffusion equation into a **Langevin equation** in order to simplify the study of correlations in time.

10.1 Random Walk and Diffusion

We already encountered the random walk process in Sec. 2.4 when we, in broad terms, made comparison between ordinary diffusion and evolutionary diffusion. Now we will develop the mathematical details. The important point is that during a time step from t to $t + \tau$, the walker's position x changes to $x + a$ with probability p_+ or to $x - a$ with probability p_-, with $p_- + p_- = 1$. During t time steps the walker will make n_+ steps to the right and n_- to the left, so $t = (n_+ + n_-)\tau$. If the walker starts at $t = 0$ from position $x = 0$, we will at time t find the walker at position $x = (n_+ - n_-)a$.

We are not really interested in walkers, however the position of the random walker can be a first attempt at modelling an observable for which a random increase or decrease

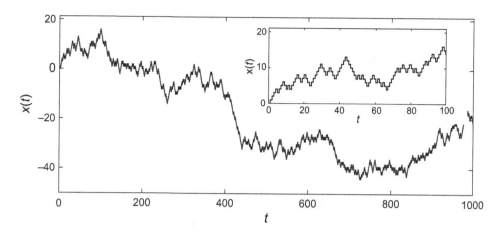

Figure 10.2 Random walk signal. The main panel shows the first 1000 time steps and the insert just the first 100 time steps.

is a reasonably effective description. We may think of the cash of a gambler, or the gains and losses of a trader or some other fluctuating data stream. At this moment the precise realisation of the random walk process is not essential. We focus on the position $x(t)$ as a realisation of a stochastically generated time signal and want to analyse its properties. An example of a trajectory of $x(t)$ for $p_+ = p_- = 1/2$ is shown in Fig. 10.2. The probability that the walker is at position $x(t)$ at time t is given by the probability of n_+ right moves and n_- left moves where n_+ and n_- are determined in terms of $x(t)$ and t as

$$x(t) = (n_+ - n_-)a \text{ and } t = (n_+ + n_-)\tau$$
$$\Downarrow$$
$$n_+ = \frac{1}{2}\left(\frac{x}{a} + \frac{t}{\tau}\right) \text{ and } n_- = \frac{1}{2}\left(\frac{x}{a} - \frac{t}{\tau}\right). \tag{10.1}$$

This is a **binomial process** and the probability is given by how many ways we can choose n_+ out of the total of $n = n_+ + n_- = t/\tau$ steps. This number is equal to the binomial coefficient $\binom{n}{n_+}$. Each step to the right occurs with probability p_+, so n_+ right steps occurs with probability $(p_+)^{n_+}$ which has to be multiplied by the probability of the n_- left-moving steps. We combine this and obtain

$$P(x,t) = \text{Prob}(x,t) = \text{Prob}\{\text{walker is at position } x = (n_+ - n_-)a \text{ at time } t = n\tau\}$$
$$= \text{Prob}(n_+, n)$$
$$= \binom{n}{n_+}p_+^{n_+} p_-^{n_-}$$
$$= \binom{n}{n_+}p_+^{n_+}(1 - p_+)^{n - n_+}. \tag{10.2}$$

When the total number of steps n, and p_+n_+ together with p_-n_- are large, the binomial distribution is well approximated by the normal distribution [140, 193]. This is an example of the Central Limit Theorem and allows us to express $P(x,t)$ as

$$P(x,t) = \text{Prob}(n_+, n) = \frac{1}{\sqrt{2\pi p_+ p_- n}} \exp\left[-\frac{1}{2}\left(\frac{n_+ - np_+}{\sqrt{np_+ p_-}} \right)^2 \right]. \qquad (10.3)$$

The advantage of this approach is that the expression in Eq. (10.2) is exact and that Eq. (10.3) is a very good approximation even for not very large values of n, p_+n_+ and p_-n_-. However, the exact approach is limited by being very particular; we are looking for approaches that can be generalised to more involved processes than a one-dimensional random walk, such as any stochastic process that involves the state of the system changing with a certain probability from one state labelled i to another labelled j. In the example of the random walker, i and j are the positions and the walker can in one time step only move to neighbour positions.

In the general case, the state of the system could be much more complicated. It could, for example, consist of the vote cast by the electorate for a parliamentary election and the time step could enumerate consecutive elections. This state could be represented by a vector

$$\boldsymbol{x} = (x_1, \ldots, x_N), \qquad (10.4)$$

where each $x_k \in \{A, B, C, ...\}$ for $k = 1, 2, \ldots, N$ denotes the vote cast for party A, B or C, etc. by voter k and N is the total number of voters.

Transitions between states correspond to the change in voting from one election to the next, i.e. to changes in $\boldsymbol{x}$ because the value of x_k changes for the voters that decide to move their support to another party. To describe this change in terms of probabilities, we will need the probability $T_{\boldsymbol{x}', \boldsymbol{x}}$ that the set of votes cast in the present election is given by $\boldsymbol{x}'$ if the set of votes in the previous election was given by $\boldsymbol{x}$.

The conservation of probability ensures that as the system evolves the probability does not disappear it only changes, with some outcomes becoming less likely while others increase in likelihood. In terms of the random walker, as time passes it becomes less likely to still find the walker near the starting point and more likely to find the walker some distance from the initial position. The conservation of probability enables us to construct a very general equation for the probability that the system is in one of its states. To make the notation more transparent, we will label a state of the system by i or j and let $T_{i,j}$ denote the probability that the system transits from state j to state i during a time step from t to $t + 1$. Finally, the probability that the system is in state i at time t will be denoted $P_i(t)$. We can now introduce the so-called **master equation**:[1]

$$P_i(t+1) = P_i(t) + \sum_{j \neq i} T_{ij} P_j(t) - \sum_{j \neq i} T_{ji} P_i(t). \qquad (10.5)$$

---

[1] When consulting the literature it is useful to know that the master equation is closely related to a similar equation called the Chapman–Kolmogorov equation [150, 465]. Here we will concentrate on the master equation because of its intuitive transparency.

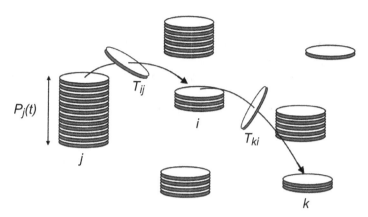

Figure 10.3 The right-hand side of the master equation represented by units (the coins) of occupancy being moved from one stack of occupancy to another.

Let us explain the meaning of this equation in general terms before we apply it to the random walker. Think of an observable that can be in a number of different states enumerated by the label i. Perhaps it is easier for a moment to imagine we have a large collection of identical systems each occupying one of the states labelled by the index i and all changing their state according to the same probabilities T_{ji}. This is depicted in Fig. 10.3, where the height of the stacks of coins indicates how many systems are in a given state at time t. There are, for example, three systems in state i. So we think of the left-hand side of the master equation $P_i(t+1)$ as the number of systems in state i at time $t + 1$. We call this the occupancy of state i. This number is related to how the systems change from one state to another during a time step. If all the systems that are in state i at time t remain there and no other systems change their state to state i, no change occurs in the occupancy of state i. This is expressed by the first term on the right-hand side of Eq. (10.5). The next two terms on the right-hand side express changes due to systems that transit from one state j to another state i. The coefficients T_{ij} denote the probability that a system in state j changes to state i during one time step. Since each of the $P_j(t)$ systems in state j makes the change from j to i with probability T_{ij}, the term $T_{ij}P_j(t)$ represents the number of systems in state j at time t that make the transition to state i during the time step from t to $t+1$. The last term represents the loss of occupancy of the label i due to transitions from i to any other state $j \neq i$.

For a given set of transition coefficients T_{ij}, the master equation allows an exact determination of how the probabilities $P_i(t)$ change with time. However, in general it can be very difficult to solve the equation analytically. Numerical iteration is in principle always possible, though for high-dimensional systems, i.e. when the configuration vector x in Eq. (10.4) has many components, a direct computational approach will be demanding. A combination of analytic approximation of the master equation and numerical analysis can then be useful, see e.g. [108, 215, 500].

To develop an understanding of how the master equation approach works, we will use it to derive an equation for the time evolution of the probability $P(x, t)$ for the one-dimensional random walker, which we already computed above in Eqs. (10.2) and (10.3). First we need the transition coefficients T_{ij}. The 'state' of the system is given by the position x of the walker and since the random steps are to neighbour sites, T_{ij} will only be different from zero for transitions from x to $x + a$ or $x - a$. We can write

$T_{x,x+a} = p_-$ representing a step to the left from position $x + a$ to position x

$T_{x,x-a} = p_+$ representing a step to the right from position $x - a$ to position x.

$$(10.6)$$

The master equation (10.5) for the random walker therefore has the form

$$P(x, t + 1) = P(x, t) + p_+ P(x - a, t) + p_- P(x + a, t) - p_- P(x, t) - p_+ P(x, t). \quad (10.7)$$

The second and third terms on the right-hand side represent walkers stepping from $x - a$ or $x + a$ onto x and the fourth and fifth terms represent walkers at x stepping to $x - a$ and to $x + a$. Since we want to sketch general mathematical methods, we turn this discrete difference equation into a differential equation, which allows the use of the methodology developed to study partial differential equations. However, we have at the same time to emphasise that when one converts discrete difference equations into differential equations, caution is needed. Let us denote the duration of each time step by τ. The left-hand side of Eq. (10.5) and Eq. (10.7) is then $P(x, t + \tau)$ and we assume that $P(x, t)$ varies slowly when we make changes to x and t. This allows us to use Taylor expansion to write

$$P(x, t + \tau) = P(x, t) + \tau \frac{\partial P(x, t)}{\partial t}$$
$$P(x + a, t) = P(x, t) + a \frac{\partial P(x, t)}{\partial a} + \frac{a^2}{2} \frac{\partial^2 P(x, t)}{\partial x^2}. \qquad (10.8)$$

Substituting these expressions into Eq. (10.7) we get

$$P(x, t + 1) - P(x, t) = p_+ P(x - a, t) + p_- P(x + a, t) - p_- P(x, t) - p_+ P(x, t)$$

$$\Downarrow \qquad\qquad (10.9)$$

$$\tau \frac{\partial P(x, t)}{\partial t} = \frac{a^2}{2}(p_+ + p_-) \frac{\partial^2 P(x, t)}{\partial x^2} - a(p_+ - p_-) \frac{\partial P(x, t)}{\partial x}.$$

To simplify the notation, we introduce δ and write

$$p_+ = \frac{1}{2} + \delta \text{ and } p_- = \frac{1}{2} - \delta, \qquad (10.10)$$

which means that δ indicates how much the walk differs from the symmetric case where right and left moves occur with equal probability. We introduce this notation in Eq. (10.9) and divide by τ to get

$$\frac{\partial P(x,t)}{\partial t} = D\frac{\partial^2 P(x,t)}{\partial x^2} - \gamma\frac{\partial P(x,t)}{\partial x}, \tag{10.11}$$

which is a diffusion equation.[2] The coefficient D of the first term on the right-hand side is called the diffusion constant and the coefficient γ of the second term is called the drift velocity. According to Eq. (10.9), they are given by

$$D = \frac{a^2}{2\tau} \quad \text{and} \quad \gamma = \frac{2a\delta}{\tau}, \tag{10.12}$$

but we must say that this will only be correct if the simple approach of the Taylor expansion in Eq. (10.8) can be trusted without any caveats, which we will see below is in fact not entirely the case. More about this in a moment.

Partial differential equations of the type in Eq. (10.12) are, for example, analysed by use of the general technique of Fourier transform outlined in Box 10.1 and explained in textbooks on applied mathematics such as [286]. The solution, derived in Box 10.1, corresponding to the random walk starting from $x = 0$ at time $t = 0$, is given by

$$P(x,t) = \frac{1}{\sqrt{4\pi Dt}} \exp\left[-\frac{(x - \gamma t)^2}{4Dt}\right]. \tag{10.13}$$

This is of the form of a normal distribution. Namely, we recall that the normal distribution of average μ and variance σ^2 is given by [193]

$$P_{\text{norm}}(x; \mu, \sigma^2) = \frac{1}{\sqrt{2\pi\sigma^2}} e^{-\frac{(x-\mu)^2}{2\sigma^2}} \tag{10.14}$$

so by comparing Eq. (10.13) and Eq. (10.14), we can conclude that the diffusion equation determines the average and variance of the position to be given by

$$\begin{aligned} \langle x \rangle &= \gamma t \\ \sigma_x^2 &= \langle x^2 \rangle - \langle x \rangle^2 = 2Dt. \end{aligned} \tag{10.15}$$

The average of the position moves linearly with time, proportional to the asymmetry between the right and left steps. And the excursions from the average given by σ are proportional to $\sqrt{t}$. As time passes, the walker will manage to reach positions ever father away from the average. The variance is time dependent even in the symmetric case $\gamma = 0$, which means that the position of the random walk is an example of a time signal following non-stationary statistics. We will return below to such signals and discuss how to characterise them and we will consider the so-called Ornstein–Uhlenbeck process, which can be thought of as a random walker attracted to its average position. This, as we will see below, turns the walk into a stationary process.

---

[2] The diffusion equation is an example of a Fokker–Planck equation. These are general partial differential equations describing the time evolution of probability densities. Many good books discuss the mathematical techniques involved in their analysis. See e.g. [150, 465].

Box 10.1 Fourier transform and solution of equations

Fourier analysis consists of writing a function $f(x)$ as a sum of, or an integral over, periodic functions. The formalism is well explained in many applied mathematics books, for example see [286]. Here we include a brief intuitive introduction. The two main formulae needed for the analysis are

$$f(x) = \int_{-\infty}^{\infty} \frac{dk}{2\pi} \hat{f}(k) e^{ikx} \qquad (10.16)$$

$$\hat{f}(k) = \int_{-\infty}^{\infty} dx f(x) e^{-ikx}. \qquad (10.17)$$

It is worth being aware that the way the factor $\frac{1}{2\pi}$ is included is done differently in the literature; sometimes it is included with the integral over k and sometimes it is split into a factor $\frac{1}{\sqrt{2\pi}}$, which is included with both the k integration and the x integration. This does not matter, but one should be careful to use the same convention for the formulae above throughout a calculation.

For people familiar with vectors, it may be helpful to think of these formulae as similar to how we write a vector $\mathbf{v}$ as a sum of the basis vectors $\mathbf{e}_x$, $\mathbf{e}_y$, etc. of the coordinate system. For example, in three dimensions we have

$$\mathbf{v} = v_x \mathbf{e}_x + v_y \mathbf{e}_y + v_z \mathbf{e}_z, \qquad (10.18)$$

where the coordinates are given by the scalar products, also referred to as the dot product:

$$v_k = \mathbf{v} \cdot \mathbf{e}_k, \text{ for } k = x, y, z. \qquad (10.19)$$

We can see that Eq. (10.16) corresponds to Eq. (10.18) if we loosely think of the integral in Eq. (10.16) as a sum

$$f(x) = \sum_k \hat{f}(k) e^{ikx}, \qquad (10.20)$$

and consider each of the functions $x \mapsto e^{ikx}$ as corresponding to a basis vector in a 'direction' labelled by k. Then the coefficient $\hat{f}(k)$ corresponds to the coordinates v_x, v_y and v_z in Eq. (10.18) and Eq. (10.17) corresponds to the scalar product in Eq. (10.19).

The inversion formulae in Eqs. (10.16) and (10.17) can be derived by use of the **Dirac delta function**, which is defined by the following formula:

$$\int_{-\infty}^{\infty} dx f(x)\delta(x - x_0) = f(x_0), \tag{10.21}$$

for any continuous function $f(x)$.

The delta function is an example of a generalised function, or a distribution, and cannot be specified in the usual way by assigning a value to each x. However, by considering different shapes for $f(x)$, it is possible to conclude that $\delta(x) = 0$ for $x \neq 0$. The delta function can be obtained as the limit of a narrow peak about $x = 0$. By choosing $f(x) = 1$, we see from Eq. (10.21) that the integral of the delta function is equal to one, so we need the peaks to become more narrow and higher as they approach $\delta(x)$ in order to keep the integral constant. Here are two examples, namely the box representation of the delta function

$$\delta_{\sqcap}(x, a) = \begin{cases} \frac{1}{a} & \text{if } -\frac{a}{2} < x < \frac{a}{2} \\ 0 & \text{otherwise} \end{cases} \tag{10.22}$$

and the Gaussian representation

$$\delta_\sigma(x) = \frac{1}{(2\pi\sigma^2)^{1/2}} e^{-\frac{x^2}{2\sigma^2}}. \tag{10.23}$$

Both $\delta_{\sqcap}(x, a)$ and $\delta_\sigma(x)$ satisfy the definition of the delta function Eq. (10.21) when we let $a \to 0$ or $\sigma \to 0$. For example

$$\int_{-\infty}^{\infty} dx f(x)\delta_{\sqcap}(x - x_0, a)$$

$$= \int_{x_0-\frac{1}{2}}^{x_0+\frac{1}{2}} \frac{1}{a} f(x) \tag{10.24}$$

$$\simeq f(x_0) + \text{terms of order in } a$$

$$\to f(x_0) \text{ for } a \to 0.$$

We can find the Fourier transform of the delta function by combining the definition in Eq. (10.21), using $x_0 = 0$, with Eq. (10.17) to conclude

$$\hat{\delta}(k) = 1. \tag{10.25}$$

Next we substitute this expression into the right-hand side of Eq. (10.16) to conclude that

$$\delta(x - x_0) = \int_{-\infty}^{\infty} \frac{dk}{2\pi} e^{-ikx}. \tag{10.26}$$

Now we can return to the Fourier transformation and show that the two expressions in Eqs. (10.16) and (10.17) are consistent. We substitute the expression for $\hat{f}(k)$ in Eq. (10.17) into Eq. (10.16) to get $f(x)$. The substitution leads to

$$\int_{-\infty}^{\infty} \frac{dk}{2\pi} \left[\int_{-\infty}^{\infty} dx' f(x') e^{-ikx'} \right] e^{ikx}$$

$$= \int_{-\infty}^{\infty} dx' f(x') \int_{-\infty}^{\infty} \frac{dk}{2\pi} e^{-ik(x'-x)} \qquad (10.27)$$

$$= \int_{-\infty}^{\infty} dx' f(x') \delta(x' - x) = f(x).$$

Next we consider how Fourier transformation can help us to solve differential equations. Consider what happens to the right-hand side of Eq. (10.16) when we differentiate with respect to the variable x:

$$f'(x) = \frac{df(x)}{dx} = \int_{-\infty}^{\infty} \frac{dk}{2\pi} (ik) \hat{f}(k) e^{ikx}, \qquad (10.28)$$

and when we different one more time we obtain

$$f''(x) = \frac{d^2 f(x)}{dx^2} = \int_{-\infty}^{\infty} \frac{dk}{2\pi} (-k^2) \hat{f}(k) e^{ikx}, \qquad (10.29)$$

because $i^2 = -1$. Comparing with the expression in Eq. (10.16) for the Fourier transforms, we conclude that the transforms of the first and second derivatives are given by

$$\widehat{f'}(k) = ik\hat{f}$$
$$\widehat{f''}(k) = -k^2 \hat{f}. \qquad (10.30)$$

We notice that differentiation translates to multiplication by the argument when we move to the Fourier transforms.

Now we applied these transformations to solve an equation similar to Eq. (10.11). Assume $g(x)$ is given and we want to find a function $f(x)$ which is a solution to

$$A \frac{d^2 f}{dx^2} + B \frac{df}{dx} = g(x), \qquad (10.31)$$

where A and B are constants. We Fourier transform the equation by replacing the Fourier transform of each of the terms $\frac{d^2 f}{dx^2}$, $\frac{df}{dx}$ and $g(x)$ to get

$$A(-k^2) \hat{f}(k) + B(ik) \hat{f}(k) = \hat{g}(k), \qquad (10.32)$$

which we rearrange to conclude that

$$\hat{f}(k) = \frac{\hat{g}(k)}{-Ak^2 + iBk}. \qquad (10.33)$$

So if we can compute the Fourier transform of $g(x)$, we just need to substitute Eq. (10.33) into Eq. (10.16) and we can determine $f(x)$ by doing the following integral:

$$f(x) = \int_{-\infty}^{\infty} \frac{dk}{2\pi} \frac{\hat{g}(k)}{-Ak^2 + iBk} e^{ikx}. \tag{10.34}$$

This method is relevant to the solution of Eqs. (10.11) and (10.68). Though in two slightly different ways which have to do with the initial conditions. First we notice that both equations involve functions of two variables, so instead of the ordinary differential equation in Eq. (10.31) we are dealing with partial differential equations of two variables, such as $f(x,t)$. This is a minor difference because we can just transform in each of the variables. Instead of Eq. (10.16) we use

$$f(x,t) = \int_{-\infty}^{\infty} \frac{dk}{2\pi} \int_{-\infty}^{\infty} \frac{d\omega}{2\pi} \hat{f}(k,\omega) e^{i(kx+\omega t)}. \tag{10.35}$$

When analysing Eq. (10.68) no initial condition is applied. The equation is assumed to hold for all $t \in (-\infty,\infty)$ and we can transform the equation in both variables and express the solution in a form similar to Eq. (10.34).

When analysing Eq. (10.11) we need to incorporate the initial condition at time $t = 0$ and therefore Fourier transform in the x variable only and solve the resulting ordinary differential equation in the t variable. This is done in the following way.

We focus first on the x dependence in diffusion Eq. (10.11) and can think of the left-hand side $\partial P / \partial t$ as the function $g(x)$ in Eq. (10.11). We solve the equation for each fixed value of t. After Fourier transforming in the x variable we obtain the equation

$$\frac{\partial \hat{P}(k,t)}{\partial t} = (-Dk^2 - \gamma ik)\hat{P}(k,t). \tag{10.36}$$

We make use of the fact that the solution to the equation $f'(x) = Cf(x)$, where C is a constant, is $f(x) = A \exp(Cx)$, where A is another constant. With this in mind, we conclude that

$$\hat{P}(k,t) = A \exp[-(Dk^2 + i\gamma k)t]. \tag{10.37}$$

The constant A is determined by the initial condition. We assume that the random walkers start from $x=0$ at time $t=0$. This means that at $t=0$ all probability is concentrated at $x=0$. We can mathematically express this initial condition by representing $P(x,0)$ by the Dirac delta function $\delta(x)$. This means we want $\hat{P}(k,0)$ to be the Fourier transform of $\delta(x)$ and we saw above right after Eq. (10.26) that $\hat{\delta}(k)=1$ and we therefore must choose $A=1$. To obtain $P(x,t)$ we use the expression in Eq (10.37) to write

$$P(x,t) = \int_{-\infty}^{\infty} \frac{dk}{2\pi} \hat{P}(k,t) e^{ikx}$$

$$= \int_{-\infty}^{\infty} \frac{dk}{2\pi} \exp[-(Dk^2 + i\gamma k)t] e^{ikx} \qquad (10.38)$$

$$= \int_{-\infty}^{\infty} \frac{dk}{2\pi} e^{-(Dk^2 + i\gamma k)t + ikx}.$$

We want to evaluate this integral by use of the formula

$$\int_{-\infty}^{\infty} dz e^{-az^2} = \sqrt{\frac{\pi}{a}}, \qquad (10.39)$$

for a Gaussian integral. This can be done by the following rearrangement:

$$-(Dk^2 + i\gamma k)t + ikx = -a(k - b)^2 + c. \qquad (10.40)$$

We determine the coefficients a and b on the right-hand side by squaring the bracket and comparing terms proportional to k^2, k and the k-independent term. The result is

$$a = Dt$$

$$b = \frac{i(x - \gamma t)}{2Dt} \qquad (10.41)$$

$$c = ab^2 = -\frac{(x - \gamma t)^2}{4Dt}.$$

Next we substitute Eq. (10.40) into Eq. (10.38) and have

$$P(x,t) = e^c \int_{-\infty}^{\infty} \frac{dk}{2\pi} e^{-a(k-b)^2} = \frac{1}{2\pi} \sqrt{\frac{\pi}{a}} e^c. \qquad (10.42)$$

We get the last equality by the substitution $z = k - b$ and Eq. (10.39). The final result is accordingly

$$P(x,t) = \frac{1}{\sqrt{4\pi Dt}} e^{-\frac{(x-\gamma t)^2}{4Dt}}, \qquad (10.43)$$

which is the result in Eq. (10.13).

Autocorrelation Function and Power Spectrum

Fourier analysis is also helpful when analysing correlations in a signal. According to the Wiener–Khinchin theorem, the autocorrelation function is the Fourier transform of the power spectrum of the signal, where the power spectrum is the absolute value squared of the Fourier transform, see e.g. Sec. 21.4 in [286] and [65] for mathematical details.

Here we present a rough sketch indicating this relationship. Consider a signal $f(t)$ and for simplicity we imagine that the average value of the signal has already been subtracted such that $\langle f(t) \rangle = 0$. Let us imagine we measure the signal $f(t)$ at time instances

$$t = -\frac{T}{2} = -\frac{N}{2}\Delta, \ldots, -2\Delta, \Delta, 0, \Delta, 2\Delta, \ldots, \frac{T}{2} = \frac{N}{2}\Delta.$$

We can then express the correlation function, see Eq. (6.55), in the following way:

$$
\begin{aligned}
C(t) &= \langle f(t_0)f(t_0 + t) \rangle \\
&= \frac{1}{N} \sum_{n=-N/2}^{N/2} f(n\Delta)f(n\Delta + t) \\
&\simeq \frac{1}{T} \int_{-\infty}^{\infty} dt_0 f(t_0)f(t_0 + t) \\
&= \frac{1}{T} \int_{-\infty}^{\infty} dt_0 \int_{-\infty}^{\infty} \frac{d\omega}{2\pi} \hat{f}(\omega)e^{it_0\omega} \int_{-\infty}^{\infty} \frac{d\omega'}{2\pi} \hat{f}(\omega')e^{i(t_0+t)\omega'} \\
&= \frac{1}{T} \int_{-\infty}^{\infty} \frac{d\omega}{2\pi} \int_{-\infty}^{\infty} \frac{d\omega'}{2\pi} \hat{f}(\omega)\hat{f}(\omega')e^{i\omega't} \int_{-\infty}^{\infty} dt_0 e^{i(\omega+\omega')}.
\end{aligned}
\tag{10.44}
$$

We make use of Eq. (10.26) and for simplicity neglect the factor $1/T$ so we can write

$$C(t) = \int_{-\infty}^{\infty} \frac{d\omega}{2\pi} \hat{f}(\omega)\hat{f}(-\omega)e^{-i\omega t}. \tag{10.45}$$

To make the right-hand side appear in the same form as Eq. (10.16) we substitute $\omega \mapsto -\omega$. Finally we notice from Eq. (10.17) that for a real function $f(t)$ we have $\hat{f}(-\omega) = \hat{f}^*(\omega)$ where the star $*$ denotes complex conjugation. The result is that Eq. (10.45) can be rewritten as

$$C(t) = \int_{-\infty}^{\infty} \frac{d\omega}{2\pi} |\hat{f}(\omega)|^2 e^{i\omega t} \tag{10.46}$$

and we can conclude that the Fourier transform of $C(T)$ is given by $\hat{C}(\omega) = |\hat{f}(\omega)|^2$.

If the power spectrum behaves like

$$|\hat{f}(\omega)|^2 \sim \frac{1}{\omega} \tag{10.47}$$

we talk about a $1/f$ (one-over-f) signal or $1/f$ fluctuations because the power spectrum depends on the frequency ω as one over the frequency. To see why such signals have received special interest [37, 351, 490] we make the substitution $u = \omega t$ in Eq. (10.46):

$$C(t) = \frac{1}{t} \int_{-\infty}^{\infty} \frac{du}{2\pi} \left| \hat{f}\left(\frac{u}{t}\right) \right|^2 e^{iu}. \tag{10.48}$$

If we neglect questions about convergence of the integral and simply substitute the $1/\omega$ expression in Eq. (10.47) for $|\hat{f}(\omega)|^2$, we notice that the right-hand side becomes independent of t. This is not entirely correct; doing a more accurate treatment, including careful handling of the convergence of the integral, shows that $C(t)$ decays extremely slowly with t indicative of very long correlations in time.

We recall that to make use of the analytic techniques available for differential equations, we replaced the discrete difference equation (10.7) by the differential equation (10.9). We say we have taken the continuum limit. Although the derivation in terms of Taylor expansion appears fairly innocent, it does tacitly assume that $P(x, t)$ changes smoothly from x to $x + a$ and from t to $t + \tau$, otherwise we are not justified in neglecting terms proportional to the higher powers of a and τ. To check how well the approximations involved represent the original random walk probability, we can compare the expression in Eq. (10.13) to the expression in Eq. (10.3). The latter expression we recall is applicable after many random walk steps. This limit may correspond to the continuum approximation since when replacing the discrete random walk steps by a continuous variable x we are forced to consider variations in x that are large compared to the step size a and variations in time t that are large compared to the time step τ. On this scale, changes in $P(x, t)$ may be relatively smooth. There are many mathematical subtleties when going from difference equations to differential equations, see e.g. Chap. 3 in [150] and [465]. Here we will take a pragmatic attitude and examine if we can make the expression in Eq. (10.13) match the one in Eq. (10.3).

First we need to keep in mind that since x is now treated as a continuous variable, $P(x, t)$ is not directly a probability but rather a probability density which gives us the probability for x in the following way:

$$\text{Prob\{walker's position is in the interval } [x, x + dx]\} = P(x, t)dx. \tag{10.49}$$

So we want to compare $P(x, t)dx$ with $\text{Prob}(n_+, n)dn_+$ in Eq. (10.3), also treating n and n_+ as continuous variables. We express x in terms of n_+ and t, making use of Eq. (10.1) to identify $n_+ = \frac{1}{2a}dx$ and arrive at

$$P(x, t)dx = \text{Prob}(n_+, n)dn_+ = \frac{1}{\sqrt{8\pi \frac{p_+ p_- a^2}{\tau} t}} \exp\left[-\frac{(x - \frac{a}{\tau}(2p_+ - 1)t)^2}{8 \frac{p_+ p_- a^2}{\tau} t} \right] dx. \tag{10.50}$$

We recall that this is a valid approximation when n, np_+ and Np_- are large coinciding with the regime for which we expect the continuum approximation to be valid. We can determine D and γ in Eq. (10.13) by insisting that the expression for $P(x, t)$ in Eq. (10.13) derived from the diffusion equation coincides with the expression in Eq. (10.50). This comparison tells us that we need to make the following identifications:

$$\gamma = \frac{a}{\tau}(2p_+ - 1) = \frac{2a\delta}{\tau}$$

$$D = \frac{2a^2}{\tau}p_+ p_- = \frac{2a^2}{\tau}\left(\frac{1}{4} - \delta^2\right). \tag{10.51}$$

This slightly more cautious handling of the continuum limit leads to the same expression for the drift velocity as did our simpler first approach in Eq. (10.12). But the diffusion coefficient differs when $\delta \neq 0$, i.e. when $p_+ \neq p_-$. Since these expressions are obtained by a comparison in the relevant region of variables, we should use these when we want to make use of the diffusion equation (10.11) to analyse random walks.

10.2 First Passage and First Return Times

In this section we use the established formalism to discuss some important aspects of widespread importance to time series from complex systems, namely return and first passage times, and correlations in time.

We start with return times and first passage times. We think of some time signal $X(t)$ and assume it behaves like a random walk. Then Eq. (10.11) describes the time evolution of the probability that the signal $X(t)$ assumes a specific value $X(t) = x$ at time t. In many situations it is of interest to know the probability distribution of the first passage time. This is the time t it takes the signal to change from a value x_0 at t_0 to reach for the first time another value x at time $t_0 + t$. We could think of $X(t)$ as representing the temperature and ask for the distribution of times until we reach a temperature x degrees higher than the one we measure today. Or if we think of finance and ask when will the Dow Jones index have decreased, or increased, by a certain amount.

Equally important is the first return time. This is the time t it takes the signal for the first time to return to a given value x_0. That is, $X(t_0) = x_0$ and then again for the first time $X(t_0 + t) = x_0$. Return times are of general interest for stochastically fluctuating signals since it is a concise way to summarise crucial aspects of the statistics of the signal. The return time could, for example, represent the time an investor starting out with x_0 capital returns to this same amount. We encountered the return of a random walker when we discussed the distribution of the progeny of branching processes, see Sec. 5.2. An introduction requiring relatively little mathematical background can be found in [361] and for a careful recent study of first return and first passage times for signals described by general one-dimensional Fokker–Planck with a discussion of relation to voter models see [30].

To be precise, we define

$$P_{1\text{pass}}(t) = \text{Prob}\{X(t + t_0) = x_0 \text{ for the first time } t \text{ given } X(t_0) = x_0\}. \tag{10.52}$$

To express $P_{1\text{pass}}(t)$ in terms of the solution to Eq. (10.11) we need to relate the flow of probability, or flow of walkers if we interpret $P(x,t)$ in Eq. (10.11) as representing the density of a collection of random walkers moving along the x-axis. To describe the flow, we introduce the probability current, or more concretely the current of walkers. In the language of density of walkers, the current $J(x,t)$ is defined as the number of walkers that cross a position x at time t.

Since walkers cannot leave the x-axis, the change in the number of walkers $(P(x, t + \delta t) - P(x,t))\delta x$ in a small interval $[x, x + \delta x]$ about position x during a small time interval δt is caused by walkers moving in, or out, of the region $[x, x + \delta x]$. This number is decreased by the net number of walkers crossing out of the interval at position $x + \delta x$ and increased by the net number entering at position x during the time interval δt, i.e.

$$(P(x, t + \delta t) - P(x, t))\delta x = -J(x + \delta x, t)\delta t + J(x, t)\delta t. \tag{10.53}$$

We divide this equation by δx and δt and take the limits $\delta x \to 0$ and $\delta t \to 0$ to obtain the relation

$$\frac{\partial P(x, t)}{\partial t} = -\frac{\partial J(x, t)}{\partial x}. \tag{10.54}$$

Comparing this expression for $\partial P / \partial t$ with Eq. (10.11) allows us to conclude that

$$J(x, t) = -D\frac{\partial P(x, t)}{\partial x} + \gamma P(x, t) + \text{constant}, \tag{10.55}$$

where the last term is a constant of integration. We choose the constant to be equal to zero. We do this to ensure that $J(x, t) = 0$ when the drift velocity $\gamma = 0$ and $\partial P / \partial x = 0$. The motivation for this is that when $\partial P / \partial x = 0$, the density of walkers, $P(x, t)$, is independent of x and this together with zero drift velocity implies that the average number of walkers passing from the left to right will equal the number passing from right to left at any position x.

We can use the current $J(x, t)$ to compute the first passage and the first return distributions. We imagine walkers start from a position $x_0 > 0$ and whenever they reach $x = 0$, they are removed. We say we have an absorbing boundary condition at $x = 0$. This is exactly the case if $x(t)$ corresponds to the amount of cash owned by a gambler. The gambler must leave the game when running out of cash. Similarly, if $x(t)$ is a measure of the number of individuals of a certain species, when $x(t) = 0$ the species is extinct and can of course not be recreated spontaneously. For this configuration the current to the left from a position $x = 0^+$ just to the right of $x = 0$ will correspond to walkers arriving at $x = 0$ for the first and only time.[3] Hence, we need to compute the solution $P(x, t)$, which corresponds to all walkers starting from x_0 at time $t = 0$, and then determine the corresponding current $-J(0^+, t)$. The current $-J(0^+, t)$ will correspond to the walkers that reach $x = 0$ at time t for the first time.

Now we are ready to analyse the first passage and first return probability. We will concentrate on the symmetric case $p_+ = p_-$ and make use of the solution in Eq. (10.13) with the diffusion coefficient given by $D = a^2 / (2\tau)$ and $\gamma = 0$ since $p_+ = p_- = 0 \Rightarrow \delta = 0$. We recall that the distribution $P(x, t)$ in Eq. (10.13) corresponds to initially having the walker, or collection of walkers, placed at position $x = 0$ at time $t = 0$. We notice that starting the walkers at $t = 0$ from a position different from zero, $x = x_0$, simply means that we need to change the origin from $x = 0$ to x_0 along the x-axis and the probability at later times corresponding to starting from $x(0) = x_0$ will be given by

---

[3] For the discrete walk depicted in Fig. 10.1, $x = 0^+$ corresponds to the position $x = a$, the first location to the right of $x = 0$.

$$P(x,t) = \frac{1}{\sqrt{4\pi Dt}} \exp\left[-\frac{(x-x_0)^2}{4Dt}\right].$$ (10.56)

We can use this solution to construct the solution we need in order to be able to determine the return time and first passage time distribution.

We are going to start the walkers from position $x_0 > 0$ and whenever a walker reaches $x = 0$ the walker will be removed. This ensures that when the walker reaches $x = 0$, it does it for the first time. We say that the boundary at $x = 0$ is absorbing. If we do not absorb the walker when it reaches $x = 0$ it may wander further about and keep returning to the origin multiple times. The absorbing boundary condition ensures that walkers arrive at $x = 0$ for the first and only time.

Mathematically we represent the walkers being removed at $x = 0$ by the condition that $P(0,t) = 0$ for all times t. So we need a solution to Eq. (10.13) subject to the initial condition that walkers start from x_0 and the boundary condition $P(0,t) = 0$. But the expression in Eq. (10.56) is clearly not zero for $x = 0$. We can, however, make use of the solution in Eq. (10.56) to construct a solution with the correct boundary condition. The method is widely use when dealing with differential equations and is called the method of image charges. The name comes from electrodynamics and the idea is that positive and negative charges can cancel each other. So, if we place image charges at adequate positions as a mathematical aid, we can make the solution we are looking for vanish where necessary. In our case we combine the solution in Eq. (10.56) with a solution where walkers of 'negative' density start from $x = -x_0$ at time $t = 0$. This means we use

$$P(x,t) = \frac{1}{\sqrt{4\pi Dt}}\left(\exp\left[-\frac{(x-x_0)^2}{4Dt}\right] - \exp\left[-\frac{(x+x_0)^2}{4Dt}\right]\right).$$ (10.57)

When we substitute $x = 0$, the two exponentials cancel each other and we have $P(0,t) = 0$. As we let t become small, each of the two Gaussians becomes more narrow and their heights at $x = \pm x_0$ increase. In the limit of $t \to 0$ we have

$$\lim_{t\to 0} \frac{1}{\sqrt{4\pi Dt}} \exp\left[-\frac{(x-x_0)^2}{4Dt}\right] = \delta(x-x_0)$$

$$\lim_{t\to 0} -\frac{1}{\sqrt{4\pi Dt}} \exp\left[-\frac{(x+x_0)^2}{4Dt}\right] = -\delta(x+x_0).$$ (10.58)

We conclude that $P(x,t)$ solves Eq. (10.13) and satisfies both the required initial and boundary condition. Figure 10.4 shows how this solution evolves in time. Initially the distribution is normalised $\int_0^\infty P(x,0) = 1$, but as time passes $\int_0^\infty P(x,t) < 1$ because walkers are lost at the $x = 0$ absorbing boundary.

Now, from the solution in Eq. (10.57), we can compute the flow, also denoted the current, of walkers corresponding to the solution in Eq. (10.57):

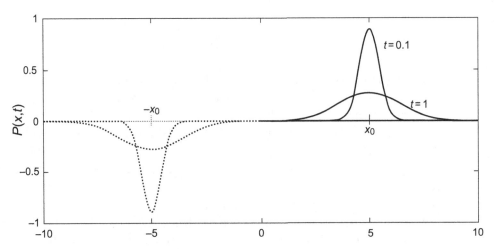

Figure 10.4 Random walk on the half axis $x > 0$ starting from position $x_0 = 5$ and with an absorbing boundary condition at $x = 0$. The solid curves correspond to the probability $P(x, t)$ for two different times $t = 0.1$ and $t = 1$. The walk is not accessing the region $x < 0$ and the dotted curves on the left-hand side do not describe probabilities but are included to show the linear combination in Eq. (10.57) ensuring that $P(0, t) = 0$ for all times.

$$J(x, t) = -D \frac{\partial P(x, t)}{\partial x}$$
$$= \frac{2D}{\sqrt{\pi}(4Dt)^{3/2}} \left[(x - x_0) e^{\frac{(x-x_0)^2}{4Dt}} - (x + x_0) e^{\frac{(x+x_0)^2}{4Dt}} \right]. \tag{10.59}$$

The distribution of first passage time to $x = 0$ is accordingly

$$P_{1\text{pass}}(t) = J(0^+, t)$$
$$= \frac{x_0}{2\sqrt{\pi}(Dt)^{3/2}} e^{-\frac{x_0^2}{4Dt}} \propto t^{-3/2} \text{ for } t \gg \frac{x_0^2}{4D}. \tag{10.60}$$

To obtain the distribution of first return times $P_{1\text{ret}}(t)$, we choose $x = 0$ and $x_0 = \epsilon$, where $0 < \epsilon \ll 1$ is a position just to the right of $x = 0$. This means that arriving at x is essentially identical to returning to x_0. We have

$$P_{1\text{ret}}(t) = \frac{\epsilon}{2\sqrt{\pi}(Dt)^{3/2}} e^{-\frac{\epsilon^2}{4Dt}} \propto t^{-3/2} \text{ for } t \gg \frac{\epsilon^2}{4D}. \tag{10.61}$$

We observe that for the diffusion equation in Eq. (10.11), which is equivalent to the random walk, the probability for the first passage and the first return time exhibit the same $t^{-3/2}$ behaviour for large times. Different behaviours are possible for different types of diffusion. For example, with different step probabilities on different domains, see e.g. [30, 361].

We also recall that our derivation explains the behaviour of the return times used when relating branching processes to random walks in Sec. 5.2.

10.3 Correlations in Time

So far we have considered how the probability of a diffusive signal, or equivalently a crowd of random walkers, spreads though space after initiation in a narrow region. We will next turn to an analysis of correlations in time of a steady-state-driven diffusive signal and again phrase our discussion in terms of the diffusion equation with a slight generalisation allowing a continuous source of walkers.

The diffusion equation (10.11) describes the time dependence of the probability that the signal, or position of the walker, is at x at time t when the probability changes because of flow from one value x to another, but with no external input beside the initial condition. Let us now imagine that some external effect can add or subtract an amount $g(x, t)$ of probability to $P(x, t)$. To include such an effect we need to add an extra term to the right-hand side of Eq. (10.11) and therefore get

$$\frac{\partial P(x,t)}{\partial t} = D\frac{\partial^2 P(x,t)}{\partial x^2} - \gamma\frac{\partial P(x,t)}{\partial x} + g(x,t). \qquad (10.62)$$

The term $g(x, t)$ is called a source term since it acts as a positive or negative supply of probability. If we assume $g(x, t)$ to represent a stochastic process corresponding to random changes to $P(x, t)$, the equation is called a Langevin equation. Combining in one equation the smooth deterministic behaviour represented by derivatives with abrupt stochastic fluctuations was first considered by Paul Langevin in his analysis of Brownian motion, i.e. a small particle suspended in water. Detailed explanations can be found in Sec. 15.5 of [362] and Chap. 9 in [465] and a slightly more cautious mathematical introduction is given in Chap. 4 in [150].

The Langevin approach is appealing whenever one is able to separate the evolution of a system into a regime of slow smooth variation from another regime of fast erratic variation. Langevin considered an equation of the form

$$m\frac{d^2 x}{dx^2} = F(x) + \chi(t). \qquad (10.63)$$

Here the left-hand side represents the acceleration of a particle with position $x(t)$. The first term on the right-hand side represents the deterministic Newtonian forces acting on the particle. This can be gravity or electrical forces. The term $\chi(t)$ is a stochastic force, which Langevin considered to be the effect of the erratic bombardment of the suspended particle by zillions of water molecules. Hence, on the time scale of the variation of the position $x(t)$, the stochastic term $\chi(t)$ is taken to change instantaneously and to assume values drawn from a random distribution with zero average. Mathematically this is represented by the specification that variance of the stochastic term acting at two different time instances t and t' is zero:[4]

$$\langle \chi(t)\chi(t')\rangle = \kappa\delta(t - t'), \qquad (10.64)$$

---

[4] One may rightly be sceptical of introducing a stochastic non-differential term $\chi(t)$ in the differential equation and care is needed, see [150]. Nevertheless, solution of the Langevin equation by means of integration is possible.

Figure 10.5 Cars/particles diffusing up and down a motorway stretching from $x = -\infty$ to $x = \infty$. At $x = 0$ an intersection allows the vehicles to enter or leave the motorway. We are interested in the time variation of the number of vehicles, $N(t)$ at $x = x_0$.

where κ is a constant multiplying the Dirac delta function introduced in Box 10.1. Though the Langevin formalism is not entirely mathematically rigorous, it offers a very appealing intuitive approach.

We will illustrate the use of the Langevin equation by two examples. The first is inspired by observed memory effects in the traffic through cities [344, 434] and information traffic [436]. The second, the **Ornstein–Uhlenbeck process**, is of very broad relevance such as e.g. to finance [47].

First we look at correlations in traffic flow. For concreteness, imagine a motorway stretching from $x = -\infty$ to $x = \infty$. At $x = 0$ vehicles can enter or leave at an intersection. We will develop a model for the time evolution of the density of cars $n(x, t)$ at position x at time t. We assume that the cars can leave or enter our system at $x = 0$ only, see Fig. 10.5.

We imagine that the density of cars is so high that the traffic is jammed and moves in a jerky stop–start fashion. Furthermore, we assume there is the same amount of traffic in both directions of the motorway, i.e. no net drift velocity, so $\gamma = 0$, see Eq. (10.55).

It therefore seems reasonable that the on-average net current will be from higher to lower density and we assume that the net current of cars is given by

$$J(x, t) = -D\frac{\partial n(x, t)}{\partial x}. \tag{10.65}$$

Next we use that for $x \neq 0$ the change with time in the local density is due to the flow of cars. So the change during a brief time interval δt in the number of cars in a small interval $[x, x + \delta x]$ about x is caused by the flow along the road and given by

$$[n(x, t + \delta t) - n(x, t)]\delta x = [-J(x + \delta x, t) + J(x, t)]\delta t. \tag{10.66}$$

We let $\delta x \to 0$ and $\delta t \to 0$ and write

$$\frac{\partial n(x, t)}{\partial t} = -\frac{\partial J(x, t)}{\partial x}. \tag{10.67}$$

We substitute Eq. (10.65) into Eq. (10.67), add the source term and arrive again at a diffusion equation

$$\frac{\partial n(x, t)}{\partial t} = D\frac{\partial^2 n(x, t)}{\partial x^2} + g(x, t). \tag{10.68}$$

The source term $g(x, t)$ will represent cars entering or leaving the motorway. Hence $g(x, t)$ will be equal to zero except when the position x is equal to the location of an intersection.

We can solve this equation again by use of the Fourier transform in a way similar to our solution of Eq. (10.11), except that this time we will transform in both variables x and t, see Box 10.1, so we write

$$n(x, t) = \int_{-\infty}^{\infty} \frac{dk}{2\pi} \int_{-\infty}^{\infty} \frac{d\omega}{2\pi} \hat{n}(k, \omega) e^{i(kx+\omega t)}, \tag{10.69}$$

which we substitute into Eq. (10.68) to obtain an expression for $\hat{n}(k, \omega)$ in terms of the Fourier transform of the drive $\hat{g}(k, \omega)$:

$$\hat{n}(k, \omega) = \frac{\hat{g}(k, \omega)}{i\omega + \gamma k^2}. \tag{10.70}$$

We substitute Eq. (10.70) into Eq. (10.69) to conclude

$$n(x, t) = \int_{-\infty}^{\infty} \frac{dk}{2\pi} \int_{-\infty}^{\infty} \frac{d\omega}{2\pi} \frac{\hat{g}(k, \omega)}{i\omega + \gamma k^2} e^{i(kx+\omega t)}. \tag{10.71}$$

To proceed we have to specify $g(x, t)$ and its Fourier transform $\hat{g}(k, \omega)$. Our aim is to study time correlations and we choose to look at the correlations in the deviation away from the average of the density of cars, $N(t)$, at some position x_0 some distance away from an interchange at $x = 0$, see Fig. 10.5. So we take as our time signal

$$N(t) \equiv n(x_0, t) - \langle n(x_0, t') \rangle_{t'}. \tag{10.72}$$

As explained in Box 10.1, the autocorrelation function of a time signal can be obtained from the Fourier transform of the so-called power spectrum, which is the absolute value squared of the Fourier transform of the considered time signal. This means we have to compute the Fourier transform of $N(t)$:

$$\hat{N}(\omega) = \int_{-\infty}^{\infty} dt N(t) e^{-i\omega t}$$

$$= \int_{-\infty}^{\infty} dt\, n(x_0, t) e^{-i\omega t} - \langle n(x_0, t') \rangle_{t'} \int_{-\infty}^{\infty} dt\, e^{-i\omega t}$$

$$= \int_{-\infty}^{\infty} dt \int_{-\infty}^{\infty} \frac{d\omega'}{2\pi} \int_{-\infty}^{\infty} \frac{dk}{2\pi} \frac{\hat{g}(k, \omega')}{i\omega' + \gamma k^2} e^{ikx_0 + i\omega' t} e^{-i\omega t} - \langle n(x_0, t') \rangle_{t'} \int_{-\infty}^{\infty} dt\, e^{-i\omega t}$$

$$= \int_{-\infty}^{\infty} \frac{dk}{2\pi} \frac{\hat{g}(k, \omega)}{i\omega + \gamma k^2} e^{ikx_0} - \langle n(x_0, t') \rangle_{t'} 2\pi \delta(\omega). \tag{10.73}$$

We used that the integral over t produces a delta function in $\omega - \omega'$ (see Box 10.1) to perform the integral over ω'. Since we are interested in the dependence of the power spectrum for $\omega > 0$, we neglect the second term in Eq. (10.73), which, because of the delta function, will only contribute to $\hat{N}(\omega)$ at $\omega = 0$.

This is how far we can go without further assumptions concerning the nature of the drive $g(x, t)$. Since this source term is meant to represent vehicles entering and leaving at position $x = 0$, we will now use

$$g(x, t) = \delta(x)\chi(t) \Rightarrow \hat{g}(k, \omega) = \hat{\chi}(\omega). \tag{10.74}$$

We then have that for $x \neq 0$ the source $g(x, t) = 0$ and at $x = 0$ the temporal variation in the flow onto and away from the 'motorway' is given by $\chi(t)$. From Eq. (10.73) we get

$$\hat{N}(\omega) = \hat{\chi}(\omega) \int_{-\infty}^{\infty} \frac{dk}{2\pi} \frac{e^{ikx_0}}{i\omega + \gamma k^2} = \frac{\hat{\chi}(\omega)}{2\pi\gamma} \int_{-\infty}^{\infty} dk \frac{e^{ikx_0}}{i\frac{\omega}{\gamma} + k^2}. \tag{10.75}$$

This integral can easily be computed by contour integration in the complex k-plane.[5] Notice that the denominator can be written as $(k - k_+)(k - k_+)$, where

$$k_\pm = \pm \sqrt{\frac{\omega}{\gamma}} \exp\left(\frac{3\pi}{8} i\right) = \pm \frac{1}{\sqrt{2}} \sqrt{\frac{\omega}{\gamma}} (-1 + i).$$

For $x > 0$, close the contour in the upper half plane and pick up the residue at k_+. The power spectrum is finally calculated as the absolute value squared of $\hat{N}(\omega)$:

$$|\hat{N}(\omega)|^2 = \frac{|\hat{\chi}(\omega)|^2}{4\gamma\omega} e^{-\sqrt{\frac{2\omega}{\gamma}} x_0}. \tag{10.76}$$

The power spectrum of the density fluctuations is clearly influenced by the power spectrum of $\chi(t)$. Let us assume that vehicles enter and leave at the intersection in a totally uncorrelated manner, perhaps not a totally unrealistic assumption.

This assumption implies that the correlation function of $\chi(t)$ is a delta function

$$\langle \chi(t_1)\chi(t_2) \rangle \propto \delta(t_1 - t_2). \tag{10.77}$$

Recall that we get the power spectrum of $\chi(t)$ as the Fourier transform of the correlation function, so we will have $|\hat{\chi}(\omega)|^2 = $ constant. In this case

$$|\hat{N}(\omega)|^2 \propto \frac{1}{\omega} e^{-\sqrt{\frac{2\omega}{\gamma}} x_0}. \tag{10.78}$$

For frequencies so small that $\sqrt{\frac{2\omega}{\gamma}} x_0 < 1$, we have $\exp\left(-\sqrt{\frac{2\omega}{\gamma}} x_0\right) \simeq 1$ and therefore

$$|\hat{N}(\omega)|^2 \propto \frac{1}{\omega} \quad \text{for} \quad \omega < \frac{\gamma}{2x_0^2} \equiv \frac{1}{2T_{\text{diff}}}. \tag{10.79}$$

Here we introduced the time scale $T_{\text{diff}} = x_0^2/\gamma$. This is the characteristic time it takes for particles, undergoing diffusion with a diffusion constant γ, to move from $x = 0$ to $x = x_0$.

At the bottom of Box 10.1, it is explained that a power spectrum inversely proportional to the frequency corresponds to very long-time correlations. This is called $1/f$, for one over frequency, behaviour.

Such long temporal correlations are observed in very many and diverse situations: the light intensity from quasars, the ocean current, the pitch or pressure fluctuations in speech and music [351], the flow of traffic, the fluctuations in the resistivity of an electric conductor [490], in time series from finance and economics [435], traffic fluctuations in cities [344, 434], transport on information networks [436] and many more.

---

[5] Contour integration is part of the theory of functions of complex variables and clear explanations can be found in Sec. 18.3 of [286] or Sec. 3.1 of [209].

Figure 10.6 Signal swapping between two possible values at constant rate ν.

Is the model we have sketched above able to explain the observed $1/f$ correlations in all these many different systems? No, probably not. Surface-driven diffusion does not seem to be central to all these situations. The question whether a general explanation for $1/f$ exists is still an open one, but self-organised criticality, see Sec. 12.1, was suggested as a general explanation [37, 214, 354] of how $1/f$ signals may be generated.

We have considered $1/f$ fluctuations here for mainly two reasons. It is a fascinating problem which is commonly encountered in complex systems and our discussion illustrates how one can use stochastic differential equations to go beyond equilibrium statistical mechanics to analyse temporal emergent behaviour.

To place the $1/f$ signal in perspective, we consider a switching signal like the one analysed in Sec. 6.6.2 once again. This time we think of a time signal $f(t)$ that can assume two values $f(t) = \pm A$ and assume that the probability of the signal swapping from one value to the other during a small time interval dt is given by νdt. See Fig. 10.6.

The autocorrelation function of this signal is, according to Eq. (6.97), given by

$$C(t) = A^2 e^{-2\nu|t|}. \tag{10.80}$$

Here we have included the absolute value of t because for a stationary signal the autocorrelation function is even. In Eq. (6.97) we only considered positive values of the argument.

We get the power spectrum as the Fourier transform of $C(t)$, see Box 10.1. So

$$
\begin{aligned}
|\hat{f}(\omega)|^2 &= \int_{-\infty}^{\infty} dt\, C(t) e^{-i\omega t} \\
&= \int_{-\infty}^{\infty} dt\, A^2 e^{-2\nu|t|} e^{-i\omega t} \\
&= A^2 \left(\int_{-\infty}^{0} dt\, e^{-2\nu(-t)} e^{-i\omega t} + \int_{0}^{\infty} dt\, e^{-2\nu t} e^{-i\omega t} \right) \\
&= A^2 \left(\left[\frac{e^{2\nu t - i\omega t}}{2\nu - i\omega} \right]_{-\infty}^{0} + \left[\frac{e^{-2\nu t - i\omega t}}{-2\nu - i\omega} \right]_{0}^{\infty} \right) \\
&= A^2 \left(\frac{1}{2\nu - i\omega} + \frac{1}{2\nu + i\omega} \right) \\
&= \frac{A^2}{4\nu^2 + \omega^2} = \frac{1}{4} \frac{(\tau A)^2}{1 + (\frac{\tau\omega}{2})^2}.
\end{aligned}
\tag{10.81}
$$

We introduced $\tau = 1/\nu$, which is the typical time scale over which to expect one swap of the signal.

The power spectrum of the switching signal decays as ω^{-2} for $\omega \gg 1/\tau$, in contrast to the $1/\omega$ decay for the time signal from the diffusive traffic model, and as $\omega \ll 1/\tau$ the power spectrum of the switching signal approaches a constant. This corresponds to the fact that a frequency-independent power spectrum is the Fourier transform of a δ function in time, i.e. a correlation function that decays instantaneously. At frequencies low compared to $1/2\tau$, correlations are lost, killed by the multiple switches that can occur for times large compared to τ.

10.4 Random Walk with Persistence or Anti-persistence: Hurst Exponent

The above analysis in terms of the autocorrelation function and Fourier transform assumes that the considered time series, or process, is stationary. That is, the statistical properties of the process do not change with time. This is obviously often *not* the case. A simple counterexample is an ordinary random walk for which e.g. the second moment x^2 is time dependent. Time series from real systems such as EEG recording from the brain [174], financial time series [167, 498], temperature records [374], etc. are examples of non-stationary time series. To handle such time series one needs alternative approaches, the **Hurst exponent** is one such example.

The Hurst exponent was introduced to analyse fluctuations in water levels on the Nile [138] and has found broad application in neuroscience, see e.g. [110] for a recent study of resting-state fMRI across the adult lifespan. It is also used to determine the long-range correlations in the earth's surface temperature [374] and for an application to time series of stock prices see e.g. [299].

We will briefly discuss the Hurst exponent [138] and use a random walk with some degree of either persistence or anti-persistence between the step increments as an illustration.

Consider a time signal consisting of the accumulated contributions

$$X(t) = \sum_{k=1}^{t} s_k. \tag{10.82}$$

We can think of $X(t)$ as the position of a random walker after t steps given by s_k with $k = 1, 2, \ldots, t$. Assume that the increments are independent and identically distributed (i.i.d.) according to a probability density $P(s)$ and the moments

$$\langle s \rangle = \sum_s s P(s)$$

$$\langle (\Delta s)^2 \rangle = \sum_s (s - \langle s \rangle)^2 P(s) \tag{10.83}$$

exist. Then it follows from the Central Limit Theorem that

$$\langle X(t) \rangle = \langle s \rangle t$$
$$\langle (\Delta X)^2(t) \rangle = \langle (\Delta s)^2 \rangle t, \tag{10.84}$$

where $\Delta X = X - \langle X \rangle$.

The time dependence for signals of the form in Eq. (10.82) can be very different when s_k are not stationary or i.i.d. The Hurst exponent H is defined to characterise such behaviour and is given by

$$\frac{\langle (\Delta X)^2(t) \rangle}{\langle (\Delta s)^2 \rangle} \propto t^{2H}. \tag{10.85}$$

So for a random walk with stationary increment distribution, i.e. when the expressions in Eq. (10.84) are satisfied, $H = 1/2$. For correlated time signals the Hurst exponent may differ from 1/2. Values of $H > 1/2$ correspond to signals which possess some degree of persistence, whereas $H < 1/2$ may be related to anti-persistence. We will illustrate this by a simple correlated random walk example below.

Originally Hurst suggested the exponent H as a way to characterise long-time behaviour and devised the so-called rescaled range, see [138], analysis to determine H as defined by Eq. (10.85). Alternatively, people often use a definition more applicable to short time [228]. Namely

$$\langle (X(t) - X(t_0))^2 \rangle \propto (t - t_0)^{2H}. \tag{10.86}$$

If the correlations in the process are homogeneous from short to long time, the two definitions are equivalent. The second definition can be used as a phenomenological way to discuss changes in the correlations of a time signal in terms of a time-dependent Hurst exponent $H(t)$ computed by means of Eq. (10.86) for short time differences $t - t_0$.

We now define a random walk signal $X(t)$ with persistence or anti-persistence acting over a single time step. Let

$$X(t + 1) = X(t) + \Delta(t) \quad \text{with} \quad \Delta(t) \in \{-1, 1\}. \tag{10.87}$$

and assume

$$\Delta(t) = \begin{cases} \Delta(t - 1) & \text{with probability } p \\ -\Delta(t - 1) & \text{with probability } 1 - p \end{cases}. \tag{10.88}$$

For $p > 1/2$ the increment will tend to be in the same direction as the previous increment, i.e. persistence. For $p < 1/2$ the tendency will be to make an increment in the direction opposite to the direction of the previous increment, so anti-persistence. And for $p = 1/2$ we have the ordinary random walk with uncorrelated increments. See Fig. 10.7 for simulated realisations of the process.

The position $X(t)$ and the increments $\Delta(t)$ become correlated because the temporal evolution of $\Delta(t)$ determines $X(t)$. The process is accordingly described by the joint probability density $P(X, \Delta, t)$. It is straightforward to write down the master equation controlling the time evolution of $P(X, \Delta, t)$.

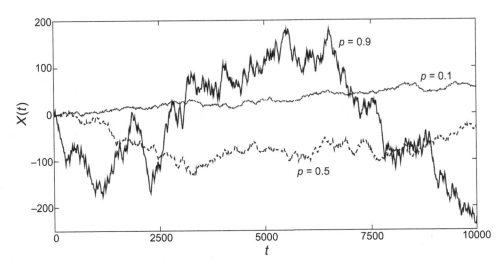

Figure 10.7 Simulated time series for the random walker defined by $X(t+1) = X(t) + \Delta(t)$ and Eq. (10.88). Each curve corresponds to a different value of the persistence probability p.

Recall the general form of the master equation in (10.5). To establish this equation for our walk we need to identify the transitions which the pair X, Δ can undergo. These are

$$
\begin{aligned}
(X - 1, 1) &\mapsto (X, -1) \quad \text{with probability } 1 - p \\
(X - 1, 1) &\mapsto (X, 1) \quad\;\; \text{with probability } p \\
(X + 1, -1) &\mapsto (X, -1) \quad \text{with probability } p \\
(X + 1, -1) &\mapsto (X, 1) \quad\;\; \text{with probability } 1 - p.
\end{aligned}
\tag{10.89}
$$

Hence we arrive at the following set of coupled equations:

$$
\begin{aligned}
P(x, -1, t+1) &= (1 - p)P(X - 1, 1, t) + pP(x + 1, -1, t) \\
P(x, 1, t+1) &= pP(X - 1, 1, t) + (1 - p)P(x + 1, -1, t).
\end{aligned}
\tag{10.90}
$$

This set of equations is easy to solve by exact numerical iteration. To obtain some analytic understanding of the effect of the time dependence of the increments, we consider a mean-field version of the master equation for the process in Eqs. (10.87) and (10.88). The mean-field approximation in this case consists of decoupling the probability densities for $X(t)$ and $\Delta(t)$. We use the notation $P_X(x, t)$ and $P_\Delta(\delta, t)$, where by $x \in \mathbb{Z}$ and $\delta \in \{-1, 1\}$, denote the values assumed by the stochastic variables X and Δ, respectively. We have

$$
P_X(x, t+1) = P_X(x - 1, t)P_\Delta(1, t) + P_X(x + 1, t)P_\Delta(-1, t) \tag{10.91}
$$

and

$$
\begin{aligned}
P_\Delta(1, t) &= pP_\Delta(1, t - 1)) + (1 - p)P_\Delta(-1, t - 1) \\
P_\Delta(-1, t) &= (1 - p)P_\Delta(1, t - 1)) + pP_\Delta(-1, t - 1).
\end{aligned}
\tag{10.92}
$$

We introduce the following simple notation:

$$f(x,t) := P_X(x,t)$$
$$g(\Delta,t) := P_\Delta(\delta,t),$$
(10.93)

to obtain the more handy expression

$$f(x,t+1) = [pf(x-1,t) + (1-p)f(x+1,t)]g(1,t-1)$$
$$+ [(1-p)f(x-1,t) + pf(x+1,t)]g(-1,t-1).$$
(10.94)

Be defining the matrix

$$\mathbf{M} = \begin{Bmatrix} p & 1-p \\ 1-p & p \end{Bmatrix},$$
(10.95)

we can write the equation for $g(\Delta,t)$ in the form

$$\begin{pmatrix} g(1,t-1) \\ g(-1,T-1) \end{pmatrix} = \mathbf{M} \begin{pmatrix} g(1,t-2) \\ g(-1,t-2) \end{pmatrix} = \mathbf{M}^{t-1} \begin{pmatrix} g(1,0) \\ g(-1,0) \end{pmatrix}.$$
(10.96)

Here is an outline of how to solve Eqs. (10.94) and (10.96):

(1) Observe that $|g(1,t) - g(-1,t)| \to 0$ when $t \to \infty$.
(2) Diagonalise Eq. (10.96) for $g(\Delta,t)$.
(3) The solution for $g(\Delta,t)$ is

$$g(1,t-1) = \frac{1}{2} + \beta\lambda^{t-1}$$
$$g(-1,t-1) = \frac{1}{2} - \beta\lambda^{t-1},$$
(10.97)

where $\beta = \frac{1}{2}(g(1,0) - g(-1,0))$ and $\lambda = 2p - 1$.

(4) Now assume $p \geq 1/2$ and solve for $f(x,t+1)$ in the continuum approximation

$$\frac{\partial f}{\partial t} = \frac{1}{2}\frac{\partial^2 f}{\partial x^2} - 2\beta\lambda^t \frac{\partial f}{\partial x}$$
(10.98)

by Fourier transform in x and conclude

$$f(x,t) = \frac{1}{\sqrt{2\pi t}} \exp\left[-\frac{1}{2t}(x + h(t))^2\right],$$
(10.99)

where $h(t) = \gamma\tau(\exp[-t/\tau] - 1)$ with $\tau = -1/\ln(\lambda)$ and $\gamma = 2\beta$.

(5) The asymptotic behaviour for $t \to \infty$:

$$\langle x \rangle(t) \simeq -h(t) \to \gamma\tau$$
$$\langle x^2 \rangle(t) = t + h^2(t) \to (\gamma\tau)^2.$$
(10.100)

And for $t \ll 1$:

$$\langle x \rangle(t) \simeq \gamma t$$
$$\langle x^2 \rangle(t) \simeq t + (\gamma t)^2.$$
(10.101)

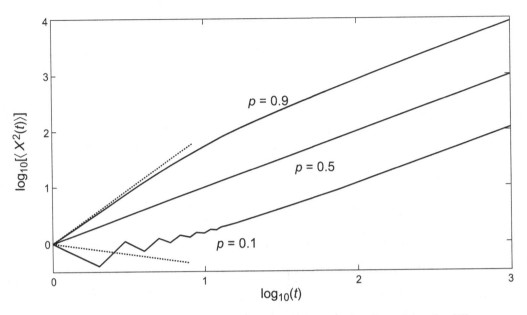

Figure 10.8 Log–log plot of the square of the displacement as a function of time for different values of p from a direct simulation of the process defined in Eqs. (10.87) and (10.88) averaged over 10^6 realisations. The slopes of the dotted straight lines are about 1.8 and -0.3 approximately. The conspicuous kinks in the curve corresponding to $p = 0.1$ happen at time steps 1, 2, 3, etc. and are due to the strong anti-persistence.

The anti-persistent case $p < 1/2$ is more difficult because $\Delta(t)$ will keep changing sign. This makes inapplicable the assumption of smooth behaviour in time underlying the above assumed continuum approximation.

Figure 10.8 shows simulated results for the square displacement. Note that if we use the slopes of the dotted lines to obtain an empirical estimate of the short-time Hurst exponent defined in Eq. (10.86) we will have $H \simeq 0.9$ for $p = 0.9$ and $H = 1/2$ for $p = 1/2$, and finally $H \simeq -0.15$ for $p = 0.1$. But the negative Hurst exponent, or $p = 0.1$, is of course not really obtained according to Eq. (10.86) since the curve for $\langle X^2(t) \rangle$ does not allow us to define a proper tangent from which to obtain an estimate of H.

However if we consider the long-time behaviour, simulations will conclude that $H = 1/2$ in all three cases, in agreement with the analytic behaviour in Eq. (10.100). This corresponds to the fact that the correlations between the time signal $x(t_1)$ and $x(t_2)$ at two different times are generated by the correlations of the consecutive increments only. Therefore, as the separation between t_1 and t_2 increases, the correlations between $x(t_1)$ and $x(t_2)$ disappear. We would have to induce long-time correlations in the increments in order to obtain a Hurst exponent different from $1/2$ at long times. See e.g. [138].

10.5 Stationary Diffusion: Ornstein–Uhlenbeck Process

Although non-stationary time signals certainly occur frequently and exhibit behaviour of the type discussed in the previous sections, stationary random signals are also encountered. We will describe how the non-stationary ordinary random walk can be turned into a stationary process in a way that appears natural. The formalism was introduced by Ornstein and Uhlenbeck to study problems in physics which can be phrased in a way equivalent to a diffusing particle subject to forces pulling it towards the origin [460]. The formalism is of very broad relevance to situations where an observable experiences a tendency to revert towards its average.

The generality of the approach can be indicated by the following few examples. For a particular careful application to finance, see [47]. An example of an application to neuronal modelling can be found in [456]. Analysis of the time involved in cognitive decisions was considered in [186] and in ecology the analysis was used in [126] to model the primary geographical range, the so-called home range, of a species where it finds most of its resources and spends most of its time.

Now we will discuss the Ornstein–Uhlenbeck process and use it as an example of how, from an intuitive stochastic Langevin equation, we can derive a Fokker–Planck equation and then use standard techniques for partial differential equations to determine the probability distribution.

First we return to the ordinary random walk. For simplicity we consider a symmetric walk for which the probability of making a step to the right is equal to the probability of making a step to the left: $p_+ = p_- = 1/2$. The position of the walker $x(t + \tau)$ at time $t + \tau$ can be expressed in terms of the position $x(t)$ at time t as

$$x(t + \tau) = x(t) + \chi(t), \tag{10.102}$$

where χ is a random variable, which with equal probability assumes the value a or $-a$, recall Fig. 10.1. The equation will lead to the flow of probability described by the master equation (10.7), or diffusion equation (10.11), which we know leads to a non-stationary process. By subtracting $x(t)$ and, somewhat casually, taking the continuum limit we can write Eq. (10.102) as

$$\frac{dx(t)}{dt} = \sigma \chi(t), \tag{10.103}$$

where we introduced the constant σ with unit of inverse time to balance dimensions.

This is an example of the Langevin equation encountered above in relation to Eq. (10.62) in the sense that we have a deterministic smooth derivative on the left-hand side and a stochastic function on the right-hand side. The Ornstein–Uhlenbeck process adds an extra term to the right-hand side of Eq. (10.103), which encourages the walker to move towards zero, namely

$$\frac{dx(t)}{dt} = -\eta x + \sigma \chi(t). \tag{10.104}$$

Here, coefficient $\eta \geq 0$ sets the strength of the restoring force. Whenever $x > 0$, the term $-\eta x$ will, if the velocity $dx/dt > 0$, slow down the speed away from zero. If $dx/dt < 0$, the term will increase the speed towards zero. For $x < 0$ we have a similar effect. We are interested in the probability density $P(x, t)$, which describes the probability $P(x, t)dxdt$ of finding the walker in an infinitesimal interval dx about the position x during an infinitesimal time interval dt about the time t if the walker started from $x = 0$ at time $t = 0$.

To determine $P(x, t)$ we need a master equation like the one in Eq. (10.7). To derive this equation we use Eq. (10.104) and consider the discrete random walk, recall Fig. 10.1. The term $\chi(t)$ makes the walker step from position x to either position $x + a$ or $x - a$ and will again lead to the last four terms on the right-hand side of Eq (10.7). We need to determine the effect of the term $-\eta x$ on the flow of probability. The term will make walkers in position x move to position $x - a$ with a likelihood given by ηx, which will subtract probability $\eta x P(x, t)$ from $P(x, t)$. The term will furthermore make walkers in position $x + a$ move to position x with a likelihood given by $\eta(x + a)$, thereby adding probability $\eta(x + a)P(x + a, t)$. In summary, we have

$$
\begin{aligned}
P(x, t + \tau) = P(x, t) &+ \eta\tau[(x + a)P(x + a, t) - xP(x, t)] \\
&+ \frac{\sigma\tau}{2}[P(x + a, t) + P(x - a, t) - 2P(x, t)].
\end{aligned}
\tag{10.105}
$$

From this equation we establish the Fokker–Planck equation

$$
\frac{\partial P}{\partial t} = \eta a \frac{\partial[xP(x, t]}{\partial x} + \frac{\sigma a^2}{2} \frac{\partial^2 P(x, t)}{\partial x^2}.
\tag{10.106}
$$

The solution corresponding to the initial condition concentrated at $x = 0$ at time $t = 0$ is

$$
P(x, t) = \left(\frac{\nu}{2\pi D[1 - \exp(-2\nu t)]}\right)^{1/2} \exp\left(-\frac{\nu x^2}{2D[1 - \exp(-2\nu Dt)]}\right),
\tag{10.107}
$$

where $D = \sigma a^2/2$ and $\nu = \eta a$. The average and the second moment are given by

$$
\begin{aligned}
\langle x \rangle &= 0 \\
\langle x^2 \rangle &= \frac{D}{\nu}(1 - e^{-2\nu t}).
\end{aligned}
\tag{10.108}
$$

We notice that, in contrast to the ordinary random walk, the second moment for the Ornstein–Uhlenbeck process converges to the constant value $\langle x^2 \rangle = D/\nu$ as time goes to infinity.

In the next section we return to evolutionary diffusion, see Sec. 2.4, as another general example of diffusive dynamics, though one that differs from the ordinary diffusion, or random walk, by being able to generate a kind of clustering in space and time.

10.6 Evolutionary Dynamics and Clustering

In Sec. 2.4 we considered qualitatively how reproduction with mutation, though similar to diffusion, differs in the way a population clusters as the dynamics lets the reproducing agents visit the various available types. In this section we will describe in some mathematical detail the similarities and differences between standard diffusion, see Sec. 10.1, and evolutionary diffusion.

We consider the stochastic dynamics of the so-called Moran process [301]. The population size N is kept constant by the dynamics. A time step from t to $t+1$ consists of first choosing with uniform probability an agent and duplicating this agent to allow the offspring to differ from the parent due to possible mutations. Next, choose another agent at random with uniform probability and remove the agent.

For simplicity we assume that each agent is characterised by a 'type' index x. During the reproduction, a mutation may occur with probability p_{mut} which makes the value of x_{off} of the offspring differ from the x value of the parent by a small amount. We want to follow the evolution of the population of types as the dynamics unfolds. To do this we define $n(x, t)$ as the number of agents of type x at time t.

The process is depicted in Fig. 10.9 and summarised below.

Reproduction: Choose a random agent, say of type x. With probability $p_{mut}/2$ place an offspring at type $x + \delta$ or $x - \delta$ and with probability $1 - p_{mut}$ place the offspring at type x, i.e.

$$\begin{aligned}
&\text{With probability } 1 - p_{mut}: \quad n(x, t) \mapsto n(x, t) + 1 \\
&\text{With probability } p_{mut}/2: \quad n(x - \delta, t) \mapsto n(x - \delta, t) + 1 \\
&\text{With probability } p_{mut}/2: \quad n(x + \delta, t) \mapsto n(x + \delta, t) + 1.
\end{aligned} \tag{10.109}$$

Death: Remove a random agent, say of type x_d, hence $n(x_d, t) \mapsto n(x_d, t) - 1$.

Figure 10.9 A sketch of the evolutionary birth–death dynamics. An agent at x duplicated at time t, a mutation occurs and at time $t + 1$, the offspring ends up at position $x_{off} \neq x$. The agent at position x_d dies at time t. The net effect is that an agent moves from position x_d to position x_{off} from time t to $t + 1$.

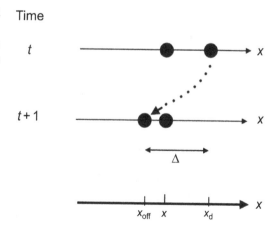

We will refer to x as the position of the agent. The net effect of the death–birth event is to move an agent from position x_d to position x_{off}, see Fig. 10.9. We can summarise the on-average changes to $n(x, t)$ by the following equation:

$$n(x, t+1) = n(x, t) + \frac{n(x - \delta, t)}{N} \frac{p_{mut}}{2} + \frac{n(x + \delta, t)}{N} \frac{p_{mut}}{2} + \frac{n(x, t)}{N}(1 - p_{mut}) - \frac{n(x, t)}{N + 1}.$$

$$(10.110)$$

This equation expresses the number of agents at position x at time t as the number at the previous time step t, the first term on the right-hand side, plus the on-average increase due to the reproduction minus the decrease due to the death. The second term on the right-hand side represents that the population at position x will increase by 1 if an agent at position $x - \delta$ reproduces and the offspring ends up at position x because of a mutation. The probability that an agent at position $x - \delta$ is selected for reproduction is $n(x - \delta, t)/N$, since each agent is equally likely to be chosen. The factor $p_{mut}/2$ represents the probability that the mutation moves the offspring from the parent position $x - \delta$ to position x by adding the amount δ to the position of the parent. The third term on the right-hand side represents an increase at position x due to reproduction with mutation at position $x + \delta$ and the fourth term corresponds to an agent at position x reproducing without any mutation. Finally, the last term represents the probability that an agent at position x is removed. Since, after reproduction, the population consists of $N + 1$ agents, an agent at position x is selected with probability $n(x, t)/(N + 1)$.

We can move the first term $n(x, t)$ to the left side of the equation, collect the remaining terms on the right-hand side and assume $N \gg 1$, so we can neglect the difference between N and $N + 1$ to obtain

$$n(x, t+1) - n(x, t) = \frac{p_{mut}}{2N}[n(x - \delta, t) + n(x + \delta, t) - 2n(x, t)].$$

$$(10.111)$$

This equation is equivalent to Eq. (10.7) for the symmetric random walk with $p_+ = p_-$, so our analysis so far suggests that the evolutionary process will move agents about amongst their types in a way essentially identical to how random walkers diffuse. From this observation we should expect that a population of agents all initially placed at $x = 0$ will smoothly spread out along the x-axis as described by the time dependence of the Gaussian peak in Eq. (10.13) with $\gamma = 0$ corresponding to $p_+ = p_- = 1/2$, see Eq. (10.51).

However, simple computer simulations reveal significant differences between the evolutionary diffusion and ordinary random walk diffusion, see Fig. 10.10. Initially, all $N = 10^4$ agents are located at $x = 0$ and the distribution $n(x, t)$ is shown at two later times $t_1 = 10^3 N$ and $t_2 = 4 \times 10^4 N$. We can think of N single time steps as one generation, since after this number of updates on average each of the N agents will have been selected precisely once for a reproduction or a death event. We denote the time measure in generations by capital T, i.e. $T_1 = 10^3$ and $T_2 = 4 \times 10^4$. We notice that, for evolutionary diffusion, at T_1 the distribution $n(x, t_1)$ is still centred around the initial position $x = 0$, similar to the distribution for ordinary random walk diffusion. But at the later time T_2, the agents undergoing evolutionary dynamics have moved as a

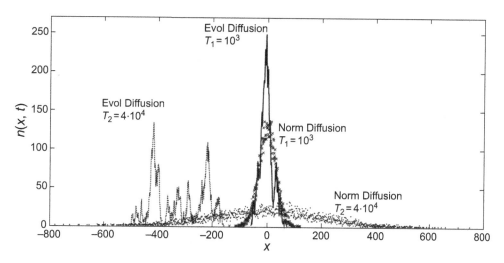

Figure 10.10 Evolutionary versus normal diffusion. For both processes, initially, all $N = 10^4$ agents are located at $x = 0$. The solid curve indicates the distribution of particles undergoing the Moran evolutionary diffusive process after $T_1 = 10^3$ generations and the dotted curve shows the same distribution at the later time $T_2 = 4 \times 10^4$. The crosses and dots show the corresponding distributions for particles undergoing ordinary diffusion. In all cases $p_{\text{mut}} = 1/2$.

cluster to the left while the ordinary random walkers are still distributed symmetrically about $x = 0$ by a broad shallow peak.

The evolutionary dynamics generates a kind of effective cohesion amongst the agents that enables them to move about like a cluster. In different realisations, the cluster may break up into different subclusters. However, the distribution of a single realisation of N agents starting from $x = 0$ never approaches the Gaussian bell shape centred about $x = 0$ [256, 257].

Since the on-average argument leading to Eq. (10.111) suggests that evolutionary diffusion and ordinary random walk diffusion should be identical, we conclude that the observed differences must be related to differences beyond the average level, or in other words the difference must have to do with the way the stochastic fluctuations act in the two cases.

The diffusion of usual random walkers is self-averaging in the sense that a single temporal realisation of a large number of agents approaches the same distribution as obtained by averaging over many realisations of the process. This is in contrast to the evolutionary diffusive process for which the time evolution of a large number of agents exhibits the 'wandering peak' behaviour just described, whereas the histogram of $n(x, t)$ averaged over many realisations approaches a Gaussian peak centred about the initial position $x = 0$ of the agents. The ensemble average obtained by averaging over many realisations of populations of evolving agents all starting from $x = 0$ but subject to different sequences of stochastic choices, implemented by different sequences of random numbers, behaves identically to the diffusion of ordinary random walkers, see Fig. 10.11.

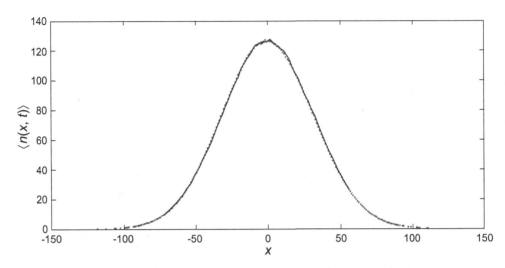

Figure 10.11 Ensemble averaged distributions for evolutionary versus normal diffusion. For both processes, initially all $N = 10^4$ particles are located at $x = 0$. The solid curve indicates the distribution of particles undergoing the Moran evolutionary diffusive process after 1000 generations. The crosses and dots show the corresponding distributions for particles undergoing ordinary diffusion. The histograms are averaged over 1000 realisations. In both cases $p_m = 1/2$.

A difference between averaging over many different realisations and the statistics of the time evolution of a single realisation of many evolving agents is encountered in various complex systems and as mentioned in Sec. 2.4, of particular importance to economics [340] and to systems with inherent disorder [112].

The difference between the two types of diffusion can be analysed by a more careful handling of the stochastic fluctuations than we did above when deriving Eq. (10.111). By use of **field-theoretic methods** one is able to conclude that the evolutionary diffusion in the limit of large N is described effectively by a Langevin equation of the form

$$\frac{\partial n(x,t)}{\partial t} = D\frac{\partial^2 n(x,t)}{\partial x^2} + A\sqrt{n(x,t)}\chi(x,t), \qquad (10.112)$$

whereas ordinary random walk diffusion is described by a Langevin equation of the following form:

$$\frac{\partial n(x,t)}{\partial t} = D\frac{\partial^2 n(x,t)}{\partial x^2} + B\sqrt{n(x,t)}\frac{\partial \chi(x,t)}{\partial x}. \qquad (10.113)$$

Here A and B are constants. The difference between the two equations consists in the derivative of the noise χ in Eq. (10.113), which corresponds to the diffusing particles being conserved. This is of course not the case for the evolutionary diffusion of the particles, or agents, since these are able to reproduce and die.

The functional form of the Langevin equation (10.112) corresponding to evolutionary diffusion, or Moran process, tells us [257] that evolutionary diffusion belongs to a large class of processes called super-Brownian motion, which is a certain type of

Markov process studied in probability theory, see e.g. [303], and for example of relevance to various interacting particle systems [118].

The behaviour of the clusters of evolutionary diffusion depends on the dimension of the type space. Different approaches, see e.g. the discussion and references in [256, 257], all conclude that two dimensions play a special role. For dimensions one and two, the size of the clusters expands with time. In one dimension the cluster size increases as the square root of time and in two dimensions the cluster size increases as the logarithm of time. In three and higher dimensions the cluster reaches a stationary size. The size varies from one realisation to another and this size is power law distributed [200].

A careful mathematical analysis by Derrida and Peliti [104], with a focus on the genealogy using methods of stochastic dynamical systems theory, relates the formation of clusters to the concept of quasi-species [122].

The formation of structure in type space, even when all individuals are subject to identical selection pressure, is the focus of the Hubble model [204]. For a concise mathematical discussion, see e.g. [285].

10.7 Master Equation, Coarse Graining and Free Energy

We now describe how the master equation 10.5 can be related to the probability relating to the stationary state described at what is called a **coarse-grained level**. The coarse-grained level of description consists of combining the microsates into groups.

Think of throwing two dice. Let D_1 and D_2 denote the number of eyes shown by dice 1 and dice 2, respectively. At the microscopic level we will specify which dice shows what for a specific number of eyes d_1 and d_2, i.e. a state will be labelled $D_1 = d_1$ and $D_2 = d_2$. A coarse-grained description could be to specify the sum of the eyes shown as $D = D_1 + D_2$. This implies that the coarse-grained state $D = 5$ contains the microstates: $(D_1, D_2) = (1, 4), (2, 3), (3, 2), (4, 1)$.

We will see that rather generally, the probabilities at the coarse-grained level can be expressed in terms of a so-called free energy. A function which we hurry to stress may have nothing to do with any energy, though in physics where the formalism first appeared, real physical energies and the real physical temperature are involved. But people make use of the terminology even in situations where they do not have a physical energy nor the physical temperature in mind. A very famous example consists of Karl Friston's *free-energy principle*, proposed as a unified theory of the brain [147]. This description makes use of notions such as temperature and free energy, but they do not refer to the thermal temperature of the brain, which for a healthy person is about 37°C, nor to the amount of physical energy in the brain.

We start from the master equation, which for convenience we write again here in the form

$$P_A(t+1) = P_A(t) + \sum_{B \neq A} T_{AB} P_B(t) - \sum_{B \neq A} T_{BA} P_A(t). \qquad (10.114)$$

We consider the states A and B to specify the configurations uniquely at some microscopic level and assume that the transition probability from state A to state B can be expressed as

$$T_{BA} = \frac{1}{1 + \exp(\beta \Delta U_{BA})}, \tag{10.115}$$

in terms of what we, with a borrowing from physics, will call a potential function $U(A)$, so $\Delta U_{BA} = U(B) - U(A)$ and a parameter $\beta > 0$. For each state A we assume we can ascribe a corresponding value $U(A)$ of the potential and will then demonstrate that the probability weights defined as

$$P_A = \frac{e^{-\beta U(A)}}{Z} \tag{10.116}$$

satisfy the stationary master equation. For this reason we will assume that these probabilities describe the system when it has relaxed to the stationary state, also referred to as the equilibrium state.[6] That the probabilities in Eq. (10.116) satisfy the stationary version, $P_A(t+1) = P_A(t)$, of the master equation is seen in the following way. We first show that Eq. (10.116) together with Eq. (10.115) ensures detailed balance. Detailed balance refers to the probability flow between two states expressed as

$$T_{BA}P_A(t) = T_{AB}P_B(t). \tag{10.117}$$

When this balance is satisfied for all times t, it follows from the master equation that the probabilities are independent of time. We see this from the following short rearrangement of Eq. (10.114):

$$\begin{aligned}
P_A(t+1) - P_A(t) &= \sum_{B \neq A} T_{AB}P_B(t) - \sum_{B \neq A} T_{BA}P_A(t) \\
&= \sum_{B \neq A} T_{BA}P_A(t) - \sum_{B \neq A} T_{BA}P_A(t) = 0,
\end{aligned} \tag{10.118}$$

where the second equality is correct when detailed balance Eq. (10.117) holds, which we now show is the case for the transition probabilities in Eq. (10.115) and probabilities in Eq. (10.116). These two equations imply that we have the following equivalent relations:

$$T_{BA}P_A(t) = T_{AB}P_B(t)$$

$$\updownarrow$$

$$T_{BA}^{-1}P_A(t) = T_{BA}^{-1}P_A(t)$$

$$\updownarrow \tag{10.119}$$

$$(1 + \exp(\beta \Delta U_{BA})P_A(t) = (1 + \exp(\beta \Delta U_{AB})P_B$$

$$\updownarrow$$

$$e^{-\beta U_A} + e^{-U_B} = e^{-\beta U_B} + e^{-U_A}.$$

[6] This is correct whenever it is possible to show that the master equation reaches the stationary state and that the stationary equation, which satisfies $P_i(t) = P_i(t+1)$, has a unique solution.

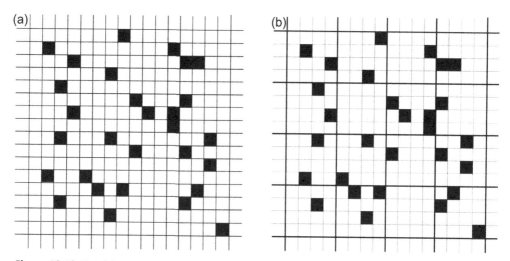

Figure 10.12 Particles on a lattice. In panel (a) the configuration is specified by recording exactly which squares are occupied. In panel (b) we group the squares together in sets of 4×4 and we only keep track of the total number of occupied sites within each 4×4 group.

Since the last equation clearly is correct, we conclude that detailed balance is satisfied.

The probabilities in Eq. (10.116) correspond to specific individual microscopic configurations. If the model represents atoms attached to positions on a surface, this would correspond to specifying precisely where the atoms are located. Or if the model refers to residences of agents, this detailed microscopic description would correspond to specifying the exact address occupied.

In many situations we are not able to know exactly the details at the microscopic level and will instead consider the probabilities at an appropriate coarse-grained level. To make the discussion transparent, let us consider particles placed on a two-dimensional lattice. See Fig. 10.12. A square can either be occupied (black) or empty (white), and at the microscopic level we will specify exactly which squares are occupied and which are not. The black squares could represent an atom being attached or an address being occupied. A more coarse level of description could consist of only specifying the local density of occupied sites. In Fig. 10.12 this is indicated by panel (b). The squares are grouped together in blocks of 4×4 and we only specify the density of occupied sites within a block, disregarding the exact position of occupied sites.

Mathematically, this corresponds to the following. A microscopic state, or configuration, of the system is specified by a vector

$$\boldsymbol{\sigma} = (\sigma_1, \sigma_2, \ldots, \sigma_N), \tag{10.120}$$

where $\sigma_i = 1$ if site number i is occupied and $\sigma_i = 0$ if unoccupied for $i = 1, 2, \ldots, N$. At the coarse-grained level we introduce the local density as

$$\rho_q = \frac{\sum_{i \in B_q} \sigma_i}{M}, \tag{10.121}$$

where B_q, $q = 1, 2, \ldots, Q$ denote the Q blocks of the course-grained description. And M is equal to the number of sites in a block, so we have $N = QM$. At the coarse-grained level the state of the system is specified by the vector

$$\boldsymbol{\rho} = (\rho_1, \rho_2, \ldots, \rho_Q). \tag{10.122}$$

We now further assume that the potential function $U(A)$ of a state A in Eq. (10.116) can be expressed as a sum over microscopic occupancy potentials ϵ_k for each site $k = 1, 2, \ldots, N$:

$$U(A) = \sum_{k=1}^{N} \epsilon_k, \tag{10.123}$$

as is e.g. the case for the Ising model, see Sec. 6.2, and that a coarse-grained potential $u(\rho_q)$ can be defined as

$$\rho_q u(\rho_q) = \frac{1}{M} \sum_{i \in B_q} \epsilon_i. \tag{10.124}$$

The potential $U(A)$ can then be expressed as $U(A) = M \sum_q \rho_q u(\rho_q)$, so we can rewrite the probabilities in Eq. (10.116) for the states $\boldsymbol{\sigma}$ as

$$P(\boldsymbol{\sigma}) = \frac{1}{Z} \exp\left[-\beta M \sum_{q=1}^{Q} \rho_q u(\rho_q) \right]. \tag{10.125}$$

To obtain the probabilities for the coarse-grained states, we need to sum the microscopic probabilities in Eq. (10.116) over all the states $\boldsymbol{\sigma}$ which correspond to one coarse-grained state $\boldsymbol{\rho}$. In state $\boldsymbol{\sigma}$ we denote the number of sites occupied in block q by m_q. There are $W(m_q) = \binom{M}{m_q}$ possible ways of occupying m_q of the M squares in block q. We let $\rho_q = m_q/M$, i.e. the local density of occupied sites, and have that the total number of micro-configurations, $W(\boldsymbol{\rho})$, corresponding to the densities specified by $\boldsymbol{\sigma}$ is given by

$$W_{\text{Tot}}(\boldsymbol{\rho}) = \binom{M}{m_1}\binom{M}{m_2} \cdots \binom{M}{m_Q} = \Pi_{q=1}^{Q} W(\rho_q). \tag{10.126}$$

The probability that the system is in a coarse-grained state ρ is therefore

$$P(\rho) = \frac{1}{Z} W_{\text{Tot}}(\boldsymbol{\rho}) \exp\left[-\beta M \sum_{q=1}^{Q} \rho_q u(\rho_q) \right]. \tag{10.127}$$

We can relate the factors $W(\rho_q)$ to the concept of entropy. Each of the $W(\rho_q)$ block micro-states, corresponding to the coarse-grained state $\boldsymbol{\rho}$, occurs with the same probability according to Eq. (10.125). By reference to Eq. (6.23) we can define the entropy of the state ρ_q of block q as

$$s(\rho_q) = \ln W(\rho_q). \tag{10.128}$$

With this notation we can write the probabilities for the coarse-grained states as

$$P(\boldsymbol{\rho}) = \frac{1}{Z} \exp \left[-\beta M \sum_{q=1}^{Q} \rho_q u(\rho_q) + \sum_{q=1}^{Q} s(\rho_q) \right]. \tag{10.129}$$

Stirling's approximation formula for factorials allows us to write for large N the logarithm $\ln N! \approx N \ln N - N$. Assume that $M \gg 1$, $m_q \gg 1$ and $M - m_q \gg 1$, then we can write

$$\begin{aligned} s(\rho_q) = \ln W(\rho_q) &= \ln \binom{M}{m_q} \\ &= \ln \frac{M!}{m_q! \, (M - m_q)!} \\ &\approx -M[\rho_q \ln \rho_q + (1 - \rho_q) \ln(1 - \rho_q)]. \end{aligned} \tag{10.130}$$

Recall that the Shannon entropy in Eq. (6.11) for a binomial variable with probabilities p and $1 - p$ is

$$S_{\text{bino}} = -p \ln p - (1 - p) \ln(1 - p). \tag{10.131}$$

Therefore, the last expression in Eq. (10.130) has the form of the sum of the Shannon entropy for a binomial for each of the M sites in a block, each occupied or empty with probability $p = \rho_q$ and $1 - p = 1 - \rho_q$, respectively.

Combining these expressions we conclude that the coarse-grained probabilities can be written in the form

$$P(\rho) = \frac{1}{Z} \exp[-\beta M F(\rho)], \tag{10.132}$$

with the definition

$$F(\boldsymbol{\rho}) = \sum_{q=1}^{Q} f(\rho_q) = \sum_{q=1}^{Q} \left[\rho_q u(\rho_q) - \frac{1}{\beta} s(\rho_q) \right]. \tag{10.133}$$

Similar expressions are derived in thermodynamics and statistical mechanics, which has made people borrow the terminology from there and call $f(\rho)$ the free energy. As should be clear from our derivation, this formalism is applicable outside of physics and applies whenever a coarse graining can be done on the probabilities in Eq. (10.116) using expressions of the form Eq. (10.124).

The functional form for the probabilities in Eq. (10.132) tells us that for large M the most likely coarse-grained states ρ will be those which minimise the total free energy $F(\rho)$. This we see in the following way. Let ρ_{min} denote the configuration that minimises $F(\rho)$ and let ρ_1 be some other state with a value $F(\rho_1) > F(\rho_{\text{min}})$. The ratio of the probabilities for ρ_{min} and ρ_1 is given by

$$\frac{P(\rho_1)}{P(\rho_{\text{min}})} = \exp(-\beta M[F(\rho_1) - F(\rho_{\text{min}})]) \to 0 \text{ as } M \to \infty. \tag{10.134}$$

We will make use of this formalism when we study segregation in Sec. 11.2.

Summary: Stochastic dynamics can emerge as a consequence of the web of interdependencies in a complex system. Although the system transitions from one state to another stochastically, the probabilities describing the likelihood of observing a specific state evolve according to the deterministic master equations.

- The random walk, with or without memory, is a particular important example of stochastic dynamics.
- The walk can be studied by use of diffusion equations.
- The character of the walk can be described in terms of return times and first passage times, and the nature of the correlations in time.
- Evolutionary dynamics can be similar to random walks but differs in essential ways by supporting a degree of cluster formation.
- By coarse graining the stationary master equation, we can connect one level of description to another and in doing so we encounter probabilistic weight functions similar to the free energy of statistical mechanics.

10.8 Further Reading

Introductory level:

The book *The Almighty Chance* [502] by Zeldovich, Ruzmaikin and Sokoloff contains a large collection of examples of how stochasticity underlies many phenomena and introduces mathematical analysis of these.

The basic mathematics is lucidly presented in *Essentials of Stochastic Processes* [119] by Rick Durrett.

Intermediate level:

The book *Stochastic Processes in Physics and Chemistry* [465] by N. G. van Kampen is very comprehensive and a classic textbook amongst mathematical scientists.

The readable *A Guide to First-Passage Processes* [361] by Sidney Redner offers an excellent way to become familiar with these aspects of stochastic processes.

Advanced level:

Mathematical details are carefully presented in the *Handbook of Stochastic Methods for Physics, Chemistry and the Natural Sciences* [150] by C. W. Gardiner.

Various applications, including to finance, are included in the mathematically rigorous *Stochastic Differential Equations: An Introduction with Applications* [320] by Bernt Øksendal.

10.9 Exercises and Projects

Exercise 1
Show that the normalisation constant A in Eq. (10.57) is given by

$$\frac{1}{A} = 2 \int_0^{\frac{x_0}{\sqrt{4Dt}}} e^{-z^2} dz.$$

Generalise the expression in Eq. (10.57) to the non-symmetric random walker for which $p_+ \neq p_-$ and derive expressions for the distribution of first passage and first return times.

Exercise 2
Let $n(x, t)$ denote the density of random walkers at position $x \in \mathbb{Z}$ at time $t \in \mathbb{Z}$. At each time step t a walker makes a move $x \mapsto x + \Delta$, where $\Delta = +1$ or $\Delta = -1$ with probability $1/2$.

(a) First derive the discrete time and space mean-field equation that describes the time evolution of $n(x, t)$. Assume the continuous time and space approximation, i.e. let $t, x \in \mathbb{R}$, and derive the partial differential equation according to which $n(x, t)$ evolves.

(b) Now assume that walkers enter and leave the x-axis according to the rate function $g(x, t)$. By use of the Fourier transform

$$n(x, t) = \int_{-\infty}^{\infty} \frac{d\omega}{2\pi} \int_{-\infty}^{\infty} \frac{dk}{2\pi} \hat{n}(k, \omega) e^{ikx + i\omega t}$$

express $n(x, t)$ in terms of the Fourier transform of $g(x, t)$.

Consider the time signal $N(t)$ describing the evolution of the density of walkers at position $x_0 > 0$, i.e. $N(t) = n(x_0, t)$. Assume that walkers can only enter or leave the system at $x = 0$ and that they do so in an uncorrelated manner. This is represented by assuming

$$g(x, t) = \delta(x)\chi(t),$$

where $\chi(t)$ is a white noise signal.

(c) Show that

$$\hat{N}(\omega) = \int_{-\infty}^{\infty} \frac{dk}{2\pi} \frac{\hat{\chi}(\omega)}{i\omega + \frac{1}{2}k^2} e^{ikx_0}.$$

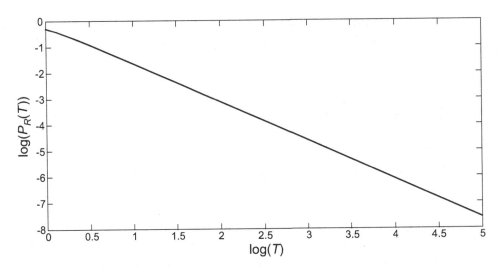

Figure 10.13 Log–log plot of the return time distribution $P_R(T)$.

(d) Show that for $\omega \ll 1/(2x_0)^2$ the signal $N(t)$ has a $1/f$ power spectrum in question (a).

Exercise 3

Consider a random walker on the set of non-negative integers $\{0, 1, 2, \ldots\}$. The probability of a step to the right is p_+ and of a step to the left is p_-, and $p_+ = p_- = 1/2$. Assume that the site $x = 0$ is an absorbing sink (i.e. the probability of moving from $x = 0$ to $x = 1$ is zero) and that $P(1, 0) = 1$, i.e. at time zero the walker starts at position $x = 1$.

(a) Express the probability $P_R(T)$ that the walker arrives at zero for the first time in terms of $P(1, t)$. *Note:* $P_R(T)$ is called the first return time probability.
(b) Iterate the master equation for $P(x, t)$ in order to numerically determine $P_R(T)$ and determine the functional dependence on T for large T. You can compare your result to the graph in Fig. 10.13.
(c) Next consider the continuum limit of the master equation and determine the probability for first return at time T from the gradient (current) at position $x \to 0^+$.

[*Note:* A good reference is S. Redner: *A Guide to First-Passage Processes*, Cambridge University Press 2001, Chap. 1. The book *Elements of Random Walk and Diffusion Processes* by O. C. Ibe, Wiley 2013, is also useful.]

Exercise 4

Consider a two-valued signal $f(t) \in \{-A, A\}$. Assume that the probability that $f(t)$ switches during the time interval dt is constant and given by νdt.

(a) Sow that the autocorrelation function

$$C(t) = \langle f(t_0)f(t_0 + t)\rangle - \langle f(t_0)\rangle \langle f(t_0 + t)\rangle,$$

where the angular brackets denote average over t_0, is given by $C(t) = A^2 \exp(-2v|t|)$.
(b) Find the power spectrum, $S(\omega)$, of the signal $f(t)$ by Fourier transform of the autocorrelation derived in (a).
(c) Establish the behaviour of $S(\omega)$ for $\omega \to 0$ and for $\omega \to \infty$

Project 1 – Non-stationary time signals

Consider a discrete-time random walker moving on the set $X = \{na|n \in \mathbb{Z}\}$, where a is the spacing between the discrete positions. Let p_+, p_0 and p_- denote the probability that the walker from time t to $t + \delta$ makes a move to the right, makes no move, makes a move to the left.

(a) Simulate the process and extract an effective Hurst exponent from the behaviour of $\langle x(t)^2\rangle$ for small values of t.
(b) Establish the master equation for

$$P(x, t) = \text{the probability that a walker starting at } x = 0,$$
$$\text{at time } t = 0 \text{ is at position } x \text{ at time } t.$$

(c) By numerical iteration, find $P(x, t)$ and the moments $\langle x(t)\rangle$ and $\langle x^2(t)\rangle$ for, say $p_+ = p_- = 0.4$, for $p_+ = p_- = 0.1$ and for $p_+ = 0.3, p_- = 0.2$
(d) For $p_+ = p_-$, in the limit $a \to 0$ and $\delta \to 0$ derive a PDE for $P(x, t)$.
(e) From the behaviour of $\langle x(t)^2\rangle$ determine the Hurst exponent.

Project 2 – Hurst exponent from the rescaled range analysis

Consider an ordinary random walk-like signal $\chi(t)$ given by the sum of increments ξ_i, i.e.

$$\chi(t) = \sum_{i=1}^{t} \xi_i.$$

Assume e.g. that all ξ_i are uniformly distributed on $(-1, 1)$.
   Use simulations to determine the Hurst exponent from the rescaled range analysis as explained in pp. 149–154 of [138]. Namely:

(a) Calculate the mean over the time span τ

$$\langle \xi \rangle_\tau = \frac{1}{\tau} \sum_{i=1}^{\tau} \xi_i.$$

(b) For $t \in [0, \tau]$ define a new variable measuring the accumulated deviation from the mean over the time period τ:

$$X(t, \tau) = \sum_{i=1}^{t} [\xi_i - \langle \xi \rangle_\tau].$$

(c) Now, for the internal $[0, \tau]$, measure the width of the range covered by the variable X by introducing

$$R(\tau) = \max_{1 \le t \le \tau} X(t, \tau) - \min_{1 \le t \le \tau} X(t, \tau).$$

(d) Calculate the standard deviation of the increments over the interval $[0, \tau]$. The tradition in Hurst analysis is to denote this standard deviation by S:

$$S(\tau) = \left(\frac{1}{\tau} \sum_{i=1}^{\tau} [\xi_i - \langle \xi \rangle_\tau]^2 \right)^{1/2}.$$

(e) Finally, determine the Hurst exponent H from the τ dependence of the ratio R/S. This is done by a fit to the form

$$R(\tau)/S(\tau) \propto \tau^H.$$

It is illuminating to study the above procedure for different distributions of the increments x_i. You may also find it illuminating to introduce correlations between the increments x_i. You can e.g. introduce some degree of persistence or anti-persistence.

Project 3 – Diffusion equation and time correlations

Let $n(x, t)$ be determined by the diffusion equation

$$\frac{\partial n(x, t)}{\partial t} = \gamma \frac{\partial^2 n(x, t)}{\partial x^2} + \beta \frac{\partial n(x, t)}{\partial x} + g(x, t)$$

where $g(x, t)$ is a given function.

(a) Use Fourier transform to express $n(x, t)$ in terms of the Fourier transform

$$\hat{g}(k, \omega) = \int_{-\infty}^{\infty} dx \int_{-\infty}^{\infty} dt e^{-i(kx + \omega t)} g(x, t).$$

(b) Assume that $\gamma = 0$ and that $\hat{g}(k, \omega) = \hat{g}(k/\omega)$, i.e. a function of the ratio k/ω. Show that

$$n(x, t) = \frac{1}{t} F\left(\frac{x}{t}\right)$$

for some function F.

(c) Assume $\gamma \neq 0$ and $\beta = 0$. Consider

$$N(t) = \int_{-\infty}^{\infty} dx n(x, t).$$

Express $N(t)$ in terms of the Fourier transform of $g(x, t)$.

(d) Determine $\hat{g}(k, \omega)$ such that the autocorrelation function $C(t)$ of $N(t)$ is proportional to the δ-function $\delta(t)$.

(e) Assume $|\hat{g}(0, \omega)|^2 = |\omega|^\alpha$, where α is a real constant such that $1 < \alpha < 2$. Find the t-dependence of the autocorrelation function $C(t)$. Ignore convergence issues.

11 Agent-Based Modelling

> **Synopsis:** In agent-based modelling the dynamics of the components, i.e. the agents, is defined in terms of rules for the agents' response to their surroundings.

The term **agent-based modelling** (ABM) corresponds to a certain interpretation of the components of a model, namely that each component has a degree of integrity and a range of actions. Some authors will only use the term ABM if the agents are assumed to represent human actors, see e.g. [64]. Though frequently we use the term even if the component, i.e. the agents, are more abstract entities that possess a repertoire of actions, which the agent executes, in response to some kind of input. This type of modelling presumably originated with von Neumann's work on self-replicating automata [477].

The capacity of present computers has made direct simulation of ABMs very efficient and the modelling strategy is finding widespread use as a way to improve our understanding through numerical experiments. New monographs and essay collections appear regularly. Here we just mention a few recent publications. A broad overview of models and applications is given in [305]. Applications to social science are discussed in [410] and [152] focuses on economics. An extensive recent review by computer scientists of available software platforms for ABM is given in [4].

We will not be concerned with the details of how to efficiently simulate ABMs, but will instead go through a few examples to illustrate how this modelling approach allows one to study the emergent collective behaviour arising when various types of elementary agents interact through some simple mechanisms.

We will focus on the type of structures that emerge at the systemic level and some mathematical methods to analyse the models. And we will look at how different regimes of entirely different behaviour emerge when one changes either the dynamics of the individual agent or tunes their interaction.

In the next section we will consider models that address how flocking can occur even without a central controlling unit, i.e. no conductor. These models are inspired by and relevant to flocking of birds and schools of fish. We notice that the dynamics of agents able to generate spatial structure is similar to the formation of clusters as a result of evolutionary diffusion considered in Sec. 2.4.

After flocking, we turn our attention to one of the famous ABMs in sociology, namely the Schelling model of segregation, and discuss the original model and how it has been developed to address many different types of segregation.

The last two sections of this chapter look at large number of agents undergoing evolutionary dynamics. We look at the Tangled Nature model which makes use of simple evolutionary dynamics of agents with replication rates determined by the network of interactions amongst the co-evolving agents. The model produces a phenomenology that exhibit, many realistic features similar to what, for example, is observed in long-time macro-evolution, ecology, cultural evolution, sustainability and the framework has also been developed to study opinion dynamics.

11.1 Flocks of Birds or Schools of Fish

Everyone has probably been astonished watching large numbers of birds, or fish, moving through the air, or sea, as one huge flexible coherent body. Several spectacular movies are available on the web, see e.g. [3, 464]. There is no indication that this highly coordinated motion of huge numbers of individuals is brought about by some central control. The evidence is that the flock motion emerges as the collective effect of the response of each individual to the motion of its fellow birds or fish. Flocking, or herding, is not only found amongst birds and fish, but in various forms amongst collections of completely different types of organisms from bacteria [503] to mammals, e.g. springbok [57]. Herding is even found amongst financial traders [330]. The fact that herding in various forms occurs in such a wide range of situations suggests we should develop models that enable us to pinpoint some basic general mechanisms. For a comprehensive review of collective motion and flock formation, see [471].

Here we will consider models inspired mainly by flocking of birds, though the mechanisms studied are sufficiently general that the models may be of relevance more broadly. Since no central command appears to guide the herd formation, it is natural to look for a mechanism based on how each individual responds to the motion in its neighbourhood. In 1996 Vicsek and collaborators published a model [470] that caught broad attention. The model assumes that one bird directs its flight according to what it experience nearby birds do. This assumption is implemented in the model by the following algorithm.

Consider a fixed number N of points, the birds, with position $\mathbf{r}_i$ and velocity $\mathbf{v}_i$ with $i = 1, 2, \ldots, N$. A bird with velocity $\mathbf{v}_i(t)$ at time t will, in the next instance $t + \Delta t$, have moved to the position

$$\mathbf{r}_i(t + \Delta t) = \mathbf{r}_i(t) + \mathbf{v}_i(t)\Delta t. \tag{11.1}$$

The influence of the surrounding birds is represented by the mechanism that determines the velocity $\mathbf{v}_i(t)$. For simplicity, all birds are assumed to move with the same fixed speed $v = |\mathbf{v}_i|$ for $i = 1, 2, \ldots, N$. But the direction of the velocity can change with time and is determined by the direction of the average velocity of the birds within a distance R_0 of bird number i. To model some amount of inaccuracy or stochasticity in the bird's ability to determine the motion of the surrounding birds, a random deviation is added. In two dimensions the direction of $\mathbf{v}_i$ is given by

$$\theta(t) = \langle \theta(t) \rangle_{R_0} + \eta \chi, \tag{11.2}$$

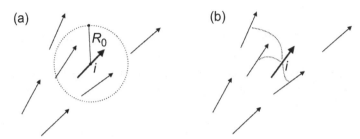

Figure 11.1 Velocities of a flock of birds. In panel (a), bird i determines its velocity as the average direction of neighbour birds within a distance R_0. In panel (b), instead of estimating distances, the bird determines its velocity from the average motion of the n_c nearest neighbours. In the figure $n_c = 3$.

with the random deviation $\chi \in [-\pi, \pi]$. See Fig. 11.1. The parameter $\eta \in [0, 1]$ determines how much the direction of the velocity is made to differ from the local average. If $\eta = 0$, the bird's direction is deterministically determined as the average direction of motion within distance R_0. In this case we will expect that birds once brought together in a dense flock will continue to move as a coherent block in a common direction. If $\eta = 1$, the term $\eta\chi$ will make any direction equally likely and the birds move at random and independently of each other. This suggests that as the parameter η is tuned from zero to one, the collective motion of the N birds will change.

Inspired by the analysis of the 2d XY model in Sec. 6.7, we may expect that the stochastic term in Eq. (11.2) can be considered as being similar to the temperature and therefore that a transition between order and disorder will occur for a certain specific strength of η. This turns out not to be entirely correct, instead the transition from disordered motion to ordered motion occurs through an intermediate stage in which regions of ordered and disordered motion coexist [158].

This two-stage onset of ordered collective motion is related to how the density of the birds influences their ability to 'agree' on a common direction of motion. To understand how the density affects the alignment of the birds, we need to take into account that in the original Vicsek model birds are assumed to pay attention to other birds within the distance R_0 and ignore birds farther away. Therefore, in regions of the flock where birds are clumped together, i.e. the density is high, a bird will be influenced by a larger number of nearby birds than is a bird in less dense regions.

Now we recall that the onset of synchronisation in the Kuramoto model, see Sec. 7.1, occurs when the strength of the coupling between rotors passes a threshold and also that synchronisation in random networks is aided by increased connectivity [339]. These two observations suggest that regions of high bird density will be able to 'synchronise', i.e. enter into aligned ordered motion, for higher values of the noise parameter η than is possible in more sparse regions of the flock. This region of coexistence between ordered and disordered motion is similar to the coexistence at a so-called first-order equilibrium phase transition, such as the coexistence between water and ice at $0°C$.

The question is how relevant the details of the transition to ordered motion is for flocks of real birds or fish, or whatever flocking phenomenon we are interested in. It has been suggested [158] that the inhomogeneous phase may be of relevance to experiments on molecular motors [380], but perhaps of less relevance to real birds where at least sometimes the transition appears to happen abruptly across the entire flock, as e.g. observed in studies of jackdaw [265].

How can we alter the Vicsek model to obtain a homogeneous transition? The assumption that birds keep track of fellow birds within a certain distance is perhaps the assumption of the model that seems most unrealistic. This will require a very rapid and accurate way for the birds to judge distance. Studies of starlings indicate that these birds at least do not use the metric distance to surrounding birds, but rather the topological distance and determine their flight by observing the six or seven nearest birds [41]. This means that the number of interactions used by a bird to determine its orientation is independent of density variations across the flock. Using topological distance instead of metric distance makes the flock of birds much more similar to the 2d XY model, or the usual Kuramoto model, and we would accordingly expect that the onset of ordered motion becomes a sharp transition, which is confirmed by detailed model studies of the Vicsek model with a fixed number of nearest-neighbour interactions [159].

That ordered collective motion can emerge from the distributed, rather than centrally controlled action of the members of a flock, is shared between models that differ in the details of how the agents respond to their surroundings. Hence the fact that no leader is needed can be seen as a robust result established by minimalistic modelling.

It is found that specific model assumptions concerning the individual bird's motion and how it depends on the surrounding birds can influence how the coherent motion is established and also influence the properties of the motion of the flock. One example is the dependence on metric versus topological distance, discussed above, leading to two different types of transitions to ordered motion.

Specific properties, such as how the density varies across a flock, also depend on details of the model. Perhaps surprisingly, the density of real starling flocks has been observed to be larger near the perimeter than in the centre [42]. To address such density variation, Lewis and Turner [261] have developed a non-metric model in which the direction a bird chooses is related to the bird's visual experience as it looks through the flock. This way, the model produces an inward tendency of the motion of the birds at the edge of the flock and an outward tendency of the birds in the centre of the flock, resulting in density variations consistent with observation.

It has been suggested by Charlesworth and Turner [82] that it is possible to relate the assumed algorithms governing the motion of the birds, or in general agents, to underlying cognitive principles. The authors demonstrate that collective motion comparable to observed animal behaviour can be obtained from a new kind of maximisation principle where the agents determine their motion in order to maximise the richness of the visual environment they experience.

11.2 Models of Segregation

We mentioned the Schelling model [386, 387] in the introduction, see Sec. 1.1, as an example of how difficult it can be to directly infer the global systemic behaviour from the preferences of the individual agents. In this section we will develop the mathematics which highlights the generality of the mechanism behind the segregation one observes when simulating the Schelling model. The original version of the model is intuitive and easy to formulate as an algorithm, which we will do below. The model is also straightforward to simulate, but analytic analysis is complicated. Hence we will first mention Schelling's original definition and then present a generalised formulation that is based on the spirit of Schelling's model while being much more approachable analytically.

In his original model, Schelling [386, 387] reduced the dynamics of social segregation to a lattice consisting of two types of agents, A and B, resident on an $L \times L$ square lattice. The only attribute of an agent is its tolerance level τ, which determines how satisfied an agent is at its current location. Agent i checks its neighbourhood to determine how many of its neighbours are of the same type as itself and determines the ratio

$$r_i = \frac{\text{number of neighbours of same type}}{\text{total number of neighbours}}. \qquad (11.3)$$

If the agent finds that $r_i < \tau$ it moves to one of the nearest empty sites which will satisfy the agent's tolerance level. If $r_i \geq \tau$ the agent is willing to remain. Simulations show that a transition occurs for a specific value τ_{seg} of the tolerance level. If $\tau < \tau_{\text{seg}}$ no segregation occurs and one observes an inhomogeneous mixture of the two types of agents in all regions of the lattice. Whereas if agents only feel comfortable when they are part of a neighbourhood containing a high density of their own type, that is for $\tau > \tau_{\text{seg}}$, the lattice splits into regions of pure A occupancy and other regions of pure B occupancy. Simulations find $\tau_{\text{seg}} \approx 1/3$.

This algorithm has been implemented in multiple ways using e.g. different definitions of neighbourhoods or different network structures and by choosing different representations of the tolerance of the agents. An overview of models and a unified mathematical approach can be found in [369].

Here we will follow the approach of [172], which makes use of the master equation analysis presented in Sec. 10.7. The model considered is slightly different from Schelling's original model, though the focus is the same, namely a study of conditions under which the population on a lattice segregates. Like in the original Schelling model, an agent will want to move if the density around him does not satisfy some criterion. In contrast to the original mode, a dissatisfied agent will move stochastically with a certain transition probability which is a function of the configuration of agents on the lattice. Since the choice of the agent is stochastic, a dissatisfied agent may sometimes remain in its present location. The stochastic update appears more realistic than a strict deterministic one, as it mimics how different people tend to react differently under identical conditions.

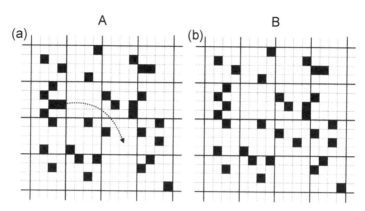

Figure 11.2 Two sections of the $L \times L$ lattice divided into neighbourhoods each containing M sites ($M = 16$ in the figure). Panel (a) depicts configuration A before the move of the agent to its new location in configuration B, depicted in panel (b).

Furthermore, since segregation relates to how the density of agents varies across the lattice, it is useful to divide the lattice into blocks each consisting of a number M of lattice sites. This way we can consider the density in each block and define a coarse-grained variable for each type of agent as the density ρ of a particular type of agent in a block – the number of that type of agent divided by M. We will present the formalism for just one type of agent. Though with one type only, the model is unable to address the segregation between two types of agents, it does still allow a segregation transition to occur. At this transition configurations change from a homogeneous agent density to configurations in which agents cluster in high-density regions, leaving other regions with very low agent density. Even in the single-agent version the segregation transition is driven by fluctuations in the agent density.

The details of the model are as follows. Consider a two-dimensional square lattice, see Fig. 11.2, consisting of $L \times L$ sites. The lattice is divided into Q neighbourhoods each of M lattice sites. Assume m of the M sites are occupied by agents, the local density of agents is then $\rho = m/M$. Instead of specifying the exact position of each agent, we will just be concerned with how many agents reside in a given block. At this coarse-grained level of description the configuration of the system is given by the vector of densities

$$\boldsymbol{\rho} = (\rho_1, \rho_2, \ldots, \rho_Q).$$

Each agent is assigned a function $u(\rho)$ which depends on the density ρ in the neighbourhood occupied by the agent. This function is called a utility function in sociology and in physics a potential. In addition a global utility is defined as

$$U_{gl} = M \sum_{q=1}^{Q} \rho_q u(\rho_q). \tag{11.4}$$

The agent decides stochastically whether to move or to stay. The agents move according to the following transition probability between configuration A and B:

$$T_{BA} = \frac{1}{1 + \exp(\beta \Delta U_{\mathrm{mov}})}, \tag{11.5}$$

where β is a parameter that controls the degree of stochasticity.

We define

$$\begin{aligned} \Delta U_{\mathrm{mov}} &= \Delta u_{\mathrm{ag}} + \alpha(\Delta U_{\mathrm{gl}} - \Delta u_{\mathrm{ag}}) \\ &= (1 - \alpha)\Delta u_{\mathrm{ag}} + \alpha \Delta U_{\mathrm{gl}}. \end{aligned} \tag{11.6}$$

This expression mixes the change Δu_{ag} experienced by the moving agent with the global changes $\Delta U_{\mathrm{gl}} - \Delta u_{\mathrm{ag}}$ experienced by all the other agents. The weighting of the mix between the individual agent and the others is controlled by the parameter $\alpha \in [0, 1]$. When $\alpha = 0$ only the experience of the moving agent matters, while for $\alpha = 1$ the move is determined by the changes experienced by the other agents.

We have the following limiting behaviour:

- $\beta \to 0$ the probability $T_{BA} \to 1/2$, the agent relocates or stays with equal probability.
- $\beta \to \infty$ the agent's decision becomes deterministic, determined by the sign of ΔU_{mov}. When $\Delta U_{\mathrm{mov}} > 0$ we have $\lim_{\beta \to \infty} T_{BA} = 0$ so the agent stays. While for $\Delta U_{\mathrm{mov}} < 0$ we have $\lim_{\beta \to \infty} T_{BA} = 1$ so the agent moves for certain.

For all values of β, transitions for which $\Delta U_{\mathrm{mov}} < 0$ are favoured, since we have $T_{BA} > 1/2$ for $\Delta U_{\mathrm{mov}} < 0$ while $T_{BA} < 1/2$ for $\Delta U_{\mathrm{mov}} > 0$.

In Sec. 10.7 it is demonstrated that in the stationary state the probability of finding the system in configuration ρ is given by

$$P(\rho) = \frac{1}{Z} e^{-\beta F(\rho)}, \tag{11.7}$$

when we are able to express the weights ΔU_{mov} in the transition probabilities in Eq. (11.5) as a change in a function $F(\rho)$ of the density ρ. So if we can find such a function $F(\rho)$ and write

$$\Delta U_{\mathrm{mov}} = F(\rho_B) - F(\rho_A), \tag{11.8}$$

then the probability that the system is in a configuration with densities ρ is given by Eq. (11.7).

We know from Sec. 10.7 that $F(\rho)$ contains a part which we call the entropy, which is determined by the number of ways $\binom{M}{m}$ the density $\rho = m/M$ in a block of M sites can be realised. Motivated by this we write

$$F(\rho) = G(\rho) - \frac{1}{\beta} S(\rho). \tag{11.9}$$

We introduce the notation $F(\rho) = M f_{\text{tot}}$ and assume a linear relationship between the total f_{tot} and the individual blocks and introduce

$$f_{\text{tot}}(\rho) = \sum_{q=1}^{Q} f(\rho_q) \tag{11.10}$$

and

$$f(\rho) = g(\rho_q) - \frac{1}{\beta} s(\rho_q). \tag{11.11}$$

The entropic part $s(\rho_q)$ was found in Sec. 10.7 and has the form

$$s(\rho) = -\rho \ln \rho - (1 - \rho) \ln(1 - \rho). \tag{11.12}$$

We need to determine the potential part[1] $G(\rho)$. To do this we compute the change ΔU_{mov} caused by the move of one agent. According to the definition in Eq. (11.4), the global part of the utility function is already a function of ρ only. But the change in utility of the moving agent needs consideration. Let the agent initially be in a block i (for initial) and the density of this block before the agent moves be denoted by ρ_i. And let the agent move to block f (for final) with a density ρ_f after the agent has arrived. This means that the density vectors of the initial state A and the final state B are given by

$$\rho_A = \left(\rho_1, \ldots, \rho_i, \ldots, \rho_f - \frac{1}{M}, \ldots, \rho_Q \right)$$
$$\rho_B = \left(\rho_1, \ldots, \rho_i - \frac{1}{M}, \ldots, \rho_f, \ldots, \rho_Q \right). \tag{11.13}$$

The change in the agent's utility can be expressed as

$$\Delta u_{\text{ag}} = u(\rho_f) - u(\rho_i). \tag{11.14}$$

The change in the global utility is, according to Eq. (11.4), only affected by the changes in the density of block i and f, which are given by

- density block i is $\rho_i - 1/M$ after the agent has left;
- density block f is $\rho_f - 1/M$ before the agent arrives.

To compute ΔU_{gl} we assume $M \gg 1$ and Taylor expand $u(\rho_i - 1/M)$ to order $1/M$ about ρ_i and $u(\rho_f - 1/M)$ about and ρ_f to get

$$\Delta U_{\text{gl}} = M[(\rho_i - 1/M)u(\rho_i - 1/M) - \rho_i u(\rho_i) + u(\rho_f) - (\rho_f - 1/M)u(\rho_f - 1/M)]$$
$$= u(\rho_f) + \rho_f u'(\rho_f) - [u(\rho_i) + \rho_i u'(\rho_i)]. \tag{11.15}$$

Combining Δu_{ag} and ΔU_{gl} to get ΔU_{mov} we conclude

$$\Delta U_{\text{mov}} = (1 - \alpha)\Delta u_{\text{ag}} + \alpha \Delta U_{\text{gl}}$$
$$= u(\rho_f) - u(\rho_i) + \alpha[\rho_f u'(\rho_f) - \rho_i u'(\rho_i)]. \tag{11.16}$$

---

[1] This terminology is borrowed from physics, though all we are doing is analysing stochastic processes of general applicability.

This is the left-hand side of Eq. (11.8). Next we demonstrate that the following function reproduces this expression when substituted on the right-hand side of Eq. (11.8). We define

$$g(\rho) = \alpha\rho u(\rho) + (1-\alpha)\int_0^\rho u(t)dt \tag{11.17}$$

and the function

$$G(\boldsymbol{\rho}) = M\sum_{q=1}^{Q}g(\rho_q). \tag{11.18}$$

We have

$$\Delta G = G(\boldsymbol{\rho}_B) - G(\boldsymbol{\rho}_A)$$

$$= \alpha\left[\left(\rho_i - \frac{1}{M}\right)u\left(\rho_i - \frac{1}{M}\right) - \rho_i u(\rho_i) + \rho_f u(\rho_f) - \left(\rho_f - \frac{1}{M}\right)u\left(\rho_f - \frac{1}{M}\right)\right]$$

$$+ (1-\alpha)\left[\int_0^{\rho_i - \frac{1}{M}}u(t)dt - \int_0^{\rho_i}u(t)dt + \int_0^{\rho_f}u(t)dt - \int_0^{\rho_f - \frac{1}{M}}u(t)dt\right]$$

$$= \alpha\left[\left(\rho_i - \frac{1}{M}\right)u\left(\rho_i - \frac{1}{M}\right) - \rho_i u(\rho_i) + \rho_f u(\rho_f) - \left(\rho_f - \frac{1}{M}\right)u\left(\rho_f - \frac{1}{M}\right)\right]$$

$$+ (1-\alpha)\left[\int_{\rho_f - \frac{1}{M}}^{\rho_f}u(t)dt - \int_{\rho_i - \frac{1}{M}}^{\rho_i}u(t)dt\right]. \tag{11.19}$$

To proceed we again expand about ρ_i and ρ_f, both in the penultimate line and the lower limit dependence of the integrals in the last line, and arrive at

$$\Delta G = u(\rho_f) - u(\rho_i) + \alpha[\rho_f u'(\rho_f) - \rho_i u'(\rho_i)], \tag{11.20}$$

from which we conclude that $\Delta G = \Delta U_{\text{mov}}$ by comparison with Eq. (11.16).

We recapitulate. To determine the distribution of agents in the configurations generated by this dynamics, we consider the coarse-grained level and determine the probabilities $P(\boldsymbol{\rho})$ for configurations given by a set of block densities $\boldsymbol{\rho} = (\rho_1, \rho_2, \ldots, \rho_Q)$. We make use of the coarse-graining formalism in Sec. 10.7 to determine the probabilities $P(\boldsymbol{\rho})$ given by Eq. (11.7) as the stationary solution of the master equation corresponding to the transition probabilities in Eq. (11.5).

So in summary we have demonstrated that in the stationary state, the density across the different neighbourhoods is distributed according to

$$P(\boldsymbol{\rho}) = \frac{1}{Z}e^{-\beta M f_{\text{tot}}(\boldsymbol{\rho})}, \tag{11.21}$$

where

$$f_{\text{tot}}(\boldsymbol{\rho}) = \sum_{q=1}^{Q}f(\rho_q) = \sum_{q=1}^{Q}\left[g(\rho_q) - \frac{1}{\beta}s(\rho_q)\right] \tag{11.22}$$

and we introduce the function $f(\rho) = g(\rho) - s(\rho)/\beta$ and

$$g(\rho) = \alpha\rho u(\rho) + (1 - \alpha) \int_0^\rho u(t)dt$$
$$s(\rho) = -\rho \ln \rho - (1 - \rho) \ln(1 - \rho).$$

(11.23)

According to Eq. (11.21), in the limit of large M the most likely configuration ρ_m is the one that minimises $f(\rho)$ under appropriate constraints. We assume that the average block density across the system, ρ_0, is given and therefore minimise under the constraint

$$\frac{1}{Q} \sum_q \rho_q = \rho_0.$$

(11.24)

To do this we introduce a Lagrange multiplier, see Box 6.1, and look for the configurations that minimise

$$J = f(\rho) + \lambda \left(\frac{1}{Q} \sum_q \rho_q - \rho_0 \right),$$

(11.25)

which implies that we need to solve the Q equations

$$\frac{\partial J}{\partial \rho_q} = \frac{\partial f}{\partial \rho_q} + \lambda = 0.$$

(11.26)

We are particularly interested in the possible existence of a transition from a homogeneous state where $\rho_q = \rho_0$ for all the blocks $q = 1, \ldots, Q$ to an inhomogeneous state i which the system separates into two types of blocks, one with density ρ_1 and another with density ρ_2. For this configuration of two densities only we have

$$f(\rho) = \gamma \left(g(\rho_1) + \frac{1}{\beta} s(\rho_1) \right) + (1 - \gamma) \left(g(\rho_2) + \frac{1}{\beta} s(\rho_2) \right),$$

(11.27)

where γ is the fraction of blocks at density ρ_1. Now Eq. (11.26) tells us that ρ_1 and ρ_2 are determined as the solutions of

$$\frac{\partial f}{\partial \rho_1} = \frac{\partial f}{\partial \rho_2} = -\lambda.$$

(11.28)

This equation determines ρ_1 and ρ_2 as functions of λ. We can then eliminate λ in terms of the overall average density ρ_0 by use of Eq. (11.24), i.e.

$$\gamma \rho_1(\lambda) + (1 - \gamma)\rho_2(\lambda) = \rho_0.$$

(11.29)

To illustrate how segregation may or may not happen depending on the parameter regime, we will consider the following example. Assume

$$u(\rho) = a\rho + b,$$

(11.30)

where a and b are constants which we will choose to make the example clear and simple. According to Eq. (11.23) this gives us

$$g(\rho) = \alpha u(\rho) + (1 - \alpha) \int_0^\rho u(t)dt$$

$$= (1 + \alpha)\frac{a}{2}\rho^2 + b\rho,$$

(11.31)

and by use of the expression for $s(\rho)$ in Eq. (11.23) we arrive at

$$f(\rho) = g(\rho) - \frac{1}{\beta}s(\rho)$$

$$= (1 + \alpha)\frac{a}{2}\rho^2 + b\rho + \frac{1}{\beta}[\rho \ln \rho + (1 - \rho)\ln(1 - \rho)].$$

(11.32)

The derivative needed for Eq. (11.28) is given by

$$f'(\rho) = (1 + \alpha)a\rho + b + \frac{1}{\beta}\ln\frac{\rho}{1 - \rho},$$

(11.33)

and we can write Eq. (11.28) as

$$\beta[(1 + \alpha)a\rho + b] = \ln\frac{1 - \rho}{\rho} - \beta\lambda.$$

(11.34)

If the overall density ρ_0 is given, which could be the case for a real application, we would have to solve this equation for each value of λ and then select the relevant λ from Eq. (11.29). To keep everything simple, we work in the opposite direction and simply assume $\lambda = 0$, and solve Eq. (11.34) to find the solutions for the densities ρ_1 and ρ_2. After that we determine the fraction γ of neighbourhoods having density ρ_1 from Eq. (11.29).

To make the illustrative mathematical analysis particularly simple, we choose[2] $\alpha = 1/2$, $a = -\frac{4}{3}$ and $b = 1$. Project 1 in this chapter considers different parameters. Figure 11.3 shows the left and the right-hand side of Eq. (11.34) for different values of β.

When Eq. (11.34) has two solutions ρ_1 and ρ_2, the fraction γ of blocks of density ρ_1 and the fraction $1 - \gamma$ of blocks with density ρ_2 is determined from Eq. (11.29), which for a given overall density ρ_0 determines γ as

$$\gamma = \frac{\rho_2 - \rho_0}{\rho_2 - \rho_1}.$$

(11.35)

Clearly, to have a solution $\gamma \in [0, 1]$ we need the average density to be between ρ_1 and ρ_2.

The analysis of the stationary state of the agent dynamics shows that the density across the system will be homogeneous if the parameter β is sufficiently small. Recall from the discussion just below Eq. (11.5) that for small values of β, the agents relocate with a near-uniform probability independent of their preferences as given by the value of their utility function. The system ends up being homogeneous, because the variation in density experienced by the agents for small values of β has little effect on their

---

[2] One realises that this choice of parameters simplifies the analysis after a little bit of fiddling around.

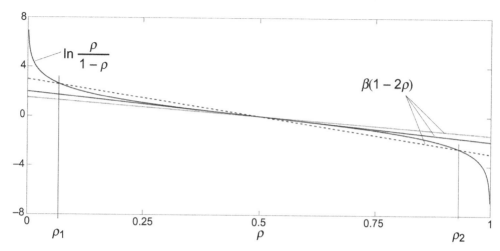

Figure 11.3 The left-hand and right-hand side of Eq. (11.34) for $\lambda = 0$ and parameters $\alpha = 1/2$, $a = -4/3$ and $b = 1$ and different values of β. The three straight lines correspond to $\beta = 3/2$ (dotted), 2 (solid) and 3 (dashed), respectively. One notices that as the slope of the straight line becomes steeper with increasing β, the straight line crosses the solid curve twice rather than once. This corresponds to Eq. (11.34) acquiring two roots.

decision to relocate. As β is increased, the transition probability Eq. (11.5) becomes more strongly dependent on the changes in utility related to an agent's move, which means that a difference between agents in cells of different density becomes important and the segregation can occur as a result.

Figure 11.4 shows how only one solution to Eq. (11.34) exists for values of $\beta < \beta_c$ and we see how two local minima develop for $\beta > \beta_c$. The solutions of Eq. (11.34) are shown in Fig. 11.5 as a function of β.

Although the functional form we have used for our example given by Eq. (11.30) was chosen for mathematical simplicity, the phenomenology is still representative of other choices of functional forms. We have so far only discussed one type of agent, which at first may appear very different from the original interest of Schelling in which agents of two types move according to the density of their own type compared to the density of the other. However, the formalism described here can also be used in the case of multiple types of agents [172] by appropriate definition of the utility functions.[3]

The approach of the Schelling model continues to be applied to many different kinds of sociological segregation. Often simulations are used to study agents that make their move stochastically based on rules relevant to a particular focus. The literature is immense and growing; here we will just mention a few recent examples.

---

[3] When consulting the paper [172] one needs to be aware that the authors use opposite signs for the transition probability in Eq. (11.5), which means that they maximise the weight function $F(\rho)$ rather than minimising, as done in our analysis. We used the sign convention of the original statistical mechanics literature, see Sec. 10.7.

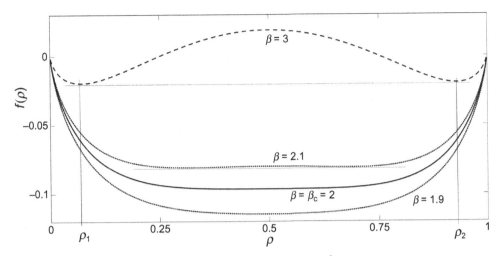

Figure 11.4 The function $f(\rho)$ in Eq. (11.32) for $\alpha = 1/2$, $a = -\frac{4}{3}$ and $b = 1$ and different β values. For $\beta < \beta_c$ the function $f(\rho)$ has only one minimum. As β increases beyond β_c, the shape of $f(\rho)$ changes and two minima appear, indicated at ρ_1 and ρ_2 for the case $\beta = 3$.

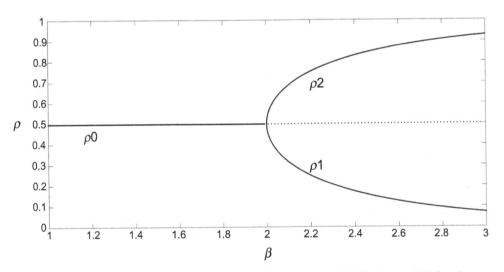

Figure 11.5 The solutions to $f'(\rho) = 0$, where $f(\rho)$ is given by Eq. (11.32) for $\alpha = 1/2$, $b = 1$, $a = -4/3$. One notices that as β is increased above $\beta_c = 2$, two new solutions appear. For $\beta < \beta_c$ the solution $\rho = 1/2$ changes from corresponding to a minima of $f(\rho)$ to becoming the position of a local maxima and the two new solutions represent the position of two minima.

Complexity science is based on the observation that similar emergent behaviour can appear in seemingly very different systems because they share common processes, in this perspective it is natural to study how social segregation compares with segregation found elsewhere, e.g. in physics. This is studied in [474], where the authors compare social segregation to different types of cluster formation amongst various types of molecules. A version of the Schelling model was developed in [375] with the aim of studying how the rate of influx of migration relates to the response of the existing population, and [409] focuses on how actual venues such as shopping malls, churches, restaurants, etc. of a city can influence segregation in a city.

When we have real human agents in mind, we are well aware that their attitude and preferences may change with time as a result of experiences. This is addressed in [461], where the tolerance level of the agents is allowed to adapt as they experience the local environment. The study concludes that adaptation leads to a polarisation of tolerance levels.

The adaptive aspect will be the theme of the next section, in which we look at the details of the Tangled Nature model, or framework.

11.3 The Tangled Nature Model

We already looked at evolution and emergence in Sec. 3.4. Now we turn to the mathematical implementation in the form of an agent-based model. The model is called Tangled Nature, or TaNa for short,[4] because the agents are entangled and their ability to act determined by a web of mutual interdependences. This way the model emphasises how the viability of a species in an ecosystem depends on both the biotic and abiotic environment or, if we think of e.g. finance and economics, how the success of one company or trader depends on the actions of the other actors in the market. The objective of the model is to see how well phenomena observed in systems undergoing evolutionary dynamics can be captured by the simple quintessential processes included in the model. These can be summarised as: mutation-prone reproduction occurring with probabilities controlled by the biotic and abiotic environment, and random death. For potential applications think of macroevolution, evolutionary ecology and, cultural evolution for example.

In the simplest version of the Tangled Nature model [216], agents of different types are allowed two actions, namely to reproduce or to die. An offspring may differ from its parent if a mutation occurs. The model focuses on the emergent structures in type space and the mode of the systemic-level dynamics [142, 413], e.g. smooth or abrupt.

In this section we first define the model mathematically and present some of its phenomenological behaviour. Next we relate the model to observations in real systems. We will discuss aspects of macro evolution such as punctuated equilibrium, see Sec. 2.7 and [169] and the gradual slowing down in the extinction rate since the Cambrian explosion [310, 311]. Furthermore, the behaviour of the distribution of types in the space of all possible agent types can be compared to features of ecosystems such as the species abundance distribution and the species area relation [20, 255].

---

[4] Some authors use other names and abbreviations, e.g. TNM, but since TNM is a standard classification system for cancer tumours it may cause less confusion if we stick to the original label TaNa.

The entanglement of the agents is mathematically captured in the following way. The type of an agent α is specified by the agent's label $\mathbf{S}_\alpha$. This label identifies the position of a node in a network of interactions. To illustrate the idea, let us assume that the node $\mathbf{S}_\alpha$ corresponds to a monkey. This node is linked with a certain strength $J(\mathbf{S}_\alpha, \mathbf{S}')$ of interaction to all other types of organisms $\mathbf{S}'$ which influence the monkey's livelihood. For example, such types as lions, $\mathbf{S}_{\text{lion}}$, and bananas, $\mathbf{S}_{\text{banana}}$, and we will want the strength of interaction $J(\mathbf{S}_\alpha, \mathbf{S}_{\text{lion}})$ to subtract from the monkey's ability to flourish while the strength $J(\mathbf{S}_\alpha, \mathbf{S}_{\text{banana}})$ must sustain the monkey as a type.

There are different ways of choosing the coupling matrix $J(\mathbf{S}, \mathbf{S}')$. In principle, data could help to quantify the interaction strength between different types. This can perhaps be done to some extent e.g. for microbial systems by experiments like those of Rivett and Bell [367], but will of course not be possible to do for all existing and potential types. Instead, we can assume for modelling purposes that the coupling matrix consists of numbers drawn from a random distribution.

We capture the beneficial and detrimental influences between the different types by defining the reproduction probability, plotted in Fig. 11.6, for agents of type $\mathbf{S}$:

$$p_{\text{off}}(\mathbf{S}, t) = \frac{1}{1 + \exp[-H(\mathbf{S}, t)]}, \tag{11.36}$$

in terms of a weight function[5]

$$H(\mathbf{S}, t) = \frac{k}{N(t)} \sum_{\mathbf{S}'} J(\mathbf{S}, \mathbf{S}') n(\mathbf{S}', t) - \mu N(t). \tag{11.37}$$

Here k is a constant used to scale the strength of the couplings $J(\mathbf{S}, \mathbf{S}')$. If we take $J(\mathbf{S}, \mathbf{S}')$ to be random numbers distributed on e.g. the interval $[-1, 1]$, the resulting strengths after multiplication by k are distributed on $[-k, k]$. Or if $J(\mathbf{S}, \mathbf{S}')$ is distributed according to a normal distribution $\mathcal{N}(0, 1)$ we have that $kJ(\mathbf{S}, \mathbf{S}')$ is distributed according to $\mathcal{N}(0, k^2)$. The total number of agents at time t is denoted by $N(t)$ and the number of agents of type $\mathbf{S}$ at time t by $n(\mathbf{S}, t)$. The abundance $n(\mathbf{S}', t)/N(t)$ can be used as a measure of how likely it is that an agent of type $\mathbf{S}$ will encounter the effect of an agent of type $\mathbf{S}'$.

So the first term on the right-hand side of Eq. (11.37) represents the accumulated effect of all the types which influence the viability of agents of type S. The second term represents that resources are limited. This is done by subtracting from H an amount proportional to the total number of agents sharing these resources. The coefficient μ represents how much the agents are limited by the resources. The larger the value of μ, the more $H(\mathbf{S})$ is decreased every time an agent is added and $N(t)$ increased. In the terminology of ecology, μ relates to the carrying capacity. If the agents represented some kind of physical molecules, μ would be related to the chemical potential.

---

[5] We could also call $H(\mathbf{S}, t)$ a utility function, see Sec. 11.2, or a potential, see Sec. 10.7. We use the term weight function to follow the literature on the Tangled Nature model. The intuition behind this term is that $H(\mathbf{S}, t)$ determines the probabilistic weight, or frequency, with which the reproduction of type $\mathbf{S}$ occurs.

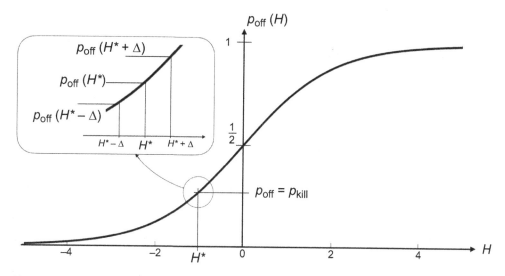

Figure 11.6 The offspring probability is given in Eq. (11.36). For a type to have a stable population, its set of interactions needs to allow the offspring probability to be approximately equal to the killing probability, which means the weight function H of stable types must be near H^*. The insert indicates the asymmetry relating to the convexity of $p_{off}(H)$, see the discussion around Eq. (11.47).

The similarity between the reproduction probability in Eq. (11.6) and the transition probability in Eq. (10.115) suggests that we should be able to find the probability of occupations in type space by the method derived in Sec. 10.7 and used to analyse the Schelling model in Sec. 11.2. However, this is not directly possible because the function $H(\mathbf{S})$ differs from the utility function considered in Secs. 10.7 and 11.2 by not depending on the local density in type space only. The coupling matrix $J(\mathbf{S}, \mathbf{S}')$ can make $H(\mathbf{S})$ depend on the density in type space around types that are positioned in type space far from $\mathbf{S}$.

Now we turn to the labelling of the types given by $\mathbf{S}$. These label nodes in the network of interactions between the different types and can be defined in any manner convenient for the specific context. One possible choice is to use the definition

$$\mathbf{S} = (S_1, S_2, \ldots, S_L), \text{ with } S_i = \pm 1. \tag{11.38}$$

This allows us, in a somewhat loose way, to think of $\mathbf{S}$ as a genome and the individual components S_i as genes, which can be in one of two alleles. Of course this idea should not be taken too seriously, but this choice makes it easy to keep track of the effect of mutations, as we now discuss.

The details of the dynamics can be implemented in many ways. One can consider sexual or asexual reproduction, or allow single or multiple offspring. Here we will look at asexual reproduction in which a reproducing agent is replaced by two offspring. This is done in the following way. At each time step an agent $\mathbf{S}$ is selected at random with uniform probability among all the $N(t)$ agents. The agent S is removed and two offspring S^1 and S^2 produced with probability $p_{\text{off}}(S, t)$. The components of each of the offspring are given according to

$$S_i^k = \begin{cases} S_i & \text{with probability } 1 - p_{\text{mut}} \\ -S_i & \text{with probability } p_{\text{mut}} \end{cases}, \qquad (11.39)$$

for each of the two: $k = 1, 2$. Note that the probability that the offspring is equal to the parent, i.e. that *none* of the 'genes' mutates, is equal to $p_{\text{mut}}^{(0)} = (1 - p_{\text{mut}})^L$.

After a reproduction update, an update representing death is performed in the following way. An agent is again selected at random with uniform probability and the selected agent is removed with probability p_{kill}, independent of type and time. Of course, real biological organisms have different probabilities of dying at a given instant, for instance because some have a much longer lifespan than others. The division time for bacteria is typically between some minutes or some hours, redwoods live for time spans between 500 and 700 years. Hence a more realistic modelling would let p_{kill} depend on the type and on the surrounding environment at a given time. However, death often occurs accidentally independent of the type. A very fit person leaving the gym could be run over by a bus. This suggests that it is sensible to only include the differences relating to type in p_{off} and not in p_{kill}.

As a result of repetitive execution of the reproductive and killing time step, the occupancy $n(S, t)$ of the types will change.

Reproduction Events

The occupancy of type S can increase when an agent of type S reproduces without any mutations, in which case $n(S, t) \mapsto n(S, t) + 1$. The occupancy of type S may also increase due to type $S' \neq S$ reproducing and producing a mutant equal to S, which leads to $n(S, t) \mapsto n(S, t) + 1$. When exactly one of the offspring is a mutant, we have $n(S, t) \mapsto n(S, t)$ and if both offspring are mutants $n(S, t) \mapsto n(S, t) - 1$, since we recall that the parent of type S is lost.

Killing Events

The effect of these are simpler. If a type S agent is selected for removal, we have $n(S, t) \mapsto n(S, t) - 1$.

Since the only actions permitted to the agents are reproduction and death, and since death occurs with the same probability for all agents, the reproduction probability summarises all the different conditions that may influence the viability of a given type of agents. From the definition of p_{kill} and $H(S, t)$ in Eqs. (11.36) and (11.37) we see that the reproductive ability of type S may change with time as a result of the changes to the occupancy of the types connected through $J(S, S')$.

It is now clear that the Tangled Nature framework shifts the focus away from intrinsic properties of the agents and instead characterises agents' behaviour, or viability, through their relations to other co-exiting agents. Clearly this is aligned with the spirit of complexity science and resonates in particular with Whitehead's suggestion quoted in Chap. 1 'that scientists should concentrate on multi-perspective networks of relationships, rather than on the behaviour of the aggregated atomic unit' [343, 493] and can also be seen as a mathematical implementation of Darwin's last paragraph in his book [94] *On the Origin of Species* from 1859, where he says:

It is interesting to contemplate a tangled bank, clothed with many plants of many kinds, with birds singing on the bushes, with various insects flitting about, and with worms crawling through the damp earth, and to reflect that these elaborately constructed forms, so different from each other, and dependent upon each other in so complex a manner, have all been produced by laws acting around us.

Of course the abilities of agents are not entirely determined by their web of interrelations and an intrinsic component characterising the agents' ability to reproduce can easily be included by adding a term to $H(\mathbf{S})$ of the form $E(\mathbf{S})$, which depends on the type label $\mathbf{S}$ only and represents the intrinsic properties of type $\mathbf{S}$ to obtain a weight function

$$H_{\text{tot}}(\mathbf{S}, t) = \frac{n(\mathbf{S}, t)}{N(t)} E(\mathbf{S}) + H(\mathbf{S}), \tag{11.40}$$

where the second term is given by Eq. (11.37). If there are no interactions between the different types, i.e. $k = 0$, so

$$H_{\text{tot}}(\mathbf{S}, t) = \frac{n(\mathbf{S}, t)}{N(t)} E(\mathbf{S}) - \mu N(t), \tag{11.41}$$

then the type $\mathbf{S}_{\text{max}}$ which has the largest $E(\mathbf{S})$ will typically rapidly outcompete other types since its reproduction rate $p_{\text{off}}(H_{\text{tot}})$ is the largest. In the end, this dominating type will be the only one alive in the ecosystem and its occupancy will be given by $n(\mathbf{S}_{\text{max}}) = N(t)$. From the balance between reproduction and annihilation, we have then

$$p_{\text{off}}(\mathbf{S}_{\text{max}}) = \dot{p}_{\text{kill}}$$

$$\Downarrow$$

$$N(t) = \frac{1}{\mu} \left[E(\mathbf{S}_{\text{max}}) - \ln \frac{p_{\text{kill}}}{1 - p_{\text{kill}}} \right]. \tag{11.42}$$

Hence, without the interactions nothing much of interest happens.

When the network of interdependencies is included, $k > 0$, the model exhibits a wealth of emergent phenomena. For example, when the strength k of the interrelation term in Eq. (11.37) is gradually increased from zero, one finds [254] that alliances between types with $E(\mathbf{S}) < E(\mathbf{S}_{\text{max}})$ can form in which a set of types happens to have favourable mutual interactions $J(\mathbf{S}, \mathbf{S}')$, which are able to increase the weight function H for the members of this set, and therefore their reproduction probability, sufficiently to make such communities as, or even more, viable than the type $\mathbf{S}_{\text{max}}$.

The combination of intrinsic and network viability, as expressed in Eq. (11.40), has also been used, see [27, 28], to study entropic effects of the evolutionary process, as represented by the TaNa model. This study investigates how types and their environment may co-evolve towards more favourable conditions for life. The aim is to relate the TaNa model to the **Gaia hypothesis** [270], which suggests that the response of the biosphere to the physical surroundings contains positive feedback loops, which help to stabilise the ecosystem.

Let us now neglect the intrinsic type specific term and focus on how co-evolution as given by Eqs. (11.36) and (11.37) reproduces a number of observed phenomena. The dynamics leads to changes in the occupancy in type space. See Fig. 3.8 for a depiction of the fixed underlying network, panel (a), in which the types $\mathbf{S}$ are the nodes and two nodes in this network $\mathbf{S}_1$ and $\mathbf{S}_2$ are linked if $J(\mathbf{S}_1, \mathbf{S}_2) \neq 0$. The reproduction and annihilation will make the occupancy of the nodes vary with time. If the reproduction cannot keep up with the killing, a type may go extinct, so $n(\mathbf{S}, t) = 0$ for that type. When mutations occur, types which previously were unoccupied may become occupied and for that type $n(\mathbf{S}, t) > 0$. Panel (b) in Fig. 3.8 indicates the dynamical network. The nodes in this dynamical network consist of the occupied, $n(\mathbf{S}, t) > 0$, types and nodes are linked only when the product $n(\mathbf{S}_1, t)J(\mathbf{S}_1, \mathbf{S}_2)n(\mathbf{S}_2, t) \neq 0$.

In what manner do we expect this network of occupied types to change, smoothly or abruptly? And what kind of networks do we expect the dynamics to generate? Will their structure adapt to make them more stable?

We will look at how adaptation is possible. The mutations described in Eq. (11.36) occur at random, as is assumed to be the case amongst real organisms. Mutations can change the occupancy of the types connected to type $\mathbf{S}$. As a consequence, the term $\sum_{\mathbf{S}'} J(\mathbf{S}, \mathbf{S}')n(\mathbf{S}', t)$ in the weight function H_S for type $\mathbf{S}$ will change, say $H_S \mapsto H_S + \Delta$. Let us assume $\Delta > 0$; this will increase the offspring probability to $p_{\text{off}}(H_S + \Delta)$. But since the mutations are random and since we assume the 'hardwired' underlying coupling matrix is random, we will expect that a fluctuation leading to $H_S \mapsto H_S - \Delta$ occurs with equal probability. This means that the fluctuations caused by mutations are equally likely to increase the offspring production as they are to decrease it, and therefore it seems like no net adaptive effect can result.

However, even if the increases or decreases in H_S are of equal size and are equally likely, they do not lead to a symmetric effect on $p_{\text{off}}(H_S)$. The reason for this is the shape of the offspring probability in Eq. (11.36). To see this look at Fig. 11.6. Types $\mathbf{S}$ which are linked through a set of interactions $J(\mathbf{S}, \mathbf{S}')n(\mathbf{S}', t)$ that are able to make the offspring probability balance the depletion of the population will satisfy

$$p_{\text{off}}(P_{\text{mut}}^{(0)})^2 + \text{back-flow} = p_{\text{kill}} + p_{\text{off}}(1 - P_{\text{mut}}^{(0)})^2. \qquad (11.43)$$

The equation expresses the balance between the probability that the population of type $\mathbf{S}$ increases by one, the left-hand side, and the probability that the population decreases by one, the right-hand side. The first term on the left-hand side expresses the probability that offspring are produced without any mutation. We have introduced

$$P_{\text{mut}}^{(0)} = (1 - p_{\text{mut}})^L, \qquad (11.44)$$

which is the probability that none of the L 'genes' of $\mathbf{S}$ change during the replication. The factor $P_{\text{mut}}^{(0)}$ needs to be included twice in Eq. (11.43) since neither offspring needs to be non-mutant to make the population of this type increase by one. The second term on the left represents the probability that population is added to type $\mathbf{S}$ due to mutations on one of the neighbour sites of $\mathbf{S}$. The term is given by

$$\text{back-flow} = \sum_{\mathbf{S}' \in \text{nb}(\mathbf{S})} p_{\text{off}}(\mathbf{S}')n(\mathbf{S}', t)p_{\text{mut}}, \qquad (11.45)$$

which sums up the probability that a single mutation on a neighbour site occurs in the direction leading to $\mathbf{S}$. We will for the moment neglect this term, assuming p_{mut} is small, and solve the balance equation (11.43) to get

$$p_{\text{off}}(2P_{\text{mut}}^{(0)} - 1) = p_{\text{kill}}. \qquad (11.46)$$

Since $P_{\text{mut}}^{(0)} \simeq 1$, we conclude that to have an approximately constant population size, a type needs $p_{\text{off}} \simeq p_{\text{kill}}$. In Fig. 11.6 we denote by H^* the value of H for which $p_{\text{off}}(H) = p_{\text{kill}}$ and our analysis tells us that the types with stationary population size will have H-values near H^*. In this region of H-values the offspring probability is convex[6] and we have

$$p_{\text{off}}(H + \Delta) - p_{\text{off}}(H) > p_{\text{off}}(H) - p_{\text{off}}(H - \Delta). \qquad (11.47)$$

This means that for adaptation to occur, we need $p_{\text{off}}(H)$ to be concave in the region around H^*. This is confirmed by simulations, see Sec. 12.6.5 in [406] where it is shown that if one uses a piecewise linear $p_{\text{off}}(H)$, i.e. without convex sections, no adaptation occurs. We can consider this as a mathematical representation of Darwin's considerations at the end of Chap. 5 in [94]:

But when a species with any extraordinarily-developed organ has become the parent of many modified descendants–which on my view must be a very slow process, requiring a long lapse of time–in this case, natural selection may readily have succeeded in giving a fixed character to the organ, in however extraordinary a manner it may be developed. Species inheriting nearly the same constitution from a common parent and exposed to similar influences will naturally tend to present analogous variations, and these same species may occasionally revert to some of the characters of their ancient progenitors. Although new and important modifications may not arise from reversion and analogous variation, such modifications will add to the beautiful and harmonious diversity of nature.

Whatever the cause may be of each slight difference in the offspring from their parents–and a cause for each must exist–it is the steady accumulation, through natural selection, of such differences, when beneficial to the individual, that gives rise to all the more important modifications of structure, by which the innumerable beings on the face of this earth are enabled to struggle with each other, and the best adapted to survive.

The TaNa model indicates that had Darwin used the language of mathematics, he could have phrased his considerations succinctly by suggesting that adaptation relies on the offspring probabilities being convex.

The conclusion is that the benefit type $\mathbf{S}$ experiences due to an increase in H caused by a mutation amongst the types influencing it, through the J-couplings, will be able to

---

[6] The graph of a convex function turns upwards, e.g. $f(x) = x^2$.

outweigh the detrimental effect it suffers due to an equal size reduction in H. We can see this as a kind of collective symmetry breaking of time in which the microscopic symmetry of individual mutations is broken and replaced by a macroscopic systemic time. At this macroscopic level the accumulated effect of all the individually undirected mutations brings about a preferred direction towards better-adapted configurations of types.

Perhaps the most conspicuous effect of the collective adaptation is the increased ability to support a larger total population size $N(t)$. Averaged over many evolutionary realisations, simulations show that the increase is logarithmic in time, $N(t) \propto \log(t)$. Here, time is defined as the number of generations. This is equivalent to the usual definition of time in Monte Carlo simulations, where one defines a time step for a system of N components as N random individual updates. For the TaNa model it is natural to think of $N(t)/p_{\text{kill}}$ individual reproduction and killing attempts as one generation and use this as our time. The point is that on average, after $N(t)/p_{\text{kill}}$ random attempts all agents present at time t have been removed.

Although averaging over many realisations of the dynamics leads to a smooth time depending on $N(t)$, the individual realisation of the dynamics is not smooth. As depicted in Fig. 11.7, the systemic-level dynamics is intermittent and consists of metastable configurations in type space that suddenly become unstable, leading to a short period of hectic rearrangement, driven by the mutations. During the hectic rearrangement, the reproductive dynamics creates new mutations and extinctions of types and in this way scans through many possible combinations of sets of co-existing types. After some relatively short period of searching through configurations, a new metastable configuration is established.

These metastable configurations are called quasi-evolutionary stable strategies (q-ESS) with a reference to evolutionary game theory where evolutionary stable strategies (ESS) consist of configurations which are stable against new mutations [283]. The q-ESS configurations consist of a core of a relatively small number of highly occupied types surrounded by a cloud of weakly occupied types which mainly receive their population as mutants from the core.[7] The essential difference between the q-ESS and ESS (and the related Nash equilibrium [280]) is that the q-ESS only involve a small part of the entire space and they do not correspond to a global optimisation over all possible configurations of types. A q-ESS becomes unstable because of the production of new mutations; the result is that reproduction and extinction will rearrange the current q-ESS into a new one, which on average is more stable than the previous. The arrival of new destabilising mutants is equivalent to a change of environment in evolutionary game theory or the introduction of new strategies in ordinary game theory [280]. Hence the q-ESS can be viewed as the set of transient states the system moves thorough on its way towards a globally stable configuration which has been tested against all possible mutations. This final state would then be an ESS. Simulations

---

[7] The core and cloud terminology was introduced in [55] to express how the highly occupied types are surrounded by neighbour types of small and erratic occupancy.

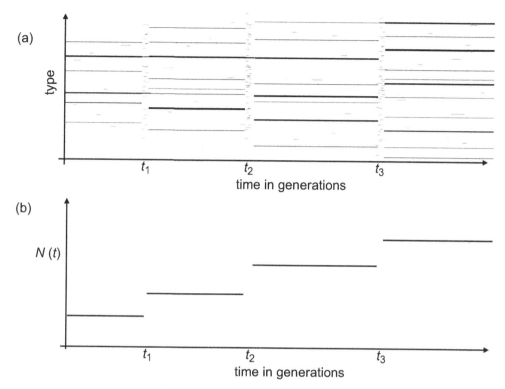

Figure 11.7 Sketch of one realisation of the dynamics of the occupancy in type space. Panel (a) indicates which of the 2^L possible different types are occupied at a given time. Panel (b) depicts the time dependence of the total number of organisms. We imagine that all the types have been labelled by a number between 1 and 2^L indicated along the y-axis in the top panel. A type's occupancy is indicated by the thickness of the line at the level of the y-axis corresponding to its type label. Types may be occupied for longer or shorter periods. The metastable configuration consists of a few core types with high occupancy surrounded by a cloud of barely occupied types. These mainly consist of types which receive their population as mutants from the core types. Every so often the meta-stable configurations suddenly become unstable and a hectic transition phase appears in which the occupancy changes rapidly. This is indicated at time t_1, t_2 and t_3.

indicate that the time it takes to reach this asymptotic state grows exponentially with L and therefore becomes beyond reach in simulations and even more so for real systems if L can be thought of as proportional to the genome length measured in number of genes.

As just mentioned, the adaptive process described above enables consecutive q-ESS configurations to become increasingly more stable, leading to an increase in their duration given by the time span $\Delta t = t_{k+1} - t_k$ between the transition at time t_k, at which the q-ESS is established, and the time t_{k+1}, where it becomes unstable [17, 406]. As a consequence of the increase in stability it is found that the rate with which the transitions occur decreases as $1/t$. This type of statistics suggests that the transitions are triggered by some kind of fluctuation following so-called record dynamics, see Sec. 12.2.

A rough idea about the time scale of the duration of the q-ESS can be obtained by a mean-field reduction of the stochastic high-dimensional dynamics to a deterministic map for the time evolution of the average value of the weight function H. This map turns out to contain **tangent map intermittency**, and this permits some insight into the scale of the lifespan of the q-ESS, see Sec. 12.3.

Every time a transition from one q-ESS to the next occurs, extinction of a large number of types happens, which is followed by creation of new types. This behaviour is qualitatively similar to the intermittency observed in the fossil record and denoted as punctuated equilibrium [169]. Moreover, the slow decay of the rate of transitions mimics the observed decay of the extinction rate observed in macroscopic evolution [310, 311].

The systemic adaptation is driven by mutants that produce sets of occupied types knitted together in a web of interactions $J(S_1, S_2)n(S_2, t)$, allowing the first term in Eq. (11.37) to become more positive or mutualistic. This means that, despite the balance between reproduction and annihilation, fixing the value of H at $H \approx H^*$, see Fig. 11.6, the adaptation enables the total population, i.e. the term $\mu N(t)$ in Eq (11.37), to increase. See Fig. 11.7.

It is difficult analytically to analyse the instability and transitions in full detail. In Sec. 13.2 we will describe a linear stability approach which, combined with simulations, gives some insights including a method to forecast the transitions a few generations in advance. Here we will look at a simple calculation which suggests that the transitions are collective in nature. To do this we focus on one specific well-occupied type $\tilde{S}$ and ask whether the occupancy $n(\tilde{S}, t)$ can be stable if we assume that the occupancies of all other types $S' \neq \tilde{S}$ remain constant independent of time, i.e. $n(S', t) = n_{S'}$ for some constants $n_{S'}$.

To analyse the stability of $\tilde{S}$, we make use of the same considerations as we did around Eq. (11.43) to note that the occupancy $n(S, t) \mapsto n(S, t) + 1$ with probability $p_{off}(S)(P^{(0)}_{mut})^2$ plus a contribution from the mutation back-flow, which we will again neglect. The population of type S decreases, i.e. $n(S, t) \mapsto n(S, t) - 1$ with probability $p_{kill} + p_{off}(S)(1 - P^{(0)})^2$. We can therefore write a mean-field equation for the time dependence as

$$\frac{\partial n(S, t)}{\partial t} = \left[p_{off}(S)(P^{(0)}_{mut})^2 - p_{kill} - p_{off}(S)(1 - P^{(0)}_{mut})^2 \right] \frac{n(S, t)}{N(t)}. \tag{11.48}$$

Let us assume that the occupancy of $n(\tilde{S}, t) = \tilde{n}$ is constant in time, i.e.

$$\frac{\partial n(\tilde{S}, t)}{\partial t} = 0$$

and therefore we have

$$\left[p_{off}(\tilde{S})(P^{(0)}_{mut})^2 - p_{kill} - p_{off}(\tilde{S})(1 - P^{(0)}_{mut})^2 \right] = 0.$$

Now we ask what happens if the occupancy of type $\tilde{S}$ is perturbed by an amount $\Delta(t) \neq 0$, so

$$n(\tilde{S}, t) = \tilde{n} + \Delta(t). \tag{11.49}$$

If $\Delta(t)$ grows with time, the occupancy of $\tilde{S}$ is unstable, whereas if $\Delta(t)$ decreases with time, the type is stable. We substitute the expression for $n(\tilde{S}, t)$ in Eq. (11.49) into Eq. (11.48) and Taylor expand $p_{\text{off}}(S)$ about $\tilde{n}$ to first order in Δ:

$$p_{\text{off}}(S) = p_{\text{off}}(\tilde{n}) + \Delta(t) \frac{d}{dn} p_{\text{off}}(\tilde{n}). \tag{11.50}$$

Using the notation

$$
\begin{aligned}
a &= 2P_m^{(0)} - 1 \\
\tilde{p} &= p_{\text{off}}(\tilde{n}) \\
A &= \frac{d}{d\tilde{n}} p_{\text{off}}(\tilde{n}),
\end{aligned}
\tag{11.51}
$$

we find

$$\frac{\partial \Delta}{\partial t} = aA \frac{\tilde{n}}{\tilde{N}} \Delta. \tag{11.52}$$

Here $\tilde{N}$ denotes the total number of agents corresponding to $n(\tilde{S}, t) = \tilde{n}$. Equation (11.52) is of the form $f'(t) = Bf(t)$, which has the solution $f(t) = f(0) \exp(Bt)$. If $B < 0$ the function $f(t)$ goes to zero with increasing t, hence to have a stable type $\tilde{S}$ we need $aA\tilde{n}/\tilde{N} < 0$. Clearly the factor $\tilde{n}/\tilde{N}$ is positive. For small mutation rates p_{mut} the factor a will also be positive. Now we show that when the type $\tilde{S}$ is a well-occupied core type, the factor A will be negative.

We have

$$A = \frac{d}{d\tilde{n}} p_{\text{off}}(\tilde{n}) = \frac{dH}{d\tilde{n}} \frac{dp_{\text{off}}}{dH}. \tag{11.53}$$

We know that $p_{\text{off}}(H)$ is a monotonously increasing function, see Eq. (11.36) and Fig. 11.6, so the derivative $dp_{\text{off}}/dH > 0$. Furthermore, since we assume $n(S, t) = \text{constant}$ for $S \neq \tilde{S}$, it follows using $J(\tilde{S}, \tilde{S}) = 0$ that

$$
\begin{aligned}
\frac{dH}{d\tilde{n}} &= \frac{d}{d\tilde{n}} \left[\frac{k}{N} \sum_{S'} J(S, S') n(S', t) - \mu N \right] \\
&= -\frac{1}{N} \left[\frac{k}{N} \sum_{S'} J(S, S') n(S', t) + \mu N \right].
\end{aligned}
\tag{11.54}
$$

For a well-occupied type the term $\frac{k}{N} \sum_{S'} J(S, S') n(S', t)$ will be positive. We see this by referring back to the initial state. Simulations show that if we start from a set of $N(0) = N_0$ agents placed at randomly selected types, $N(t)$ decreases together with the number of different types. This happens because the random non-adapted set of interactions between the types is unable to counter the $-\mu N(t)$ term in Eq. (11.36) to establish the balance between reproduction and annihilation expressed by Eq. (11.43). After a transient period the diversity decreased to at most a few types and $N(t)$ will have decreased to a value $N(t) = N_{min}$ sufficiently small that even without interactions, $H = -\mu N_{min}$ makes $p_{off}(H)$ big enough to satisfy the balance in Eq. (11.43). Now reproduction sets in and associated mutants will initiate the adaptive search for configurations of occupied types which possess a web of interactions that allow some occupied types (the core) to have $\frac{k}{N}\sum_{S'} J(S, S')n(S', t) > 0$, consistent with the fact that $N(t)$ increases with time to exceed N_{min}.

The conclusion is that the adaptive dynamics will produce configurations where the fluctuations about the core types follow the dynamics of Eq. (11.52) with $A < 0$ and therefore perturbations about the stationary state will decrease. This stability against fluctuations in a single type's occupancy, keeping the occupancy of the other types fixed, resembles the Nash equilibrium or the ESS of game theory and evolutionary game theory [194].

The lesson we learn from this analysis is that the transitions between different metastable configurations, or q-ESS states, must involve some degree of collective rearrangement. Simulations show that transitions commonly are triggered by the appearance of one, or more, mutants with a favourable set of interactions to the existing core [55]. In Sec. 12.3 we will look at how to forecast transitions when dynamics supports the creation of such new disruptive types.

We will close this section with a brief discussion of how the TaNa phenomenology compares with observation. The model in the form presented here is too abstract to allow a detailed quantitative comparison, but at a stylistic qualitative level the model behaves in many ways similar to observations. We mentioned above that the punctuated equilibrium, and gradual slowing down of the extinction rate observed in macro evolution, compares with the intermittency and decreasing transition rate in the TaNa model.

The model is also able to reproduce various ecological observations, such as an increase in mutualism between the extant types [20, 248], see e.g. [75, 431] for discussions relating to real systems. Ecologists use the so-called species abundance distribution (SAD) to describe how many species have a given size in terms of number of individuals belonging to the species. In real ecosystems one frequently observed the SAD to have a shape well parameterised by a log-normal distribution. Similar behaviour is reproduced by the TaNa model, see [20].

When ecologists count the number of different species, $N_s(A)$, encountered in a given area of size A, a species area relation (SAR) of the form $N_s(A) \propto A^z$ is commonly reported with an exponent z which can depend on the genera and geography considered. When the TaNa model is generalised to include mobility of the agents on a lattice, comparable relations are found between the number of different types and the area they reside in [255].

In the schematic framework of the TaNa model, dynamics of different types of agents embedded in a web of mutual interactions encapsulate the dynamics of many different classes of complex systems. Therefore the framework has been generalised to analyse a broad range of situations where agents co-evolve. For more details and references see [216], here we just indicate the breadth of the approach by listing a few examples. Cultural evolution is considered in [313], the dynamics of the UN's sustainability indices is modelled in [467, 468]. Opinion dynamics and polarisation are modelled in [357]; this version of the Tangled Nature framework uses agents placed in a network of social interactions which influence the opinion formation. The model addresses in particular the relationship between the structure of the social network and whether polarisation or consensus emerges.

Evolutionary aspects of economics was analysed in [368, 443]. Formation of food webs and ecosystem structure was investigated in [306, 366] and bacterial resistance was studied in [390].

Summary: Agent-based modelling, combined with computer simulations, provides an efficient way of analysing the consequences at the aggregate level of the actions assumed to be available to the individual agents. We looked at three examples of structure-forming models:

- Variations of the Vicsek model of emergent flock formation by agents moving according to an equation of motion.
- Variations of the Schelling model of emergent segregation by agents moving according to social preference.
- The Tangled Nature model of co-evolving interacting agents giving rise to intermittent collective adaptive non-stationary dynamics.

11.4 Further Reading

Introductory level:

A basic crisp introduction to ABM is included in the brief book *Complexity: A Very Short Introduction* [196] by John Holland.

The book *Think Complexity* [113] by Allen Downey is a Python-based starting point for simulations, including ABMs, of models of complex behaviour. Concerning Chap. 9, the reader needs to be aware that $1/f$ spectra are not observed in the BTW sandpile model, see Sec. 12.1.

Intermediate level:

A non-technical scientific and very readable discussion of agent-based modelling is given in *Emergence: From Chaos to Order* [195] by John Holland.

Straightforward explanations of how to construct computer codes for ABMs can be found in Chap. 19 in Hiroki Sayama's book *Introduction to the Modeling and Analysis of Complex Systems* [379].

The practicalities of ABM are thoroughly presented in *The Handbook of Agent Based Modelling* [305] by Craig Mounfield.

Modelling of human behaviour, including a multitude of applications, is done in the elegant book *Agent-Based Modelling and Geographical Information Systems: A Practical Primer (Spatial Analytics and GIS)* [89] by Andrew Crooks, Nicolas Malleson, Ed Manley and Alison Heppenstall.

Advanced level:

ABMs from an engineering perspective are discussed in the 13 papers in the book *Empirical Agent-Based Modelling – Challenges and Solutions: Volume 1. The Characterisation and Parameterisation of Empirical Agent-Based Models* [414] edited by Alexander Smajgk and Olivier Barreteau.

The paper 'Agent-based models for the emergence and evolution of grammar' [425] discusses an intriguing study of language formation.

11.5 Exercises and Projects

Exercise 1

Consider a population consisting of individuals that are either of opinion A or opinion B. Initially the entire population is of opinion A. New information arrives, which is scrutinised by everyone and after doing this, each individual changes their opinion from A to B with probability p. Each individual has q_c neighbours.

(a) Let P_∞ denote the probability that a randomly chosen person belongs to an infinite (system-spanning) cluster of B individuals. Determine in mean field the value p_c of p at which P_∞ becomes non-zero.

(b) Determine the mean-field exponent, β, describing the behaviour of $P_\infty \propto (p - p_c)^\beta$ in the critical regime above p_c.

Now assume that the population can be considered to live on a one-dimensional square lattice of infinite extent.

(c) Determine the probability, $P(s)$, that a randomly chosen individual belongs to a B-cluster of size s.

(d) Determine the average size of clusters of B people when visiting clusters by choosing a person at random as done in (c).

(e) Determine the exponent γ which describes the behaviour of the average size of B-clusters as the critical value of p is approached.

Exercise 2

Consider the merger process described in [472]. The process is controlled by the merger probability

$$p_{meger} = p(1 + n_A)^{\alpha},$$

where $\alpha = 3/2$ is considered in the paper.

(a) Simulate the process for $\alpha \neq 3/2$, say $\alpha = 1$ and $\alpha = 2$, and compare your result to the result in Fig. 3 of the paper.
(b) Derive the set of mean field master equations given in Eqs. (3.2), (3.3) and (3.4).
(c) For different values of α, solve the mean-field master equations by numerical iteration starting from an initial population of 2000 agents.

Exercise 3

To better understand the details of the analysis in Sec. 11.2, reproduce Fig. 11.8 which corresponds to a slightly different choice of parameters, namely $a = -1.48$ instead of $a = -4/3$, than used for Fig. 11.4, leading to a less symmetric case. The dashed straight line in Fig. 11.8 touching the dashed curve representing $f(\rho)$ for $\beta = 2.8$ indicates how one may graphically solve Eq. (11.28). Use this to determine the segregation transition to determine the roots of Eq. (11.28) as a function of β.

Exercise 4

Simulation of a population of N agents. Each agent i is allocated a willingness parameter $w_i \in [0, 1]$ and possesses one of two opinions A or B. Focus on two choices for the distribution of w_i.

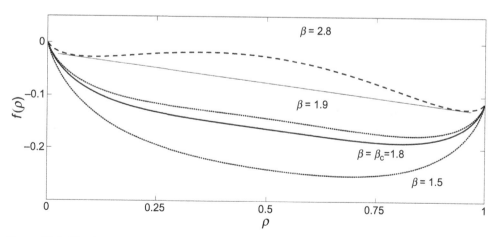

Figure 11.8 The function $f(\rho)$ in Eq. (11.32) for $\alpha = 1/2$, $b = 1$, $a = -1.48$ and different β values. Show that for these parameters $\beta_c \simeq 1.8$.

(i) All agents have $w_i = 1$.

(ii) A fraction f of agents are allocated $w_i = 0$ and the rest have $w_i = 1$.

The opinion of the agents evolves according to the following steps:

(1) Choose with uniform probability (i.e. $1/N$) an agent i_0.

(2) Initially allocate opinion A or B with probability $1/2$ to each of the agents.

(3) Select again at random an *odd* number n of other agents.

(4) The opinion of agent i_0 is now updated with probability w_{i_0} to be equal to the majority opinion of the group of n agents.

Now simulate the dynamics by repeating steps (3) and (4) and study the evolution of the distribution of opinions and compare case (i) and (ii) for various choices of f, n and N. It is of particular interest to know how reaching consensus, i.e. all agents end up with the same opinion, depends on f and n.

It is also of interest to investigate how the dynamics behaves for different ways of assigning the parameters w_i.

Another interesting question is how a small group of agents with low value of w_i can influence the dynamics. To do this, start from an initial configuration where a small part of the population is allocated opinion A and a low value of w_i, while the rest of the population is allocated opinion B and higher values of w_i.

Project 1 – Dynamics of agents on a network

Consider a population of N agents on a network. Let $nb(i)$ denote the set of neighbours of agent i, i.e. if $j \in nb(i)$ then j is a neighbour of i. The opinion x_i of an agent can assume one of the three values $\{-1, 0, 1\}$. Think of $x_i = -1$ as against, $x_i = 0$ as undecided and $x_i = 1$ as for. The sum of the opinions of the neighbours of i is given by

$$\tilde{x}_i = \sum_{j \in nb(i)} x_j.$$

We introduce the following function $F : \mathbb{R} \mapsto (0, 1/2)$:

$$f(t) = \frac{1}{2} \frac{1}{1 + e^{-t}}.$$

We assume that the exchange of ideas and arguments between the members of the population is such that the probability $P(x_i)$ that a randomly chosen agent has opinion x_i is given by the following table:

x_i	$P(x_i)$
-1	$f(-\beta\tilde{x}_i)$
1	$f(\beta\tilde{x}_i)$
0	$1 - f(-\beta\tilde{x}_i) - f(\beta\tilde{x}_i)$

(a) Derive an expression for the average opinion value $\langle x_i \rangle$ of agent i in terms of $f(\tilde{x}_i)$.

Assume all agents i have the same number of neighbours $q = |nb(i)|$ and make use of a mean-field (MF) approximation consisting of the following replacement:

$$\tilde{x}_i \mapsto q \langle x_i \rangle = q \langle x \rangle.$$

(b) Derive an MF equation for $\langle x \rangle$.
(c) Show that the MF equation allows for two regimes separated by a certain value β_c. For $\beta < \beta_c$ the only solution of the MF equation is $\langle x \rangle = 0$, whereas for $\beta > \beta_c$ a non-zero solution also exists.
(d) Determine β_c in terms of q.
(e) Expand the MF equation in the vicinity above the critical point in order to determine the critical exponent controlling how $\langle x \rangle$ depends on $\Delta = \beta - \beta_c$ for $\Delta > 0$.

We now turn to the dynamic evolution of the opinions of the agents. To simplify our analysis we only want to keep track of the number of agents $A(t)$ with opinion -1, the number $C(t)$ of undecided agents and the number of agents $B(t)$ with opinion 1. To do this we establish first the master equation in continuous time for the probability $P(x, t)$ that a randomly chosen agent has a certain opinion $x \in \{-1, 0, 1\}$ at time t. We assume that the flow between the three groups is given by the following rates (see also Fig. 11.9): the flow rate from the undecided group to either group A or group B is equal to ω. The rate from A to C is γ and from B to C is η. There is no direct flow between A and B. The total number of agents, $N = A(t) + B(t) + C(t)$, is conserved.

(f) Show that the master equation for $P(x, t)$ leads to a set of coupled autonomous equations for $A(t)$, $B(t)$ and $C(t)$ of form similar to e.g.

$$\frac{dA}{dt} = \omega C - \gamma A.$$

Figure 11.9 Sketch of the flow between the population groups.

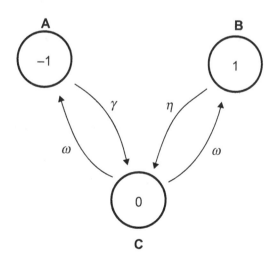

Figure 11.10 Sketch of the flow between the population groups.

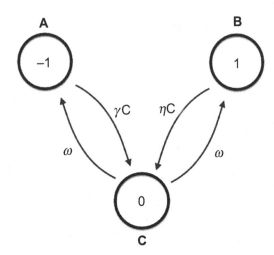

(g) Determine the size of the population groups A, B and C at the fixed point of the set of equations in (f).

(h) Show that the fixed point is attractive and determine the characteristic time of approach towards the fixed point.

Let us next imagine that the 'undecided' group C has an essential effect on the agents in groups A and B. We take that effect into account by assuming that the rate with which agents flow from either group A or group B into group C is proportional to the number of agents in group C. See Fig. 11.10.

(i) Write down the relevant master equation and show that the set of autonomous equations for $A(t)$, $B(t)$ and C is of form similar to e.g.

$$\frac{dA}{dt} = \omega C - \gamma AC.$$

(j) Analyse the autonomous equations and show that there is a change in the dynamics of the system at

$$N_c = \omega \left(\frac{1}{\gamma} + \frac{1}{\eta} \right).$$

Namely, for $N > N_c$ a non-zero number of agents never make up their mind, whereas for $N < N_c$ all agents end up in group A or B.

Project 2 – Opinion dynamics and cultural evolution

The voter model, see e.g. [141], and the Axelrod model of cultural trait dynamics [31] have inspired very many agent-based studies of the dynamics of opinions and cultural evolution. This project starts from the voter and Axelrod model and then looks at generalisations which include aspects of evolutionary dynamics of entangled agents.

(a) Describe the dynamics of the voter model as e.g. presented in [141].

(b) What are the salient features of the Axelrod model [31]?

(c) On the basis of (a) and (b), briefly describe the structure and dynamics of the Tangled Worldview Model of Opinion Dynamics presented in [357].

(d) According to [357], what are the conditions under which opinions polarise?

(e) Why is the dynamics of the Tangled Worldview Model intermittent?

(f) Briefly relate the cultural evolution model in [313], called the Tangled Axelrod Model, to the voter model and the original Axelrod model.

(g) Why is the dynamics of the model in [313] intermittent?

(h) How would you rate the Tangled Worldview Model and the Tangled Axelrod Model according to the O'Keeffe–Einstein propositions?

12 Intermittency

Synopsis: We discuss high and low-dimensional, as well as stationary and non-stationary, mathematical formalisms leading to intermittent dynamics.

In Sec. 2.7, we described in general terms how the dynamics at the level of the individual components can be smooth and steady while the same dynamics considered at the aggregated systemic level may appear very different. In this chapter we will consider the mathematical description of mechanisms that can lead to jerky, or intermittent, systemic dynamics.

We first consider the very influential idea of self-organised criticality (SOC) suggested in 1989 by Bak, Tang and Wiesenfeld [37]. They proposed that the components of complex systems commonly undergo dynamics consisting of steady gradual loading, followed by sudden relaxation when a certain threshold is exceeded. The collective dynamics of such systems will exhibit calm periods while load is being applied, followed by dramatic release of accumulated internal strain. This abrupt release will occur in the form of some kind of avalanche, similar to the dynamics observed in mountains when the snow, built up over many years, suddenly crashes down the mountain in a snow avalanche. SOC assumes the intermittent dynamics to be stationary, i.e. no evolutionary changes of the system, over long time scales. Plenty of examples of SOC behaviour have been suggested, including avalanches in granular piles, earthquakes, brain activity and many more [214, 354].

After SOC, we will consider record-driven intermittency in which the abrupt non-stationary systemic relaxation is triggered by fluctuations amongst the components that manage to release pent-up strain every time a fluctuation bigger than any hitherto experienced one occurs, i.e. when a record fluctuation happens. Hence the name record dynamics. Examples have been suggested to include relaxation of magnetic materials, colloids, superconductors and hungry ants [404, 405, 406].

Finally we will consider how the theory of deterministic dynamical systems may be used to understand intermittency. This will involve a macroscopic control parameter, which follows dynamics that can be captured by a mathematical map containing a kind of bottleneck called tangent map intermittency. Mean-field analysis will be used to derive the deterministic tangent map from the stochastic dynamics at the level of the components.

12.1 Self-Organised Criticality

The field of SOC was introduced by simulations of a very simple algorithm, which the authors, Bak, Tang and Wiesenfeld [37], suggested would resemble the dynamics of sand sliding down the side of a sandpile. But more importantly, the paper expected that the dynamics of the algorithm, consisting of a slow external load generating intermittent bursts of relaxation, would be of very general relevance and explain why many time signals exhibit $1/f$ power spectra and why fractal structures are so widely observed.

It is curious to point out that real sandpiles behave differently than the computer model and that the $1/f$ spectrum reported for the computer model turned out[1] to be $1/f^2$, for details see e.g. [214]. This did not prevent the suggestion that the dynamics of complex systems self-generates critical behaviour to raise significant interest across the sciences and continues to do so [354, 484].

We have in Secs. 3.1 and 6.3 mentioned that exactly at the critical point where a transition between two different states of organisation occurs, correlations can be long-ranged. This corresponds to the existence of spatial and temporal organisation characterised by power-law-dependent correlation functions, which implies scale invariance and relates to fractal structures. For equilibrium thermodynamic systems to reach the critical point, a parameter, such as the temperature, needs to be externally tuned to the exact value corresponding to the transition point.

As SOC implicates, it was suggested that a critical state would appear all by itself as a direct consequence of the dynamics of the complex system. Moreover, it was also suggested that instead of looking at correlation functions, the critical state and the long-range correlations would be manifest through intermittent abrupt release of strain leading to events similar to avalanches. That the systems self-organise to a scale-free, or critical, state was assumed to be indicated by distribution functions of event sizes that follow power-law functional forms instead of the usual Gaussian peaked shape.

The interest in SOC originates in the fact that intermittent burst dynamics is observed in a long range of phenomena such as forest fires, rain showers, earthquakes, brain activity, solar flares, extinction events and more. And in all these cases the size of the events, i.e. the area of burnt forest, amount of water precipitated, etc., are found, at least approximately, to follow power laws.

However, it is still in most cases not entirely clear how close the observed broad distributions of event sizes compare with power laws and exactly how relevant the mechanisms suggested by SOC models are to the scenarios in nature. This simply means that more work is needed to improve our understanding of the models, see e.g. [325, 327], and of the observed indicators of criticality in real systems, see e.g. [100].

In the next subsections we will briefly present two central models of SOC. An overview of some of the early work can be found in [214] and a comprehensive and very detailed presentation of the first couple of decades of SOC research is given in [354].

---

[1] Unintentionally, according to Chao Tang, the authors reported the square root of the power spectrum instead of the power spectrum (C. Tang, personal communication, APS March meeting, 1990, Los Angeles).

12.1.1 Sandpile Models

The model used to introduce SOC consists of a d-dimensional hypercubic lattice of linear size L. A dynamical variable $z(\mathbf{r})$ is assigned to each of the L^d lattice sites $\mathbf{r}$. Various versions of update algorithms for $z(\mathbf{r})$ have been considered, which all share the same idea that $z(\mathbf{r})$ remains constant if its value is below a certain threshold value z_{thr} and changes by redistributing some, or all, of $z(\mathbf{r})$ to neighbour sites when $z(\mathbf{r}) > z_{thr}$, see Fig. 12.1.

In one dimension these update rules correspond to interpreting $z(\mathbf{r})$ as the absolute value of a slope given by $z(r) = |h(r+1) - h(r)|$, where $h(r)$ is the height of the pile at position r. Note in Fig. 12.1 that the height of the pile decreases with increasing r, so $z(r) = |h(r+1) - h(r)| = h(r) - h(r+1)$. In one dimension, adding a grain to the height at position r, i.e. $h(r) \mapsto h(r) + 1$, will increase the absolute slope on site r, namely

$$z(r) = h(r) - h(r+1) \mapsto (h(r) + 1) - h(r+1) = z(r) + 1.$$

Moving a grain from position r to position $r+1$ leads to the following redistribution of downwards slopes, see panel (b) in Fig. 12.1:

$$z(r-1) = h(r-1) - h(r) \mapsto h(r-1) - [h(r) - 1] = z(r) + 1$$
$$z(r) = h(r) - h(r+1) \mapsto [h(r) - 1] - [h(r+1) + 1] = z(r) - 2$$
$$z(r+1) = h(r+1) - h(r+2) \mapsto [h(r+1) + 1] - h(r+2) = z(r+1) + 1.$$

In higher dimensions the slope is given by the gradient of the height variable $\nabla h(\mathbf{r})$, i.e. a vector with direction, and the redistribution of height would most naturally be assumed to occur along $\nabla h(\mathbf{r})$. For simplicity, the higher-dimensional versions of the

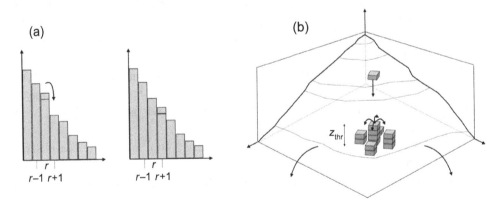

Figure 12.1 Sketch of granular pile. The update algorithm of the z-variable in rules R1 and R2 is inspired by how the slope changes in the one-dimensional case in panel (a). We imagine the slope at position r has become too steep and a grain topples from position r to $r+1$. Consider the downslope at an arbitrary position k to be given by the difference in height at location k and its neighbour $k+1$ to the right. As an effect of the relocation from location r to $r+1$, the downslope at r decreases and the downslope at the two neighbour locations $r-1$ and $r+1$ increases. The sketch in panel (b) indicates how, in the two-dimensional case, a grain added at a random position may trigger a relocation onto the four neighbour sites according to rules R1 and R2 (see the text).

model neglect this, so the meaning of $z(\mathbf{r})$ becomes more symbolic and it is safest simply to think of the algorithm as representing the dynamics of a local 'strain variable' which remains stable until its value surpasses a certain threshold.

Hence we will think of the sandpile model as a metaphor and not worry about how the model exactly relates to real granular piles. But the idea that an amount of buildup can locally be sustained up to a certain threshold, after which abrupt release occurs, relates convincingly to situations we are familiar with e.g. in earthquakes, snow on mountain sides or firing of neurones. A very careful and detailed discussion of the sandpile model in various variations is given in [354].

Since the gross features of the phenomenology in terms of intermittent release of relaxation events of various sizes do not depend on the exact details of the definition, and since the relation to real piles of sand turned out to be at most metaphorical, we will illustrate here the philosophy of the 'sandpile' class of models of SOC by describing a more straightforward model introduced by Zhang [504].

The Zhang model considers a variable $E(\mathbf{r}) \in \mathbb{R}$ which is thought of as representing in general terms the 'energy' stored on site $\mathbf{r}$, where $\mathbf{r}$ again represents sites on a d-dimensional hypercubic lattice of linear dimension L. Energy is added to sites chosen at random with uniform probability in small packages of size $\delta \in [0, E_{max}]$. As long as $E(\mathbf{r}) < E_{max}$ nothing happens, but when $E(\mathbf{r})$ exceeds E_{max} the site becomes unstable and all the energy of the unstable site is redistributed to the neighbour sites.

The dynamics in two dimensions is summarised by the following two rules:

R1 *Loading*

Choose at random with uniform probability a site $\mathbf{r} \in [0, L] \times [0, L]$ and add energy

$$E(\mathbf{r}) \mapsto E(\mathbf{r}) + \delta \tag{12.1}$$

which represents the buildup of energy, or strain, on site $\mathbf{r}$.

R2 *Relaxation*

If $E(x, y) > E_{max}$ then redistribute according to

$$\begin{aligned} E(x, y) &\to 0 \\ E(x \pm 1, y \pm 1) &\to E(x \pm 1, y \pm 1) + E(x, y)/4. \end{aligned} \tag{12.2}$$

Boundary Conditions

If the relaxing site (x, y) is on the boundary of the system, the energy that should have been added to those neighbour sites which are outside the square $[0, L] \times [0, L]$ is lost.

The model is driven by two different types of dynamics. The driving is considered to represent some external macrodynamics that slowly builds up the strain in the system. One can think of the very slow relative motion of tectonic sheets adding stress on an earthquake fault, leading to internal deformation and strain. This part of the dynamics consists of sequential repetition of R1 until a site appears with $E(\mathbf{r}) > E_{max}$.

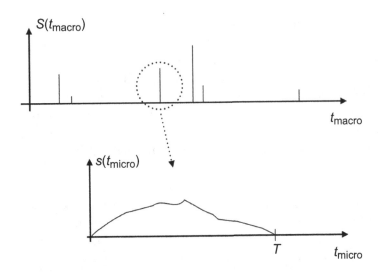

Figure 12.2 Sketch of the dynamics unfolding at the macro and micro time scales. Top panel indicates the dynamics at the macro time scale. At this scale the individual avalanches appear as instantaneous spikes of activity. The height of the spike is equal to the total number of relaxations involved in the avalanche. The bottom panel zooms in on one particular avalanche and indicates its progress on the micro time scale.

When a site appears with $E(\mathbf{r}) > E_{max}$, we say that an avalanche has been triggered and we change to microdynamics. In the analogy with earthquakes, this part of the dynamics corresponds to the rapid release of energy during the quake. In the model this mode of the dynamics consists of applying R2 by use of parallel updates. This is done by sweeping through the entire lattice and making a copy $E(\mathbf{r}, t+1) = E(\mathbf{r}, t)$. Next, for each $\mathbf{r}$, use the copy $E(\mathbf{r}, t)$ to update $E(\mathbf{r}, t+1)$ according to R2. That is, if a site has $E(\mathbf{r}, t) > E_{max}$ then set $E(\mathbf{r}, t+1) = 0$ and $E((x \pm 1, y \pm 1), t+1) = E((x \pm 1, y \pm 1), t) + E(\mathbf{r}, t)/4$.

The dynamics is intermittent if we look at the release of energy on the macro time scale, see Fig. 12.2. A macro time step is equal to one addition event, i.e. one application of R1. Most often, when load is added, the E value of the receiving site remains below E_{max} so no further rearrangement of $E(\mathbf{r})$ is needed. This means that the relaxation activity on the system is intermittent, with long periods where load is gradually added without any relaxation. But these quiet periods will be interrupted, when the load brings a site variable $E(\mathbf{r})$ above E_{max}. When this happens we switch to the micro time scale and apply R2 until all sites are brought below E_{max} once again. We count the total number, T, of parallel sweeps needed to obtain $E(\mathbf{r}) < E_{max}$ on all sites and the total number of sites visited, S, using R2 during this process. We call T the duration of the avalanche and S its size.

Viewed on the macro time scale, the repetitive application of R2 to relax the avalanche is considered to be instantaneous. In the earthquake analogy, years separate earthquakes which themselves last seconds or at most minutes.

When $E(\mathbf{r}) < E_{max}$ is re-established on all sites, we have reached a new stable configuration. The micro time variable t_{micro} counts the number of parallel sweeps and we let $s(t_{micro})$ denote the number of sites relaxed by rule R2 in sweep number t_{micro}. This means that for each avalanche of duration T and size S we have

$$S = \sum_{t_{micro}=1}^{T} s(t_{micro}). \tag{12.3}$$

The SOC literature has been particularly interested in the probability distributions $P(T)$ of durations of avalanche relaxation events and the probability distribution $P(S)$ of the size of the relaxation event. In a large number of various SOC models, evidence mainly from simulations shows that these distributions resemble power laws $P(T) \sim T^{-\alpha}$ and $P(S) \sim S^{-\tau}$ for some range of S and T. No particular parameter has to be tuned to obtain this behaviour. All that is needed is that the driving is slow compared to the relaxation. This is ensured by allowing full relaxation, i.e. the application of rule R2 when needed, in between each driving step, rule R1.

In this sense the system's dynamics organises itself into the state characterised by the power laws, hence we can say it is self-organising. And since the distributions at least resemble power laws and we know that power laws may correspond[2] to the absence of a particular scale as found in the critical state of equilibrium systems (see Secs. 2.1, 3.1 and 3.2), it was suggested in [37] that this type of avalanche dynamics consists of self-organisation to a critical state.

12.1.2 Mean-Field Analysis

It is simple to use mean-field theory to study how the load and relaxation dynamics may generate events that are distributed according to power laws [85]. It is very illuminating to think of the progression of the avalanches as a branching process, see Chap. 5. As usual when we make a mean-field description, we neglect the actual geometry and estimate instead what to expect is happening on average. For simplicity we apply R1 with $\delta = 1$ and also assume E_{max} to be an integer. To gain insight about how the dynamics defined in R1 and R2 can lead to power laws, we use a generalised version of R2. In order to make mean field applicable, we need to destroy the spatial correlations. To do this we replace the nearest-neighbour sites in R2 by a new set of randomly selected sites each time a site becomes unstable. So we replace R2 by

R2-MF *Relaxation*
If $E(\mathbf{r}) \geq E_{max}$ then redistribute according to

$$\begin{aligned}
E(\mathbf{r}) &\to E(\mathbf{r}) - E_{max} \\
E(\mathbf{r}_q) &\to E(\mathbf{r}_q) + 1, \text{for } q = 1, 2, \ldots, q_{max},
\end{aligned} \tag{12.4}$$

---

[2] See [422] Chap. 14 concerning how power laws can appear in systems that are not critical in the sense of equilibrium statistical mechanics, i.e. without any diverging correlations.

where $\mathbf{r}_q$ denotes q_{max} randomly chosen sites, which each receive a unit of load. Let $q_{max} = \nu E_{max}$; if $\nu < 1$ the system loses an amount $E_{max} - \nu E_{max}$ every time R2-MF is applied. This allows us to study the consequence of non-conservative, $\nu < 1$, update rules compared with a conservative $\nu = 1$ rule.

In the language of branching processes, we consider an unstable site to be a node. When the load of this site is redistributed according to R2-MF, a number $q = 0, 1, \ldots, q_{max}$ of new nodes may be created. Namely, each receiving node that becomes unstable. This view allow us to consider the application of R1 as the creation of the root node. The consecutive application of R2-MF will then create the ensuing generations of the branching process, creating a branching tree as sketched in Fig. 5.1.

The duration T of the avalanche is equal to the number of generations up to extinction of the branching tree and the size of the avalanche S is equal to the total progeny of the root node. To determine the statistics of the avalanches in terms of a branching process, we need to determine the branching probabilities and the average branching ratio.

The branching probability can be expressed in terms of the probability density $P(E)$. Here $P(E)$ denotes the probability that the dynamical variable $E(\mathbf{r})$ of an arbitrarily chosen site $\mathbf{r}$ assumes the value E. Since a site becomes unstable when $E(\mathbf{r}_{us}) \geq E_{max}$, the q_c packages distributed when an unstable site relaxes will create new unstable sites whenever they hit sites with E-values equal to $E_{max} - 1$. The probability, p_n, that n new unstable sites are produced is given by

$$p_n = \binom{q_{max}}{n}[P(E_{max} - 1)]^n[1 - P(E_{max} - 1)]^{q_{max}-n}. \tag{12.5}$$

The average branching ratio μ is given by

$$\mu = \langle n \rangle = \sum_{n=0}^{q_{max}} np_n = q_{max}P(E_{max} - 1). \tag{12.6}$$

The probabilities $P(E)$ for $E < E_{max}$ can be determined from the master equation in the stationary state. We know the master equation from Eq. (10.5). The master equation expresses the change in $P(E)$ with time. With our present notation we have

$$P(E, t + 1) = P(E, t) + \sum_{E'}[T_{E,E'}P(E', t) - T_{E',E}P(E, t)], \tag{12.7}$$

and we need to find the transition probabilities $T_{E,E'}$. During a single micro time step, a site will increase its E-value by an amount equal to the number of times it is chosen as a (random) neighbour for a relaxing unstable site. The probability that a site is chosen as neighbour is equal to $\frac{1}{N-1} \simeq \frac{1}{N}$, where N denotes the number of sites in the system. That is, the probability that a site is chosen to be neighbour to k over-critical sites is equal to N^{-k}. In the limit of $N \to \infty$ we can neglect the possibility $k > 1$. In this approximation a site can at most receive one unit of sand during a micro time step, so the transition probabilities will connect E-values that differ precisely by one unit. When a site relaxes, it induces a change in the E-value of q_{max} other sites, accordingly we have

$$T_{E,E'} = \frac{q_{max} n_t}{N} \delta_{E-1,E'} \text{ for } 0 < E - 1 < E_{max}, \tag{12.8}$$

where n_t is the number of times R2-MF has been applied during the sweep across the system at micro time step t, leading to the redistribution of $q_{max} n_t$ units, and $\delta_{i,j}$ as usual denotes the Kronecker δ-function.[3] There is one more transition coefficient, namely

$$T_{0,E_{max}} = 1, \tag{12.9}$$

representing that sites with $E = E_{max}$ will relax to $E = 0$ in the next micro time step. The master equation can now be written as

For $1 \leq E \leq E_{max} - 1$: $\quad P(E, t+1) - P(E, t) = \dfrac{q_{max} n_t}{N}[P(E-1, t) - P(E, t)]$

For $E = 0$: $\quad P(0, t+1) - P(0, t) = -\dfrac{q_{max} n_t}{N} P(0, t) + P(E_{max}, t)$

For $E = E_{max}$: $\quad P(E_{max}, t+1) - P(E_{max}, t) = \dfrac{q_{max} n_t}{N} P(E_{max} - 1, t) - P(E_{max}, t).$

$$\tag{12.10}$$

Next we look at the stationary state in which the probability is independent of time, i.e.

$$P(E, t+1) = P(E, t) \equiv P(E)$$

and we can from Eq. (12.10) conclude that

For $1 \leq E \leq E_{max} - 1$: $P(E-1) = P(E)$ $\tag{12.11}$

and further that

$$P(E_{max}) = \frac{q_{max} n_t}{N} P(0) \text{ and } P(E_{max}) = \frac{q_{max} n_t}{N} P(E_{max} - 1). \tag{12.12}$$

Now we use the normalisation condition

$$\sum_{E=0}^{E_{max}} P(E) = 1 \tag{12.13}$$

to conclude that

For $0 \leq E < E_{max}$: $P(E) = \dfrac{1}{E_{max} + \frac{q_{max} n_t}{N}} \simeq \dfrac{1}{E_{max}},$ $\tag{12.14}$

where the last approximate equality holds in the limit of large N, assuming that as the system size is increased n_t grows more slowly than N.

From Eq. (12.6) we obtain the branching ratio $\mu = \nu$. We recall from Chap. 5 that the distribution of progeny is given by

$$P(S) \propto S^{-3/2} \exp\left(-\frac{S}{S_0(\nu)}\right), \tag{12.15}$$

where $S_0(\nu) \propto (1 - \nu)^{-2}$.

---

[3] $\delta_{i,j} = 0$ when $i \neq j$ and $\delta_{i,j} = 1$ when $i = j$.

We can conclude that for avalanche sizes $S < S_0$, the avalanche size distribution behaves like a power law, but the exponential cutoff at $S_0(\nu)$ is *independent* of the system size N. So only when $\nu = 1$, and the redistribution during relaxation in R2-MF conserves the load variable E, does the model support a state with no cutoff on the power-law behaviour. This conclusion agrees with results from simulations of the mean-field dynamics defined by combining R1 with R2-MF, see [85].

We can also, by use of the result in Eq. (5.42), conclude that for $\nu = 1$ the distribution of durations behaves like a power law $P(T) \sim T^{-\alpha}$ with exponent $\alpha = 2$.

The specific values $\tau = 3/2$ and $\alpha = 2$ will be applicable to all systems for which the dynamics reduces to mean-field dynamics that can be mapped to an uncorrelated branching process. But in general we expect different values for the exponents when correlations in space and time are essential.

Inspired by the experience from the theory and phenomenology of equilibrium critical phenomena, the expectation was that the exponents characterising the power laws, such as τ and α, would be the same for classes of models and classes of real systems. The reason for this is that power laws in equilibrium systems apply to the behaviour at length scales large compared to the separation of the components, i.e. to scales corresponding to the accumulated action of many components. To illustrate the idea, we can think of waves on water. The dynamics of the waves is very different from the dynamics of the individual water molecules. When waves role across the oceans, the individual water molecules stay essentially at the same location and simply perform a circular motion, which combines to generate the ruling waves. And the hydrodynamics of waves only depends on a few material parameters such as density and viscosity, not on the details of the molecules of the liquid.

At large scale the accumulated action obliterates the dependence on details. The result is that the critical behaviour classified by the exponents is determined by general properties such as symmetries and dimension [59, 164]. This limits greatly the number of sets of exponents. Each set of exponents is said to describe a universality class. The mean-field exponents $\tau = 3/2$ and $\alpha = 2$ describe one such class relevant to systems with negligible correlations. When correlations are of importance, equilibrium critical systems are described by different exponents corresponding to one of the other few classes. We will discuss in the next subsection how the situation turned out to be more complicated for the SOC models.

12.1.3 Lessons from Sandpile Models

In this subsection we briefly look at the way scale-free behaviour is realised in SOC. The power laws encountered in the models, and in the real systems considered to exhibit SOC behaviour, are harder to analyse than is usually the case for equilibrium systems in a critical state and the existence of universality classes is found to be much harder to settle.

To check if e.g. a histogram for $P(S)$ obtained from simulations, or from observed data, exhibits power-law behaviour, one will first check if a plot of $\log[P(S)]$ against

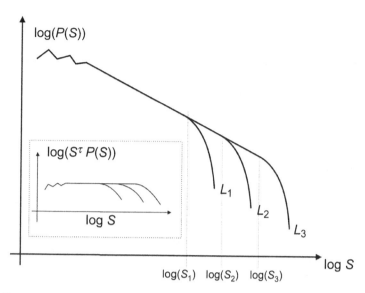

Figure 12.3 Sketch of log–log plots of histograms of avalanche size distributions for systems of different linear sizes $L_1 < L_2 < L_3$. Main panel: we see deviation from power-law behaviour for small values of S, and for large values of S. If the deviation at large values is due to finite system size, the point of departure from power law must move to the right when the linear size of the system, L, is increased, as indicated. If the power-law exponent is τ, the linear part of the log–log plot must become horizontal when plotting $\log(S^\tau P(S))$ vs. $\log S$, as indicated in the insert.

$\log S$ shows that $\log[P(S)]$ depends linearly on $\log S$. This is because, if $P(S) = AS^{-\tau}$ we have $\log[P(S)] = \log A - \tau \log S$. The size of the system will limit the maximally possible observed avalanche size, hence we cannot expect the power-law relation $P(S) = AS^{-\tau}$ to hold for arbitrary large values of S on a finite system. Nor do we, as discussed above (see end of Sec. 12.1.2), expect the scale-free behaviour corresponding to the power law to hold for short length scales, i.e. for small values of S. So, when dealing with a concrete realisation, we only expect a straight line in log–log plots over a finite region, see Fig. 12.3.

If the observed deviation from power-law behaviour is determined by the system size, the upper value of S at which $P(S)$ deviates from the straight line, denoted S_i with $i = 1, 2, \ldots$ in Fig. 12.3, will move to larger values when the linear size, L_i, of the system is increased. So to check if the size of the system is the only factor limiting the range of observed power law, it is essential to be able to compare histograms for different system sizes. Furthermore, the value of the unknown exponent τ can be systematically estimated by finding the value of τ that makes the product $S^\tau P(S)$ fit a *horizontal* straight line over the widest range of S values for each system size L, see insert in Fig. 12.3. The behaviour of the type indicated in the figure is typical of critical behaviour in equilibrium systems and if e.g. we find $S_i \propto L_i^D$ for some exponent $D > 0$ we will expect that as L is increased, $P(S)$ behaves as a power law over an ever-increasing range of S values.

On the other hand, if the cutoff S_i does not move to larger values with increasing L, we conclude that a characteristic scale, in addition to the system size, exists and hence that we are not observing true critical behaviour.

Determining exponents through a careful analysis of the size dependence is called finite size scaling analysis and is essential for a systematic understanding of the behaviour to be expected in the limit of large systems [354]. The appearance of ever-larger regions of scale-free power-law behaviour is observed in equilibrium critical systems when increasing the system size and only a few classes of power-law exponents are found. Each class contains systems that differ in detail but share symmetries and dimensionality.

For SOC systems the situation turned out to be much more subtle, see [354]. The dependence on system size can, for SOC systems, be more intricate than that described here and it is sometimes found that what appear to be minor differences in the definition of models lead to different values for the power-law exponents. It is still not entirely clear in what way observed approximate power-law behaviour relates to critical behaviour in the sense of long-range correlation functions with slow algebraic dependence on separation as discussed in Sec. 6.4. We will return to this point in connection with the forest fire model presented in the next section. But first we want to discuss the consequences for the merit of the kind of simple models traditionally used to study SOC.

Although the intermittent release of activity observed in models of SOC dynamics in most cases does not allow accurate prediction of universality classes and associated exponents with the precision known in equilibrium statistical mechanics, models are being used to analyse very broad ranges of phenomena and identify similarities and differences in behaviour.

There is a huge and growing literature applying ideas inspired by SOC and we can only list a few examples. Applications to phenomena in earth systems, such as earthquakes and forest fires, are described in [100, 190, 457]. A range of examples from plasma physics and solar flares can be found in [291, 397].

As an illustration, we mention one example of the application of the philosophy of SOC in which a load and relaxation model has been applied to develop a phenomenology of brain dynamics. In a series of papers by de Arcangelis and collaborators, a model with dynamical rules similar in spirit to the definition in R1 and R2 in Eqs. (12.1) and (12.2) has been used to relate observed brain dynamics[4] to avalanche activity [101]. The model is used to discuss the effect of the brain's modular structure on the power laws observed in the avalanche distribution [373], and even to demonstrate how avalanche dynamics can be used as the basis for learning in a neural network [294].

The model used by de Arcangelis and collaborators [101] is somewhat more involved than the rules R1 and R2 above, because it includes a representation of the plasticity known to exist in the brain. This is done by including a site variable, the action potential v_i of a neurone, corresponding to the variable $E(\mathbf{r})$ in Eqs. (12.1) and (12.2), together with an additional variable, the conductance variable g_{ij}, across the synopsis between neurone i and neurone j. The dynamics of v_i and g_{ij} are coupled together. When the action potential v_i exceeds a certain threshold $v_{\max}$, the neurone i 'fires', which means that its value v_i is set to zero and 'charge' is redistributed to neighbour sites according

---

[4] In Ref. [101] the data reference given for Fig. 3 is a misprint, and should read [35], i.e. our [315].

to the conductance between site i and its neighbours. The flow of current, on the other hand, leads to an increase in the conductance variable g_{ij} across the synapses carrying the current. These updates are repeated, i.e. an avalanche is unfolding, until all action potentials are below $v_{\max}$. When an avalanche finishes, the plasticity of the couplings between neurones is again taken into account. At this stage an equal amount is subtracted from *all* conductance variables g_{ij} corresponding to the average increase experienced by the conductance variables active during the avalanche. In this way the strength of the inactive synapses variable g_{ij} will reduce relative to the active ones. When a synapse variable g_{ij} falls below a certain σ_t value, it is put to zero, i.e. $g_{ij} < \sigma_t \Rightarrow g_{ij} \mapsto 0$.

Although this modelling framework is extremely simple compared to the intricate dynamics of the brain, it has nevertheless been able to suggest some crucial features needed for the phenomenology to match the observations. For example, the necessity of a fraction of inhibitory synapsis is indicated by the fact that the power spectrum of the model dynamics only matches observed brain dynamics when inhibition is included in the model [267]. An inhibitory synapse is represented by a connection between neurones i and j where the relaxation of v_i leads to a decrease in v_j rather than the usual increase.

Let us summarise. Simple load and relax dynamics as originally introduced in the context of self-organisation to a critical state is commonly seen to produce approximate power-law behaviour if not exact critical behaviour and the SOC framework has inspired models, such as the model of brain dynamics by de Angelis and collaborators, that do very well in terms of the three O'Keeffe–Einstein propositions discussed in Chap. 4.

So, if SOC models do not produce perfect self-similarity and power laws, the question is in what way they are described by long-range correlations in the form of algebraically decaying correlation functions, i.e. the hallmark of critical behaviour as defined in equilibrium statistical mechanics, see Sec. 6.4. We will discuss this in the next section by looking at the forest fire model of SOC.

12.1.4 Forest Fire Model

The model describes spreading of some attribute among agents [114]. If the agents are trees and the attribute is fire, we can think of the dynamics of the model as fires ravaging a forest. If the agents are people and the attribute an infectious decease, we can think of the model as a contribution to epidemiology, in which case we will consider the attribute to be the infection spreading through the population. Here we will discuss the model from the perspective of self-organisation towards critical behaviour and how simple model dynamics is able to reproduce statistical aspects of the phenomenology of intermittent activity seen in very different systems, in particular rain and brain.

We will also use the model to look at how long-range correlations can suggest a critical state despite the event distributions only following power laws approximately. The model and its power law and scaling properties have been intensely studied, and a detailed description is given in [354].

We define the model in the following way. Each site of a d-dimensional cubic lattice, or each node of a network, can be in one of three states. The site can be occupied by a tree T, or by a tree on fire F, or the site can be empty E. The dynamical update rules are as follows and are applied in parallel.

R1 *External drive.* A site is selected at random with uniform probability. If occupied by a tree, the tree catches fire with probability f and becomes a fire site: $T \mapsto F$. One can think of this as a tree being hit by lightning, or in epidemiological terms as an infection arriving from outside the population.

R2 *Activity spreading.* A tree site, neighbour to a fire site, becomes a fire site: $T \mapsto F$.

R3 *Relaxation.* A fire site becomes an empty site: $F \mapsto E$. When we think of fire as an infection spreading, this update corresponds to an infected recovering and staying immune until the next time rule R2 produces a tree, i.e. a new susceptible, on the site.

R4 *Internal loading.* Empty sites become occupied by a tree with probability p: $T \mapsto F$. In the epidemiological interpretation, this corresponds to the arrival of a new susceptible individual.

The separation of the time scale of the slow external drive from the internal fast dynamics is ensured by choosing $0 \ll f \ll p \ll 1$. These relations can be made explicit by the following protocol.

After a fire has burned out, R1 is applied until yet again a tree catches fire, at which point one repeats the application of R2 and R3 to let the fire burn out and so forth. The dynamics can be implemented in more optimal ways for simulation purposes, see [354, 355].

The intermittent dynamics generated by the model has been used to discuss very different systems. For example, measles dynamics at the Faroe Islands was considered in [363]. Because the islands are very remote and the population small, a burst of infection will run through the population and die out before a new infection arrives by ship or plane. And very similar dynamics in which the spreading R2 only occurs with a certain probability was implemented on the human connectome to model bursts of brain activity in [178].

Brain activity was further analysed [433] along the same lines as an analysis of precipitation [341] and it was demonstrated that the forest fire model exhibits behaviour similar to that observed in brain and rain [325]. More details will be given below.

The distribution of the sizes of the fires follows approximate power laws, but with an unusual dependence on the size of the system. We mentioned above, see Fig. 12.3, that in equilibrium critical phenomena distributions approach power laws with higher accuracy when the system size is increased. Similarly, when one tunes the control parameter towards the critical state, the range of power-law behaviour also grows. The forest fire exhibits a region of approximate power-law behaviour, which grows as the separation between the time scale of lightning, given by $1/f$, and the time scale of tree growth, given by $1/p$, is increased, i.e. when $\theta = p/f$ increases. We should expect increasing θ to bring the system closer to the critical region and therefore more accurate power-law behaviour as θ is increased. But unexpectedly, though the region over which approximate power-law behaviour is observed increases, the accuracy of the power law decreases as θ increases [171, 354, 355]. Therefore we are led to the conclusion that strict scale invariance characterised by power-law behaviour is not happening in the model.

A somewhat disappointing finding if exact power-law behaviour is taken to be the core of SOC. However, when we look at observational data, we seldom have event distributions described by very accurate power laws. Hence we can ask if the dynamics of an SOC model like the forest fire model allows us to reproduce other features of the dynamics of real systems. When we look at the way in which the dynamics favours configurations close to critical behaviour, we do see similarities between the simulated forest fire dynamics and the observed dynamics of precipitation [341] and brain activity [433].

So let us compare three types of activity: precipitation in the form of rain, neuronal firing as monitored by an fMIR scanner and repeated applications of rules R2, R3 and R4 in the forest fire model. In each case we choose a parameter that relates to the ability of a state of the system to support bursts of activity. We think of this parameter as the control parameter, similar to the temperature in a thermal equilibrium phase transition. In the case of precipitation, the humidity w of the atmosphere is taken as the control parameter plotted along the x-axis in Fig. 12.4. When analysing the fMRI signal from the brain, the parameter x is taken to be the number of voxels in the fMRI scan with a signal above a certain value and for the forest fire model we choose x to be the density of sites occupied by trees.

In analogy with equilibrium phase transitions we have to identify a quantity that can be used as an order parameter, generically denoted $\langle M(x) \rangle$ in Fig. 12.4. For the Ising model the order parameter is the average value of all the spin variables, see Eq. (6.31) and Fig. 6.4. In the case of rain the order parameter is the average precipitation

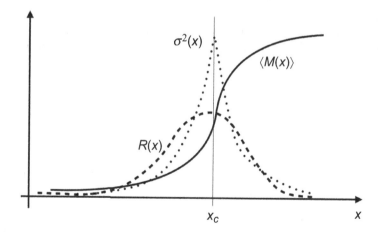

Figure 12.4 Sketch of the behaviour of a system driving itself into a region around a critical onset. The x-axis indicates the value of a relevant control parameter x. At the value $x = x_c$ long-reaching correlations appear. The state of the system is characterised by some order parameter $\langle M(x) \rangle$ and its x dependence is described by the solid curve. The variance of $M(x)$ is indicated by the dotted line. The dashed line indicates the probability that the system observed at a random instance in time will have a given value of the x parameter.

rate $\langle P \rangle(w)$ at a given humidity w. The study of brain activity [433] uses the largest connected cluster of active voxels as order parameter, as is usually done in the theory of **percolation**, see Sec. 8.4 and Project 1 of Chap. 8 and [84]. The behaviour of the forest fire can be compared to the rain and brain behaviour by choosing as order parameter the largest cluster of trees [325, 326].

All three systems exhibit behaviour qualitatively like that sketched in Fig. 12.4, where we see a gradual increase in the order parameter. Recall that in equilibrium phase transitions the control parameter, for example the temperature, is controlled from outside the system. In the three systems we are looking at here, the value of x is determined by the dynamics of the systems. So as the dynamics pushes through the available configurations, the control parameter x, i.e. humidity, number of active voxels or number of trees, will change. The curve $R(x)$ in Fig. 12.4 indicates the 'residence time', which is given by the probability $R(x)$ that when observing the system at a random instant in time, one will find that the control parameter has the specific value x.

It is striking that though the order parameter does not change from zero to non-zero at a precise value of the control parameter x (as is the case in equilibrium transitions, see Fig. 6.4), nevertheless for all three systems the order parameter changes over a certain region which is also a region of increased fluctuations, or susceptibility, as indicated by the curve $\sigma(x)$, and it is the region in which the system is most likely to be found. In all three cases the probability distributions describing the events (shower sizes, connected cluster of brain activity and connected clusters of trees) exhibit approximate power laws as reported in [341, 354, 433].

That the behaviour in this region may also be related to some degree of critical behaviour in the sense of long-range correlations is seen in simulations of the forest fire model. Since clusters on fire are assumed to burn down instantaneously, one defines for each site i a variable $S_i = 1$ if the site is occupied by a tree and $S_i = 0$ if the site is empty. The correlation function similar to that defined in Eq. (6.55) for this variable can then be simulated and although the dependence on system size is intricate, the correlation function does approximatively follow algebraic decay; see [325, 327] for more details and references.

12.2 Record Dynamics

The dynamics of SOC intermittency consists of slow external driving punctuated by rapid internal relaxation. The systems are assumed to be in stationary states, meaning that after some transient the properties of the system and its statistics are time independent. In this section we will look at a different route to intermittent dynamics driven by internal fluctuations which trigger a change in some of the system properties every time a record fluctuation occurs. That is a fluctuation which is bigger than any hitherto encountered fluctuation.

The resulting emergent collective dynamics is called record dynamics and was first developed as a model of so-called charge density wave dynamics [407] and has subsequently been used to analyse dynamics in a variety of systems. The abrupt intermittent relaxations are called quakes and on average bring about a slight increase in the stability, which on the other hand makes the time between quakes increase. Formulated differently, the probability that a quake occurs at time t is found to decrease like $1/t$. That is, a gradual decay in activity, such as e.g. described by the Omori aftershock law [462]. To a very good approximation, the record controlled quake process can be considered a Poisson process in logarithmic time. This means that when the ordinary time axis is replaced by the logarithm of time $\tau \equiv \ln t$, the process becomes stationary and the probability that a quake event occurs in a short logarithmic time interval $\Delta \tau = \tau_2 - \tau_1$ depends on the difference τ only and not explicitly on the value τ_1 and τ_2.

Record dynamics have been used to analyse physical systems out of equilibrium, such as spin glasses, magnetic fields in superconductors and colloidal relaxation. It has been applied to biology to model micro and macro evolution (punctuated equilibrium) and the behaviour of foraging ants [404, 406]; cultural evolution has also been investigated within this framework [29, 313].

In the next subsections we will first, in some detail, describe the mathematics behind record dynamics with the aim of highlighting how the resulting statistics is very general and essentially independent of the fluctuations at the microscopic level of the components. After having presented the basics of the mathematics, we will discuss a few applications.

12.2.1 Statistics of Records

A very comprehensive discussion of the mathematical theory of records is given in [309], and details can also be found in [404, 406]. Here we will give an intuitive presentation of the most basic facts. Consider a stochastic variable X. The only requirement is that X can assume infinitely many values, so we will think of X as being distributed on the real numbers $\mathbb{R}$. Let $P_X(x)$ denote the probability density function for X; we then have that the accumulated density is given by

$$F(x) = \text{Prob}\{X < x\} = \int_{-\infty}^{x} dx P_X(x). \tag{12.16}$$

Next we construct a sequence of independent equally distributed random numbers by drawing a number from X at each time step $0, 1, 2, \ldots$:

$$x_0, x_1, x_2, \ldots, x_t, x_{t+1}, \ldots. \tag{12.17}$$

We want to find the probability, $P(t)$, that the number drawn at time step t, i.e. x_t, is a record. This will be the case if $x_t > x_q$ for $q = 0, 1, 2, \ldots, t-1$. We are not interested in the specific value x_t of X at the time step t, we just want it to be a record. So to compute the probability for this event, we need to sum over all values of x_t consistent with x_t being a record. Therefore $P(t)$ is given by

$$P(t) = \int_{-\infty}^{\infty} \text{Prob}\{\max\{x_0, x_1, \ldots, x_{N-1}\} < x\} P_X(x) dx. \tag{12.18}$$

Since the numbers drawn at different time steps are independent, we can express the probability inside the integral in terms of the accumulated distribution in the following way:

$$\text{Prob}\{\max\{x_0, x_1, \ldots, x_{t-1}\} < x\} = \text{Prob}\{x_0 < x \text{ and } x_1 < x \ldots \text{ and } x_{t-1} < x\}$$
$$= \text{Prob}\{x_0 < x\}\text{Prob}\{x_1 < x\} \cdots \text{Prob}\{x_{t-1} < x\}$$
$$= [F(x)]^t. \tag{12.19}$$

We substitute this expression into Eq. (12.18) and get

$$P(t) = \int_{-\infty}^{\infty} [F(x)]^t P_X(x) dx. \tag{12.20}$$

This integral can be computed without specifying how X is distributed. We know in general, see Eq. (12.16), that $dF(x)/dx = P_X(x)$, so

$$P(t) = \int_{-\infty}^{\infty} [F(x)]^t \frac{dF}{dx} dx, \tag{12.21}$$

and we can make the substitution

$$z = F(x) \text{ and } dz = \frac{dF(x)}{dx} dx \tag{12.22}$$

and have for the integration boundaries

$$x = -\infty \rightarrow z = F(x) = 0 \text{ and } x = \infty \rightarrow z = F(x) = 1.$$

We combine these results to write

$$P(t) = \int_0^1 z^t dz = \left[\frac{z^{t+1}}{t+1}\right]_0^1 = \frac{1}{t+1}. \tag{12.23}$$

We can also understand this result from the point of view that since the values $x_0, x_1, \ldots, x_t$ are independent, the largest value amongst them is equally likely to be obtained at any of the $t + 1$ draws and therefore occurs at time step t with probability $1/(t+1)$.

We emphasise the generality of the result in Eq. (12.18). Namely that the record occurs as t decreases like $1/(t+1)$, no matter how X is distributed. All that is needed is for X to be able to assume infinitely many different values, otherwise at some time step one will draw the largest possible X value and then no more records can be obtained after that.

Let us turn to the more general probability that n records occur at specific time steps in a sequence of t time steps. As per the definition we consider the first draw x_0 at time $t = 0$ to be a record and denote the times of the consecutive $n - 1$ records by $0 < t_1 < \cdots < t_{n-1} \le t$. We denote the probability for the n records at $0, t_1, \ldots, t_{n-1}$ by $P_n(t; 0, t_1, \ldots, t_{n-1})$. Although the derivation is somewhat more involved than the analysis above leading to Eq. (12.23), we can again compute the

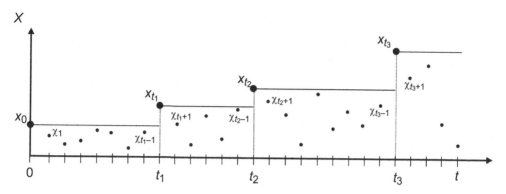

Figure 12.5 Sketch of records in a sequence of uncorrelated and equally distributed random numbers. Three records, in addition to the one at time $t = 0$, are seen at time t_1, t_2 and t_3.

probability in terms of integrals over the values $x_0, x_{t_1}, \ldots, x_{t_{n-1}}$ assumed by X at the record times $0, t_1, \ldots, t_{n-1}$. In the integrals below, we denote the values of X assumed at the time steps between record times by χ_t, see Fig. 12.5. So the sequence of assumed X values is given by

$$x_0, \chi_1, \ldots, \chi_{t_1-1}, x_{t_1}, \chi_{t_1+1}, \ldots, \chi_t$$

and we have

$$P_n(t; 0, t_1, \ldots, t_{n-1}) = \int_{-\infty}^{\infty} dx_{n-1} P(x_{n-1}) \int_{-\infty}^{x_{n-1}} dx_{n-2} P(x_{n-2})$$
$$\int_{-\infty}^{x_{n-2}} dx_{n-3} P(x_{n-3}) \cdots \int_{-\infty}^{x_2} dx_1 P(x_1) \int_{-\infty}^{x_1} dx_0 P(x_0)$$
$$\text{Prob}\{\chi_1 < x_0, \chi_2 < x_0, \ldots, \chi_{t_1-1} < x_0\} \tag{12.24}$$
$$\text{Prob}\{\chi_{t_1+1} < x_{t_1}, \chi_{t_1+2} < x_1, \ldots, \chi_{t_2-1} < x_1\}$$
$$\cdots$$
$$\text{Prob}\{x_{t_{n-1}+1} < x_{t_{n-1}}, \ldots, x_n < x_{t_{n-1}}\}.$$

In a similar way to our steps from Eq. (12.18) to Eq. (12.20), we introduce the accumulated probability function and subsequently introduce the substitution introduced in Eq. (12.22):

$$P_n(t; 0, t_1, \ldots, t_{n-1}) = \int_{-\infty}^{\infty} dx_{n-1} P(x_{n-1}) \int_{-\infty}^{x_{n-1}} dx_{n-2} P(x_{n-2})$$
$$\int_{-\infty}^{x_{n-2}} dx_{n-3} P(x_{n-3}) \cdots \int_{-\infty}^{x_2} dx_1 P(x_1) \int_{-\infty}^{x_1} dx_0 P(x_0)$$
$$F(x_0)^{t_1-1} F(x_1)^{t_2-t_1-1} \cdots F(x_{n-1})^{t-t_{n-1}}$$
$$= \int_0^1 dz_{n-1} z_{n-1}^{t-t_{n-1}} \int_0^{z_{n-1}} dz_{n-2} z_{n-2}^{t_{n-2}-t_{n-3}-1} \cdots \int_0^{z_2} dz_1 z_1^{t_2-t_1-1} \int_0^{z_1} dz_0 z_0^{t_1-1}$$
$$= \frac{1}{(t+1)t_{n-1}t_{n-2} \cdots t_1}. \tag{12.25}$$

From this expression we can find the total probability $P_n(t)$ that precisely n records occur during the sequence of time steps $0, 1, 2, \ldots, t$ by summing over the specific record times $t_1, t_2, \ldots, t_{n-1}$. We have

$$P_n(t) = \sum_{0 < t_1 < t_2 < \cdots < t_{n-1} \leq t} P_n(t; 0, t_1, \ldots, t_{n-1}). \tag{12.26}$$

This sum cannot be reduced exactly to a closed form, but for $1 \ll n \ll t$ it is to a good approximation given by the following simple expression, see e.g. Sec. 14 in [309] or Sec. 5.2 in [406]:

$$P_n(t) = \frac{1}{t} \frac{(\ln t)^{(n-1)}}{(n-1)!} \quad \text{with } n = 1, 2, 3, \ldots. \tag{12.27}$$

We recall that the Poisson distribution of a stochastic variable $X = k \in \{0, 1, 2, \ldots\}$ with average λ is given by

$$\text{Prob}\{X = k\} = \frac{\lambda^k e^{-k}}{k!}, \tag{12.28}$$

and notice that if we introduce $\mu_t = \ln t$, we can write Eq. (12.27) as a Poisson distribution

$$P_n(t) = \frac{\mu_t^{n-1} e^{-\mu_t}}{(n-1)!} \quad \text{with } n = 1, 2, 3, \ldots. \tag{12.29}$$

Here the average number of records during the time steps $0, 1, \ldots, t$ is given by $\mu_t = \ln t$. Note that because we count the first event x_0 as the first record, we should let $n = k + 1$ when we compare with the Poisson distribution.

In summary, we have found that the records in a sequence of independent equally distributed random numbers can to a good approximation[5] be considered to be a Poisson process when we substitute the ordinary time t by its logarithm $\ln t$.

From this we can conclude that the average number of records, μ_t, grows linearly with the logarithm of time, i.e. $\mu_t = \ln t$, and therefore we have for the rate with which records happen on average:

$$\frac{d\mu_t}{dt} = \frac{1}{t}, \tag{12.30}$$

so a gradual decrease of activity. Furthermore, since the variance of a Poisson process is equal to its average, we can conclude that the variance in the number of records during time t is given by $\sigma^2(t) = \ln t$.

For an ordinary Poisson process for which an event occurs with probability νdt during an infinitesimal time interval dt, the average number of events during a time interval t is $\mu = \nu t$ and we know, see Eq. (6.95), that for a Poisson process with fixed rate ν, the probability that no event occurs during the time interval t is given by

$$P_0(t) = e^{-\nu t}. \tag{12.31}$$

---

[5] For a test of the accuracy of the approximation, see [404].

Since we can consider the record process as a Poisson process in the 'time' variable $\tau = \ln t$, we have $\mu = \tau$ and hence we can consider that the process in logarithmic time is a Poisson process with rate $\nu = 1$. So the probability that no records occur during the logarithmic time interval $\Delta \tau$ is, according to Eq. (12.31), given by $P_0(\Delta \tau) = e^{-\Delta \tau}$ or, expressed in terms of the probability for the logarithms of two consecutive record times, $\tau_{k-1} = \ln t_{k-1}$ and $t_k = \ln t_k$:

$$\text{Prob}\{\tau_k - \tau_{k-1} < \Delta \tau) = e^{-\Delta \tau}. \tag{12.32}$$

Written slightly differently:

$$\text{Prob}\left\{\ln\left(\frac{t_k}{t_{k-1}}\right) < s\right\} = e^{-s}. \tag{12.33}$$

After having established some of the mathematical properties of the statistics of records in sequences of uncorrelated identically distributed numbers, we need to address the question of why the resulting non-stationary dynamics may be of relevance to real systems. We will consider this in the next subsection.

12.2.2 Spin Glasses, Superconductors, Ants and Evolution

It is not immediately obvious how drawing uncorrelated random numbers and monitoring the sequence of draws at which a new record value is obtained relates to the dynamics of many-component complex systems. Let us consider systems stability and how it relates to the dynamics of the parts.

To make our considerations concrete, we start with an example from physics describing a situation of competitive frustration. Think e.g. of a situation where the interaction between pairs of components may carry more or less strain corresponding to more or less energy. The Ising model in Sec. 6.2 can be used as a simple illustration. The total energy of all components, the Ising spins, is given by a sum of the energy of each pair of interactions, which we can write in a more general form than the one considered in Sec. 6.2.

Let us assume the total energy is given by

$$H[S_1, \ldots, S_N] = -\sum_{i=1}^{N} \sum_{j \in \text{nb}(i)} J_{ij} S_i S_j. \tag{12.34}$$

Here the sum over j runs over the set of neighbours $\text{nb}(i)$ of spin i. Because the energy of different pairs contains different factors $J_{ij} = J_{ji}$, the way the energy depends on the precise assignment of $S_i = -1$ or $S_i = 1$ for each spin i is very complicated. If we let J_{ij} be random numbers drawn from a distribution symmetrical about zero, e.g. from the interval $[-1, 1]$, we can have situations of so-called frustration. A simple version of this is illustrated in the left panel in Fig. 12.6. To minimise the pair energy of S_1 and S_2 we need $S_1 = S_2$, but then the third spin S_3 is unable to minimise the interaction energy $J_{13} S_1 S_3$ simultaneously with minimising the energy stored in $J_{23} S_2 S_3$ (since to

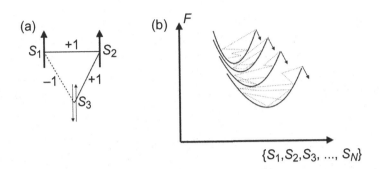

Figure 12.6 Panel (a) example of frustration. The couplings between the spins are $J_{12} = +1$, $J_{23} = +1$ and $J_{13} = -1$. Three Ising spins coupled together in a way that prohibits spin S_3 to simultaneously minimise its interaction energy with S_1 and S_2 if S_1 and S_2 have minimised their interaction energy. Panel (b) contains a sketch of the hierarchical structure of consecutively deeper valleys of the free energy controlling the relaxation.

minimise the first term $J_{13}S_1S_3$ we need $S_3 = -S_1$ while to minimise the second term $J_{23}S_2S_3$ we need $S_3 = S_2$).

   This kind of frustration between pairwise requirements leads to the energy given by Eq. (12.34) having a dependence on the configuration $\{S_1, S_2, \ldots, S_N\}$ that corresponds to a hierarchical mountain landscape with valleys within valleys [404], as depicted in the right panel of Fig. 12.6. This is not entirely correct; the structure that really matters is the free energy $F = E - TS$, which combines the configurational energy with the multiplicity of the configuration as measured by the entropy, see e.g. Secs. 6.7.2 and 10.7.

   We now imagine that our spin system has been started out at high temperature corresponding to a large amount of internal energy. We then suddenly lower the external temperature. To move towards equilibrium, the spin system needs to lower the free energy, which it does by lowering its internal energy by trying to minimise the pair terms in the Hamiltonian in Eq. (12.34). The thermal fluctuations will kick the configurations around in the local valleys in Fig. 12.6. Most of the time the fluctuations will be too small to bring the configuration out of the local valley, but once in a while a fluctuation will occur that is big enough to overcome the barrier of the particular valley. This will be a record fluctuation, since had a fluctuation that big, or bigger, occurred earlier, the configuration would already have escaped from this valley at that instant.

   We have used the language of the Ising system to describe how hierarchical release of internal pent-up strain can occur when the relaxation of one part of the interaction leads to stress on other parts. This situation is in general terms familiar from various situations and we know it from daily life as 'you cannot please everyone'. Think of opinion dynamics. To please one of my acquaintances I might want to agree to serve beef, but will then frustrate those of my friends who favour veganism.

   Or in the context of evolution, think of the multitude of contradictory interdependencies species in a food web are subject to. For example, is it best to be little and agile and able to escape rapidly, or is it better to be big and strong and able to stand one's ground or even attack.

As yet another example, think of the dynamics of the fault lines between the tectonic plates. The earthquakes caused by the jerky motion of the plates may be related to the fact that when a local section of the fault moves to release some part of the internal strain, this motion is likely to increase the strain between other parts.

So we conclude that at least intuitively, it is possible that fluctuation-driven record dynamics may be of relevance to a range of real systems. We will finish the section by briefly looking at some examples and focus on what may be learned from realising that a specific system and its intermittent dynamics may be analysed from the viewpoint of record dynamics.

The simplest way to identify if the behaviour of an intermittent time signal is consistent with record dynamics is to plot the number μ_t of events and the variance σ_t^2 of the number of events during a time interval of length t against $\ln t$. Since record dynamics behaves like a Poisson process when considered as a function of the *logarithm* of time, both μ_t and σ_t^2 need to be linear functions of $\ln t$ if record dynamics is behind the observed dynamics. This is the case e.g. for the abrupt transitions observed in the Tangled Nature model of evolution and the abrupt changes in models of magnetic relaxation in superconductors [17]. The same is observed in experiments on colonies of foraging ants, Fig. 2 in [365] is particularly illuminating.

We recall that the statistics of records is independent of the statistics of the underlying fluctuations. This implies that the observed behaviour, if driven by record dynamics, must be robust against changes at the microscopic level. Accordingly, we will expect that when the microscopic fluctuations are generated by thermal fluctuations, the observed intermittent relaxation should essentially be independent of the temperature. This is a very strong requirement. Thermal dynamics is in general expected to depend exponentially on the temperature, since configurations are visited with probabilities given by the Boltzmann weights in Eq. (6.7). The Boltzmann weights lead to the so-called Arrhenius activation law, stating that the rate with which a thermal system overcomes an energy barrier ΔE is proportional to $\exp(-\Delta E/k_B T)$. So thermal relaxation is expected to depend exponentially on temperature. In contrast, magnetic relaxation in superconductors has been observed in experiments to depend only very weakly on the temperature over a very broad range of temperatures. This, together with the behaviour of μ_t and σ_t^2 mentioned above, suggests that record fluctuations may be behind the observed weak temperature dependences [321].

We already mentioned that the $1/t$ time dependence of the Omori aftershock law [462] could suggest that the aftershocks are triggered by record fluctuations of some kind. If this is the case, we would expect that the robustness of record dynamics against the details of the microscopic fluctuations makes the aftershock statistics independent of details such as the nature of the fault and the size of the preceding earthquake. This seems at least approximatively to be the case [95].

The distinction between the ever occurring fluctuations kicking the system about within a valley, see Fig. 12.6, and the record fluctuations which cause a move to a deeper level of stability suggests different behaviour should be observed depending on the time scale of an observation. At time scales too short to capture the record events, behaviour essentially identical to equilibrium is expected, while at longer time scales the effect of the infrequent big record fluctuations should lead to a different behaviour.

Examples of such behaviour can be seen e.g. in simulations of a spin glass in contact with a heat reservoir. A spin glass is a disordered magnetic system, which shares dynamical properties with ordinary glass, such as the material windows and drinking vessels are made of. Both spin glass and ordinary glass are out of thermodynamic equilibrium and undergo very slow relaxation towards the equilibrium state. Proper equilibrium is only reachable at elevated temperatures. At short times the energy fluctuations exchanged between the spin glass and the heat reservoir are Gaussian distributed [403]. This is what we expect, since the heat exchanged is the sum of many random microscopic parcels and therefore the Central Limit Theorem tells us that the energies will be Gaussian distributed as long as correlations between the individual microscopic parcels are negligible. When the observation time is extended to allow sampling of the infrequent record events, the distribution of energy exchange changes form and develops a region with an exponential rather than a Gaussian shape. This is an illustration of how systems undergoing record dynamics may appear to be in equilibrium when observed on short time scales. Clearly it is of significant practical relevance to know what constitutes the time scale over which equilibrium behaviour is to be expected, and how long one should expect to wait before a record fluctuation kicks the system to a new quasi-equilibrium state. We will return to the question of estimating the time scale associated with the abrupt transitions in the next section and in the next chapter we will consider the important but difficult question of how to forecast abrupt events in complex systems.

To finish this section we will, in the context of evolutionary ecology and ant colonies, ask what consequences record dynamics has for stability. This framework implies that, at least predominantly after a transition a new transition can only be released if a fluctuation of larger size occurs. In this way we expect that an old ecosystem, which is the result of many past evolutionary transitions, is more stable than a young one. This may be correct, at least in the sense that the fossil record seems to indicate that the rate of mass extinctions has been decreasing over time spans of millions of years [310, 311]. Whether we can expect increasing stability with age of the local ecosystem is not clear. We do not know what constitutes a big or a small fluctuation in this case, and we may expect that external factors such as those caused by human activity are more likely to disturb the local ecology than are internal fluctuations.

If we apply the same line of thought to the experiment on the foraging ants [365], we could in principle check if the ant colony becomes more stable the more ants have left the nest by comparing the effect of applying a pertinent perturbation at early and late times. Perhaps one might add a drop of ant repellent and check if nests which experience many ants leaving tend to be affected less than nests where only a few ants have left at the time the repellent is applied.

12.3 Tangent Map Intermittency

The previous examples of mechanisms leading to intermittent dynamics, SOC and record dynamics, are concerned with the emergent collective dynamics of very many interacting components. In this section we describe an approach from the theory of deterministic dynamical systems [391], in which intermittency is obtained from a discrete map of a single variable. The intermittency arises when the identity is nearly tangential to the map, see Fig. 12.7. The relevance of such maps to high-dimensional stochastic systems depends on whether a robust macroscopic degree of freedom emerges, which is able to capture the dominant dynamics, see Sec. 2.2. One example of this is the paradigmatic Pomeau–Manneville map [348], which was derived to represent intermittency in weakly turbulent fluid dynamics, i.e. for dynamics of very high dimension.

Tangent map intermittency was, in 1983, suggested by Procaccia and Schuster as a universal mechanism of $1/f$ noise. The same phenomenon which inspired SOC in 1987 [37]. It is not in general straightforward to connect in detail a high-dimensional complex system with the one-dimensional map considered by Procaccia and Schuster [353]. However, this was done e.g. for the Tangled Nature model, see Sec. 11.3. Using mean-field approximations, a map for the average of the weight function H was derived and a time scale for the intermittency was obtained by combining the mean-field analysis with data from simulations, see [107].

We now describe how a time scale for the quiet periods of the intermittent dynamics can be extracted from the map. Assume that the state of our system is well described by a single variable x_t captured at discrete integer time steps $t \in \mathbb{Z}$.

We assume the iterative relation $x_{t+1} = f(x_t)$ and that the tangent of the map $f(x)$ at a point x_c is parallel with and close to the identity $y = x$ as indicated in Fig. 12.7. The closeness of the identity to the map implies that many iterations of the map are needed to get through the bottleneck between the map and the identity. This is indicated by the dotted line staircase in the left panel in Fig. 12.7. Note that as we iterate the map $x_{t+1} = f(x_t)$, the value of x at time $t + 1$ becomes the argument of $f(x)$ in the next

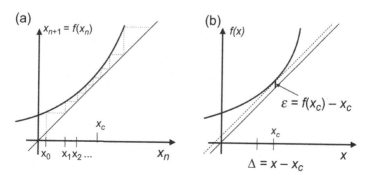

Figure 12.7 Panel (a) iterative map. Panel (b) bottleneck between identity and map.

iteration. Hence the horizontal lines stretching from $y = f(x_t)$ out to the identity $y = x$. In the bottleneck, many iterations are needed to make $y = f(x_t)$ change significantly.

Since the tangent at x_c is assumed to be parallel to the identity, we have $f'(x_c) = 1$ and Taylor expanding to second order about x_c we can write

$$f(x) = f(x_c) + (x - x_c) + \frac{1}{2}f''(x_c)(x - x_c)^2. \tag{12.35}$$

We denote the distance between the identity and $f(x)$ at x_c by ϵ, so $f(x_c) = x_c + \epsilon$. Next we introduce Δ to denote the deviation from x_c and write $x = x_c + \Delta$ and introduce $\beta = f''(x_c)$, so we get

$$x_{t+1} = f(x_t)$$
$$\Downarrow$$
$$x_{t+1} = x_c + \epsilon + (x_t - x_c) + \beta(x_t - x_c)^2 \tag{12.36}$$
$$\Downarrow$$
$$\Delta_{t+1} - \Delta_t = \epsilon + \beta\Delta_t^2.$$

We can easily get an estimate of how long the variable x_t spends in the bottleneck region about x_c by making use of a continuum approximation. To do this we use Taylor expansion, i.e. $\Delta_{t+1} = \Delta_t + \frac{d\Delta}{dt}$, and obtain the differential equation

$$\frac{d\Delta}{dt} = \epsilon + \beta\Delta^2. \tag{12.37}$$

We can rearrange the equation to obtain

$$\int \frac{d\Delta}{\epsilon + \beta\Delta^2} = \int dt$$
$$\Downarrow$$
$$t + t_0 = \frac{1}{\sqrt{\epsilon\beta}}\arctan\left(\sqrt{\frac{\beta}{\epsilon}}\Delta\right) \tag{12.38}$$
$$\Downarrow$$
$$\Delta = \sqrt{\frac{\epsilon}{\beta}}\operatorname{tg}[\sqrt{\epsilon\beta}(t + t_0)].$$

We see that the time scale for the change in Δ is given by $1/\sqrt{\epsilon\beta}$, or to be slightly more specific we compute the time it takes Δ to change from $-\infty$ to ∞. Recall that $\operatorname{tg}(-\pi/2) = -\infty$ and that $\operatorname{tg}(\pi/2) = \infty$. So an upper bound on the time it takes x_t to move through the bottleneck is given by the time difference between the time t_- and t_+ given by

$$\sqrt{\epsilon\beta}(t_- + t_0) = -\frac{\pi}{2} \quad \text{and} \quad \sqrt{\epsilon\beta}(t_+ + t_0) = \frac{\pi}{2}. \tag{12.39}$$

So our estimate of the time x_t spends in the bottleneck is

$$T = t_+ - t_- = \frac{\pi}{\sqrt{\epsilon\beta}}. \tag{12.40}$$

We see that T becomes large if $f(x_c)$ is close to the tangent, corresponding to ϵ small, or if the curvature of the map around x_c is small, corresponding to $\beta = f''(x_c)$ small.

Obviously, we will need to repeat the passage through the bottleneck in order to produce an intermittent signal. This will be the case if the map has a shape to the right of x_c that sends x_t back to the left of x_c. The following function is an example of such a map:

$$f(x) = \begin{cases} x^2 + \frac{1}{4} + \epsilon & \text{if } 0 \le x < \frac{3}{4} \\ \frac{3}{16x} + 0.0725 & \text{if } \frac{3}{4} \le x \le 1 \end{cases}. \tag{12.41}$$

The map in Eq. (12.41) is just meant as an illustration and does not represent any known phenomenon from the real world. Perhaps a more realistic way to obtain repeated passages through the bottleneck region is by assuming that the map to the right of x_c is chaotic in a way that leads to x_t being kicked back below x_c, see [353]. Another mechanism could be random reinjection, where some random mechanism maps $x_t > x_c$ back to $x_t < x_c$. This scenario has been suggested to be relevant to the Tangled Nature model [106]. For this model the map $x_{t+1} = f(x_t)$ describes the evolution of the average $\langle H \rangle$ of the weight function in Eq. (11.37) [107]. See Fig. 12.8. The passage through the bottleneck corresponds to the extended metastable q-ESS epochs, which are separated by the hectic search through the type space leading to the establishment of a new q-ESS. Since the value of H is attracted to the region about H^*, see Fig. 11.6, it seems natural to assume that reinjection happens in the map describing the evolution of $\langle H \rangle$.

Moreover, the adaptive dynamics makes on average each q-ESS slightly more stable than the previous one, which corresponds to the map controlling $\langle H \rangle$ evolving its shape towards smaller values of ϵ and β in Eq. (12.40). When the relevant parameters for the map are extracted from simulations, it is found that the description in terms of a tangent map is able to estimate the duration of the q-ESS with reasonable accuracy [107].

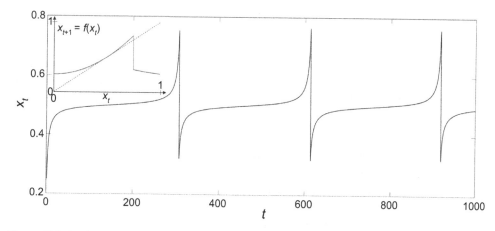

Figure 12.8 Main panel shows the iterated time dependence $x_{t+1} = f(x_t)$ of the map in the insert defined in Eq. (12.41).

Summary: We have described a number of complementary mechanisms which lead to intermittent dynamics. We distinguished between three prominent examples:

Self-organised criticality focuses on stationary statistics of high-dimensional systems and emphasises power laws. Although approximate power laws, and hence systems exploring regions in the vicinity of a critical point, may be most relevant.

Record dynamics can apply to systems undergoing relaxation. The emphasis is on systems with very many components frustrated in their mutual interaction. The rate of the intermittent abrupt events decreases inversely with time.

Tangent map intermittency in principle a low-dimensional description, but can be of relevance to high-dimensional systems if the systemic dynamics is controlled by an emergent robust collective variable. The tangent map combined e.g. with mean-field analysis may be helpful when estimating intermittent time scales.

12.4 Further Reading

General audience:

An entertaining autobiographical explanation of how self-organised criticality was conceived is given in the book *How Nature Works: The Science of Self-Organized Criticality* [34] by Per Bak, one of the inventors of SOC.

The book *The Structure of Evolutionary Theory* [169] by Stephen Jay Gould discusses intermittency in the form of punctuated equilibrium. The book is at a high scientific level but requires no particular background.

Intermediate level:

A relatively brief introduction to the original ideas behind the suggestion of self-organised criticality is given in the book *Self-Organized Criticality: Emergent Complex Behavior in Physical and Biological Systems* [214] by Henrik Jeldtoft Jensen.

Record dynamics and its applications to a very broad array of situations is discussed in *Stochastic Dynamics of Complex Systems: From Glasses to Evolution* [406] by Paolo Sibani and Henrik Jeldtoft Jensen.

Intermittency in the context of dynamical systems, and in particular tangent intermittency, is discussed with great clarity in Chap. 4 of the book *Deterministic Chaos: An Introduction* [391] by Heinz Georg Schuster.

Advanced level:

Dynamical systems intermittency is discussed at a mathematically advanced and thorough level in the book *New Advances on Chaotic Intermittency and its Applications* [127] by Sergio Elasker and Ezequiel del Río.

A thoroughly comprehensive and careful exposition of self-organised criticality is given in the book *Self-Organised Criticality: Theory, Models and Characterisation* [354] by Gunnar Pruessner.

12.5 Exercises and Projects

Exercise 1
The time between events in SOC models is typically exponentially distributed, indicating no correlations between events. This is often in contrast to observed behaviour. Solar flares is an example of this. Read [180] and discuss how the OFC, despite its lack of strictly critical behaviour, may be of relevance to solar flare dynamics.

Exercise 2
Simulate the Manna model on a square lattice. The model is defined in the following way, see [276, 354]. Each site can contain zero or one particle. New particles are added to randomly chosen sites. If the site is empty, the added particle remains at the site. If the site was already occupied, an avalanche is triggered consisting of redistributing *both* particles to randomly chosen neighbour sites. This redistribution of particles is repeated by parallel updates until all sites again contain either zero or one particle. When this happens the avalanche is considered to have stopped and a new particle is added to a random site. Numerically determine the distribution of avalanche sizes, durations and time between avalanches.

Exercise 3
Carefully go through the arguments leading to Eq. (12.23). Next derive Eq. (12.27).

Exercise 4
Consider the logistic map

$$x(t+1) = f(x(t))] = rx(t)(1 - x(t)).$$

Derive analytically an expression for the fixed point x^* given by $x^* = f(x^*)$ and the point x_{tan} for which $f(x)$ has a tangent parallel to the identity, i.e. $f(x_{\mathrm{tan}}) = 1$. Use the expression in Eq. (12.40) to estimate the number of iterations needed to pass through the bottleneck.

Choose $r = 1 + \Delta$. For $\Delta = 0.1, 0.01$ and 0.001, numerically iterate the map and determine the number of iterations, T, needed to go from $x(0) = 10^{-5}$ to $X(T) = x^* \pm 10^{-5}$; compare your results with the results obtained from Eq. (12.40).

Exercise 5
Machine learning, chaos and intermittency are discussed in [237]. Start with the abstract and discussion sections and then skim read the paper with the aim of making a brief overview of how machine learning can be used to develop models of real chaotic systems.

Project 1 – Self-organised branching processes: mean-field theory for avalanches
The project studies the paper [501]. Recall that the avalanche dynamics evolves on two time scales. The first, denoted by t, counts the numbers of energy units added to the pile. In between each addition of an energy unit, the relaxation dynamics is allowed to run. This is the branching process depicted in Fig. 1 of [501]. The relaxation probability p is kept constant while an avalanche spreads, which is described as a branching process. However, the adding of energy units and their escape at the boundary make the relaxation probability vary on the time scale t, hence Eq. (1), which is considered to represent the time dependence of p on the time scale of energy addition.

(a) Write down the branching probabilities for the process defined on p. 4071, bottom of the right column.
(b) Compute the generator function for the process and determine the average branching ratio from the generator. For which value p_c of p is the process critical?
(c) Let τ_{ex} denote the time to extinction. By use of generator formalism show that the probability

$$P(\tau_{\text{ex}} > n) \propto \exp\left(-\frac{n}{n_0}\right), \quad \text{where } n_0 \in \mathbb{R}^+.$$

(d) Let the average branching ratio be of the form $\mu = 1 - \Delta$. Show that for $\Delta \to 0^+$ the scale $n_0 \to \infty$ algebraically as

$$n_0 \propto \Delta^{-a}$$

and determine the value of the exponent a.
(e) Explain why Eq. (1) in [501], though not in any way an exact description, nevertheless can be seen as a reasonably effective description of the probability $p(t)$. Equations (1) and (3) assume that one unit of energy leaves the system for each of the $\sigma_n(p, t)$ relaxing boundary sites. One might perhaps find it more natural to assume that each boundary site lost two units of energy, given the description of the process at the top of the left column on p. 4072. How would the fixed point value p_c of Eq. (3) change if we assumed that each relaxing boundary site lost two energy units? What is the value of p_c in this case when $n \to \infty$?
(f) Simulate numerically the process described in Eq. (1) and reproduce curves similar to those in Fig. 2 of [501].

(g) Assume $p(0) = 0.1$. Demonstrate numerically that the time t^* at which $p(t)$ reaches the asymptotic value $p(t^*) \simeq 1/2$ approximately depends exponentially on n and show how this dependence can be expected from an estimate based on the slope of $p(t)$ at $t = 0$.

(h) Study numerically the process for different values of n and produce plots like in Fig. 3 of [501]. Explain why the exponent τ is expected to satisfy $\tau = 3/2$.

(i) Demonstrate numerically that $D(s)$ in Fig. 3 falls off exponentially for large values of s.

(j) Consider $p < p_c$. Use generator formalism to explain why, for fixed value of p, the time average of $\sigma_n(p, t)$ is given by

$$\langle \sigma_n(p, t) \rangle = (2p)^n. \tag{1}$$

Now consider the limit $n \to \infty$ and compute the average of $Y_\infty = \sum_{n=0}^{\infty} \sigma_n$ first by summing Eq. (1) and next from the generator for Y_∞.

(k) Show that the variance $\text{Var}\, \sigma_n = \langle [\sigma_n(p, t)]^2 \rangle - \langle \sigma_n(p, t) \rangle^2$ in the limit of $p \to p_c^-$ is given by

$$\text{Var}\, \sigma_n = n.$$

Within this approximation calculate the correlation function

$$C(r) = \langle S_1 S_r \rangle - \langle S_1 \rangle^2$$

by use of a continuum Poisson process.

Project 2 – Epidemiology, record dynamics and tangent maps

We consider the outbreak of a transmittable disease in an infinite population. The disease is characterised by its ability $\eta \in [0, 1)$ to infect. Each individual has a specific level of resistance R distributed on the interval $[0, 1]$. The resistance of an individual is independent of the other individuals and constant in time, except that an infected person becomes immune, i.e. R for that individual becomes equal to 1. Once a year an infection of strength η drawn as i.i.d. from $[0, 1)$ hits all individuals in the population and those with $R < \eta$ become infected.

(a) We treat time as a discrete variable measured in years. Assume a person succumbs to the disease at year t_1. What is the probability that the next incident occurs at t_2?

(b) Consider the number of outbreaks Q that have occurred between two years t_1 and t_2. Assume $Q \gg 1$ and compute to a good approximation the probability that $Q = q$.

Now we consider some details of the spreading of the disease. The probability that an infected person A directly transmits the disease to k other people is given by

$$p_k = (1 - e^{-\alpha})e^{-\alpha k}, \quad k = 0, 1, 2, \ldots \text{ and } \alpha > 0. \tag{1}$$

(c) Determine the average number μ of direct transmissions caused by the infected person A. This is the first generation of people infected by A.

(d) Now consider the tree of infections originating from a given infected person A. Use the relation

$$g_{Z_n}(s) = g(g_{Z_{n-1}}(s))$$

between the generator for the size of generation n and $n-1$ and the generator for the p_k to determine the ratio between the average number of infected, $\langle Z_n \rangle$ and $\langle Z_m \rangle$, in generations n and m.

Let $\Pr(T)$ denote the probability that the infection originating from person A is still spreading in the population after T generations of infections.

(e) Determine α in Eq. (1) such that the ratio $\Pr(T_1)/\Pr(T_2)$ only depends on T_1/T_2.

In a big urban area, of total fixed population N, the fraction x_n of people infected by some disease is measured every week, $n = 1, 2, 3, \ldots$. Assume that x_n evolves according to the map

$$x_{n+1} = f(x_n).$$

Assume that $f(x)$ generates intermittency by having a region nearly tangential to the identity.

(f) Make use of the theory of intermittent tangent maps to derive an expression that estimates the number of iterations during which x_n is quasi-static.
(g) Assume that

$$f(x) = \begin{cases} x^2 + \frac{1}{4} + \epsilon & \text{for } 0 \leq x \leq \frac{3}{4} \\ \frac{3}{16}\frac{1}{x} + 0.0725 & \text{for } \frac{3}{4} < x \leq 1 \end{cases}.$$

Plot x_n as a function of n for $n = 0$ to $n = 1000$ starting from $x_0 = 0.01$ for the following values of $\epsilon = 10^{-2}, 10^{-3}$ and 10^{-4}.
(h) Use the result of question (f) to estimate the duration between the intermittent bursts of infection for each of the values of ϵ mentioned in question (g) and comment on how these estimates compare with the behaviour observed in the graph of question (g).

13 Tipping Points, Transitions and Forecasting

> **Synopsis:** We discuss forecasting of the transitions accompanying the intermittent dynamics of complex systems. Co-evolutionary dynamics is particularly challenging.

The term 'tipping point' indicates an irreversible, relatively sudden, change in the state of a system. Such tipping points can be part of the intrinsic intermittent dynamics or they may occur as a consequence of changes to external parameters. We discussed the behaviour in general terms in Sec. 3.5. The development of reliable methodologies to forecast these abrupt system-altering events such as ecosystem collapse, climate change, earthquakes, onset of epileptic seizure or financial crashes is very important, very difficult and still a wide open field [151, 202, 432, 483].

In this chapter we will just mention two mathematical strategies for analysing the change in systems resilience before a transition occurs. The first assumes that some external control parameter is changing in a way that leads to a decrease in the stability and eventually the appearance of a non-reversible tipping point. The analysis assumes that the state of the considered system is captured by at least one robust collective variable, e.g. the total population of a species, and seeks to forecast an approaching transition from an increase in the fluctuations of this variable as the external conditions are changing [383, 384].

The second approach we will look at analyses the stability of high-dimensional complex dynamics which undergo transitions caused by the intrinsic dynamics despite the external conditions remaining fixed. This analysis investigates the appearance of an unstable direction in state space [76, 243, 347]. We will discuss how the overlap between the current state and the most unstable direction in state space can be used as an early warning signal [76, 347].

13.1 Externally Induced Transitions

We look at some robust collective parameter, $x(t)$. It could be the total population in a given region [385]. We assume that initially the parameters of the system are such that $x(t)$ oscillates in time about a stable value; we sketched this behaviour in Fig. 3.9. We are not interested in a ball rolling in a potential, which would be described by Newton's equation, but rather we have in mind that $x(t)$ evolves in time as the net effect of the sum of many individual effectively stochastic changes – accidental death and birth events if we think of $x(t)$ being the total population size. It is therefore natural to assume that the time evolution of $x(t)$ from generation to generation can be captured by a random walk which is subject to some stabilising effect. Hence we can make use of the analysis of the Ornstein–Uhlenbeck process in Sec. 10.5.

To illustrate the behaviour we look at the discrete version and assume that $x(t)$ evolves according to

$$x(t+1) = x(t) - \eta(t)x(t) + \sigma\chi(t), \tag{13.1}$$

see Eq. (10.104). We have assumed that the coefficient $\eta(t)$ of the restoring term is time dependent. This is thought of as representing the change in external parameters which makes the bowl in Fig. 3.9 more shallow.

By way of illustration, we assume the following time dependence:

$$\eta(t) = \begin{cases} \eta_0 & \text{if } 0 \le t \le t_0 \\ \eta_0(1 - \frac{t-t_0}{T}) & \text{if } t_0 < t < t_0 + T, \\ 0 & \text{if } t_0 + T \le t \end{cases} \tag{13.2}$$

where η_0 is a positive constant. Initially the external conditions are kept fixed and $\eta(t)$ is independent of time until time t_0 is reached. During the next time interval of duration T, the external conditions change, making $\eta(t)$ linearly decrease until $t_0 + T$, at which time the restoring force disappears and x_t from then on becomes an ordinary random walker.

Figure 13.1 indicates the behaviour. We know from Eq. (10.108) that for time-independent η the Ornstein–Uhlenbeck random walker is confined within a region which in the continuum approximation is proportional to the ratio σ/η.

We see, as expected, in Fig. 13.1 that the amplitude of fluctuations away from the average $\langle x(t) \rangle = 0$, and the associated return times, increase as the restoring force decreases. It has been suggested [383] to use the increase in the fluctuations and the ever slower return to the average as precursors from which one can try to forecast the transition to zero restoring force $\eta(t) = 0$, which happens in our example at $t = 4000$.

The success of applying this approach to real systems will depend on how well the effect of the approaching transition is captured by one, or a few, macroscopic parameters. It may also depend on how distinct the onsets of fluctuations are and how easily they can be distinguished from the fluctuations present even before the restoring force starts to weaken. Although potential applications are listed e.g. in [384], it is still not clear how successful the focus on increased fluctuations and slowing down is in predicting sudden transitions in real systems. We will, in the next section, see that the intrinsic dynamics of some classes of complex systems may cause abrupt tipping points which are not foreshadowed by slowing down or a distinct increase in fluctuations.

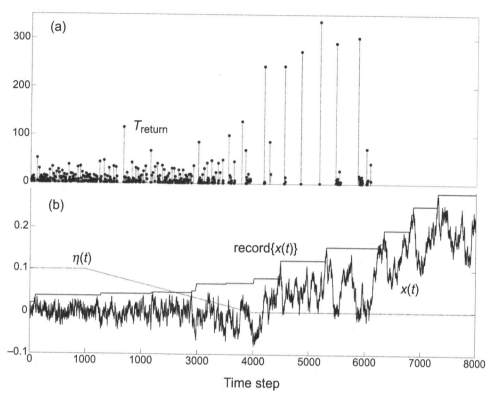

Figure 13.1 Realisation of the Ornstein–Uhlenbeck process with time-dependent coefficient η. The top panel (a) contains the time since last crossing of $x(t) = 0$ as a function of the time step given along the x-axis at the bottom of panel (b). In the bottom panel (b) the solid curve represents the position $x(t)$ of the random walker in Eq. (13.1). The dotted line describes the time evolution of the coefficient $\eta(t)$ of the restoring force term given in Eq. (13.2). The solid staircase curve is the record value of x_t up to time t. Parameters are $\eta = 0.1$, $\sigma = 0.01$, $t_0 = 10^3$ and $T = 3 \times 10^3$.

13.2 Intrinsic Instability

We have noted previously a number of times, e.g. see Sec. 3.5, Chap. 12 and Sec. 11.3, that intermittent transitions may be intrinsic and therefore can occur even when the external conditions remain constant. Hence, to construct early warning signs of such transitions, we need to identify indicators of change in the system's stability that may occur even without any change in the surroundings.

As a first step, let us look at the temporal evolution of two populations $x(t)$ and $y(t)$. The x population evolves according to the following Ornstein–Uhlenbeck process with a time-independent $\eta > 0$ and $\bar{X} > 0$ given by

$$x(t+1) = [x(t) - \eta(x(t) - \bar{X})] \exp(-\alpha y(t)) + \sqrt{x(t)}\chi(t) \tag{13.3}$$

and the restoring term exponentially damped, $\alpha > 0$, by coupling to a growth process controlling the y population according to

$$y(t+1) = \begin{cases} y(t) + \nu x(t) & \text{with prob. } p \\ y(t)(1+\beta) + \sqrt{y(t)}\chi(t) & \text{with prob. } 1-p \end{cases}, \qquad (13.4)$$

where we assume the two constants ν and β to be positive. We can loosely think of $x(t)$ and $y(t)$ as being proportional to the size of the populations of two types X and Y. Type Y may become populated as a result of a mutation from population X, which occurs with probability p. If we think of an example from economics, perhaps X could represent traditional bookshops and Y represent an online trader like Amazon. As long as $y(t) = 0$, type X follows an Ornstein–Uhlenbeck process except we have multiplied the noise term χ by the square root of the size of the populations. This is a simplistic way to indicate that the noise arises from fluctuations in the death and birth processes underlying the evolution of the type.[1] Namely, if there are no members of the type around, there will be no birth nor any deaths, so in this case the noise should not be able to change the population size. The same holds for type Y.

As soon as Y becomes populated due to a mutation from X, the Y population grows exponentially. This corresponds to a type that fits very well into the given environment and as it grows it has a detrimental effect on X, modelled by the exponential term $\exp(-\alpha y(t))$ in Eq. (13.3).

Figure 13.2 contains the results of a simulation where initially there is no Y population present, so the X population oscillates as we expect for an Ornstein–Uhlenbeck process about the mean $\bar{X}$. Since, according to Eq. (13.4), the Y population becomes populated with probability p after some time we will have $y(t) > 0$ and the population will start to grow. Hence the trajectory in Fig. 13.2 moves into the xy-plane. At first the X population continues to fluctuate about the value $\bar{X}$, but after some time the Y population will have grown sufficiently large that the factor $\exp(-\alpha y(t))$ in Eq. (13.3) will cause the decline of the X population. The motion away from the stable fluctuations about $x = \bar{X}$ described by Eq. (13.3) for $y(t) = 0$ can be anticipated by linear stability analysis of the fixed points of the non-stochastic part of Eqs. (13.3) and (13.4).

To study the stability we subtract $x(t)$ and $y(t)$ from both sides of the equations and replace $x(t+1) - x(t)$ and $y(t+1) - y(t)$ by dx/dt and dy/dt. Next we introduce

$$F(x,y) = [x - \eta(x - \bar{X})]\exp(-\alpha y) - x \qquad (13.5)$$
$$G(x,y) = \beta y.$$

With this notation we can write the non-stochastic part of the continuum time version of Eqs. (13.3) and (13.4) as

$$\frac{dx}{dt} = F(x,y)$$
$$\frac{dy}{dt} = G(x,y). \qquad (13.6)$$

The fixed points (x^*, y^*) are defined as the solutions to $F(x,y) = 0$ and $G(x,y) = 0$. They are called fixed points because if, for some t_0, we have $(x(t_0), y(t_0)) = (x^*, y^*)$, we

---

[1] See the discussion in Sec. 10.6; see Eq. (10.112), where we mentioned that field-theoretic analysis suggests the noise term should be multiplied by the square root of the population size.

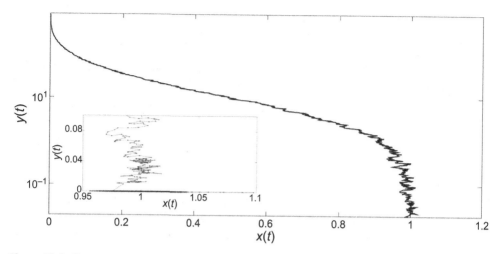

Figure 13.2 Coupled Ornstein–Uhlenbeck processes, see Eqs. (13.3) and (13.4). In the insert one sees the initial trajectory along the x-axis while $y(t)$ remains equal to zero. When the second variable $y(t)$ becomes populated, due to a mutation from x, it grows exponentially. As $y(t)$ increases, the fixed point at $(x,y) = (1,0)$ is destabilised because of Y's suppressive effect on X. See main panel. Parameters: $\eta = 10^{-1}$, $\alpha = \beta = \nu = 10^{-2}$, $\bar{X} = 1$, $p = 10^{-3}$, $(x(0), y(0)) = (1.1, 0)$.

will remain at this position for all t since according to Eq. (13.6) the time derivatives of x and y are zero. We can ask what happens if $(x(t_0), y(t_0))$ is not exactly equal to the fixed point but in its vicinity. To do this we make use of the expansion on eigenvectors similar to the discussion of Eqs. (8.118)–(8.121). See Box 8.1 concerning eigenvectors. We write $x(t) = x^* + \delta_x(t)$ and $y(t) = y^* + \delta_y(t)$ and expand $F(x,y)$ and $G(x,y)$ to first order in $\delta_x(t)$ and $\delta_y(t)$. When we make use of $F(x^*, y^*) = G(x^*, y^*) = 0$, we obtain the following equations for the time evolution:

$$\frac{d\delta_x}{dt} = \partial_x F(x^*, y^*)\delta_x + \partial_y F(x^*, y^*)\delta_y$$

$$\frac{d\delta_y}{dt} = \partial_x G(x^*, y^*)\delta_x + \partial_y G(x^*, y^*)\delta_y. \tag{13.7}$$

Or, using matrix notation:

$$\frac{d}{dt}\delta = M\delta, \tag{13.8}$$

where M is given by

$$M = \left\{ \begin{array}{cc} \partial_x F & \partial_y F \\ \partial_x G & \partial_y G \end{array} \right\}_{(x^*, y^*)}$$

$$= \left\{ \begin{array}{cc} (1-\eta)e^{-\alpha y} - 1 & -\alpha e^{-\alpha y}[x - \eta(x - \bar{X})] \\ 0 & \beta \end{array} \right\}_{(x^*, y^*)} \tag{13.9}$$

$$= \left\{ \begin{array}{cc} -\eta & -\alpha \bar{X} \\ 0 & \beta \end{array} \right\}.$$

The eigenvalues and eigenvectors of M are given by

$$\lambda_1 = \eta \quad v_1 = (1,0)$$
$$\lambda_2 = \beta \quad v_2 = \left(-1, \frac{\eta + \beta}{\alpha \bar{X}}\right). \tag{13.10}$$

When we write $\delta(t) = c_1(t)v_1 + c_2(t)v_2$ and substitute into Eq. (13.8) we find, see e.g. Eq. (8.121):

$$\delta(t) = c_1(0)e^{\lambda_1 t}v_1 + c_2(0)e^{\lambda_2 t}v_2. \tag{13.11}$$

This tells us that if we start from a position off the x-axis, i.e. with $c_2(0) > 0$, we will be pushed away from the x-axis since the eigenvalue $\lambda_2 > 0$ and therefore the y-component $c_2(0)e^{\lambda_2 t}$ of δ will increase with time. The more our initial position $(c_1(0), c_2(0))$ is aligned along the unstable direction v_2, the sooner we will move away from the fixed point at $(x^*, y^*) = (\bar{X}, 0)$.

Inspired by these considerations, we now turn to the high-dimensional formalism and describe an early warning measure for high-dimensional stochastic dynamics, which makes use of how much the configuration of the fully stochastic system is aligned along the eigenvector corresponding to the largest eigenvalue of the stability matrix of the mean-field approximation to the stochastic dynamics.

To have a concrete example in mind, we will use the Tangled Nature model of stochastic evolutionary dynamics of reproduction, mutation and killing, see Sec. 11.3. The configuration of the model is given by the vector

$$\mathbf{n}(t) = (n(\mathbf{S}^1), n(\mathbf{S}^2), \ldots, n(\mathbf{S}^{2^L})). \tag{13.12}$$

We can derive a mean-field equation for the on-average behaviour of $\mathbf{n}(t)$ by arguments similar to the analysis we did in Sec. 11.3, where we discussed stability when only one type changes its occupancy in time, see Eqs. (11.48) to (11.52). This time, however, we will allow for simultaneous changes to the occupancy of all types, so we will have a high-dimensional equation involving the vector $\mathbf{n}(t)$ and a matrix involving the probabilities governing the stochastic updates.

Our mean-field equation will, in the continuum approximation, be of the form

$$\frac{d\mathbf{n}}{dt}(t) = \frac{1}{N(t)}\mathbb{T}(\mathbf{n}(t)) \cdot \mathbf{n}(t), \tag{13.13}$$

where the matrix element of the dynamical matrix $\mathbb{T}(\mathbf{n}(t))$ is given by

$$\mathbb{T}(\mathbf{n}(t))_{ij} = [p_{\text{off}}^j(2p_{\text{mut}}^{(0)} - 1) - p_{\text{kill}}]\delta_{i,j} + 2p_{\text{off}}^j(p_{\text{mut}})^{Ld_{ij}}(1 - p_{mut})^{L(1-d_{ij})}(1 - \delta_{ij}). \tag{13.14}$$

Here p_{off}^j denotes the offspring probability of type j in the configuration $\mathbf{n}(t)$ and $p_{\text{mut}}^{(0)} = (1 - p_{\text{mut}})^L$ is the probability that no mutation occurs and $d_{ij} = \sum_{q=1}^{L} |S_q^i - S_q^j|/(2L)$ is the Hamming distance[2] between type i and type j. Finally, δ_{ij} is the usual

---

[2] That is the number of elements $S_q^i \neq S_q^j$ for $q = 1, 2, \ldots, L$.

Kronecker delta function. The derivation of this equation is straightforward, though a bit cumbersome, and follows the same analysis as used in Eqs. (11.48) to (11.52). Details can be found in [76, 346, 347].

The metastable q-ESS configurations will, in mean field, correspond to the fixed-point configurations $\mathbf{n}^*$ of Eq. (13.13) given by

$$\mathbb{T}(\mathbf{n}^*) \cdot \mathbf{n}^* = 0. \tag{13.15}$$

It is not possible analytically to find $\mathbf{n}^*$ when $\mathbf{n}(t)$ is a high-dimensional vector and $\mathbb{T}(\mathbf{n}(t))$ depends in a complicated non-linear way on $\mathbf{n}(t)$. Since we are looking for ways to forecast when to expect an approaching transition, we will assume we are able to observe the current configuration of the fully stochastically fluctuating system. During the quiescence of the q-ESS, the occupancy vector $\mathbf{n}(t)$ will only exhibit weak time dependence. We will average $\mathbf{n}(t)$ over a local time window to obtain the configuration $\bar{\mathbf{n}}_{\text{stoc}}$. We will expect this configuration to be approximately a fixed point, i.e. we need to check $\mathbb{T}(\bar{\mathbf{n}}_{\text{stoc}}) \cdot \bar{\mathbf{n}}_{\text{stoc}} \simeq 0$. If that is the case we next do linear stability analysis about $\bar{\mathbf{n}}_{\text{stoc}}$ by writing $\mathbf{n}(t) = \bar{\mathbf{n}}_{\text{stoc}} + \delta\mathbf{n}(t)$ and expanding to first order in $\delta\mathbf{n}(t)$ to get

$$\frac{d\delta\mathbf{n}}{dt}(t) = \frac{1}{N(t)}\mathbb{T}(\bar{\mathbf{n}}_{\text{stoc}} + \delta\mathbf{n}(t)) \cdot (\bar{\mathbf{n}}_{\text{stoc}} + \delta\mathbf{n}(t)), \tag{13.16}$$

which leads to the equation

$$\frac{d\delta\mathbf{n}}{dt} = \mathbb{M}(\bar{\mathbf{n}}_{\text{stoc}})\delta\mathbf{n}. \tag{13.17}$$

Here $\mathbb{M}$ is given by

$$M_{ij} = \frac{1}{N(t)}\left\{ T_{ij}(\mathbf{n}) + \sum_{q=1}^{2^L} \frac{\partial T_{iq}(\mathbf{n})}{\partial n_j}n_q \right\} \tag{13.18}$$

evaluated at $\mathbf{n} = \bar{\mathbf{n}}_{\text{stoc}}$. The stability of the fluctuations $\delta\mathbf{n}$ about the state $\bar{\mathbf{n}}_{\text{stoc}}$ is given by the eigenvalues λ_i of the matrix $\mathbb{M}(\bar{\mathbf{n}}_{\text{stoc}})$. Namely, similar to when in Sec. 8.5.3 we discussed synchronisation on networks, that is Eqs. (8.117) to (8.121), we imagine expanding $\delta\mathbf{n}$ on the eigenvectors $\mathbf{v}_i$ of $\mathbb{M}(\bar{\mathbf{n}}_{\text{stoc}})$. From Eq. (13.17) we will get

$$\delta\mathbf{n}(t) = \sum_i c_i(0)e^{\lambda_i t}\mathbf{v}_i. \tag{13.19}$$

Positive eigenvalues will force the component of $\delta\mathbf{n}$ along the corresponding eigenvectors to grow and therefore bring the system away from the metastable state $\bar{\mathbf{n}}_{\text{stoc}}$. One might therefore try to use the fluctuations of $\delta\mathbf{n}(t)$ as a warning signal of approaching instabilities. However, because of the stochastic nature of the basic dynamics, $\delta\mathbf{n}(t)$ fluctuates significantly even while the q-ESS state still possesses resilience. This is indicated in Fig. 13.3, where large excursions away from $\bar{\mathbf{n}}_{\text{stoc}}$ repeatedly occur followed by a reduction of $\delta\mathbf{n}(t)$ and a return to the same metastable configuration. One notices in particular the large peak in $|\delta\mathbf{n}(t)|$ to the left of the figure. For this reason it would be difficult to select a threshold τ (in the figure indicated by the horizontal dotted line) such that the event $|\delta\mathbf{n}(t)| > \tau$ could be used as an early warning.

Figure 13.3 Behaviour of fluctuations $\delta\mathbf{n}(t)$ and warning measure $Q(t)$ before a transition. The insert shows a blow-up just before the transition. The grey and black balls indicate data points at generations; each generation consists of $N(t)/p_{kill}$ time steps, see Sec. 11.3. For simulation details see [76]. © 2014 American Physical Society.

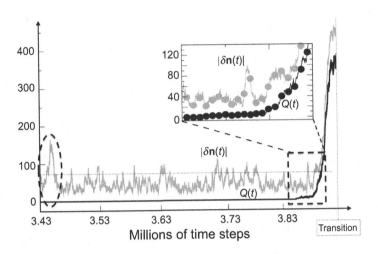

A much more reliable measure is given by the overlap between $\delta\mathbf{n}(t)$ and the normalised eigenvector $\mathbf{e}_{max}$ corresponding to the most positive eigenvalue λ_{max}. We define a warning signal $Q(t)$ from the growth predicted during the succeeding Δt time units by the expansion in Eq. (13.19) of the $\delta\mathbf{n}(t)$ component along the most unstable direction:

$$Q(t) = |e^{\lambda_{max}\Delta t}\delta\mathbf{n}(t) \cdot \mathbf{e}_{max}|. \qquad (13.20)$$

We see from Fig. 13.3 that $Q(t)$ remains very small, even when $|\delta\mathbf{n}(t)|$ fluctuates significantly away from the present q-ESS. The indicator $Q(t)$ only begins to increase when $\delta\mathbf{n}(t)$ due to mutations develops an overlap with the 'dangerous' direction $\mathbf{e}_{max}$. This means that it is possible to choose a threshold for $Q(t)$, which does not lead to false positives. The insert in Fig. 13.3 suggests that $Q(t)$ is able to predict a transition, which brings a q-ESS to an abrupt end, several generations before it suddenly occurs. This methodology is also able to forecast transitions in stochastic replicator, i.e. game-theoretic, systems [347].

For real systems it can be difficult to have enough knowledge about the controlling dynamics to be able to construct a mean-field description of the dynamics similar to Eq. (13.13). If sufficient data is available, it may be possible from time-series analysis to construct the stability matrix $\mathbb{M}$, the so-called Jacobian matrix [50]. One can then compute the largest eigenvalue and its eigenvector from this observationally determined stability matrix and combine it with observations of the state $\mathbf{n}(t)$ to determine $\bar{\mathbf{n}}_{stoc}$, which would in principle allow $Q(t)$ to be computed.

Summary: We distinguished between forecasting transitions caused by changing external conditions and transitions triggered by the intrinsic systems dynamics:

- For the first situation it has been suggested that an increase in fluctuations and a slowing down of the dynamics can be used as precursors.
- In the second situation it may be possible to construct an early warning signal from a tailored instantaneous linear stability analysis.

13.3 Further Reading

General audience:

The popular science book by Malcolm Gladwell, *The Tipping Point: How Little Things Can Make a Big Difference* [161] offers an entertaining introduction.

The book by Marten Scheffer, *Critical Transitions in Nature and Society* [382] requires no mathematical background and explains the approach discussed in Sec. 13.1.

Intermediate level:

A non-mathematical collection of scientific and scholarly papers is available in the open-access book *Addressing Tipping Points for a Precarious Future* edited by Timothy O'Riordan and Timothy Lenton [318]. Although no quantitative methodology for forecasting is suggested beyond the narrative of Sec. 13.1, the papers address a very broad range of aspects including science, culture, economics, spirituality and education.

Advanced level:

To develop an informed assessment of the approach sketched in Sec. 13.1, it is recommended to study the papers considered in Project 1 below.

13.4 Exercises and Projects

Exercise 1

Study numerically the fluctuations in the process in Eq. (13.1) with $\eta(t)$ given by Eq. (13.2). Try, from the changing time dependence of the fluctuations (the variance and the autocorrelations), to predict the transition taking place at time $t^* = 4000$ where $\eta(t)$ vanishes. You may find it difficult to obtain good forecasting efficiency.

Exercise 2

Simulate the process in Eqs. (13.3) and (13.4). Construct $Q(t)$ in Eq. (13.20) for the motion in the vicinity of the fixed point at $(x, y) = (\bar{X}, 0)$. Starting from different initial points on the x-axis near $x = \bar{X}$, study the fluctuations $|\delta(t)| = |(x(t) - \bar{X}, y(t) - 0)|$ away from the fixed point at $(\bar{X}, 0)$ and the temporal behaviour of $Q(t)$. Compare the behaviour of $|\delta(t)|$ and $Q(t)$ in terms of early warning signals.

You can, for example, consider the following parameters: $\eta = 10^{-1}$, $\alpha = \beta = \nu = 10^{-2}$, $\bar{X} = 1$, $p = 10^{-4}$.

Exercise 3

Simulate the process in Eqs. (13.3) and (13.4) in the vicinity of the point at $(\bar{X}, 0)$ and try to use as an early warning signal the logarithmic time dependence considered in Eq. (1) of [243].

Project 1 – Early warning: a literature study of current discussion

Being able to forecast approaching transitions or tipping points in complex systems is of great importance and very difficult. Since no well-established consensus concerning the most viable approach exists, this project will look at some of the discussions in the literature.

The suggestion that increase in fluctuations and slowing down [384] can be used as early warning signs has attracted significant interest, but the approach is not always successful.

(a) List the various early warning signals suggested in [243, 347, 384]. Comment on the difference between internally and externally triggered transitions and the role of stochasticity.

(b) In [73], Ornstein–Uhlenbeck theory is combined with the bifurcation approach of [384] and applied to model systems and empirical data. Give a summary of the discussion.

(c) Read [175] and explain the weaknesses the authors discuss in using increased fluctuations as precursors in finance. Relate your explanation to Fig. 13.3.

(d) It has been suggested that the forecasting approach presented in [384] can be improved by use of 'deep learning'. Outline the discussion in [74].

(e) Summarise the findings in [389], which study autocorrelations as early warning in psychopathology.

(f) Briefly explain the idea behind 'critical speeding up' as presented in [445]. Discuss how applicable this approach is in your opinion.

14   Concluding Comments and a Look to the Future

> **Synopsis:** We point out that complexity science is developing fast and that present and future scientific and societal challenges will need fundamental improvements in our ability to analyse and deal with complex emergent behaviour. It is therefore desirable to spread widely the awareness of approaches and insights from complexity science.

In this book we have considered complexity science to be the study of phenomena emerging at the aggregate level. Phenomena produced by the interactions amongst the components. Phenomena which are fundamentally different from any of the attributes of the individual components. We can think of complexity science as a study of the *collective* carried by the *in-between*. The specific properties of the individual components are secondary compared to the emphasis on the collective consequences of interconnectedness. The focus of complexity science is accordingly the study and classification of which kind of behaviour to expect from which kind of interaction amongst components. The internal workings of the components are left to the subject sciences. Complexity science looks for general 'laws' for aggregate emergent phenomena.

The progress of subject sciences, such as biology or sociology, involves an ever-increasing specialisation as deeper levels of refinement are uncovered. We see this e.g. in how a discipline like biology keeps dividing into subfields such as cell biology, molecular biology, evolutionary biology, etc.

This is in contrast to complexity science, which focuses on phenomena encountered across different fields and aims at improving our understanding by systematic studies of how the same kind of emergence may occur in situations belonging to very different fields of science. This makes complexity science inherently transdisciplinary, which raises educational and collaborative challenges both for the progress of complexity science itself and for its applications.

Firstly, the phenomenology that needs explanation by the complexity scientist comes from behaviour observed across the different subject sciences and hence in order to know which questions to address, a person working in complexity science needs to be able to absorb input from various subject sciences.

Secondly, in order for advances made in complexity science to become of use to people addressing systemic-level phenomena in specific fields of science, new insights of complexity science need to be communicated in a way that is comprehensible to experts in the various specific subject fields.

This duality between theoretical progress and application is of course not particular to complexity science, but there can be a tendency to see complexity science solely as the cross-disciplinary collaborative application of the expertise that e.g. mathematicians and physicists working together with biologists or neuroscientists can bring to bear on a specific problem. It is important to remember that foundational knowledge has to be continuously developed for the successful application and that the scrutiny of how the aggregate emerges from the individual is a scientific frontier of great current significance.

Although sometimes not fully appreciated, complexity science involves fundamental enquiries needed to establish the principles and concepts necessary for an adequate and timely worldview. Previously, fundamental science had a tendency to be seen as those subjects that dig deep and narrow, like high-energy physics, which has been considered to be the most fundamental of investigations. We need to realise that it is equally fundamental and important to understand the new realities that emerge when interdependent parts are put together. Understanding emergent phenomena at the aggregate level is not just fundamental in the sense of being intriguing and intellectually stimulating, the solution of some of the most pressing problems will need input from complexity science. We need to continue to improve our understanding of emergent aggregate behaviour and learn how to make use of these insights in our attempts to tackle urgent problems such as pollution, climate change, urbanisation, growing inequality, poverty, corruption and democratic deficit.

But no single individual is likely to be able to possess specialised insights from a broad range of subject fields and combine this with an expertise in complexity science. It therefore becomes increasingly necessary to develop a shared language that allows different experts to communicate.

It is the hope that working with the present book may help people from across the sciences to establish some shared conceptual stepping stones that can aid the flow of insights across traditional subject boundaries and thereby push complexity science forward and spread its applications.

Below is a list of some crucial open questions in complexity science.

- Quantifying and classifying emergence e.g. by use of information-theoretic approaches such as in Sec. 9.5.2.
- Developing our ability to forecast. We do not expect it to be possible to predict minute details such as we are used to when dealing with deterministic mechanics used e.g. in space travel. But we have good reason to imagine, see Chap. 13, that we may be able to understand better how to develop early warning signs of approaching systemic upheavals.
- Improving our understanding of why systems of very different types under their own dynamics settle in a regime in the vicinity of what is called a critical state, see Secs. 2.1 and 12.1.1.
- To what extent is it necessary to include higher-order interactions between components in order to understand the aggregate level? We routinely base our description of a system on adding together pairwise interactions, see e.g. Secs. 6.2 and 11.3, but we know that this is not always sufficient as discussed e.g. in [53].

- Improving our understanding of the applicability of complexity science: when is a phenomenon an example of general features of emergence and when is it a component-specific property?
- Improving our theoretical understanding of what we can call the taxonomic classification of emergent phenomena.

Let us finish our journey with the great vista perspective as it is formulated by John Donne (1572–1631), exceptional poet, scholar, theologian and more. That parts are intrinsically interconnected and that this affects our responsibility has long been realised. John Donne expressed this with exceptional clarity and beauty in his poem from 1623:

> No man is an island
> Entire of itself,
> Every man is a piece of the continent,
> A part of the main.
> If a clod be washed away by the sea,
> Europe is the less.
> As well as if a promontory were.
> As well as if a manor of thy friend's.
> Or of thine own were:
> Any man's death diminishes me,
> Because I am involved in mankind,
> And therefore never send to know for
> whom the bell tolls;
> It tolls for you.

Four hundred years after Donne's poetic narrative on the interconnectedness of the world, we can go beyond the narrative and the hunch. We start to be able to systematically study and classify emergent behaviour and we have begun to develop concepts and tools that allow us to scientifically approach, and in new ways understand, the profound world right in front of us.

It is now possible and important to take note of the lessons from complexity science and to apply its insights when dealing with ecology, neuroscience, economics, engineering, etc.

14.1 Further Reading

General audience:

As mentioned in the preface, the approach of this book has been to develop complexity science as the scientific study of emergent phenomena rather than starting from a definition of complex systems. Nevertheless, the latter approach is the endeavour of the book *What Is a Complex System?* [247] by James Ladyman and Karoline Wiesner.

A very inspiring collection of essays by people connected with the Santa Fe Institute during the period 1984–2019 can be found in the enjoyable collection *Worlds Hidden in Plain Sight* [242] edited by David Krakauer. No mathematics background is needed. The collection radiates enthusiasm and argues forcefully the need for the perspective of complexity science when dealing with some of the most urgent problems facing us.

Geoffrey West's book *Scale* [491] presents very engaging discussions of some of the most spectacular ambitions of complexity science.

Intermediate level:

Complexity science seen as the quantitative approach to the analysis of complex systems is presented with mathematical detail in the book *Introduction to the Theory of Complex Systems* [442] by Stefan Thurner, Rudolf Hanel and Peter Klimek.

In the exceptionally ambitious and profound *From Strange Simplicity to Complex Familiarity: A Treatise on Matter, Information, Life and Thought* [121], Manfred Eigen discusses how a better understanding of the physical aspects of information can help our understanding of the most complex of all phenomena: life and its origin.

Glossary

2d XY model is the technical name for a model originating in the theory of magnetism. It has become a prime example of how statistical mechanics is able mathematically to analyse emergent behaviour of relevance to large classes of very different systems.

Adjacency matrix is a matrix used to describe how nodes are linked together in a network.

Agent is in complexity science used to denote a component with some integrity, which possesses a range of possible actions. In the game-theoretic model called prisoner's dilemma, an agent can either choose to cooperate or defect. One may even think of the components of the Ising model as agents which can be in one of two states.

Agent-based model is a type of model in which the components, the agents, evolve according to a deterministic or stochastic set of rules prescribing how each agent reacts to its own state and the state of the surrounding agents and, potentially, other external factors.

Bak–Sneppen model is an elegant and very simple mathematical algorithm interpreted to model the dynamics of creation and extinction of interdependent species. The model operates at the level of species.

Binomial distribution is the probability distribution relating to the sum of binary random events. For example the distribution describes the probability of getting k heads (in any order) out of N throws.

Binomial process is a stochastic process with two outcomes. For example the repeated throwing of a coin, in which case the two possible outcomes for each throw are heads or tails. The number of heads in N throws follows the binomial distribution.

Branching process consists of a process of 'parents' producing 'offspring'. As a simple example, think of an organism that can produce zero, one, two, etc. offspring. We think of the offspring as branches from the parents. And each offspring will in turn be able to sprout new branches.

Breaking of symmetry refers to a symmetry, e.g. symmetry between up and down, being present at one level of description or in some circumstances, not being present e.g. at the systemic level or under other circumstances. Liquid water is symmetric under rotation, whereas solid ice flakes are symmetric only when rotated through $60°$.

Canonical ensemble refers in statistical mechanics to the abstract collection of all possible states of the components of a system in contact with a heat bath, i.e. at fixed temperature.

Cartesian combination of two sets, say $A = \{a_1, a_2\}$ and $B = \{b_1, b_2, b_3\}$, consists of combining all the elements of A and B to form the set $A \times B = \{(a_1, b_1), (a_1, b_2), (a_1, b_3), (a_2, b_1), (a_2, b_2), (a_2, b_3)\}$. The set $A \times B$ is called the Cartesian product of A and B.

Cellular automata are mathematical algorithms devised by Stanislaw Ulam and John von Neumann consisting of cells typically arranged on a grid. Each cell contains a dynamical variable. The time evolution of a given cell's variable is determined by the values of neighbour cells.

Centrality in the context of network theory is a measure of the importance of a node.

Central Limit Theorem is a mathematical statement from probability theory. The essence of the theorem is that when random numbers $X_1, X_2, \ldots, X_n$, all equally distributed with average μ and standard deviation σ, are added together:

$$S_n = X_1 + X_2 + \cdots + X_n,$$

then the adjusted sum

$$\frac{1}{\sigma\sqrt{n}}(S_n - n\mu)$$

will be normally distributed in the limit of large n.

Characteristic scale refers e.g. to a spatial distance or a temporal interval of particular relevance. As a means of illustration, think of the characteristic scale of commuting distance for people living in cities, which is typically of order some kilometres, whereas interplanetary space travel is measured in many tens of millions of kilometres. The characteristic lifetime of cells in the human body is roughly measured in tens of days, whereas the lifetime of the human body is measured in tens of years.

Coarse-grained level consists in lumping components together in groups in one way or another. We may e.g. describe population by use of various groupings such as employers and employees rather than in terms of each individual. Or we may describe a liquid in terms of the density of the fluid at a point in space and time rather than keeping track of each individual liquid molecule.

Collective degree of freedom refers to a single parameter used to capture the behaviour of many components. For example the centre of mass position of a solid body or the average height of an age group of young children.

Community is used in network science to denote a group of nodes, which according to some criterion form a whole.

Complexity science we take to be the systematic investigation of emergent phenomena. The aim is to characterise and understand commonalities shared by emergent phenomena in different systems and fields of research.

Configuration models is a procedure for the construction of a random network with a prescribed degree distribution.

Continuous-phase transitions: see first-order transition.

Correlation coefficient is a quantity in statistics, used as a measure of interrelatedness between two quantities. The correlation coefficient of two independent quantities, say the number of eyes of two different dice, is zero. For this reason the correlation coefficient is often taken to be indicative of possible relations between quantities. Caution is needed since the correlation coefficient can be zero between dependent quantities. See the discussion after Eq. (9.4).

Correlation length is defined as the characteristic length scale of decay of the correlation coefficient $C(r)$ between two quantities measured at different positions and is generically denoted by ξ and defined for exponential decay as $C(r) \propto \exp(-r/\xi)$.

Criticality in statistical mechanics refers to the state of a system where the correlation length is infinite. Such a state is characterised by fractal structures, power-law distributions and extreme sensibility to external forces, perturbations or stimuli.

Critical phenomena refers to the phenomenological behaviour, the critical state, at or near a critical point.

Critical point refers to the set of values of the control parameters, e.g. pressure and temperature, for which a system exhibits critical behaviour.

Critical state is a term sometimes used in a vague sense to indicate a situation which is special in some respect. In statistical mechanisms a system is said to be in the critical state if correlations between fluctuations at different positions, and/or different times, exhibit a slow algebraic decay with the separation in space and time. Spatial configurations of such states commonly exhibit fractal structure.

Critical temperature is the precise temperature at which a phase transition takes place. That is, the critical point along the temperature axis.

Curse of dimensionality refers to the fact that to determine a joint probability for N observables $X_1, \ldots, X_N$ the number of possible events increases very rapidly as $k_1 k_2 \cdots k_N$, where k_i denotes the possible number of values the observable X_i can assume. This will therefore require a very large amount of data to estimate histograms for the joint distribution of $X_1, X_2, \ldots, X_N$.

Degree can mean many things. In network theory the degree of a node is equal to the number of links attached to the node.

Diffusion process refers to a dynamical process involving stochastic motion. The motion of a grain of pollen suspended in water is a prototype example called Brownian motion after the Scottish botanist who described his observation of the motion of pollen. The huge numbers of water molecules collide with the pollen from all sides and with a large spread of velocities. The net effect of all these collisions is to knock the pollen around in random directions. We observe a diffusion process every time we watch a drop of cream slowly spread in a cup of coffee. In this case the water molecules knock the huge macromolecules of the cream about.

Dirac delta function is a so-called generalised function denoted by $\delta(x)$ introduced by the physicists Paul Dirac. One may intuitively, with some caution, think of the function as given by

$$\delta(x) = \begin{cases} 0 & \text{for } x \neq 0 \\ \infty & x = 0 \end{cases}.$$

Dynamical systems theory is a mathematical discipline that describes the time evolution of one or more particles. It makes use of differential equations and seeks to analyse their behaviour without being able to explicitly solve the dynamical equations.

Eigenvector is a term in linear algebra. If a vector v is an eigenvector for a matrix M, multiplying v by M will correspond to multiplying v by a number. This number is called the eigenvalue corresponding to the eigenvector and can be computed from the properties of the matrix M.

Emergence we will consider to denote any property or phenomenon at the level of the whole system which is produced by interactions between the constituent parts and which the parts do not possess individually.

Entropy is a term with widespread use in science. In thermodynamics, entropy is related to heat. In statistical mechanics and probability theory it relates to the probabilities of the various states a system can occupy. In the theory of coding, now called information theory, it is used to describe the amount of information carried by a string of message code. In all cases entropy is related to disorder and uncertainty. The less we know about the exact state of a system, or the exact detail of a piece of code, the higher is the entropy we will assign to it.

Equilibrium can have different meanings in different contexts. When dealing with time series, equilibrium is sometimes used to indicate stationarity, which means that the statistical properties are time independent. This meaning is relevant e.g. to punctuated equilibrium (see entry below) in which the equilibrium separating the punctuation events is considered to be essentially stationary. In statistical mechanics and thermodynamics, equilibrium refers to a precise experimental situation where the system under consideration can be considered to be at a fixed temperature.

Ergodicity refers to the ability of a system to reach every possible configuration. For example, if we throw two ordinary coins, we are able to observe all four configurations: head–head, head–tail, tail–head and tail–tail. Imagine the coins when brought together interact in a way which forces the second coin to show the same face as the first. We will then only be able to observe head–head or tail–tail and we say ergodicity has been broken.

Error threshold is a concept from evolutionary biology. Selection acts on the variety of different types produced by mutations. Those types which in a given situation are able to produce most offspring will become dominant types in the population. But if mutations occur too frequently, the offspring will always differ from the parents and the favourable trends of the parents will be lost in the population as fast as they appear. Hence no adaptation can occur and we say the mutation rate is above the error threshold in the sense that mutational errors happen too frequently.

Evolutionary adaptive dynamics consists prototypically of agents which produce offspring of some kind. The offspring may differ from the parent agent due to changes caused by mutations. Selection pressure due to external factors and/or interaction with co-evolving agents will give agents with a certain enhanced ability to reproduce in the given situation. The reproductive dynamics will therefore be able to produce a population with an increased number of agents with beneficial properties. In this kind of evolutionary adaptive dynamics the adaptation happens at the level of the population, while the properties of the individual do not change over its lifespan.

Exact differential in the mathematical theory of differentiation refers to a relationship between the total change of a function of several variables and the change given by the partial derivatives corresponding to the changes in each of the variables. If we know that a differential function $E(S, V)$ satisfies that the change dE due to a change dS and dV of the two variables is given by

$$dE = AdS + BdV,$$

then we can conclude that the partial derivatives must satisfy

$$\frac{\partial E}{\partial S} = A \text{ and } \frac{\partial E}{\partial V} = AB.$$

Excess degree is the degree of a node minus one.

Exponential growth refers to a time dependence of some quantity $f(t)$ given by an expression of the form $f(t) = A \exp(at)$, where A and a are constants. This can be written as $f(t) = A[\exp(a)]^t$. Since $\exp(a) < 1$ when $a < 0$ whereas $\exp(a) > 1$ when $a > 0$, and multiplying numbers smaller than one leads to rapid decrease whereas multiplying numbers larger than one leads to rapid increase, we say that $f(x)$ decreases exponentially fast for $a < 0$ and grows exponentially fast for $a > 0$.

Field-theoretic methods are mathematical formalisms which describe a system in terms of fields, i.e. functions of space and time such as the electric field. The methodology has been developed to become a general approach for the analysis of many types of systems including stochastic processes.

First-order phase transition is an equilibrium transition, such as from ice to water, in which the two phases (ice and water) can exist together. The correlation length of the system remains finite at the transition, in contrast to a second-order phase transition at which the correlation length diverges to infinity. The order parameter undergoes a discontinuous jump at the first-order transition point.

First passage times are the times it takes a random walker to move past a certain milestone for the first time.

First return times are the times it takes a random walk to return for the first time to a starting point.

Fitness is used in many different meanings. In biology, economics and sociology it is important to keep in mind that fitness should be seen in context and expresses a relationship between an organism, the company, or a person and the surroundings. Fitness should not be treated as an intrinsic property on a par with body weight, assets or age.

Fluctuation–dissipation relations are formulas that make a bridge between the internal fluctuations in the equilibrium system and how much change an external applied force may induce. The latter will lead to dissipation of energy.

fMRI scanner stands for functional magnetic resonance imaging. By studying the magnetic response inside the brain, the scanner is able to identify the local molecular composition. This is used to relate the local oxygen content to the neuronal activity or 'functioning'.

Fokker–Planck equations are partial differential equations for a probability distribution.

Fractals are heterogeneous geometric objects which possess repetitive structures in such a way that when a part of the fractal is magnified, the part will resemble the whole structure. Clouds, mountains and the crown of a large tree all exhibit some degree of fractal geometry. The romanesque cauliflower is a particularly elegant and regular fractal structure.

Gaia hypotheses consist of the suggestion that life develops in a symbiotic relationship with the non-biological surroundings in a way that sustains life.

Game theory consists in the mathematical investigation of agents interacting according to sets of predefined strategies. The relative benefit of one strategy against another is defined through a set of payoffs.

Generator function is a mathematical object used in many branches of mathematics. They are useful because they store mathematical information in a compact form. This information can then be extracted e.g. by evaluating the generator function at certain values of its argument or by differentiation.

Giant cluster is used in percolation theory to denote a cluster reaching across the system. The size of the cluster will grow as the size of the system is increased.

Giant component is a connected set of nodes in a network which contains a major part of all the nodes. In particular, the size of a giant cluster will increase if the size of the network is increased.

Graph Fourier transform is a generalisation of the Fourier expansion to functions defined on networks.

Half-link denotes half of the link between two nodes of a network. One half of the link is assigned to each of the two connected nodes.

Hurst exponent is a measure of how fluctuations in a time signal develop in time. An ordinary random walk has a Hurst exponent $H = 1/2$. Signals for which the direction of fluctuations tend to persist have $H > 1/2$ and signals in which fluctuations tend to swap direction have $H < 1/2$.

Ideal gas in physics is defined as a collection of non-interacting particles.

Information theory is a huge research field, which may be thought of as developing mathematical tools to help unravel what information a probability distribution may contain.

Intermittency denotes temporal evolution consisting of long quiet stretches separated by abrupt hectic activity.

Ising model is a very schematic and very successful model which originates in the theory of magnetic phase transitions and has become the prototype example of elegant succinct modelling.

Kinetic energy refers to the energy related to the motion of a body. The kinetic energy of a point particle of mass m moving with velocity v is, in Newtonian mechanics, given by $\frac{1}{2}mv^2$.

Langevin equation is a differential, or partial differential, equation which contains a stochastic noise term. For example

$$m\frac{dx^2}{dt^2} = F(x) + \chi(t)$$

where $x(t)$ may denote the position of a particle, $F(X)$ the force acting on the particle at position x and $\chi(t)$ a random force acting at time t.

Laplacian, or Laplace operator, is a term from the mathematics of calculus and denotes a sum of second derivatives of a function $f(x_1, x_2, \ldots, x_N)$ of several variables. The following is common notation. The Laplacian is denoted by Δ and the gradient by ∇, related as

$$\Delta f = \nabla \cdot \nabla f = \sum_{i=1}^{N} \frac{\partial^2 f}{\partial x_i^2}.$$

Logical gate is a physical realisation of a computational procedure or its mathematical description. In the information-theoretic context a logical gate is simply an example of a very simple function, which maps two inputs $x_1 \in \{0, 1\}$ and $x_2 \in \{0, 1\}$ to an output $x_3 = f(x_1, x_2) \in \{0, 1\}$.

Lotka–Volterra model describes the time variation of prey and predator by use of mean-field considerations. Assume the density of members of species prey is denoted by $x(t)$ and the density of the predator species is $y(t)$. The time variation is then assumed to be of the form

$$\frac{dx}{dt} = ax(t) - bx(t)y(t)$$

$$\frac{dy}{dt} = cx(t)y(t) - dy(t),$$

with a, b, c and d all being positive numbers. The term $ax(t)$ represents birth of a species x. The product term $bx(t)y(t)$ is taken to represent the likelihood of an encounter between prey and predator. The term is subtracted from the growth of x as the outcome of such an encounter may result in a member of the pre-species being eaten. This event, on the other hand, sustains the predator species represented by the term $cx(t)y(t)$. Finally, the predators die at a rate given by $dy(t)$.

Louvain algorithm is an algorithm for subdividing a network into clusters of well-connected nodes.

Magnetic moments can be thought of as a compass needle. The magnetic moment of a compass needle is itself the result of adding the effect of the magnetic moments of the molecules in the material.

Master equation is an equation for the time evolution of a probability distribution.

Maximum entropy principle is a way to determine probability distributions. The idea is that, under relevant constraints, the probability distribution to be used to describe a system is the one that maximises the entropy.

Mean field theory is a general approximative approach in which one is looking for the 'on-average' behaviour. This can be done in many ways, depending on the specific system. One can e.g. replace all the different interactions a component experiences from the rest of the system by the average effect of these interactions and then look at the evolution of statistical moments rather than looking at the time dependence of the entire probability distributions.

Microcanonical ensemble is a term in statistical mechanics, which assumes all the possible states of a closed system to occur with equal probability. Ensemble simply refers to the abstract set of the physical states of the components of the considered system. Canonical is probably used to reflect that the assumption should in general be accepted since it is a natural assumption. Why Gibbs, who introduced the term in his 1902 book on statistical mechanics [156], used the term 'micro' is not clear, but it distinguishes this ensemble from the canonical ensemble (see above).

Mixed embedding is a procedure for selecting the most important subset of past or future events from a time series in order to be able to handle the required histograms.

Mutual information is a measure of how similar the joint probability distribution $P_{X,Y}(x,y)$ of two quantities is to the product of the marginal, or individual, probability distributions $P_X(x)P_Y(y)$. The mutual information is zero only in the case $P_{X,Y}(x,y) = P_X(x)P_Y(y)$, otherwise positive.

Network considered mathematically consists of nodes and links connecting nodes. Network science and graph theory develop the mathematical description of such geometry structures and further investigate both theoretically and from data the relationship between the structure of networks, networks' dynamical evolution and dynamics taking place on network structures.

Normal distribution is an often encountered probability distribution shaped as a peak of width σ centred around the average value μ and given by the mathematical expression

$$P(x) = \frac{1}{\sqrt{2\pi\sigma^2}} \exp\left[-\frac{1}{2}\left(\frac{x-\mu}{\sigma}\right)^2\right].$$

Observables we use to denote quantities of a system that can be measured or simulated and thereby used to characterise the properties of a given system.

Order parameter is an observable quantity, which can be used to differentiate between a region of order and another region of disorder. The rigidity of a material can be used as an order parameter for the

transition between solid and melting. The crystalline ordered solid is rigid; as temperature is increased the crystalline order is lost and the rigid solid becomes a deformable liquid.

Ornstein–Uhlenbeck process is a stochastic random walk process in which the fluctuating signal experiences a pull towards its average value. This makes the process stationary in time, in contrast to ordinary diffusion.

Partial differential equation is an equation involving the partial derivatives of a function of several variables. For example, for a function of two variables $f(x,y)$ a partial differential equation will contain the partial derivatives

$$\frac{\partial f}{\partial y} \text{ and } \frac{\partial f}{\partial y}.$$

Partition function in statistical mechanics denotes the sum needed in order to ensure the probabilities add up to one. Remarkably, from this the average value of systems properties can be computed.

Percolation in mathematics and physical science refers to the process of establishing a path across a system by adding small snippets of paths together. One may think of a checkker board at which we place coins at random on the squares. We then ask which percentage of squares need to be occupied by coins in order for a path of coin-covered squares to form that stretches from one side of the board to the other. Careful discussions can be found in [84, 263, 424].

Phase can mean very different things. When dealing with periodic variation, the phase increases by a constant amount for every period. For example, think of the earth periodic circulation of the sun. For each full circumnavigation the corresponding phase can be taken to be time, in which case the phase increases by (approximately) 365 days, or we can take the variation of the angle of the line connecting the sun and the earth, which will increase by 360°. This meaning of phase should not be confused with the one used in connection with e.g. phase transition.

Phase transitions consist in the change of matter from one phase to another. For example, from liquid to solid or from magnetic to non-magnetic as a consequence of a change in temperature.

Poisson process is a fundamental type of stochastic process. As an example, think of an event that occurs with fixed probability during a time interval of a second. This could be that we have added 99 black balls and one red ball into a bag and every second we pull out one ball at random, look at its colour and return it to the bag. The process of observing the red ball is a Poisson process and the probability that the event happens during a draw is 1/100.

Punctuated equilibrium was coined by Niels Eldredge and Stephen Jay Gould as a term to describe the observation that the fossil record consists of bursts of extinction and creation events interspaced by periods of quiescence.

Quantum many-body theory is the branch of physics that investigates how interactions between quantum particles such as electrons and ions in a material bring about the properties of a material such as its electric resistance or thermal properties.

Random walk refers to a stochastic process in which the value of a quantity, $x(t)$, changes from one time step to the next by adding a random increment, i.e. $x(t + 1) = x(t) + \Delta(t)$, where $\Delta(t)$ is a number drawn at random.

Redundancy refers to some degree of duplication. Since the eyes of the opposite faces of a dice always add up to 7, I know by redundancy that when I throw a five, two eyes will be facing the table.

Renormalisation group is a mathematical technique which describes how the effective interaction between regions changes as one changes the length scale, or size of the regions. In the limit of large regions the effective interaction becomes independent of the size if we are dealing with a scale-invariant phenomenon.

Response function refers to a mathematical function, which describes the changes in a system as a response to some external change. An established formalism exists in statistical mechanics which can be used to study response functions of physic systems. We would also like to know the response function that describes the impact on the economy from a change e.g. in the interest rate.

Scale invariance means that the same structure and controlling mechanisms are independent of length or time scale. For example, think of cloud formation. Large clouds are like small clouds, except they are bigger. This leads to self-similar geometry.

Second-order or continuous phase transition is a change of order in a system at which the correlation length diverges to infinity and the order parameter changes continuously from zero in the disordered phase to non-zero in the ordered phase.

Self-similar means that the whole of an object looks like its parts. Many objects in nature are self-similar. For example, think of clouds or some plants, such as broccoli or cauliflower.

Social segregation consists in the separation e.g. geographically of a population into distinct groups according to colour, wealth, ethnicity or similar attribute.

Species turns out to be an example of an emergent entity which is difficult to define in a rigorous and precise way. Although we routinely use the term, a universally useful definition does not exist. The precise role of species in ecology and evolution is currently actively debated.

Spin glass is a term originating in the study of magnetic materials. It refers to inhomogeneous materials where the interaction between the microscopic components varies from one location to another. The dynamics of these magnetic materials have many similarities with ordinary glass, hence the name, and a new field of statistical mechanics was developed to describe spin glasses, which have had far-reaching influence in e.g. computer science by making fundamental contributions to our understanding of computational neuronal networks.

Spin quantum number refers to a quantum property of particles such as electrons or protons. It is called spin because it is possible, in a loose heuristic sense, to think of this property as similar to the momentum carried by the rotation of a spinning top. The quantum spin either assumes integer values 0, 1, 2, ... or so-called half integer values, 1/2, 3/2, 5/2, ... The remarkable finding is that statistical properties such as how many particles can be in the same physical state are determined by whether we are dealing with integer or half-integer spin particles. This leads to the emergent properties of the half and integer spin particles being dramatically different.

Spin waves are wave-like structures carried by the microscopic spins, or magnetic moments, of magnetic materials.

Stochastic variable is a sequence X_n which for each n assumes a value x_n with a certain probability $P_X(x) = \text{Prob}\{X_n = x, n\}$. If the probability $\text{Prob}\{X_n = x, n\}$ is independent of n, we are dealing with a stationary stochastic variable. For time series, the subscript n will be counting the time steps.

Superconductivity refers to the loss of electric resistance when materials are cooled to very low temperatures. It was discovered in 1911 by H. K. Onnes and it took until the 1950s to develop an understanding of super conductivity as a subtle emergent collective quantum-mechanical phenomenon.

Superfluidity is the property of certain liquids, most famously helium, to flow entirely without viscosity, i.e. without friction. It is a quantum phenomenon which happens at very low temperatures.

Symmetry breaking describes how a configuration changes from one symmetry to another. Water is isotropic, so looks the same in all directions. When water freezes and forms a snowflake, the rotational symmetry is broken down to a sixfold symmetry.

Synchronisation denotes the coordinated time evolution of many dynamical components. For example, strict synchronisation of pendulums corresponds to all the pendulums moving identically, whereas the synchronised action of musicians in an orchestra can consist of each musician playing different notes though accurately coordinated with all the other musicians.

Synergy simply means that the collective effect of a set of components is greater than a direct sum of the effect of the components individually. That is, something new is created by the collective.

Tangent map intermittency is a term from the theory of dynamical systems describing a certain mechanism for time evolution characterised by long periods of little change separated by short bursts of rapid change.

Tangled Nature model is a mathematical model of individuals undergoing co-evolutionary stochastic birth, mutation and death dynamics. The dynamics of individuals generates intermittent dynamics at the level of species.

Temporal graph signal transform is another term for graph Fourier transfer. For an explanation of this term, see the entry above.

Time to extinction for a birth–death process of non-overlapping generations, such as annual plants, is equal to the number of generations until no new descendants are produced.

Transfer entropy is an information-theoretic measure expressible in terms of Shannon entropy and conditioned Shannon entropy. It can be seen as quantifying and giving a direction to the interdependence between two variables represented by time series.

Vortices can refer to rotational motion in a liquid or to similar circular geometric structures in the 2d XY model of magnetism.

References

[1] Nobel Prize in Physics 1973. https://www.nobelprize.org/prizes/physics/1973/josephson/facts/.

[2] Nobel Prize in Physics 2010. https://www.nobelprize.org/prizes/physics/2010/press-release/.

[3] The billion sardine dance – BBC Earth. https://www.youtube.com/watch?v=Hg-NsZQFSAk, 2015.

[4] S. Abar, G. K. Theodoropoulos, P. Lemarinier and G. M. P. O'Hare. Agent based modelling and simulation tools: A review of the state-of-art software. *Computer Science Review*, 24:13–33, 2017.

[5] D. Abásolo, S. Simons, R. Morgado da Silva, G. Tononi and V. V. Vyazovskiy. Lempel–Ziv complexity of cortical activity during sleep and waking in rats. *Journal of Neurophysiology*, 113(7):2742–2752, 2015.

[6] F. Abdelnour, M. Dayan, O. Devinsky, T. Thesen and A. Raj. Functional brain connectivity is predictable from anatomic network's Laplacian eigenstructure. *NeuroImage*, 172(Feb):728–739, 2018.

[7] F. A. Razak and H. J. Jensen. Quantifying 'causality' in complex systems: Understanding transfer entropy. *PLoS ONE*, 9(6):e99462, 2014.

[8] D. M. Abrams, R. Mirollo, S. H. Strogatz and D. A. Wiley. Solvable model for chimera states of coupled oscillators. *Physical Review Letters*, 101(8):084103, 2008.

[9] D. M. Abrams and S. H. Strogatz. Chimera states for coupled oscillators. *Physical Review Letters*, 93:174102, 2004.

[10] S. Ackerman. *Discovering the Brain*. National Academic Press, 1992.

[11] A. Y. Aldhebiani. Species concept and speciation. *Saudi Journal of Biological Sciences*, 25(3):437–440, 2018.

[12] J. A. Almendral and A. Díaz-Guilera. Dynamical and spectral properties of complex networks. *New Journal of Physics*, 9:187, 2007.

[13] V. Ambegaokar, B. I. Halperin, D. R. Nelson and E. D. Siggia. Dissipation in two-dimensional superfluids. *Physical Review Letters*, 40:783–786, 1978.

[14] V. V. Ambegaokar, B. I. Halperin, D. R. Nelson and E. D. Siggia. Dynamics of superfluid films. *Physical Review B*, 21:1806–1826, 1980.

[15] J. M. Amigó, R. Dale and P. Tempesta. A generalized permutation entropy for random processes. *Chaos*, 31:013115, 2021.

[16] J. M. Amigó, R. Dale and P. Tempesta. Complexity-based permutation entropies: From deterministic time series to white noise. *Communications in Nonlinear Science and Numerical Simulation*, 105:106077, 2022.

[17] P. E. Anderson, H. J. Jensen, L. P. Oliveria and P. Sibani. Evolution in complex systems. *Complexity*, 10:49–56, 2004.

[18] P. W. Anderson. More is different. *Science*, 177(4047):393–396, 1972.

[19] P. W. Anderson. *More and Different: Notes from a Thoughtful Curmudgeon*. Word Scientific, 2011.

[20] P. Anderson and H. J. Jensen. Network properties, species abundance and evolution in a model of evolutionary

ecology. *Journal of Theoretical Biology*, 232:551–558, 2005.

[21] P. W. Anderson. *Basic Notions of Condensed Matter Physics*. Perseus, 1997.

[22] T. Andrillon, A. T. Poulsen, L. K. Hansen, D. Léger, and S. Kouider. Neural markers of responsiveness to the environment in human sleep. *Journal of Neuroscience*, 36(24):6583–6596, 2016.

[23] R. G. Andrzejak, C. Rummel, F. Mormann and K. Schindler. All together now: Analogies between chimera state collapses and epileptic seizures. *Scientific Reports*, 6:1–10, 2016.

[24] A. M. Apergis-Schoute, B. Bijleveld, C. M. Gillan, N. A. Fineberg, B. J. Sahakian and T. W. Robbins. Hyperconnectivity of the ventromedial prefrontal cortex in obsessive-compulsive disorder. *Brain and Neuroscience Advances*, 2:239821281880871, 2018.

[25] A. Arenas, A. Diaz-Guilera, J. Kurths, Y. Moreno and C. Zhou. Synchronization in complex networks. *Physics Report*, 469:93–153, 2008.

[26] A. Arenas, A. Díaz-Guilera and C. J. Pérez-Vicente. Synchronization reveals topological scales in complex networks. *Physical Review Letters*, 96(11):114102, 2006.

[27] R. Arthur and A. Nicholson. An entropic model of Gaia. *Journal of Theoretical Biology*, 430:177–184, 2017.

[28] R. Arthur and A. Nicholson. Selection principles for Gaia. *Journal of Theoretical Biology*, 533:110940, 2022.

[29] R. Arthur, A. Nicholson, P. Sibani and M. Christensen. The Tangled Nature Model for organizational ecology. *Computational and Mathematical Organization Theory*, 23(1):1–31, 2017.

[30] O. Artime, N. Khalil, R. Toral and M. San Miguel. First-passage distributions for the one-dimensional Fokker–Planck equation. *Physical Review E*, 98(4):042143, 2018.

[31] R. Axelrod. The dissemination of culture: A model with local convergence and global polarization. *The Journal of Conflict Resolution*, 41(2):203–226, 1997.

[32] L. Bachelier. Théorie de la spéculation. *Annales de l'École Normale Supérieure*, 3:21–86, 1900.

[33] G. Baglietto, E. V. Albano and J. Candia. Criticality and the onset of ordering in the standard Vicsek model. *Interface Focus*, 2(6):708–714, 2012.

[34] P. Bak. *How Nature Works: The Science of Self-organized Criticality*. Oxford University Press, 1997.

[35] P. Bak, K. Christensen, L. Danon and T. Scanlon. Unified scaling law for earthquakes. *Physical Review Letters*, 88(17):178501, 2002.

[36] P. Bak and K. Sneppen. Punctuated equilibrium and criticality in a simple model of evolution. *Physical Review Letters*, 71(24):4083–4086, 1993.

[37] P. Bak, C. Tang and K. Wiesenfeld. Self-organized criticality: An explanation of 1/f noise. *Physical Review Letters*, 59(4):381–384, 1987.

[38] A. Balanov, N. Janson, D. Postnov and O. Sosnovtseva. *Synchronization: From Simple to Complex*. Springer, 2008.

[39] D. Baldwin and J. Walker. Neurotic, stress-related and somatoform disorders. *Companion to Psychiatric Studies*, 453–491, 2010.

[40] P. Ball. *Critical Mass: How One Thing Leads to Another*. Farrar, Straus and Giroux, 2004.

[41] M. Ballerini, N. Cabibbo, R. Candelier, A. Cavagna, E. Cisbani, I. Giardina, V. Lecomte, A. Orlandi, G. Parisi, A. Procaccini, M. Viale and V. Zdravkovic. Interaction ruling animal collective behavior depends on topological rather than metric distance: Evidence from a field study. *PNAS*, 105(4):1232–1237, 2008.

[42] M. Ballerini, N. Cabibbo, R. Candelier, A. Cavagna, E. Cisbani, I. Giardina, A. Orlandi, G. Parisi, A. Procaccini, M. Viale and V. Zdravkovic. Empirical investigation of starling flocks: A benchmark study in collective animal behaviour. *Animal Behaviour*, 76:201–215, 2008.

[43] K. Bansal, J. O. Garcia, S. H. Tompson, T. Verstynen, J. M. Vettel and S. F. Muldoon. Cognitive chimera states in human brain networks. *Science Advances*, 5(4):eaau8535, 2019.

[44] A.-L. Barabási and H. E. Stanley. *Fractal Concepts in Surface Growth.* Cambridge University Press, 1995.

[45] A.-L. Barabàsi and R. Albert. Emergence of scaling in random networks. *Science*, 286(15):509–511, 1999.

[46] A.-L. Barabasi and J. Frangos. *Linked: How Everything is Connected to Everything Else and What it Means for Business, Science, and Everyday Life.* Basic Books, 2014.

[47] O. E. Barndorff-Nielsen and N. Shepard. Non-Gaussian Ornstein–Uhlenbeck-based models and some of their uses in financial economics. *Journal of the Royal Statistical Society, Series B: Statistical Methodology*, 63(2):167–241, 2001.

[48] L. Barnett, A. B. Barrett and A. K. Seth. Granger causality and transfer entropy are equivalent for Gaussian variables. *Physical Review Letters*, 103(23):238701, 2009.

[49] C. Barras. Mind maths: Your brain teeters on the edge of chaos. https://institutions.newscientist.com/article/mg21729032-100-mind-maths-your-brain-teeters-on-the-edge-of-chaos/, 2013.

[50] E. Barter, A. Brechtel, B. Drossel and T. Gross. A closed form for Jacobian reconstruction from time series and its application as an early warning signal in network dynamics. *Proceedings of the Royal Society A: Mathematical, Physical and Engineering Sciences*, 477:20200742, 2021.

[51] D. S. Bassett and J. Stiso. Spatial brain networks. *Comptes Rendus Physique*, 19(4):253–264, 2018.

[52] J. Feder. *Fractals.* Plenum, 1988.

[53] F. Battiston, E. Amico, A. Barrat, G. Bianconi, G. Ferraz de Arruda, B. Franceschiello, I. Iacopini, S. Kéfi, V. Latora, Y. Moreno, M. M. Murray, T. P. Peixoto, F. Vaccarino and G. Petri. The physics of higher-order interactions in complex systems. *Nature Physics*, 17(10):1093–1098, 2021.

[54] R. J. Baxter. *Exactly Solved Models in Statistical Mechanics.* Academic Press, 1982.

[55] N. Becker and P. Sibani. Evolution and non-equilibrium physics: A study of the Tangled Nature Model. *Epl*, 105(1), 2014.

[56] J. M. Beggs and D. Plenz. Neuronal avalanches in neocortical circuits. *Journal of Neuroscience*, 23(35):11167–1117, 2003.

[57] R. C. Bigalke. Observations on the behaviour and feeding habits of the springbok, *Antidorcas marsupialis. Zoologica Africana*, 7(1):333–359, 1972.

[58] N. Biggs. *Algebraic Graph Theory.* Cambridge University Press, 1974.

[59] J. J. Binney, N. J. Dowrick, A. J. Fisher and M. E. J. Newman. *The Theory of Critical Phenomena: An Introduction to the Renormalization Group.* Oxford Uinversity Press, 1993.

[60] T. Biyikoğlu, J. Leydold and P. F. Stadler. *Laplacian Eigenvectors of Graphs: Perron–Frobenius and Faber–Krahn Tye Theorems.* Springer, 2007.

[61] V. D. Blondel, J. L. Guillaume, R. Lambiotte and E. Lefebvre. Fast unfolding of communities in large networks. *Journal of Statistical Mechanics: Theory and Experiment*, 2008(10):P10008, 2008.

[62] S. Boccaletti, A. N. Pisarchik and C. I. del Genio. *Synchronization: From Coupled Systems to Complex Networks.* Cambridge University Press, 2018.

[63] B. Bollobás. *Modern Graph Theory.* Springer Science and Business Media, 1998.

[64] E. Bonabeau. Agent-based modeling: Methods and techniques for simulating human systems. *Proceedings of the National Academy of Sciences of the United States of America*, 99(Suppl. 3):7280–7287, 2002.

[65] E. Bonabeau and P. Lederea. On 1/f power spectra. *Journal of Physics A: Mathematical and Theoretical*, 27:L243–L250, 1994.

[66] P. Bonacich. Factoring and weighting approaches to clique identification. *Journal of Mathematical Sociology*, 113–120, 1972.

[67] P. Bonacich. Some unique properties of eigenvector centrality. *Social Networks*, 29(4):555–564, 2007.

[68] M. Bonaventura, V. Nicosia and V. Latora. Characteristic times of biased random walks on complex networks. *Physical Review E*, 89(1):012803, 2014.

[69] U. Brandes, D. Delling, M. Gaertler, R. Gorke, M. Hoefer, Z. Nikoloski and D. Wagner. On modularity clustering. *IEEE Transactions on Knowledge and Data Engineering*, 20(2):172–188, 2008.

[70] L. Brilouin. *Science and Information Theory.* Dover Publications, 2004.

[71] K. Brodmann. *Localisation in the Cerebral Cortex.* Springer, 2006.

[72] E. Bullmore and O. Sporns. Complex brain networks: Graph theoretical analysis of structural and functional systems. *Nature Reviews Neuroscience*, 10(3):186–198, 2009.

[73] T. M. Bury, C. T. Bauch and M. Anand. Detecting and distinguishing tipping points using spectral early warning signals. *Journal of the Royal Society Interface*, 17(170):20200482, 2020.

[74] T. M. Bury, R. I. Sujith, I. Pavithran, M. Scheffer, T. M. Lenton, M. Anand and C. T. Bauch. Deep learning for early warning signals of regime shifts. *Proceedings of the National Academy of Sciences*, 118(39):e2106140118, 2021.

[75] W. Cai, J. Snyder, A. Hastings and R. M. D'Souza. Mutualistic networks emerging from adaptive niche-based interactions. *Nature Communications*, 11(1):5470, 2020.

[76] A. Cairoli, D. Piovani and H. J. Jensen. Forecasting transitions in systems with high-dimensional stochastic complex dynamics: A linear stability analysis of the tangled nature model. *Physical Review Letters*, 113(26):264102, 2014.

[77] N. A. Campbell, L. A. Urry, M. L. Cain, S. A. Wasserman, P. A. Minorsky and J. B. Reece. *Biology: A Global Approach.* Pearson, 11th edition, 2017.

[78] V. Cataudella and P. Minnhagen. Simple estimate for vortex fluctuations in connection with high-T_c superconductors. *Physica C*, 166:442–450, 1990.

[79] A. Cavagna, A. Cimarelli, I. Giardina, G. Parisi, R. Santagati, F. Stefanini and M. Viale. Scale-free correlations in starling flocks. *Proceedings of the National Academy of Sciences of the United States of America*, 107(26):11865–11870, 2010.

[80] P. M. Chaikin and T. C. Lubensky. *Principles of Condensed Matter Physics.* Cambridge University Press, 1995.

[81] D. J. Chalmers. Strong and weak emergence. In P. Clayton and P. Davies (eds), *The Re-emergence of Emergence.* Oxford University Press, 2006.

[82] H. J. Charlesworth and M. S. Turner. Intrinsically motivated collective motion. *Proceedings of the National Academy of Sciences of the United States of America*, 116(31):15362–15367, 2019.

[83] D. Chorozoglou, D. Kugiumtzis and E. Papadimitriou. Testing the structure

of earthquake networks from multivariate time series of successive main shocks in Greece. *Physica A: Statistical Mechanics and its Applications*, 499:28–39, 2018.

[84] K. Christensen and N. R. Moloney. *Complexity and Criticality*. Imperial College Press, 2005.

[85] K. Christensen and Z. Olami. Sandpile models with and without an underlying spatial structure. *Physical Review E*, 48(5):3361–3372, 1993.

[86] A. Corral and F. Font-Clos. Criticality and self-organization in branching processes: Application to natural hazards. http://arxiv.org/abs/1207.2589, 2012.

[87] T. M. Cover and J. A. Thomas. *Elements of Information Theory*, 2nd edition. John Wiley & Sons, 2005.

[88] J. C. Cox, J. E. Ingersoll and S. A. Ross. A theory of the term structure of interest rates. *Theory of Valuation*, 53(2):129–151, 2005.

[89] A. Crooks, N. Malleson, E. Manley and A. Heppenstall. *Agent-Based Modelling and Geographical Information Systems: A Practical Primer (Spatial Analytics and GIS)*. SAGE Publications, 2018.

[90] T. J. Czaczkes, C. Grüter, L. Ellis, E. Wood and F. L. W. Ratnieks. Ant foraging on complex trails: Route learning and the role of trail pheromones in *Lasius niger*. *Journal of Experimental Biology*, 216(2):188–197, 2013.

[91] R. D'Andrea and A. Ostling. Challenges in linking trait patterns to niche differentiation. *Oikos*, 125:1369–1385, 2016.

[92] L. Danon, A. P. Ford, T. House, C. P. Jewell, M. J. Keeling, G. O. Roberts, J. V. Ross and M. C. Vernnon. Networks and the epidemiology of infectious disease. *Interdisciplinary Perspectives on Infectious Diseases*, 2011:284909, 2011.

[93] C. Darwin and A. Wallace. On the tendency of species to form varieties; and on the perpetuation of varieties and species by natural means of selection. *Zoological Journal of the Linnean Society*, 3(9):45–62, 1858.

[94] C. R. Darwin. *On the Origin of Species by Means of Natural Selection, or the Preservation of Favoured Races in the Struggle for Life*. John Murray, 1859.

[95] J. Davidsen, C. Gu and M. Baiesi. Generalized Omori–Utsu law for aftershock sequences in southern California. *Geophysical Journal International*, 201(2):965–978, 2015.

[96] P. Davies (ed.). *The New Physics*. Cambridge University Press, 1989.

[97] P. Davies. *The Demon in the Machine: How Hidden Webs of Information are Finally Solving the Mystery of Life*. Penguin, 2020.

[98] P. Davies and N. H. Gregersen (eds). *Information and the Nature of Reality: From Physics to Metaphysics*. Cambridge University Press, 2011.

[99] R. Dawkins. *The Selfish Gene*. Oxford University Press, 1976.

[100] L. de Arcangelis, C. Godano, J. R. Grasso and E. Lippiello. Statistical physics approach to earthquake occurrence and forecasting. *Physics Reports*, 628(March):1–91, 2016.

[101] L. de Arcangelis, C. Perrone-Capano and H. J. Herrmann. Self-organized criticality model for brain plasticity. *Physical Review Letters*, 96(2):028107, 2006.

[102] J. de Haan. How emergence arises. *Ecological Complexity*, 3(4):293–301, 2006.

[103] J. C. Delvenne, R. Lambiotte and L. E. C. Rocha. Diffusion on networked systems is a question of time or structure. *Nature Communications*, 6, 2015.

[104] B. Derrida and L. Peliti. Evolution in a flat fitness landscape. *Bulletin of*

Mathematical Biology, 53(3):355–382, 1991.

[105] S. A. di Collobiano, K. Christensen and H. J. Jensen. The Tangled Nature Model as an evolving quasi-species model. *Journal of Physics A: Mathematical and Theoretical*, 36:883–891, 2003.

[106] A. Diaz-Ruelas, H. J. Jensen, D. Piovani and A. Robledo. Relating high-dimensional stochastic complex systems to low-dimensional intermittency. *European Physical Journal: Special Topics*, 226(3):341–351, 2017.

[107] A. Diaz-Ruelas, H. J. Jensen, D. Piovani and A. Robledo. Tangent map intermittency as an approximate analysis of intermittency in a high dimensional fully stochastic dynamical system: The Tangled Nature model. *Chaos*, 26(12):123105, 2016.

[108] R. Dickman. Numerical analysis of the master equation. *Physical Review E*, 65(4):047701, 2002.

[109] D. Dolan, H. J. Jensen, P. A. M. Mediano, M. Molina-Solana, H. Rajpal, F. Rosas and J. A. Sloboda. The improvisational state of mind: A multidisciplinary study of an improvisatory approach to classical music repertoire performance. *Frontiers in Psychology*, 9(Sep):1341, 2018.

[110] J. Dong, B. Jing, X. Ma, H. Liu, X. Mo and H. Li. Hurst exponent analysis of resting-state fMRI signal complexity across the adult lifespan. *Frontiers in Neuroscience*, 12(Feb):34, 2018.

[111] P. Doreian, V. Batagelj and A. Ferligoj (eds). *Advances in Network Clustering and Blockmodeling*. Wiley-Blackwell, 2020.

[112] V. Dotsenko. *An Introduction to the Theory of Spin Glasses and Neural Networks*. World Scientific, 1995.

[113] A. B. Downey. *Think Complexity*. O'Reilly Media, 2012.

[114] B. Drossel and F. Schwabl. Self-organized critical forest-fire model. *Physical Review Letters*, 69(11):1629–1632, 1992.

[115] P. Duan, F. Yang, T. Chen and S. L. Shah. Direct causality detection via the transfer entropy approach. *IEEE Transactions on Control Systems Technology*, 21(6):2052–2066, 2013.

[116] J. A. Dunne, R. J. Williams and N. D. Martinez. Food-web structure and network theory: The role of connectance and size. *Proceedings of the National Academy of Sciences of the United States of America*, 99(20):12917–22, 2002.

[117] R. Durrett. *Branching Process Models of Cancer (Mathematical Biosciences Institute Lecture Series Book 1)*. Springer, 2015.

[118] R. Durrett, L. Mytnik and E. Perkins. Competing super-Brownian motions as limits of interacting particle systems. *Electronic Journal of Probability*, 10(2000065):1147–1220, 2005.

[119] R. Durrett. *Essentials of Stochastic Processes*. Springer, 2004.

[120] R. Durrett. *Random hraph Dynamics*. Cambridge University Press, 2007.

[121] M. Eigen. *From Strange Simplicity to Complex Familiarity: A Treatise on Matter, Information, Life and Thought*. Oxford Uinversity Press, 2013.

[122] M. Eigen, J. McCaskill and P. Schuster. Molecular quasi-species. *Journal of Physical Chemistry*, 92(24):6881–6891, 1988.

[123] A. Einstein. Theorie der Opaleszens von homogenen Flüssigkeiten und Flüssigkeitsgemischen in der Nähe des kritischen Zustandes. *Ann. Phys. (Leipzig)*, 33:1275–1298, 1910.

[124] A. Einstein. *Investigations on the Theory of the Brownian Movement*. Dover Publications, 1956.

[125] A. Einstein. *Autobiographical Notes*. Open Court Publishing Company, 1979.

[126] J. M. Eisaguirre, T. L. Booms, C. P. Barger, S. D. Goddard and G. A. Breed. Multistate Ornstein–Uhlenbeck approach for practical estimation of movement and resource selection around central places. *Methods in Ecology and Evolution*, 12(3); 507–519, 2021.

[127] S. Elaskar and E. del Río. *New Advances on Chaotic Intermittency and its Applications*. Springer, 2018.

[128] N. Eldredg and S. J. Gould *Punctuated Equilibria: An Alternative to Phyletic Gradualism*. Freeman, Cooper & Co., 1972.

[129] F. Engels. *Dialectics of Nature*. Progress Publishers, 1934.

[130] P. Erdös and A. Rényi. On random graphs. *Publicationes Mathematicae*, 6:290–297, 1959.

[131] M. Erkurt. Emergence of form in embryogenesis. *Journal of the Royal Society Interface*, 15(148):20180454, 2018.

[132] E. Estevez-Rams, R. Lora Serrano, B. Aragón Fernández and I. Brito Reyes. On the non-randomness of maximum Lempel Ziv complexity sequences of finite size. *Chaos*, 23(2), 2013.

[133] E. Estevez-Rams, R. Lora-Serrano, C. A. J. Nunes and B. Aragón-Fernández. Lempel–Ziv complexity analysis of one dimensional cellular automata. *Chaos*, 25(12):123106, 2015.

[134] P. Expert, S. De Nigris, T. Takaguchi, and R. Lambiotte. Graph spectral characterization of the XY model on complex networks. *Physical Review E*, 96(1):012312, 2017.

[135] P. Expert, T. S. Evans, V. D. Blondel and R. Lambiotte. Uncovering space-independent communities in spatial networks. *Proceedings of the National Academy of Sciences of the United States of America*, 108(19):7663–7668, 2011.

[136] P. Expert, R. Lambiotte, D. R. Chialvo, K. Christensen, H. J. Jensen, D. J. Sharp and F. Turkheimer. Self-similar correlation function in brain resting-state functional magnetic resonance imaging. *Journal of the Royal Society Interface*, 8(57):472–479, 2011.

[137] N. Farid and K. Christensen. Evolving networks through deletion and duplication. *New Journal of Physics*, 8:212, 2006.

[138] J. Feder. *Fractals*. Plenum Press, 1988.

[139] W. Feller. *An Introduction to Probability Theory and Its Applications, Vol. 1*. John Wiley & Sons, 1950.

[140] W. Feller. *An Introduction to Probability Theory and Its Applications, Vol. 1*, 3rd edition. John Wiley & Sons, 1968.

[141] J. Fernández-Gracia, K. Suchecki, J. J. Ramasco, M. San Miguel and V. M. Eguíluz. Is the voter model a model for voters? *Physical Review Letters*, 112(15):158701, 2014.

[142] W. M. Fitch and F. J. Ayala (eds). *Tempo and Mode in Evolution: Genetics and Paleontology 50 Years After Simpson*. National Academic Press, 1995.

[143] K. Ford and J. A. Wheeler. *Geons, Black Holes, and Quantum Foam: A Life in Physics*. W. W. Norton & Co., 2000.

[144] H. C. Fogedby, P. Hedegård and A. Svane. Low temperature spin wave and domain wall thermodynamics and form factors for the classical easy-plane ferromagnetic chain. *Physica*, 132B: 17–55, 1985.

[145] S. Fortunato and D. Hric. Community detection in networks: A user guide. *Physics Reports*, 659:1–44, 2016.

[146] R. E. Freeman, J. S. Harrison and S. Zyglidopoulos. *Stakeholder Theory*. Cambridge University Press, 2018.

[147] K. Friston. The free-energy principle: A unified brain theory? *Nature Reviews Neuroscience*, 11(2):127–138, 2010.

[148] P. A. Gago and S. Boettcher. Universal features of annealing and aging in

compaction of granular piles. *Proceedings of the National Academy of Sciences of the United States of America*, 117(52):33072–33076, 2020.

[149] M. Galesic and D. L. Stein. Statistical physics models of belief dynamics: Theory and empirical tests. *Physica A: Statistical Mechanics and its Applications*, 519:275–294, 2019.

[150] C. W. Gardiner. *Handbook of Stochastic Methods for Physics, Chemistry and the Natural Sciences*. Springer, 2004.

[151] H. Gatfaoui and P. de Peretti. Flickering in information spreading precedes critical transitions in financial markets. *Scientific Reports*, 9(1):106–112, 2019.

[152] D. D. Gatti, G. Fagiolo, M. Gallegati, M. Richiardi and A. Russo (eds). *Agent-Based Models in Economics: A Toolkit*. Cambridge University Press, 2018.

[153] M. Gell-Mann. *The Quark and the Jaguar: Adventures in the Simple and the Complex*. Abacus, 1994.

[154] P. Gerlee and T. Lundh. *Scientific Models: Red Atoms, White Lies and Black Boxes in a Yellow Book*. Springer, 2016.

[155] S. Gibb, R. F. Hendry and T. Lancaster (eds). *The Routledge Handbook of Emergence*. Routledge, 2019.

[156] J. W. Gibbs. *Elementary Principles in Statistical Mechanics*. The Perfect Library, 2015.

[157] E. N. Gilbert. Random graphs. *Annals of Mathematical Statistics.*, 30:1141–1144, 1959.

[158] F. Ginelli. The physics of the Vicsek model. *European Physical Journal: Special Topics*, 225(11–12):2099–2117, 2016.

[159] F. Ginelli and H. Chate. Relevance of metric-free interactions in flocking phenomena. *Physics Review Letters* 105(Oct):168103, 2010

[160] L. Giot, J. S. Bader, C. Brouwer, A. Chaudhuri, B. Kuang, Y. Li, Y. L. Hao, C. E. Ooi, B. Godwin, E. Vitols, G. Vijayadamodar, P. Pochart, H. Machineni, M. Welsh, Y. Kong, B. Zerhusen, R. Malcolm, Z. Varrone, A. Collis, M. Minto, S. Burgess, L. McDaniel, E. Stimpson, F. Spriggs, J. Williams, K. Neurath, N. Ioime, M. Agee, E. Voss, K. Furtak, R. Renzulli, N. Aanensen, S. Carrolla, E. Bickelhaupt, Y. Lazovatsky, A. Dasilva, J. Zhong, C. A. Stanyon, R. L. Finley Jr, K. P. White, M. Braverman, T. Jarvie, S. Gold, M. Leach, J. Knight, R. A. Shimkets, M. P. McKenna, J. Chant and J. M. Rothberg. A protein interaction map of *Drosophila melanogaster*. *Science*, 302(Dec):1727–1736, 2003.

[161] M. Gladwell. *The Tipping Point: How Little Things Can Make a Big Difference*. Abacus, 2002.

[162] M. Glazer and J. S. Wark. *Statistical Mechanics: A Survival Guide*. Oxford University Press, 2001.

[163] J. Gleick. *The Information: A History, a Theory, a Flood*. Fourth Estate, 2011.

[164] N. Goldenfeld. *Lectures on Phase Transitions and the Renormalization Group*. CRC Press/Taylor & Francis, 1992.

[165] R. Goldstein and M. S. Vitevitch. The influence of closeness centrality on lexical processing. *Frontiers in Psychology*, 8(Sep):1683, 2017.

[166] C. Goodnight. On multilevel selection and kin selection: Contextual analysis meets direct fitness. *Evolution*, 67(6):1539–1548, 2013.

[167] P. Gopikrishnan, V. Plerou, Y. Liu, L. A. N. Amaral, X. Gabaix and H. E. Stanley. Scaling and correlation in financial time series. *Physica A: Statistical Mechanics and its Applications*, 287(3–4):362–373, 2000.

[168] H. Goto, E. Viegas, H. J. Jensen, H. Takayasu, and M. Takayasu. Appearance of unstable monopoly state caused by selective and concentrative

mergers in business networks. *Scientific Reports*, 7(1):3–8, 2017.

[169] S. Gould. *The Structure of Evolutionary Theory*. The Belknap Press of Harvard University Press, 2002.

[170] C. J. W. Granger. Investigating causal relations by econometric models and cross-spectral methods. *Econometrica*, 37(3):424–438, 1969.

[171] P. Grassberger. Critical behaviour of the Drossel–Schwabl forest fire model. *New Journal of Physics*, 4:17.1–17.5, 2002.

[172] S. Grauwin, E. Bertin, R. Lemoy and P. Jensen. Competition between collective and individual dynamics. *Proceedings of the National Academy of Sciences of the United States of America*, 106(49):20622–20626, 2009.

[173] C. M. Grinstead and J. L. Snell. *Introduction to Pobability*. American Mathematical Society, 1997.

[174] S. Guan, R. Jiang, H. Bian, J. Yuan, P. Xu, C. Meng and B. Biswal. The profiles of non-stationarity and non-linearity in the time series of resting-state brain networks. *Frontiers in Neuroscience*, 14(Jun):493, 2020.

[175] V. Guttal, S. Raghavendra, N. Goel and Q. Hoarau. Lack of critical slowing down suggests that financial meltdowns are not critical transitions, yet rising variability could signal systemic risk. *PLoS ONE*, 11(1):e0144198, 2016.

[176] P. Haccou, P. Jagers and V. A. Vatutin. *Branching Processes: Variation, Growth, and Extinction of Populations*. Cambridge University Press, 2005.

[177] E. M. Hafner-Burton and A. H. Montgomery. Centrality in politics: How networks confer influence. *SSRN Electronic Journal*, 2012.

[178] A. Haimovici, E. Tagliazucchi, P. Balenzuela and D. R. Chialvo. Brain organization into resting state networks emerges at criticality on a model of the human connectome. *Physical Review Letters*, 110(17):178101, 2013.

[179] T. J. D. Halliday, P. Upchurch and A. Goswami. Eutherians experienced elevated evolutionary rates in the immediate aftermath of the Cretaceous–Palaeogene mass extinction. *Proceedings of the Royal Society B: Biological Sciences*, 283(1833):20153026, 2016.

[180] D. Hamon, M. Nicodemi and H. J. Jensen. Continuously driven OFC: A simple model of solar flare statistics. *Astronomy and Astrophysics*, 387:326–334, 2002.

[181] R. Hanel and S. Thurner. A comprehensive classification of complex statistical systems and an axiomatic derivation of their entropy and distribution functions. *EPL (Europhysics Letters)*, 93(2):20006, 2011.

[182] R. Hanel and S. Thurner. When do generalized entropies apply? How phase space volume determines entropy. *EPL*, 96:50003, 2011.

[183] T. E. Harris. *The Theory of Branching Processes*. Springer, 1963.

[184] G. S. Hartnett and M. Mohseni. Self-supervised learning of generative spin-glasses with normalizing flows. arXiv:2001.00585, 2020.

[185] T. Hastie and R. Tibshirani. *Elements of Statistical Learning*. Springer, 2009.

[186] R. A. Heath. The Ornstein–Uhlenbeck model for decision time in cognitive tasks: An example of control of nonlinear network dynamics. *Psychological Research*, 63(2):183–191, 2000.

[187] O. A. Heggli, J. Cabral, I. Konvalinka, P. Vuust and M. L. Kringelbach. A Kuramoto model of self–other integration across interpersonal synchronization strategies. *PLoS Computational Biology*, 15(10):e1007422, 2019.

[188] R. Hegselmann. Thomas C. Schelling and James M. Sakoda: The intellectual, technical, and social history of a model.

Journal of Artificial Societies and Social Simulation, 20(3):15, 2017.

[189] C. K. Hemelrijk and H. Hildenbrandt. Schools of fish and flocks of birds: Their shape and internal structure by self-organization. *Interface Focus*, 2(6):726–737, 2012.

[190] S. Hergarten. *Self-Organized Criticality in Earth Systems*. Springer, 2002.

[191] R. Hersh. *What is Mathematics, Really?* Oxford University Press, 1997.

[192] J. Hizanidis and N. Kouvaris. Metastable and chimera-like states in the *C. elegans* brain network. *Cybernetics and Physics*, 4(1):17–20, 2015.

[193] P. G. Hoel, S. C. Port and C. J. Stone. *Introduction to Probability Theory*. Houghton Mifflin Co., 1971.

[194] J. Hofbauer and K. Sigmond. *Evolutionary Games and Population Dynamics*. Cambridge University Press, 1998.

[195] J. H. Holland. *Emergence from Chaos to Order*. Helix Books, 1997.

[196] J. H. Holland. *Complexity: A Very Short Introduction*. Oxford Uinversity Press, 2014.

[197] P. Holme. Rare and everywhere: Perspectives on scale-free networks. *Nature Communications*, 10(1):8–10, 2019.

[198] S. Hossenfelder. The truth about scientific models. *Scientific American*, 18 June, 2020.

[199] B. Houchmandzadeh, E. Dumonteil, A. Mazzolo and A. Zoia. Neutron fluctuations: The importance of being delayed. *Physical Review E*, 92(5):052114, 2015.

[200] B. Houchmandzadeh and M. Vallade. Clustering in neutral ecology. *Physical Review E*, 68:061912, 2003.

[201] P. Hövel, A. Viol, P. Loske, L. Merfort and V. Vuksanović. Synchronization in functional networks of the human brain. *Journal of Nonlinear Science*, 30(5):2259–2282, 2020.

[202] J. Huang, T. Meng, Y. Deng and F. Huang. Catching critical transition in engineered systems. *Mathematical Problems in Engineering*, 2021:5589429, 2021.

[203] X. Huang, L. V. S. Lakshmanan and J. Xu. *Community Search over Big Graphs*. Morgan & Claypool, 2019.

[204] S. P. Hubbell. *The Unified Neutral Theory of Biodiversity and Biogeography*. Princeton University Press, 2001.

[205] B. D. Hughes. *Random Walks and Random Environments, Vol. 1 & 2*. Clarendon Press, 1995.

[206] G. Iacono, R. Massoni-Badosa and H. Heyn. Single-cell transcriptomics unveils gene regulatory network plasticity. *Genome Biology*, 20(1):110, 2019.

[207] S. Itani. A graph signal processing framework for the classification of temporal brain data. *Proceedings of the European Signal Process Conference 2020*, pp. 1180–1184, 2020.

[208] M. Jalili, A. Salehzadeh-Yazdi, S. Gupta, O. Wolkenhauer, M. Yaghmaie, O. Resendis-Antonio and K. Alimoghaddam. Evolution of centrality measurements for the detection of essential proteins in biological networks. *Frontiers in Physiology*, 7(Aug): 375, 2016.

[209] G. J. O. Jameson. *A First Course on Complex Functions*. Chapman & Hall, 1970.

[210] M. A. Javed, M. S. Younis, S. Latif, J. Qadir and A. Baig. Community detection in networks: A multidisciplinary review. *Journal of Network and Computer Applications*, 108(Sep 2017):87–111, 2018.

[211] E. T. Jaynes. Information theory and statistical Mechanics II. *Physical Review*, 108(2):171–190, 1957.

[212] E. T. Jaynes. *Probability Theory: The Logic of Science*. Cambridge University Press, 2003.

[213] E. T. Jaynes. Information theory and Statistical mechanics. *Physical Review*, 106:620–630, 1957.

[214] H. J. Jensen. *Self-Organized Criticality: Emergent Complex Behavior in Physical and Biological Systems*. Cambridge University Press, 1998.

[215] H. J. Jensen. Emergence of network structure in models of collective evolution and evolutionary dynamics. *Proceedings of the Royal Society A: Mathematical, Physical and Engineering Sciences*, 464(2096):2207–2217, 2008.

[216] H. J. Jensen. Tangled Nature: A model of emergent structure and temporal mode among co-evolving agents. *European Journal of Physics*, 40(1):014005, 2018.

[217] H. J. Jensen, R. H. Pazuki, G. Pruessner and P. Tempesta. Statistical mechanics of exploding phase spaces: Ontic open systems. *Journal of Physics A: Mathematical and Theoretical*, 51:375002, 2018.

[218] H. J. Jensen, R. H. Pazuki, G. Pruessner and P. Tempesta. Statistical mechanics of exploding phase spaces: Ontic open systems. *Journal of Physics A: Mathematical and Theoretical*, 51(37):375002, 2018.

[219] H. J. Jensen and P. Tempesta. Group entropies: From phase space geometry to entropy functionals via group theory. *Entropy*, 20:804, 2018.

[220] H. J. Jensen and P. Minnhagen. Two-dimensional vortex fluctuations in the non-linear current–vltage characteristics for high-temperature superconductors. *Physical Review Letters*, 66:1630–1633, 1991.

[221] H. J. Jensen and H. Weber. Phenomenological study of vortices in a two-dimensional XY model in a magnetic field. *Physics Reviews B*, 45:10468, 1992.

[222] P. Jiruska, M. de Curtis, J. G. R. Jefferys, C. A. Schevon, S. J. Schiff and K. Schindler. Synchronization and desynchronization in epilepsy: Controversies and hypotheses. *Journal of Physiology*, 591(4):787–797, 2013.

[223] P. A. Johnson, B. Rouet-Leduc, L. J. Pyrak-Nolte, G. C. Beroza, C. J. Marone, C. Hulbert, A. Howard, P. Singer, D. Gordeev, D. Karaflos, C. J. Levinson, P. Pfeiffer, K. M. Puk and W. Reade. Laboratory earthquake forecasting: A machine learning competition. *Proceedings of the National Academy of Sciences of the United States of America*, 118(5):e2011362118, 2021.

[224] S. Johnson. *Emergence: The Connected Lives of Ants, Brains, Cities and Software*. Penguin Books, 2001.

[225] R. H. Jones. *Reductionism: Analysis and the Fullness of Reality*. Bucknell University Press, 2000.

[226] F. Jordan. Searching for keystones in ecological networks. *OIKOS*, 99(3):607–612, 2002.

[227] L. Kang, C. Tian, S. Huo and Z. Liu. A two-layered brain network model and its chimera state. *Scientific Reports*, 9(1):14389, 2019.

[228] H. Kantz and T. Schreiber. *Nonlinear Time Series Ananlysis*. Cambridge University Press, 2000.

[229] P. Karampelas, J. Kawash and T. Özyer (eds). *From Security to Community Detection in Social Networking Platforms*. Springer, 2019.

[230] K. Kawamura, F. Parrenin, L. Lisiecki, R. Uemura, F. Vimeux, J. P. Severinghaus, M. A. Hutterli, T. Nakazawa, S. Aoki, J. Jouzel, M. E. Raymo, K. Matsumoto, H. Nakata, H. Motoyama, S. Fujita, K. Goto-Azuma, Y. Fujii and O. Watanabe. Northern Hemisphere forcing of climatic cycles in Antarctica over the past 360,000 years. *Nature*, 448(7156):912–916, 2007.

[231] M. J. Keeling and K. T. D. Eames. Networks and epidemic models.

Journal of the Royal Society Interface, 2(4):295–307, 2005.

[232] T. H. Keitt. Coherent ecological dynamics induced by large-scale disturbance. *Nature,* 454(7202):331–334, 2008.

[233] A. I. Khinchin. *Mathematical Foundation of Information Theory.* Dover Publications, 1957.

[234] E. Y. Kim, D. Ashlock and S. H. Yoon. Identification of critical connectors in the directed reaction-centric graphs of microbial metabolic networks. *BMC Bioinformatics,* 20(1):328, 2019.

[235] M. Kimmel and D. E. Axelrod. *Branching Processes in Biology: 19 (Interdisciplinary Applied Mathematics 19).* Springer, 2015.

[236] A. Kirkley, H. Barbosa, M. Barthelemy and G. Ghoshal. From the betweenness centrality in street networks to structural invariants in random planar graphs. *Nature Communications,* 9(1):2501, 2018.

[237] L.-W. Kong, H. Fan, C. Grebogi and Y.-C. Lai. Emergence of transient chaos and intermittency in machine learning. *Journal of Physics: Complexity,* 2:035014, 2021.

[238] K. Kosmidis and M. T. Hütt. The *E. coli* transcriptional regulatory network and its spatial embedding. *European Physical Journal E,* 42(3):30, 2019.

[239] J. M. Kosterlitz. The critical properties of the two-dimensional XY model. *Journal of Physics C: Solid State Physics,* 7:1046–1060, 1974.

[240] J. M. Kosterlitz and D. J. Thouless. Ordering, metastability and phase transitions in two-dimensional systems. *Journal of Physics C: Solid State Physics,* 6(7):1181–1203, 1973.

[241] C. Koutlis, V. K. Kimiskidis and D. Kugiumtzis. Identification of hidden sources by estimating instantaneous causality in high-dimensional biomedical time series. *International*

Journal of Neural Systems, 29(4):1850051, 2019.

[242] D. C. Krakauer (ed.). *Worlds Hidden in Plain Sight.* The Santa Fe Institute Press, 2019.

[243] C. Kuehn, G. Zschaler and T. Gross. Early warning signs for saddle-escape transitions in complex networks. *Scientific Reports,* 5:13190, 2015.

[244] D. Kugiumtzis. Direct-coupling information measure from nonuniform embedding. *Physical Review E,* 87(6):062918, 2013.

[245] Y. Kuramoto. Self-entrainment of a population of coupled non-linear oscillators. In H. Araki (ed.), *International Symposium on Mathematical Problems in Theoretical Physics. Lecture Notes in Physics, Vol. 39,* p. 420. Springer, 1975.

[246] Y. Kuramoto and D. Battogtokh. Coexistence of coherence and incoherence in nonlocally coupled phase oscillators. arXiv:cond-mat/0210694v1, 2002.

[247] J. Ladyman and K. Wiesner. *What is a Complex System?* Yale University Press, 2020.

[248] S. Laird and H. J. Jensen. The Tangled Nature model with inheritance and constraint: Evolutionary ecology restricted by a conserved resource. *Ecological Complexity,* 3:253–262, 2006.

[249] S. Laird and H. J. Jensen. A non-growth network model with exponential and 1/k scale-free degree distributions. *Europhysics Letters,* 76(4):710–716, 2006.

[250] S. Laird and H. J. Jensen. Correlation, selection and the evolution of species networks. *Ecological Modelling,* 209(2–4):149–156, 2007.

[251] M. Lallouache, A. S. Chakrabarti, A. Chakraborti and B. K. Chakrabarti. Opinion formation in kinetic exchange models: Spontaneous symmetry-

breaking transition. *Physical Review E*, 82(5):056112, 2010.

[252] M. J. Lamb and E. Jablonka. *Evolution in Four Dimensons*. MIT Press, 2005.

[253] L. D. Landau and E. M. Lifshitz. *Statistical Physics, Vol. 5*. Elsevier, 1980.

[254] D. Lawson, H. J. Jensen and K. Kaneko. Diversity as a product of inter-specific interactions. *Journal of Theoretical Biology*, 243(3):299–307, 2006.

[255] D. J. Lawson and H. J. Jensen. The species area relationship and evolution. *Journal of Theoretical Biology*, 241:590–600, 2006.

[256] D. J. Lawson and H. J. Jensen. Neutral evolution in a biological population as diffusion in phenotype space: Reproduction with local mutation but without selection. *Physical Review Letters*, 98(9):2–5, 2007.

[257] D. J. Lawson and H. J. Jensen. Understanding clustering in type space using field theoretic techniques. *Bulletin of Mathematical Biology*, 70(4):1065–1081, 2008.

[258] K. Lehnertz, S. Bialonski, M. T. Horstmann, D. Krug, A. Rothkegel, M. Staniek and T. Wagner. Synchronization phenomena in human epileptic brain networks. *Journal of Neuroscience Methods*, 183(1):42–48, 2009.

[259] T. M. Lenton. Gaia and natural selection. *Nature*, 439:439–447, 1998.

[260] J. M. Levine, J. Bascompte, P. B. Adler and S. Allesina. Beyond pairwise mechanisms of species coexistence in complex communities. *Nature*, 546(7656):56–64, 2017.

[261] J. M. Lewis and M. S. Turner. Density distributions and depth in flocks. *Journal of Physics: Applied Physics*, 50:494003, 2017.

[262] M. Li, Y. Han, M. J. Aburn, M. Breakspear, R. A. Poldrack, J. M. Shine and J. T. Lizier. Transitions in information processing dynamics at the whole-brain network level are driven by alterations in neural gain. *PLoS Computational Biology*, 15(10):e1006957, 2019.

[263] M. Li, R. R. Liu, L. Lü, M. B. Hu, S. Xu and Y. C. Zhang. Percolation on complex networks: Theory and application. *Physics Reports*, 907:1–68, 2021.

[264] E. H. Lieb and J. Yngvason. The mathematical structure of the second law of thermodynamics. In A. J. de Jong, D. Jerison, G. Lustig, B. Mazur, W. Schmid and S.-T. Yau (eds), *Current Developments in Mathematics 2001*, pp. 89–129. International Press, 2002.

[265] H. Ling, G. E. McIvor, J. Westley, K. van der Vaart, R. T. Vaughan, A. Thornton and N. T. Quellette. Behavioural plasticity and the transition to order in jackdaw flocks. *Nature Communications*, 10:5174, 2019.

[266] E. Lippiello, L. De Arcangelis and C. Godano. Influence of time and space correlations on earthquake magnitude. *Physical Review Letters*, 100(3):038501, 2008.

[267] F. Lombardi, H. J. Herrmann and L. De Arcangelis. Balance of excitation and inhibition determines 1/f power spectrum in neuronal networks. *Chaos*, 27:047402, 2017.

[268] F. Lombardi, H. J. Herrmann, C. Perrone-Capano, D. Plenz and L. De Arcangelis. Balance between excitation and inhibition controls the temporal organization of neuronal avalanches. *Physical Review Letters*, 108(22):228703, 2012.

[269] L. D. Lord, P. Allen, P. Expert, O. Howes, M. Broome, R. Lambiotte, P. Fusar-Poli, I. Valli, P. McGuire and Federico E. Turkheimer. Functional brain networks before the onset of psychosis: A prospective fMRI study with graph theoretical analysis. *NeuroImage: Clinical*, 1(1):91–98, 2012.

[270] J. Lovelock. *The Ages of Gaia, New Edition*. Oxford University Press, 1988.

[271] V. Lowe. *Understanding Whitehead*. The Johns Hopkins Press, 1962.

[272] Z. Lu, J. Wahlström and A. Nehorai. Community detection in complex networks via clique conductance. *Scientific Reports*, 8(1):5982, 2018.

[273] D. M. N. Maia, J. E. M. de Oliveira, M. G. Quiles and E. E. N. Macau. Community detection in complex networks via adapted Kuramoto dynamics. *Communications in Nonlinear Science and Numerical Simulation*, 53:130–141, 2017.

[274] S. Majhi, B. K. Bera, D. Ghosh and M. Perc. Chimera states in neuronal networks: A review. *Physics of Life Reviews*, 28:100–121, 2019.

[275] J. Mallet. A species definition for the modern synthesis. *Trends in Ecology & Evolution*, 10(7):294–299, 1995.

[276] S. S. Manna. Two-state model of self-organized criticality. *Journal of Physics A: Mathematical and Theoretical*, 24(7):L363, 1991.

[277] A. S. R. Manstead. The psychology of social class: How socioeconomic status impacts thought, feelings, and behaviour. *British Journal of Social Psychology*, 57(2):267–291, 2018.

[278] R. Maor, T. Dayan, H. Ferguson-Gow and K. E. Jones. Temporal niche expansion in mammals from a nocturnal ancestor after dinosaur extinction. *Nature Ecology and Evolution*, 1(12):1889–1895, 2017.

[279] L. Margulis and D. Sagan. *What is Life?* University of California Press, 1995.

[280] M. Maschler, E. Solan and S. Zamir. *Game Theory*. Cambridge University Press, 2020.

[281] P. Massobrio, L. De Arcangelis, V. Pasquale, H. J. Jensen and D. Plenz. Criticality as a signature of healthy neural systems: multi-scale experimental and computational studies. *Frontiers in Systems Neuroscience*, April, 2015.

[282] N. Masuda, M. A. Porter and R. Lambiotte. Random walks and diffusion on networks. *Physics Reports*, 716-717:1–58, 2017.

[283] J. Maynard Smith. *The Theory of Evolution*. Cambridge University Press, 1995.

[284] J. Maynard Smith and E. Szathmáry. *The Origins of Life: From the Birth of Life to the Origins of Language*. Oxford University Press, 1999.

[285] A. J. McKane, D. Alonso and R. V. Solé. Analytic solution of Hubbell's model of local community dynamics. *Theoretical Population Biology*, 65(1):67–73, 2004.

[286] D. A. McQuarrie. *Mathematical Methods for Scientists and Engineers*. University Science Books, 2003.

[287] E. S. Meckes and M. W. Meckes. *Linear Algebra*. Cambridge University Press, 2018.

[288] P. A. M. Mediano, F. E. Rosas, C. Timmermann, L. Roseman, D. J. Nutt, A. Feilding, M. Kaelen, M. L. Kringelbach, A. B. Barrett, A. K. Seth, S. Muthukumaraswamy, D. Bor and R. L. Carhart-Harris. Effects of external stimulation on psychedelic state neurodynamics. *bioRxiv*, p. 2020.11.01.356071, 2020.

[289] P. A. M. Mediano, A. K. Seth and A. B. Barrett. Measuring integrated information: Comparison of candidate measures in theory and simulation. *Entropy*, 21(1):21, 17, 2019.

[290] F. Menczer, S. Fortunato and C. A. Davis. *A First Course in Network Science*. Cambridge University Press, 2020.

[291] M. Mendoza, A. Kaydul, L. De Arcangelis, J. S. Andrade, and H. J. Herrmann. Modelling the influence of photospheric turbulence on solar flare statistics. *Nature Communications*, 5:6–9, 2014.

[292] M. Meyer, S. Havlin and A. Bunde. Clustering of independently diffusing individuals by birth and earth processes. *Physical Review E*, 54(5):5567–5570, 1996.

[293] A. M. Michalak, E. J. Anderson, D. Beletsky, S. Boland, N. S. Bosch, T. B. Bridgeman, J. D. Chaffin, K. Cho, R. Confesor, I. Daloglu, J. V. Depinto, M. A. Evans, G. L. Fahnenstiel, L. He, J. C. Ho, L. Jenkins, T. H. Johengen, K. C. Kuo, E. Laporte, X. Liu, M. R. McWilliams, M. R. Moore, D. J. Posselt, R. P. Richards, D. Scavia, A. L. Steiner, E. Verhamme, D. M. Wright and M. A. Zagorski. Record-setting algal bloom in Lake Erie caused by agricultural and meteorological trends consistent with expected future conditions. *Proceedings of the National Academy of Sciences of the United States of America*, 110(16):6448–52, 2013.

[294] L. Michiels Van Kessenich, D. Berger, L. De Arcangelis and H. J. Herrmann. Pattern recognition with neuronal avalanche dynamics. *Physical Review E*, 99(1):010302, 2019.

[295] L. Michiels Van Kessenich, L. De Arcangelis and H. J. Herrmann. Synaptic plasticity and neuronal refractory time cause scaling behaviour of neuronal avalanches. *Scientific Reports*, 6(Aug):32071, 2016.

[296] P. Minnhagen. The two-dimensional Coulomb gas, vortex unbinding, and superfluid-superconducting films. *Review of Modern Physics*, 59:1001–1066, 1987.

[297] P. Minnhagen and P. Olsson. Monte Carlo calculation of the vortex interaction for high-Tc superconductors. *Physical Review B*, 44(9):4503–4511, 1991.

[298] M. Mitchell. *Complexity: A Guided Tour*. Oxford Uinversity Press, 2009.

[299] S. K. Mitra. Is Hurst exponent value useful in forecasting financial time series? *Asian Social Science*, 8(8):111–120, 2012.

[300] A. B. Monteiro and L. Del Bianco Faria. Causal relationships between population stability and food-web topology. *Functional Ecology*, 31(6):1294–1300, 2017.

[301] P. A. P. Moran. *The Statistical Processes of Evolutionary Theory*. Clarendon Press, 1962.

[302] E. Morin. *On Complexity*. Hampton Press, 2008.

[303] P. Mörters. The average density of super-Brownian motion. *Annales de l'institut Henri Poincare (B) Probability and Statistics*, 37(1):71–100, 2001.

[304] F. Moss and S. Gielen (eds). *Neuro-Informatics and Neural Modelling*. Elsevier, 2001.

[305] C. C. Mounfield. *The Handbook of Agent Based Modelling*. Independently published, 2020.

[306] Y. Murase, T. Shimada, N. Ito and P. A. Rikvold. Random walk in genome space: A key ingredient of intermittent dynamics of community assembly on evolutionary time scales. *Journal of Theoretical Biology*, 264(3):663–672, 2010.

[307] R. Natrajan, H. Sailem, F. K. Mardakheh, M. A. Garcia, C. J. Tape, M. Dowsett, C. Bakal and Y. Yuan. Microenvironmental heterogeneity parallels breast cancer progression: A histology–genomic integration analysis. *PLOS Medicine*, 13(2):e1001961, 2016.

[308] Z. Néda, E. Ravasz, T. Vicsek, Y. Brechet and A. L. Barabási. Physics of the rhythmic applause. *Physical Review E*, 61(6 B):6987–6992, 2000.

[309] V. B. Nevzorov. *Records: Mathematical Theory*. American Mathematical Society, 2001.

[310] M. E. J. Newman and G. J. Eble. Decline in extinction rates and scale invariance in the fossil record. *Paleobiology*, 25(4):434–439, 1999.

[311] M. E. J. Newman and P. Sibani. Extinction, diversity and survivorship of taxa in the fossil record. *Proceedings of the Royal Society London Series B*, 266:1593–1599, 1999.

[312] M. E. J. Newman. *Networks: An Introduction*, 2nd edition. Oxford University Press, 2018.

[313] A. E. Nicholson and P. Sibani. Cultural evolution as a nonstationary stochastic process. *Complexity*, 21(6):214, 2015.

[314] V. Nicosia, M. Valencia, M. Chavez, A. Díaz-Guilera and V. Latora. Remote synchronization reveals network symmetries and functional modules. *Physical Review Letters*, 110(17):174102, 2013.

[315] E. Novikov, A. Novikov, D. Shannahoff-Khalsa, B. Schwartz, and J. Wright. Scale-similar activity in the brain. *Physical Review E*, 56(3):R2387–R2389, 1997.

[316] A. Nowak, R. R. Vallacher, M. Zochowski and A. Rychwalska. Functional synchronization: The emergence of coordinated activity in human systems. *Frontiers in Psychology*, 8(Jun):945, 2017.

[317] T. Nozawa, M. Kondo, R. Yamamoto, H. Jeong, S. Ikeda, K. Sakaki, Y. Miyake, Y. Ishikawa and R. Kawashima. Prefrontal inter-brain synchronization reflects convergence and divergence of flow dynamics in collaborative learning: A pilot study. *Frontiers in Neuroergonomics*, 2(Jun):686596, 2021.

[318] T. O'Riordan and T. Lenton (eds). *Addressing Tipping Points for a Precarious Future*. OUP/British Academy, 2013.

[319] M. Oizumi, L. Albantakis and G. Tononi. From the phenomenology to the mechanisms of consciousness: Integrated information theory 3.0. *PLoS Computational Biology*, 10(5):e1003588, 2014.

[320] B. Øksendal. *Stochastic Differential Equations: An Introduction with Applications*. Springer, 2005.

[321] L. P. Oliveira, H. J. Jensen, M. Nicodemi and P. Sibani. Record dynamics and the observed temperature plateau in the magnetic creep-rate of type-II superconductors. *Physical Review B*, 71(10):104526, 2005.

[322] T. D. Olszewski. Persistence of high diversity in nonequilibrium ecological communities: Implications for modern and fossil ecosystems. *Proceedings of the Royal Society B: Biological Sciences*, 279(1727):230–236, 2012.

[323] O. E. Omel'chenko. The mathematics behind chimera states. *Nonlinearity*, 31:R121–164, 2018.

[324] R. Otter. The multiplicative process. *Annals of Mathematical Statistics*, 20:206–224, 1949.

[325] L. Palmieri and H. J. Jensen. The emergence of weak criticality in SOC systems. *Epl*, 123(2):20002, 2018.

[326] L. Palmieri and H. J. Jensen. The forest fire model: The subtleties of criticality and scale invariance. *Frontiers in Physics*, 8(Sep):257, 2020.

[327] L. Palmieri and H. J. Jensen. Investigating critical systems via the distribution of correlation lengths. *Physics Review Research*, 2:013199, 2020.

[328] M. J. Panaggio and D. M. Abrams. Chimera states: Coexistence of coherence and incoherence in networks of coupled oscillators. *Nonlinearity*, 28(3):R67–R87, 2015.

[329] A. Papana, D. Kugiumtzis and P. G. Larsson. Detection of direct causal effects and application to epileptic electroencephalogram analysis. *International Journal of Bifurcation and Chaos*, 22(9):12500222, 2012.

[330] A. Park and H. Sabourian. Herding and contrarian behavior in financial markets. *Econometrica*, 79(4):973–1026, 2011.

[331] P. Paradisi, P. Allegrini, A. Gemignani, M. Laurino, D. Menicucci and A. Piarulli. Scaling and intermittency of brain events as a manifestation of consciousness. *AIP Conference Proceedings*, 1510(Jan):151–161, 2013.

[332] S. J. Park, J. W. Kim, H. J. Lee, H. J. Park, and P. Gloor. Behavioral aspects of social network analysis. In *Proceedings of the 5th International Conference on Collaborative Innovation Networks COINs15*, 2015.

[333] W. S. Parker and E. Winsberg. Values and evidence: how models make a difference. *European Journal for Philosophy of Science*, 8(1):125–142, 2018.

[334] R. Pastor-Satorras, E. Smith and R. V. Solé. Evolving protein interaction networks through gene duplication. *Journal of Theoretical Biology*, 222(2):199–210, 2003.

[335] W. Pauli. The connection between spin and statistics. *Physical Review*, 58:716, 1940.

[336] G. P. Pavlos, M. N. Xenakis, L. P. Karakatsanis, A. C. Iliopoulos, A. E. G. Pavlos and D. V. Sarafopoulos. Universality of Tsallis non-extensive statistics and fractal dynamics for complex systems. *CHAOS 2012 - 5th Chaotic Modeling and Simulation International Conference, Proceedings*, pp. 405–458, 2012.

[337] J. Pearl and D. MacKenzie. *The Book of Why: The New Science of Cause and Effect*. Penguin, 2019.

[338] K. Pearson. The problem of the random walk. *Nature*, 72:294, 1905.

[339] T. Pereira, D. Eroglu, G. Baris Bagci, U. Tirnakli and H. J. Jensen. Connectivity-driven coherence in complex networks. *Physical Review Letters*, 110(23):234103, 2013.

[340] O. Peters. The time resolution of the St Petersburg paradox. *Philosophical Transactions of the Royal Society A: Mathematical, Physical and Engineering Sciences*, 369(1956):4913–4931, 2011.

[341] O. Peters and J. D. Neelin. Critical phenomena in atmospheric precipitation. *Nature Physics*, 2(6):393–396, 2006.

[342] S. E. Petersen and O. Sporns. Brain networks and cognitive architectures. *Neuron*, 88(1):207–219, 2016.

[343] E. Peterson. The conquest of vitalism or the eclipse of organicism? The 1930s Cambridge organizer project and the social network of mid-twentieth-century biology. *The British Journal for the History of Science*, 47(2):281–304, 2014.

[344] G. Petri, P. Expert, H. J. Jensen and J. W. Polak. Entangled communities and spatial synchronization lead to criticality in urban traffic. *Scientific Reports*, 3:1798, 2013.

[345] A. Pikovsky, M. Rosenblum and J. Kurths. *Synchronization: A Universal Concept in Nonlinear Sciences*. Cambridge University Press, 2001.

[346] D. Piovani. *Analysing and forecasting transitions in complex systems* – https://spiral.imperial.ac.uk/handle/10044/1/31380. PhD thesis, 2015.

[347] D. Piovani, J. Grujić and H. J. Jensen. Linear stability theory as an early warning sign for transitions in high dimensional complex systems. *Journal of Physics A: Mathematical and Theoretical*, 49(29):295102, 2016.

[348] Y. Pomeau and P. Manneville. Intermittent transition to turbulence in dissipative dynamical systems. *Communications in Mathematical Physics*, 74(2):189–197, 1980.

[349] N. R. Poniatowski. Superconductivity, broken gauge symmetry, and the Higgs mechanism. *American Journal of Physics*, 87(6):436–443, 2019.

[350] A. Pournaki, L. Merfort, J. Ruiz, N. E. Kouvaris, P. Hövel and J. Hizanidis. Synchronization patterns in modular neuronal networks: A case study.

Frontiers in Applied Mathematics and Statistics, 5(Oct):52, 2019.

[351] W. H Press. Flicker noises in astonomy and elsewhere. *Comments on Astrophysics*, 7(4):103–119, 1978.

[352] D. De Solla Price. A general theory of bibliometric and other cumulative advantage processes. *Journal of the American Society for Information Science*, 27(5):292–306, 1976.

[353] I. Procaccia and H. Schuster. Functional renormalisation-group theory of universal 1/f noise in dynamical systems. *Physical Review A*, 28(2):1210–1212, 1983.

[354] G. Pruessner. *Self-Organised Criticality. Theory, Models and Characterisation.* Cambridge University Press, 2012.

[355] G. Pruessner and H. J. Jensen. Broken scaling in the forest-fire model. *Physical Review E*, 65(5):021903, 2002.

[356] M. I. Rabinovich, M. A. Zaks and P. Varona. Sequential dynamics of complex networks in mind: Consciousness and creativity. *Physics Reports*, 883:1–32, 2020.

[357] H. Rajpal, F. E. Rosas and H. J. Jensen. Tangled worldview model of opinion dynamics. *Frontiers in Physics*, 7(Oct):163, 2019.

[358] G. M. Ramir-Ávila, J. Kurths and J. L. Deneubourg. Fireflies: A paradigm in synchronzation. In M. Edelman and M. Sanjuan (eds), *Chaotic, Frational, and Complex Dynamics: New Insights an Perspectives*, pp. 35–64. Springer, 2017.

[359] F. A. Razak. *Mutual information based measures on complex interdependent networks of neuro data sets.* PhD thesis, 2013. https://doi.org/10.25560/11579

[360] D. Reagle and D. Salvatore. Forecasting financial crises in emerging market economies. *Open Economies Review*, 11(3):247–259, 2000.

[361] S. Redner. *A Guide to First-Passage Processes.* Cambridge University Press, 2001.

[362] F. Reif. *Fundamentals of Statistical and Thermal Physics.* McGraw-Hill, 1965.

[363] C. J. Rhodes, H. J. Jensen and R. M. Anderson. On the critical behaviour of simple epidemics. *Proceedings of the Royal Society B: Biological Sciences*, 264(1388):1639–1646, 1997.

[364] T. O. Richardson, K. Christensen, N. R. Franks, H. J. Jensen and A. B. Sendova-Franks. Group dynamics and record signals in the ant *Temnothorax albipennis*. *Journal of the Royal Society Interface*, 8(57):518–528, 2011.

[365] T. O. Richardson, E. J. H. Robinson, K. Christensen, H. J. Jensen, N. R. Franks and A. B. Sendova-Franks. Record dynamics in ants. *PLoS ONE*, 5(3):e9621, 2010.

[366] P. A. Rikvold and R. K. P. Zia. Punctuated equilibria and 1/f noise in a biological coevolution model with individual-based dynamics. *Physical Review E*, 68(3):31913, 2003.

[367] D. W. Rivett and T. Bell. Abundance determines the functional role of bacterial phylotypes in complex communities. *Nature Microbiology*, 3(7):767–772, 2018.

[368] J. D. Robalino and H. J. Jensen. Entangled economy: An ecosystems approach to modeling systemic level dynamics. *Physica A: Statistical Mechanics and its Applications*, 392(4):773–784, 2013.

[369] T. Rogers and A. J. McKane. A unified framework for Schelling's model of segregation. *Journal of Statistical Mechanics: Theory and Experiment*, 2011(7):P07006, 2011.

[370] F. E. Rosas, P. A. M. Mediano, M. Gastpar and H. J. Jensen. Quantifying high-order interdependencies via multivariate extensions of the mutual information. *Physical Review E*, 100(3):32305, 2019.

[371] J. Runge. Causal network reconstruction from time series: From

theoretical assumptions to practical estimation. *Chaos*, 28(7):075310, 2018.

[372] J. Runge, V. Petoukhov, J. F. Donges, Hlinka, N. Jajcay, M. Vejmelka, D. Hartman, N. Marwan, M. Paluš and J. Kurths. Identifying causal gateways and mediators in complex spatio-temporal systems. *Nature Communications*, 6:8502, 2015.

[373] R. Russo, H. J. Herrmann and L. De Arcangelis. Brain modularity controls the critical behavior of spontaneous activity. *Scientific Reports*, 4:4312, 2014.

[374] K. Rypdal, L. Østvand and M. Rypdal. Long-range memory in Earth's surface temperature on time scales from months to centuries. *Journal of Geophysical Research Atmospheres*, 118(13):7046–7062, 2013.

[375] A. Sahasranaman and H. J. Jensen. Rapid migrations and dynamics of citizen response. *Royal Society Open Science*, 6(3):181864, 2019.

[376] J. M. Sakoda. The checkerboard model of social interaction. *The Journal of Mathematical Sociology*, 1(1):119–132, 1971.

[377] A. Sarracino, O. Arviv, O. Shriki and L. de Arcangelis. Predicting brain evoked response to external stimuli from temporal correlations of spontaneous activity. *Physical Review Research*, 2:033355, 2020.

[378] K. Sato, Y. Ito, T. Yomo and K. Kaneko. On the relation between fluctuation and response in biological systems. *Proceedings of the National Academy of Sciences of the United States of America*, 100(Suppl. 2):14086–14090, 2003.

[379] H. Sayama. *Introduction to the Modeling and Analysis of Complex Systems*. Open SUNY Textbooks, 2015.

[380] V. Schaller, C. Weber, C. Semmrich, E. Frey and A. R. Bausch. Polar patterns of driven filaments. *Nature*, 467(Sep):73–77, 2010.

[381] M. Schartner, A. Seth, Q. Noirhomme, M. Boly, M. A. Bruno, S. Laureys and A. Barrett. Complexity of multi-dimensional spontaneous EEG decreases during propofol induced general anaesthesia. *PLoS ONE*, 10(8):1–21, 2015.

[382] M. Scheffer. *Critical Transitions in Nature and Society (Princeton Studies in Complexity Book 16)*. Princeton University Press, 2020.

[383] M. Scheffer, J. Bascompte, W. A. Brock, V. Brovkin, S. R. Carpenter, V. Dakos, H. Held, E. H. van Nes, M. Rietkerk and G. Sugihara. Early-warning signals for critical transitions. *Nature*, 461(7260):53–59, 2009.

[384] M. Scheffer, S. R. Carpenter, T. M. Lenton, J. Bascompte, W. Brock, V. Dakos, J. Van De Koppel, I. A. Van De Leemput, S. A. Levin, E. H. Van Nes, M. Pascual and J. Vandermeer. Anticipating critical transitions. *Science*, 338(6105):344–348, 2012.

[385] M. Scheffer, S. Carpenter, J. A. Foley, C. Folke and B. Walker. Catastrophic shifts in ecosystems. *Nature*, 413(6856):591–596, 2001.

[386] T. C. Schelling. Dynamic models of segregation. *Journal of Mathematical Sociology*, 1:143–186, 1971.

[387] T. Schelling. American economic association models of segregation. *The American Economic Review*, 59(2):488–493, 1969.

[388] T. Schreiber. Measuring information transfer. *Physical Review Letters*, 85(2):461–464, 2000.

[389] M. J. Schreuder, C. A. Hartman, S. V. George, C. Menne-Lothmann, J. Decoster, R. van Winkel, P. Delespaul, M. De Hert, C. Derom, E. Thiery, B. P. F. Rutten, N. Jacobs, J. van Os, J. T. W. Wigman and M. Wichers. Early warning signals in psychopathology: What do they tell? *BMC Medicine*, 18(1):269, 2020.

[390] L. S. Schulman. Bacterial resistance to antibodies: A model evolutionary study. *Journal of Theoretical Biology*, 417(Jan):61–67, 2017.

[391] H. G. Schuster. *Deterministic Chaos*. VCH Verlagsgesellschaft, 1989.

[392] F. Schwabl. *Statistical Mechanics*. Springer, 2000.

[393] M. Serafino, G. Cimini, A. Maritan, A. Rinaldo, S. Suweis, J. R. Banavar and G. Caldarelli. True scale-free networks hidden by finite size effects. *Proceedings of the National Academy of the United States of America*, 118:e2013825118, 2021.

[394] C. E. Shannon. A mathematical theory of communication. *Bell System Technical Journal*, 27(4):623–656, 1948.

[395] J. Shao, X. Wang, Q. Yang, C. Plant and C. Böhm. Synchronization-based scalable subspace clustering of high-dimensional data. *Knowledge and Information Systems*, 52(1):83–111, 2017.

[396] K. J. Sharkey. Localization of eigenvector centrality in networks with a cut vertex. *Physical Review E*, 99(1):012315, 2019.

[397] A. S. Sharma, M. J. Aschwanden, N. B. Crosby, A. J. Klimas, A. V. Milovanov, L. Morales, R. Sanchez and V. Uritsky. 25 Years of self-organized criticality: Space and laboratory plasmas. *Space Science Reviews*, 198(1-4):167–216, 2016.

[398] H. H. Shen. Core concept: Resting-state connectivity. *Proceedings of the National Academy of Sciences of the United States of America*, 112(46):14115–14116, 2015.

[399] J. Shen and B. Zheng. On return–volatility correlation in financial dynamics. *Europhysics Letters*, 88(2):28003, 2009.

[400] J. Shena, J. Hizanidis, V. Kovanis and G. P. Tsironis. Turbulent chimeras in large semiconductor laser arrays. *Scientific Reports*, 7(Feb):42116, 2017.

[401] D. W. Sherburne. *A Key to Whitehead's Process and Reality*. University of Chicago Press, 1981.

[402] D. I. Shuman, S. K. Narang, P. Frossard, A. Otega and P. Vandergheynst. The emerging field of signal processing on graphs: Extending high-dimensional data analysis to networks and other irregular domain. *IEEE Signal Processing Magazine*, 30(3):83–98, 2013.

[403] P. Sibani and H. J. Jensen. Intermittency, aging and extremal fluctuations. *Europhysics Letters*, 76(4):563–569, 2006.

[404] P. Sibani, S. Boettcher and H. J. Jensen. Record dynamics of evolving metastable systems: Theory and applications. *European Physical Journal B*, 94(1):37, 2021.

[405] P. Sibani and S. Christiansen. Non-stationary aging dynamics in ant societies. *Journal of Theoretical Biology*, 282(1):36–40, 2011.

[406] P. Sibani and H. J. Jensen. *Stochastic Dynamics of Complex Systems: From Glasses to Evolution*. Imperial College Press, 2013.

[407] P. Sibani and P. B. Littlewood. Slow dynamics from noise adaptation. *Physical Review Letters*, 71(10):1482, 1993.

[408] E. Siggiridou, C. Koutlis, A. Tsimpiris and D. Kugiumtzis. Evaluation of Granger causality measures for constructing networks from multivariate time series. *Entropy*, 21(11):e21111080, 2019.

[409] D. Silver, B. Ultan and P. Adler. Venues and segregation : A revised Schelling model. *PLoS ONE*, 16(1):e0242611, 2021.

[410] E. Silverman. *Methodological Investigations in Agent-Based Modelling: With Applications for the Social Sciences*. Springer, 2018.

[411] H. A. Simon. On a class of skew distribution functions. *Biometrika*, 42(3/4):425, 1955.

[412] H. A. Simon. The architecture of complexity. *Proceedings of the American Philosophical Society*, 106(6):467–482, 1962.

[413] G. G. Simpson. *Tempo and Mode in Evolution*. Columbia University Press, 1944.

[414] A. Smajgl and O. Barreteau (eds). *Empirical Agent-Based Modelling - Challenges and Solutions: Volume 1, The Characterisation and Parameterisation of Empirical Agent-Based Models*. Springer, 2013.

[415] S. J. M. Smith. EEG in the diagnosis, classification, and management of patients with epilepsy. *Neurology in Practice*, 76(2):069245, 2005.

[416] R. V. Solé and J. M. Montoya. Complexity and fragility in ecological networks. *Proceedings of the Royal Society B: Biological Sciences*, 268(1480):2039–2045, 2001.

[417] R. V. Solé, R. Ferrer-Cancho, J. M. Montoya and S. Valverde. Selection, tinkering, and emergence in complex networks. *Complexity*, 8(1):20–33, 2002.

[418] V. Soloviev, S. Semerikov and V. Solovieva. Lempel–Ziv complexity and crises of cryptocurrency market. *Advances in Economics, Business and Management Research*, 129:299–306, 2020.

[419] M. G. Song and G. T. Yeo. Analysis of the air transport network characteristics of major airports. *Asian Journal of Shipping and Logistics*, 33(3):117–125, 2017.

[420] W. Song, J. Wang, K. Satoh and W. Fan. Three types of power-law distribution of forest fires in Japan. *Ecological Modelling*, 196(3–4):527–532, 2006.

[421] X. Song and A. Taamouti. A better understanding of Granger causality analysis: A big data environment. *Oxford Bulletin of Economics and Statistics*, 81(4):911–936, 2019.

[422] D. Sornette. *Critical Phenomena in Natural Sciences: Chaos, Fractals and Disorder: Concepts and Tools*. Springer, 2000.

[423] D. Stauffer and S. Solomon. Ising, Schelling and self-organising segregation. *European Physical Journal B*, 57(4):473–479, 2007.

[424] D. Stauffer and A. Aharony. *Introduction to Percolation Theory*. Taylor & Francis, 1994.

[425] L. Steels. Agent-based models for the emergence and evolution of grammar. *Philosophical Transactions of the Royal Society B: Biological Sciences*, 371(1701):20150447, 2016.

[426] D. L. Stein and C. M. Newman. *Spin Glasses and Complexity*. Princeton University Press, 2013.

[427] J. V. Stone. *Information Theory: A Tutorial Introduction*. Sebtel Press, 2015.

[428] R. F. Storms, C. Carere, F. Zoratto and C. K. Hemelrijk. Complex patterns of collective escape in starling flocks under predation. *Behavioral Ecology and Sociobiology*, 73(1):10, 2019.

[429] S. Strogatz. *Sync: The Emerging Sicence of Spontaneous Order*. Penguin Books, 2004.

[430] J. Stuhlman and B. Buhler. *Georgia O'Keeffe: Circling Around Abstraction*. Hudson Hill Press, 2007.

[431] S. Suweis, F. Simini, J. R. Banavar and A. Maritan. Emergence of structural and dynamical properties of ecological mutualistic networks. *Nature*, 500(7463):449–452, 2013.

[432] D. Swingedouw, C. I. Speranza, A. Bartsch, G. Durand, C. Jamet, G. Beaugrand and A. Conversi. Early warning from space for a few key tipping points in physical, biological, and social–ecological systems. *Surveys in Geophysics*, 41:1237–1284, 2020.

[433] E. Tagliazucchi, P. Balenzuela, D. Fraiman and D. R. Chialvo. Criticality in large-scale brain fMRI dynamics unveiled by a novel point process analysis. *Frontiers in Physiology*, 3(Feb):15, 2012.

[434] M. Takayasu and H. Takayasu. 1/f Noise in a traffic model *Fractals*, 01(04):860–866, 1993.

[435] M. Takayasu and H. Takayasu. Fractals and economics. *Mathematics of Complexity and Dynamical Systems*, 0(1):512–531, 2012.

[436] M. Takayasu, H. Takayasu and T. Sato. Critical behaviors and 1/f noise in information traffic. *Physica A: Statistical Mechanics and its Applications*, 233(3–4):824–834, 1996.

[437] R. P. Taulor, A. P. Micolich and D. Jones. Fractal analysis of Pollock's drip paintings. *Nature*, 399:422, 1999.

[438] P. Tempesta. Group entropies, correlation laws and zeta functions. *Physical Review E*, 84:21121, 2011.

[439] P. Tempesta. Beyond the Shannon–Khinchin formulation: The composability axiom and the universal-group entropy. *Annals of Physics*, 365:180–197, 2016.

[440] P. Tempesta. Formal groups and z-entropies. *Royal Society A*, 472:20160143, 2016.

[441] P. Tempesta and H. J. Jensen. Universality classes and information-theoretic measures of complexity via group entropies. *Scientific Reports*, 10(1):5952, 2020.

[442] S. Thurner, R. Hanel and P. Klimek. *Introduction to the Theory of Complex Systems*. Oxford University Press, 2018.

[443] S. Thurner, P. Klimek and R. Hanel. Schumpeterian economic dynamics as a quantifiable model of evolution. *New Journal of Physics*, 12:075029, 2010.

[444] M. Tinkham. *Introduction to Superconductivity*, 2nd edition. Dover Publications, 2004.

[445] M. Titus and J. Watson. Critical speeding up as an early warning signal of stochastic regime shifts. *Theoretical Ecology*, 13(4):449–457, 2020.

[446] G. P. Tolstov. *Fourier Series*. Dover Publications, 1976.

[447] L. Tolstoy. *War and Peace*. Oxford University Press, 2017.

[448] G. Tononi. Integrated information theory of consciousness: An updated account. *Archives italiennes de biologie*, 150(4):293–329, 2012.

[449] G. Tononi, G. M. Edelman and O. Sporns. Complexity and coherency: Integrating information in the brain. *Trends in Cognitive Sciences*, 2(12):474–484, 1998.

[450] G. Tononi, O. Sporns and G. M. Edelman. A measure for brain complexity: Relating functional segregation and integration in the nervous system. *Proceedings of the National Academy of Sciences of the United States of America*, 91(11):5033–5037, 1994.

[451] V. A. Traag, L. Waltman and N. J. van Eck. From Louvain to Leiden: Guaranteeing well-connected communities. *Scientific Reports*, 9(1):5233, 2019.

[452] R. J. Trudeau. *Introduction to Graph Theory*. Dover Publications, 1976.

[453] C. Tsallis. *Introduction to Nonextensive Statistical Mechanics: Approaching a Complex World*. Springer, 2009.

[454] C. Tsallis. Possible generalization of Boltzmann–Gibbs statistics. *Journal of Statistical Physics*, 52(1/2):479–487, 1988.

[455] C. Tsallis. Beyond Boltzmann–Gibbs–Shannon in physics and elsewhere. *Entropy*, 21(7):696, 2019.

[456] H. C. Tuckwell, F. Y. M. Wan and J. P. Rospars. A spatial stochastic neuronal model with Ornstein–Uhlenbeck input current. *Biological Cybernetics*, 86(2):137–145, 2002.

[457] D. L. Turcotte. Self-organized criticality. *Reports on Progress in Physics*, 62:1377–1429, 1999.

[458] A. Turing. The chemical basis of morphogenesis. *Philosophical Transactions of the Royal Society of London B*, 237(641):37–72, 1952.

[459] L. Turnbull, M. T. Hütt, A. A. Ioannides, S. Kininmonth, R. Poeppl, K. Tockner, L. J. Bracken, S. Keesstra, L. Liu, R. Masselink and A. J. Parsons. Connectivity and complex systems: Learning from a multi-disciplinary perspective. *Applied Network Science*, 3(1):11, 2018.

[460] G. E. Uhlenbeck and L. S. Ornstein. On the theory of the Brownian motion. *Physical Review*, 36:823–841, 1930.

[461] L. Urselmans and S. Phelps. A Schelling model with adaptive tolerance. *PLoS ONE*, 13(3):e0193950, 2018.

[462] T. Utsu, Y. Ogata and R. S. Matsu'ura. The centenary of the Omori formula for a decay law of aftershock activity. *Journal of Physics of the Earth*, 43:1–33, 1995.

[463] M. P. van den Heuvel and H. E. Hulshoff Pol. Exploring the brain network: A review on resting-state fMRI functional connectivity. *European Neuropsychopharmacology*, 20(8):519–534, 2010.

[464] J. van Ijken. Flight of the starlings. *National Geographic*, 2016.

[465] N. G. van Kampen. *Stochastic Processes in Physics and Chemistry*. North-Holland Personal Library, 2007.

[466] F. Varela, J-P. Lachaux, E. Rodriguez and J. Martinerie. The brainweb: Phase synchronisation and large-scale integration. *Nature Reviews Neuroscience*, 2(Apr):229–239, 2001.

[467] P. Vázquez, J. A. Del Rio, K. G. Cedano, M. Martínez and H. J. Jensen. An entangled model for sustainability indicators. *PLoS ONE*, 10(8):e0135250, 2015.

[468] P. Vázquez, J. A. del Río, K. G. Cedano, J. van Dijk and H. J. Jensen. Network characterization of the Entangled Model for sustainability indicators: Analysis of the network properties for scenarios. *PLoS ONE*, 13(12):e0208718, 2018.

[469] P. E. Vértes, A. F. Alexander-Bloch, N. Gogtay, J. N. Giedd, J. L. Rapoport and E. T. Bullmore. Simple models of human brain functional networks. *Proceedings of the National Academy of Sciences of the United States of America*, 109(15):5868–5873, 2012.

[470] T. Vicsek, A. Czirók, E. Ben-Jacob, I. Cohen and O. Schochet. Novel type of phase transition in a system of self-driven particles. *Physical Review Letters*, 75(6):1226–1229, 1995.

[471] T. Vicsek and A. Zafeiris. Collective motion. *Physics Reports*, 517(3-4):71–140, 2012.

[472] E. Viegas, S. P. Cockburn, H. J. Jensen and G. B. West. The dynamics of mergers and acquisitions: Ancestry as the seminal determinant. *Proceedings of the Royal Society A: Mathematical, Physical and Engineering Sciences*, 470(2171):20140370, 2014.

[473] E. Viegas, M. Takayasu, W. Mirura, K. Tamura, T. Ohnishi, H. Takayasu and H. J. Jensen. Cosystems Perspective on Financial Networks. *Complexity*, 18(6):34–48, 2013.

[474] D. Vinković and A. Kirman. A physical analogue of the Schelling model. *Proceedings of the National Academy of Sciences of the United States of America*, 103(51):19261–19265, 2006.

[475] I. Vlachos and D. Kugiumtzis. Nonuniform state-space reconstruction and coupling detection. *Physical Review E*, 82(1):016207, 2010.

[476] J. W. von Goethe. *Theory of Colours*. CreateSpace Independent Publishing Platform, 2015.

[477] J. von Neumann. *Theory of Self-reproducing Automata.* University of Illinois Press, 1967.

[478] J. von Neumann and Oskar Morgenstern. *The Theory of Games and Economic Behavior: 6oth Anniversary Commemorative Edition.* Golden Keys Success, 2020.

[479] M. M. Waldrop. *Complexity: The Emerging Science at the Edge of Order and Chaos.* Simon & Schuster, 1992.

[480] D. M. Walsh. *Organisms, Agency, and Evolution.* Cambridge University Press, 2015.

[481] X. Wan. *Time series causality analysis and EEG data analysis on music improvisation.* – https://spiral.imperial.ac.uk/handle/10044/1/23956. PhD thesis, 2014.

[482] X. Wan, B. Crüts and H. J. Jensen. The causal inference of cortical neural networks during music improvisations. *PLoS ONE*, 9(12):e112776, 2014.

[483] G. Wang, Y. Li and X. Zou. Several indicators of critical transitions for complex diseases based on stochastic analysis. *Computational and Mathematical Methods in Medicine*, 2017(Cv):7560758, 2017.

[484] N. W. Watkins, G. Pruessner, S. C. Chapman, N. B. Crosby and H. J. Jensen. 25 Years of self-organized criticality: Concepts and controversies. *Space Science Reviews*, 198(1–4):3–44, 2016.

[485] A. J. Watson and J. E. Lovelock. Biological homeostasis of the global environment: The parable of Daisyworld. *Tellus B*, 35 B(4):284–289, 1983.

[486] D. Watts. *Six Degrees: The New Science of Networks.* Vintage, 2004.

[487] W. Weaver. Science and complexity. *American Scientist*, 36(4):536–544, 1948.

[488] H. Weber and P. Minnhagen. Monte Carlo determination of the critical temperature for the two-dimensional XY model. *Physical Review B*, 37:5986–5989, 1988.

[489] S. Weinberg. *The First Three Minutes: A Modern View of the Origin of the Universe.* Basic Books, 1993.

[490] M. B. Weissmann. 1/f Noise and other slow, nonexponetial kinetics in condensed matter. *Reviews of Modern Physics*, 60:537–571, 1988.

[491] G. West. *Scale: The Universal Laws of Life and Death in Organisms, Cities and Companies.* The Orion Publishing Group, 2017.

[492] A. N. Whitehead. *An Introduction to Mathematics.* Cambridge University Press, 1911.

[493] A. N. Whitehead. *Science and the Modern World.* Simon & Schuster, 1970.

[494] A. N. Whitehead. *Process and Reality (Gifford Lectures).* Macmillan USA, 1979.

[495] E. P. Wigner. The unreasonable effectiveness of mathematics in the natural sciences. *Communications on Pure and Applied Mathematics*, XIII:1–14, 1960.

[496] A. M. Wink. Eigenvector centrality dynamics from resting-state fMRI: Gender and age differences in healthy subjects. *Frontiers in Neuroscience*, 13(Jun):648, 2019.

[497] Y. Wu, R. Lu, H. Su, P. Shi and Z.-G. Wy. *Synchronization Control for Large-Scale Network Systems: 76 (Studies in Systems, Decision and Control, 76).* Springer, 2016.

[498] K. Yamada, H. Takayasu and M. Takayasu. The grounds for time dependent market potentials from dealers' dynamics. *European Physical Journal B*, 63(4):529–532, 2008.

[499] G. U. Yule. A mathematical theory of evolution, based on the conclusions of Dr. J. C. Willis. *Philosophical Transactions of the Royal Society London B*, 213(402-10):21–87, 1925.

[500] J. Zand, U. Tirnakli and H. J. Jensen. On the relevance of q-distribution functions: The return time distribution of restricted random walker. *Journal of Physics A: Mathematical and Theoretical*, 48(42):425004, 2015.

[501] S. Zapperi, K. B. Lauritsen and E. E. Stanley. Self-organized branching processes-mean-field theory for avalanches. *Physical Review Letters*, 74(22):4071–4074, 1995.

[502] Ya. B. Zeldovich, A. A. Ruzmaikin and D. D. Sokoloff. *The Almighty Chance*. World Scientific, 1990.

[503] H. P. Zhang, A. Be'er, E. L. Florin and H. L. Swinney. Collective motion and density fluctuations in bacterial colonies. *Proceedings of the National Academy of Sciences of the United States of America*, 107(31):13626–13630, 2010.

[504] Y. C. Zhang. Scaling theory of self-organized criticality. *Physical Review Letters*, 63(5):470–473, 1989.

[505] J. Zinn-Justin. *Quantum Field Theory and Critical Phenomena*, 5th edition Oxford University Press, 2021.

[506] J. Ziv and A. Lempel. A universal algorithm for sequential data compression. *IEEE Transactions on Information Theory*, 23(3):337–343, 1977.

[507] J. Ziv and A. Lempel. Compression of individual sequences via variable-rate coding. *IEEE Transactions on Information Theory*, 24(5):530–536, 1978.

Index

adaptation and evolutionary dynamics
  importance of surrounding structures to, 61
  mutation, 62–63
  network structures, 61–63
  reproducing agent, 62
adaptive evolutionary dynamics, 39–40
adjacency matrix, 179
agent-based modelling (ABM)
  computer simulation, 324
  defined, 324
  segregation models, 328–337
  Tangled Nature model, 337–349
  Vicsek model of flocking behaviour, 325–327
Anderson, Philip W., 16
anti-persistence (random walk), 302–306
Architecture of Complexity, The, 17
Aristotle, 13, 15

Bak–Sneppen model, 80
betweenness centrality in network theory, 187–188
binomial process, 281
Bohr, Niels, 15
Boltzmann weights and Shannon entropy, 116–118
branching process
  defined, 25, 93, 94
  generator functions, 95–103
  random walks, 103–106
breaking of symmetry, 16

canonical ensemble, 113
Cartesian combinations, 249
causal relations entropy estimates, 237–240
cellular automata, 17
Central Limit Theorem, 10
characteristic scale, 23–25

chimera states
  defined, 30
  synchronisation, 170–174
closeness centrality in network theory, 187
clustering and evolutionary dynamics, 309–313
collective robust degrees of freedom, 26–28
community detection in network theory, 188–196
complexity measures
  group entropy, 272–273
  information theory and emergence, 259–272
  Lempel–Ziv, 256–259
complexity science, *see also* mathematics
  aim of, 13
  and data science, 6
  defined, 18, 397
  descriptive definition, 5
  dynamics and evolution, 47
  focus of, compared to other sciences, 7
  and game theory, 6
  multidisciplinary nature of, 7
  and network science, 6
  open questions in, 398–399
  shift to focus to processes at aggregate level, 8
  synchronisation, 48
  as systematic investigation of the general patterns and structures of emergent phenomena, 6
  transdisciplinary nature of, 397
  working definition, 5
configuration model
  defined, 197
  network generation, 204–205
correlation function ranges
  exact approach, 134–139
  intuitive discussion, 139–143

correlations in time and dynamical probabilities, 297–302
course-grained level, 313–315
critical phenomena, 50
critical point
  correlation examples, 132
  correlations, 129–132
  defined, 50
  fluctuations, 132
  nature of, 125–127
  response function, 128–129
critical state, 26
critical temperature, 125

data science and complexity science, (
degree centrality in network theory, 179–184
diffusion, evolutionary, 31–33
  Ornstein–Uhlenbeck process, 307–308
  random walk and, 280–293
diffusion process, 28
diffusion process types, 31
diffusion equation, 280
dynamic network generation, 205–20
dynamical systems theory
  defined, 65
  stochastic, 279–317
dynamics and evolution intermittenc
  356–381

eigenvector centrality in network theory, 184–187
electroencephalogram (EEG), 36
emergence and complexity science
  Adam Smith and, 13
  adaptive evolutionary dynamics, 39–40
  Aristotle and, 13
  defined, 18
  evolutionary adaptive dynamics a
  14
  history of views on, 17

ideal gas, 10
information theory and, 259–272
laws between different types of
   components, 13
normal distribution, 14
process philosophy and, 13
social segregation, 12
statistical mechanics and,
   110–155
emergence and complexity science,
   properties
characteristic scale, 23–25
collective robust degrees of freedom,
   26–28
evolutionary diffusion, 31–33
network science, 35–37
structural coherence in space and
   time, 28–31
summary, 21
symmetry breaking, 33–35
temporal mode, 37–38
transitions, 34
emergence protypical model usefulness
Bak–Sneppen and Tangled Nature
   model, 80
Ising model, 79
need for simplicity in, 76–78
Schelling model, 80
emergent behaviour models
evolutionary dynamics and
   adaptation, 60–64
Ising model, 48–52
Kuramoto model, 60
mean-field model, 64–69
network models, 52–56
synchronisation, 57–60
emergent structures, summary, 47
ensembles and probability
microcanonical ensemble,
   111–113
entropy and information theory
causal relations estimates, 237–240
complexity measures, 272–273
defined, 113
degree of complexity measures,
   256–274
master equation and, 315–317
probability distribution, 245–256
time series and networks,
   241–244
error threshold, 46
evolutionary adaptive dynamics,
   14

evolutionary diffusion, 31–33,
   378
evolutionary dynamics, see also Tangled
   Nature model
evolutionary dynamics and clustering,
   309–313
exact differential, 112
excess degree, 197
exponential growth, 25

field-theoretic methods, 312
first passage times, 280, 293
first return times, 280, 293
Fokker–Planck equation, 280
forecasting tipping points, 387–394
forest fire model, 367–370
Fourier transform and solution of
   equations, 286
Fourier transform graph, 219
fractals, 26
functional magnetic resonance imaging
   (fMRI)
brain scanner, 35
correlation functions, 133
scanner, 25

game theory and complexity science, 6
generator functions
proof, 95
sizes and lifetimes, 97–103
giant cluster, 196
giant cluster spreading in networks,
   196–203
Gilbert model, 53, 55–56

half-links (configuration model), 204
hardwired network, 63
Hurst exponent, 302–306

ideal gas and emergence, 10
information science, 15
information theory and complexity
   science
background of, 230–232
degree of complexity measures,
   256–274
emergence and, 259–272
interdependence, 232–237
information theory and entropy,
   113
interdependence in information theory,
   232–237

intermittency mode, 21
intermittent dynamics
record dynamics, 370–378
self-organised criticality (SOC),
   357–370
tangent map, 379–381
Ising model
and statistical mechanics, 119–125
components of, 48
magnetic interpretation, 50
physics interpretation, 49
record dynamics examples, 375–376
usefulness of simplicity in, 79

kinetic energy, 29
Kuramoto model
synchronisation, 60
systems behaviour of, 164–170

Lagrange multipliers, 114–118
Laplacian on a network, 220–224
Lempel–Ziv complexity measure,
   256–259
Lotka–Volterra model, 64
Louvain algorithm, 192

magnetic moments, 34
magnetoencephalography (MEG), 36
Marx, Karl, 77
master equation, 280, 282, 283, 313–317
mathematics, see also complexity science
agent-based modelling, 324–349
branching process, 93–106
entropy, 237–256
intermittent dynamics, 356–381
as language, 89–90
network theory, 177–219
statistical mechanics, 110–155
stochastic dynamics and probability
   equations, 279–317
synchronisation, 163–174
tipping points, 387–394
maximum energy principle, 113, 246
mean-field model
dimensionality, 64–67
and Ising model, 119–125
forecasting, 67–69
self-organised criticality (SOC),
   361–364
microcanonical ensemble, 111–113
Millennium Bridge example, 57–58, 67
More Is Different, 16

Morin, Edgar, 17
mutation, 62–63, 68

network generation
    background of, 204
    configuration model, 204–205
    dynamic, 205–209
    steady-state evolution, 209–212
network models
    no-scale, 53–56
    scale, 52–53
    Tangled Nature model, 61–63
network models distribution
    binomial, 54, 56
    exponential growth, 54, 56
    power law, 54, 56
network science
    and complexity science, 6
    emergence and, 35–37
    entropy and, 241–244
network theory
    basic concepts, 178
    community detection, 188–196
    giant cluster spreading, 196–203
    history of, 177
    language discrepancy around, 177
    node importance measures, 179–188
network theory, dynamics
    network generation, 204–212
    random walk, 212–216
    synchronisation, 216–219
node measure in network theory
    betweenness centrality, 187–188
    closeness centrality, 187
    connectance, 179
    degree centrality, 179–184
    eigenvector centrality, 184–187
    usefulness of, 188
non-equilibrium mode, 22
normal distribution, 14

occupied network, 63
On Complexity, 17
organized complexity, 15
Ornstein–Uhlenbeck process,
    307–308

partition function, 112, 118
persistence (random walk), 302–306
phase transitions, 16
phenomenology
    complexity science, 397

Tangled Nature model, 348–349
preferential attachment, 55–56
probability distribution
    entropy and, 245–256
    stochastic, 279–317
process philosophy, 13
progeny size and generator functions,
    99–102
punctuated equilibrium, 21, 39, 80

quantum many-body theory, 16

random walk
    branching trees and, 103–106
    defined, 31
    and diffusion, 280–293
    network theory, 212–216
    with persistence or anti-persistence,
        302–306
record dynamics
    background of, 370–371
    statistics of records, 371–375
record dynamics examples
    ants, 378
    evolution, 378
    Ising model, 375–376
    superconductivity, 377–378
reproducing agent, 62, 93
resting state network, 37

sandpile models, 358–361, 364–367
scale-invariant phenomena, 50
Schelling model
    defined, 12
    detail usefulness, 80
    segregation models, 328–337
segregation models, 328–337
self-organised criticality (SOC)
    background of, 357
    forest fire model, 367–370
    mean-field analysis, 361–364
    sandpile models, 358–361, 364–367
Shannon entropy and Boltzmann
    weights, 116–118
Shannon–Khinchi axioms, 246
Simon, Herbert A., 17
simplicity in modelling
    building blocks not always evident,
        77–78
    and complex phenomena, 76
    in sociology and economics, 76–78
    Marx on, 77

Tolstoy on, 77
Smith, Adam, 13
social segregation and complexity
    science, 12
spin glass model, 121
spin quantum number, 89
spin waves, 144
stationary diffusion, 307–308
statistical mechanics
    2d XY model, 155
    correlation ranges, 133–143
    critical point and, 125–127
    ensembles, 110–119
    Ising model, 119–125
steady-state evolution in network
    generation, 209–212
structural coherence in space and time,
    28–31
superconductivity
    defined, 35
    record dynamics examples, 377–378
    synchronisation and, 58
symmetry breaking, 33–35
synchronisation, 21, 30, 48
    chimera states, 170–174
    defined, 163
    Kuramoto model, 164–170
    network theory, 216–219
    as structure in time, 57–58
    superconductivity, 58

tangent map intermittency, 379–381
Tangled Nature model, see also
    evolutionary dynamics
    adaptation and, 60–64
    agent-based modelling (ABM),
        337–349
    killing events, 340–348
    phenomenology, 348–349
    reproduction events, 340
    time dependence, 23
    transitions, 68–69
    usefulness of, 80
temporal graph signal transform,
    219
temporal mode
    intermittency, 21
    non-equilibrium, 22
    overview, 37–38
time to extinction, 102–103
tipping points

externally induced transitions,
   387–388
intrinsic instability, 389–394
Tolstoy, Leo, 77
toy models, 75
transfer entropy, 236
transition, 34
2d XY model

mathematical details,
   148–153
overview, 143
vortex unbinding, 152–154

up–down symmetry breaking, 34

Vicsek model of flocking behaviour,
   325–327

vortex unbinding
   2d XY model, 152–154
   superconductors, crystals,
      154–155
vortices, 144

Weaver, Warren, 15